# BIOLOGICAL OXIDATION

BY

## CARL OPPENHEIMER, M.D., Ph. D.

AND

## KURT G. STERN, Ph. D.

WITH THE COLLABORATION
OF W. ROMAN, Ph. D.

**1939**

Springer-Science+Business Media, B.V.

Softcover reprint of the hardcover 1st edition 1939

ISBN 978-94-017-5835-2          ISBN 978-94-017-6291-5 (eBook)
DOI 10.1007/978-94-017-6291-5

# PREFACE.

Two years ago, when I began to write the section dealing with desmolases of the Supplement to the 5th edition of my book „Die Fermente und ihre Wirkungen", I was confronted with the necessity of integrating an immense material. It became obvious that this would be impossible without the guidance of a theory capable of encompassing the diverse facts and phenomena. In spite of the many attempts made in that direction, no comprehensive theory was available. I was therefore compelled to develop such a guiding theory. It was clear that any such theory would have to reconcile the apparently contradictory views of WARBURG and WIELAND. While this would have been a very difficult task in 1926, when the last complete edition of „Die Fermente" appeared, such a synthesis appeared feasible in 1937 owing to the modifications which the theories of WARBURG and WIELAND had experienced in the meantime (by these workers and their collaborators themselves) and owing to the contributions made by the pure theory (HABER, WILLSTÄTTER, MICHAELIS) as well as by the study of specific enzyme reactions (THUNBERG, School of HOPKINS, KEILIN, v. EULER, KUHN, SZENT-GYÖRGYI, etc.). I should like to emphasize that the theory of biological oxidation as evolved on this foundation was and still is considered a tool made for the task of arranging the experimental material in an orderly and logical fashion. The subsequent period of writing the special parts of the „Supplement" has satisfied me that this aim has been attained. This unitary theory comprises the XVII. Main Part of the „Supplement" to „Die Fermente", which is now complete (The Hague, Dr. W. JUNK). In this work I enjoyed the valuable collaboration of Dr. W. ROMAN (London) who assisted in the writing of several chapters of the General Part and also wrote largely the section dealing with the descriptive chemistry of the enzyme system. Thus, Dr. ROMAN has also contributed to the present book.

My wish to take the general theory of biological oxidation out of the larger treatise and to present it to the biochemists in the form of a monograph was realized thanks to the willingness of Dr. KURT G. STERN to cooperate in the undertaking. The present book is the product of our collaboration.

CARL OPPENHEIMER.
THE HAGUE.

# PREFACE.

As an experimental worker in the field of biological oxidation I felt keenly the lack of a comprehensive theoretical treatment of the subject. Therefore I welcomed the idea of Professor C. OPPENHEIMER to develop the chapter on biological oxidation in his Supplement to „Die Fermente" into a monograph. Given a considerable latitude in the execution of the plan, I have endeavored to render the text less dependent on other expositions. In doing so it was understood, however, that for fuller information the reader should still be referred to the „main work" by OPPENHEIMER (1926), the Supplement, and to the original publications.

Besides making numerous alterations throughout the text, I have to assume responsibility for the following sections which were either newly added or completely rewritten: Redox Potentials in Heterogeneous Systems, Affinity and Rate of Reaction, Semiquinones as Intermediate Steps of Oxidation-Reduction Systems, Photochemistry of the Respiratory Ferment, Chemistry of the Hemin Enzymes, Copper Proteins, Yellow Enzymes, Carboxylase, Cozymase and other Pyridine Coenzymes, Protein Bearers of Coenzymes, Quinoid Mesocatalysts, Fumaric Acid Catalysis and Citric Acid Cycle. A large number of new references has been added to the bibliography which has been completely rearranged. An attempt has been made to incorporate the more significant contributions up to the end of 1938 when the manuscript was concluded. In order to keep the volume of the book within reasonable limits, the detailed discussion of certain aspects of cell respiration such as the effect of external factors and the mechanism of the PASTEUR-MEYERHOF effect had to be omitted with regret. For a recent review of the latter phenomenon reference is made to the comprehensive article by D. BURK (157b).

A number of colleagues and friends assisted generously in this undertaking. Dr. DEAN BURK helped to disentangle the intricate questions of terminology and symbols presented by the section on Energetics and Potentials. Dr. R. A. SHIPLEY read the entire manuscript and suggested numerous changes pertaining to style and language. Drs. W. ROMAN, K. SALOMON and J. L. MELNICK shared the task of proof reading. To these and other colleagues who offered suggestions and criticisms, I am greatly indebted. I also wish to thank my wife, Else, for her assistance in the arrangement of the bibliography.

KURT G. STERN,
Laboratory of Physiological Chemistry,
Yale University School of Medicine,
New Haven, Connecticut.

# Biological Oxidation.

———

## Table of Contents

# Special Part

# General Part *).

## A. Introduction.

A comprehensive theory of biological oxidation catalysis is being developed on the basis of two theories which only a few years ago were considered incompatible. We refer to the theories of OTTO WARBURG and of HEINRICH WIELAND. The battle cries ,,activation of hydrogen'' and ,,activation of oxygen'' have lost much of their significance due to the progress made in our understanding of the underlying mechanisms. Strictly speaking, neither an activation of hydrogen nor of oxygen occurs in biological oxidation. Both terms are eliminated in an e x a c t description of the mechanism. However, they may be retained as a matter of convenience for the purpose of describing certain important relationships.

If a mere description is intended it would be more correct to call the better known biological oxidations d e h y d r o g e n a t i o n catalyses instead of o x i d a t i o n catalyses. In all of these processes — a few rather obscure instances excepted — a transfer of h y d r o g e n takes place. Theoretically it does not matter whether molecular oxygen or another chemical compound functions as the acceptor of the hydrogen. This conclusion is important because it adds the concept of t h e o r e t i c a l uniformity of the mechanism of such reactions to the recognition of the b i o l o g i c a l uniformity of dehydrogenation without oxygen (a n o x y b i o s i s) and dehydrogenation with oxygen (o x y b i o s i s). Theoretically it is of secondary importance, whether the acceptor of the hydrogen is a u t o x i d i z a b l e, i.e. capable of further transfer of the hydrogen to molecular oxygen as the acceptor; or whether it is not autoxidizable, i.e. capable of hydrogen transfer to other chemical compounds; or whether the system may utilize both types of acceptors. The point of primary importance is the hydrogen transfer itself, representing the disturbance of an equilibrium whereby the hydrogen of an organic compound, called the d o n a t o r, is first labilized and then shifted to another compound, the a c c e p t o r, in accordance with a thermodynamic potential. If the acceptor is molecular oxygen, ultimate oxidation via hydrogen peroxide to water takes place; if the acceptor is another chemical compound, anoxybiontic reactions occur which we call oxido-reductions.

Processes of both kinds may occur spontaneously, i.e. without catalysis, if the thermodynamic potential driving the general reaction

$$DH_2 + Acc \rightarrow D + AccH_2 \quad \ldots \ldots \ldots \ldots \ldots \ldots \quad (1)$$

is not opposed by kinetic hinderances. In case the reaction proceeds too slowly it may be catalysed by inserting new reactants which speed up the process and which thereby function as c a t a l y s t s. The general equation becomes:

---

*) The whole subject has been reviewed in detail by v. EULER (288) and by OPPENHEIMER (905) Important special aspects are treated in the publications (1169, 1171, 1373, 1374, 974, 105, 106 903, 904, 1302, 804).

$$DH_2 + Cat \rightarrow D + Cat\ H_2 + Acc \rightarrow D + Cat + AccH_2 \quad . \quad . \quad . \quad .(2)$$

It follows that in each case the catalyst must represent a reversible oxidation-reduction system. It must be able to accept readily hydrogen from the donator in accordance with the natural thermodynamic potential and to give it off just as readily to an acceptor of higher potential. By virtue of this cyclic change from the oxidized to the reduced state and back to the oxidized state the catalyst eliminates the inherent kinetic obstacles in the system and promotes the completion of reaction (1) in accordance with the laws of thermodynamics.

Later it will be shown that there exist certain relationships between thermodynamics and kinetics. Here it may suffice to mention that a thermodynamic potential which is too steep may inhibit the reaction and may require the interposition of oxidation-reduction systems of intermediary potential range as catalysts. Certain products of sugar decomposition, for instance, which possess a strongly negative potential are unable to yield hydrogen directly to molecular oxygen or to the strongly positive hemin systems. A whole series of catalysts with graded potentials is required in this case to permit the hydrogen transfer from the sugar to oxygen via the hemin systems.

The reaction scheme (2) represents the general mechanism of dehydrogenation catalysis. In order to achieve complete oxidation it is necessary for the hydrogen to be ultimately received by an autoxidizable redox system which in turn reduces oxygen to hydrogen peroxide:

$$DH_2 + Cat \rightarrow D + Cat\ H_2 + O_2 \rightarrow D + Cat + H_2O_2.$$

Hydrogen peroxide is then transformed into water either by simple ("catalatic") decomposition or by subsequent hydrogenation.

The greatest obstacle preventing a merging of the theories of WARBURG and WIELAND was WIELAND's conception that dehydrogenation catalysts were specific only with respect to the donator but not with regard to the acceptor (hydrogen peroxide excepted (p. 34)). In other words, once the hydrogen is "activated", it should go to any acceptor including molecular oxygen. Subsequent experiments from the school of WIELAND, notably those of BERTHO, have shown that the dehydrogenases of WIELAND are not only "donator-specific" but also "acceptor-specific"; only such catalysts which have an affinity for oxygen, and are therefore autoxidizable systems, may effect terminal oxidation. The postulate of WARBURG that only certain systems can effect the ultimate oxidation is therefore justified. Today we know that the most important autoxidizable catalyst of living cells is the respiratory ferment of WARBURG.

The reconciliation of the two theories includes the polemics concerning certain definitions. Strictly speaking there exists neither primary "activation" of hydrogen nor of oxygen. If we express oxido-reduction in terms of electron changes, the "activation" of the hydrogen consists in the labilization of hydrogen and subsequent liberation of an hydrogen ion (proton); the corresponding electron is taken up by the acceptor which is thereby reduced. The "activation" of the oxygen, on the other hand, consists in the charging of the oxygen molecule with two electrons which serve as the points of attachment of the liberated hydrogen ions:

$$DH_2 + O_2 \rightarrow D + 2\ H^+ + 2\ \varepsilon + O_2 \rightarrow D + {}^-O{-}O^- + 2\ H^+ \rightarrow D + H_2O_2*).$$

---

*) The symbol $^-O{-}O^-$ is meant solely to indicate the charging of the oxygen; it does not imply a suggestion as to the physical meaning of this type of charge.

Both processes are coupled inseparably: Ultimate oxidation is achieved by simultaneous „activation" of both hydrogen and oxygen.

It may be of interest to mention briefly at this point, that formerly „activation of oxygen" was conceived quite differently and that WARBURG initially reckoned with such an activation by his iron systems. We refer to the process of activation via metal peroxides. Such a primary activation of oxygen by a rearrangement of the inert molecule $O_2$ to —O—O— may undoubtedly occur in certain instances, e.g. during the catalytic hydrogenation of molecular oxygen. In this way inorganic and organic substances of strongly oxidizing character may arise. However, this process is certainly not of decisive importance in biological reactions. Firstly, it is not known whether a catalysis is possible at all in this manner or whether only induced oxidations are thereby carried out. Furthermore, there is little indication that such reactions are essential for vital oxidation, e.g. of double bonds of unsaturated fatty acids.

In principle, therefore, the biological decomposition of metabolites may be uniformly described as consisting of a sequence of identical reactions. Hydrogen is transferred from one donator to the first acceptor, and from the latter to another and a third and a fourth acceptor, until in one case the hydrogen, still in organic linkage, comes to rest in a stable system, e.g. in the form of lactic acid or ethyl alcohol, presenting the picture of pure Anoxybiosis; or else, until at one point molecular oxygen comes into play with the aid of an autoxidizable system. In this case the hydrogen ends up as water after it has gone through the intermediary stage of hydrogen peroxide. This is the case of complete Oxybiosis, i.e. of complete combustion.

The successive hydrogen transfer proceeds in a continuous fashion. The conception of catalysts themselves as nothing but interposed acceptors which receive hydrogen and hand it on to the next acceptor is important. Formerly we used to speak of decomposition products and of steps of decomposition; we considered the catalysts simply as promotors of such reactions. Today we know that the catalysts represent acceptors in the same sense as the decomposition products, the only distinction being by definition that the catalysts are not eliminated in the course of the process but are constantly regenerated by the oxidation-reduction cycle. This suggests a novel possibility: at a certain position of the equilibrium, e.g. at a certain potential level, a decomposition product arising in the course of metabolism may temporarily function as a catalyst by virtue of its character as a reversible redox system, until the conditions of environment change and the substance becomes once more a metabolite undergoing further degradation. This persisting of a "metabolic catalyst" may be static or dynamic in nature; a certain amount of this substance may persist for a certain length of time or else it may be rapidly decomposed and an equivalent amount may arise anew in the decomposition process. We are dealing here with the class of non-enzymatic catalysts of the living cell. They are non-enzymatic because they are simple compounds without a colloidal bearer. However, there are transitions from non-enzymatic catalysts over doubtful intermediary forms to classical enzymes, e.g. the dehydrogenases or the respiratory ferment. Examples of non-enzymatic cell catalysts (see the review by KISCH 571)) are natural quinoid pigments like pyocyanine, chlororaphine, toxoflavin, and also fumaric acid (SZENT-GYÖRGYI 1126)), ascorbic acid, or thiol compounds like glutathione. If it is shown that such a compound is only active in combination with a colloidal bearer, the system becomes an enzyme by definition. Such intermediary forms are called "vitazymes" and "hormozymes" by EULER (details on p. 87).

Considering biological oxidation from the standpoint of energetics we find that the hydrogen in the course of its transfer follows a path prescribed by the thermodynamic potential: starting at a high reduction potential level it sinks to a low level, loosing constantly free energy and performing as much work as possible in accordance with the energy requirements of cell metabolism. The energy is liberated by the chain of coupled reactions which we call oxido-reduction processes. Dehydrogenation, i.e. separation of hydrogen from a donator, in the physiological temperature range is almost always an endothermal process. Only its coupling with hydrogenation yields an over-all reaction with liberation of energy *) (See OPPENHEIMER **902**), p. 1287).

As far as these individual reactions may be conceived as isothermal and reversible processes, their free energy changes may be expressed in terms of redox potentials. Considerable data have been obtained permitting a certain insight into the sequence of oxido-reductions. The process may begin at strongly negative potentials, i.e. with substances of high reducing power like products formed from sugars. Redox systems of suitable potential range are provided as catalysts. They hand over the hydrogen to redox systems of less negative potential. Gradually the normal potential of zero (at ph 7) is reached and passed, and the positive range is entered.

In anoxybiontic metabolism, no high potential levels are attained. The transition to redox systems of strongly positive potential is usually preparatory to the participation of molecular oxygen, i.e. to oxybiosis. Here the last intermediary systems of positive potential range, especially the cytochromes and the respiratory ferment come into play. The latter, being an autoxidizable system, is capable of yielding the hydrogen to molecular oxygen, thereby terminating the chain. The last step liberates the large amount of free energy corresponding to the combustion of hydrogen to form water. This, in the last analysis, appears to be the ultimate aim of cell metabolism.

The uniform character of biological breakdown finds it counterpart in the uniform character of the enzymes promoting it, of the Oxidoreducases. WIELAND, in placing emphasis on the detachment of hydrogen from a specific donator, calls them Dehydrases, lately occasionally also Hydrokinases. The classical term Oxidases would appear doomed to sink into oblivion, unless an attempt is made to fill the old name with a new and theoretically sound meaning. We have seen that those systems which promote the hydrogenation of molecular oxygen must be autoxidizable; in some cases, e.g. in that of the respiratory ferment of WARBURG, they will react exclusively with molecular oxygen. Such enzymes as well as those which act by primary activation of oxygen via peroxide formation might perhaps properly be called oxidases. Such a course would involve the renaming of certain enzymes which formerly were designated as oxidases, e.g. succinic oxidase. It may, then, be advisable to differentiate between dehydrases (dehydrogenases) and oxidases as the two subgroups of the hydrokinases, using their acceptor specificity as the criterion.

While oxidoreductases of all kinds cover the main chain of events concerned with the hydrogen transfer and hydrogen fixation, other enzymes, too, form a part of the desmolytic mechanism. Firstly, an enzyme is required for the production of carbon dioxide from preformed molecules resulting from extensive dehydrogenation. In the case of the plant cell and of the microorganisms this is achieved with the help of

---

*) At high temperatures the equilibrium is shifted. The same metal catalysts which hydrogenate at low temperatures act now as dehydrogenases and liberate hydrogen. Bacteria may likewise form free hydrogen. The enzymes responsible for this phenomenon have only recently been studied.

carboxylase. In the case of the animal cell the origin of carbon dioxide as an end-product of metabolism is still partly obscure. Furthermore, we find catalase in all respiring cells. It has been claimed that catalase has the task of destroying hydrogen peroxide which arises as an intermediate product of metabolism. This is purely hypothetical in view of the fact that no formation of hydrogen peroxide in the cells of higher organisms has as yet been demonstrated. In addition, some auxiliary systems are required in desmolysis. Certain products of dehydrogenation reactions which can no longer be attacked in a direct manner must be hydrated previous to further breakdown. The best known Hydratase is the strictly specific Fumarase which transforms fumaric acid into malic acid. Also in the case of intermediary formed aldehydes hydratation represents a very important phase of desmolysis; the Cannizzaro reaction with aldehydes is a dismutation to alcohol and carboxylic acid. It is not quite clear as yet whether the enzymes concerned here, the mutases, are simple hydratases or whether they are also oxidoreducases which dehydrogenate one molecule of aldehyde at the expense of another aldehyde molecule as the donator. There exist other enzymes for special functions: Aspartase which catalyses the equilibrium reaction

Fumaric acid + Ammonia ⇄ Aspartic acid,

enzymes promoting the formation and the dissolution of carbon - carbon bonds.e.g. Carboligase and Aldolase (at first called Zymohexase) which catalyzes typical aldol formation from aldehydes up to a certain equilibrium point, especially the equilibrium between phosphorylated trioses and hexoses. Finally, there are some complex enzyme systems in bacteria which operate with molecular hydrogen (Hydrogenase, etc.) and with free nitrogen (Nitrogenase), e.g. in the process of nitrogen fixation.

Anoxybiosis and Oxybiosis, Fermentation and Respiration, are not two different processes but only two stages of one process. Which of these two stages exists in a cell depends on external factors (presence of oxygen) and internal factors (presence and activity of autoxidizable catalysts like the respiratory ferment). The phenomenon of the Pasteur-Meyerhof effect, i.e. the suppression of fermentation upon admitting oxygen to cells capable of respiration, is evidence of the existance of a coupling link between oxybiosis and anoxybiosis. This effect has been interpreted by Meyerhof as a measure of energetic economy: many more substrate molecules must be split in anaerobiosis than in aerobiosis to obtain the same amount of free energy. Upon admittance of oxygen a part of the large amount of energy made available by the oxidation of a few split products is utilized for the resynthesis of the rest to the original substrate (carbohydrate). Recent work (cf. Burk 157b)) indicates that the oxygen tension rather than the respiration is the decisive factor in the suppression of fermentation under aerobic conditions. Furthermore, in most instances, the decrease in the formation of fermentative end products in air appears to be due to an actual diminution of fermentation rather than to an increased resynthesis.

At least as far as the breakdown of carbohydrates is concerned, the uniformity in principle of the metabolism of all types of cells, whether animal, plant or microorganism, may be accepted as proved.

# B. Theories of Oxidoreduction.

## I. General.

### 1)  Historical:

In order to understand the simplification which has been achieved in the theory of oxidoreduction in recent years it will be necessary to sketch briefly the fundamental conceptions which form the basis of these developments.

WIELAND has built his theory of catalytic dehydrogenation around the fact that the catalytic forces attack primarily the hydrogen of the substrate. It is rendered labile and thus enabled to follow the trend of the thermodynamic potential in such a manner that the hydrogenation of the acceptor yields more energy than is required for the dehydrogenation of the substrate (donator). It is assumed that during this process donator as well as acceptor are adsorbed on the surface of the catalyst. The mechanism is essentially the same whether the reaction is brought about by unspecific or mildly specific catalysts of known constitution, e.g. metal surfaces, or by specific biological catalysts, i.e. enzymes. It is further assumed that the specificity of these catalysts, called dehydrases (hydrokinases) depends entirely on the structure of the organic radical containing the hydrogen subject to labilization. Therefore the dehydrases are named after the effective donator which may be for instance an aldehyde, succinic acid, hydroxy acids, or a purine base. No acceptor specificity is assumed. Any thermodynamically permissible acceptor, also oxygen, will do. Only a few acceptors are excepted in the case of normal biological catalysts, e.g. peroxides with the inclusion of hydrogen peroxide and disulfides. This concept encompasses the ultimate oxidation of hydrogen to water with the intermediate formation of hydrogen peroxide.

The validity of this scheme for the state of a n o x y b i o s i s has never been seriously questioned. The fight between the schools of WIELAND and WARBURG revolved around the question of the ultimate oxidation and of oxidation by oxygen in general. Originally, WARBURG did not recognize an "ultimate" oxidation; the oxygen was thought to attack the substrate immediately in all types of biological oxidations. The primary phenomenon was the "activation" of oxygen, occurring with the aid of iron. The iron content of the organic catalyst was taken as the only important feature. Molecular oxygen cannot attack organic compounds directly: always the ferrous iron of the catalyst is converted into a state of higher oxidation. The latter represents "oxygen in active form" which is capable of oxidizing the substrate. The proof for the decisive importance of iron was seen in the specific inhibition of the reaction with molecular oxygen by hydrocyanic acid. Iron forms stable complexes with this inhibitor. [Certain instances of non-inhibition by HCN have already been observed by WARBURG, for instance in the case of the respiration of the alga Chlorella.] WARBURG rejected WIELAND's idea of an oxidation of "activated" hydrogen by "non-activated" oxygen, and the activation of hydrogen altogether. He admitted only the possibility of an unspecific "activation" of the entire substrate molecule by the surface forces of the catalyst.

However, if we look into this matter from a chemical point of view such a non-specific activation can hardly be anything else but the provision of "available" hydrogen; WARBURG's model reactions (e.g. the oxidation of cysteine on charcoal) are actually dehydrogenations. The mere fact that many catalytic oxidations with molecular oxygen are highly specific with respect to the substrate is sufficient proof that the primary step in the catalysis occurs within the substrate molecule, i.e. at the hydrogen atoms of the substrate. The oxidation of succinic acid to fumaric acid by molecular oxygen, as catalyzed by succinic dehydrogenase, is a suitable example. But even if the necessity for an "activation" or labilization of hydrogen for the case of oxidation by molecular oxygen is admitted, the original theory of WIELAND is unable to explain this "ultimate" oxidation. The postulate of the free acceptor choice by "activated" hydrogen is untenable. The claim of WARBURG that a new type of catalysis with the participation of iron intervenes at this stage was irrefutable. WIELAND was unable to explain satisfactorily the effect of hydrocyanic acid. The same holds for the special role of the peroxidases. The interpretation by WIELAND that they are probably phenol dehydrogenases which show an acceptor specificity for hydrogen peroxide punctures his postulate of free choice of acceptors. Finally, the well known fact that during fermentation in presence of air the "active" hydrogen of fermentation does not combine with oxygen but with the thermodynamically much weaker acceptor acetaldehyde refutes the claim that activated hydrogen may readily and under all conditions combine with molecular oxygen.

Several years ago it was pointed out by OPPENHEIMER 902) that WIELAND's theory is deficient in this respect and that WARBURG was justified in postulating a special mechanism for oxidation by molecular oxygen, namely catalysis via iron systems. OPPENHEIMER made an attempt to reconcile the two theories by assuming that both reactants, hydrogen as well as oxygen, must be "activated" in order to react with each other. The activation by iron according to WARBURG was then considered to be the most important form of oxygen activation. However, it was thought possible that any other form of "activated" oxygen would do, e.g. peroxides like hydrogen peroxide or organic peroxides as well as the quinoid "respiratory pigments" of PALLADIN. This view was soon afterwards accepted in principle by HOPKINS 497) and by RAPKINE and WURMSER 963).

Subsequent developments have changed this conception of the subject with respect to details. The question as to how oxygen is activated is no longer to the point inasmuch as WARBURG has provided a quite different interpretation. Furthermore, WIELAND has given up his theory of the free choice of acceptor; instead he believes that the dehydrogenases are only able to transfer hydrogen to oxygen if they have a special affinity for oxygen as well as for hydrogen, i.e. if they act in a sort of ternary system. This development of fundamental importance will be dealt with in detail on p. 34.

## 2) General Theory of Catalysis by Oxido-Reduction.

Before we undertake to draw a picture of this type of catalysis it is appropriate to emphasize that there are two essential reservations in the title. Firstly, it does not read "oxidative" catalysis but purposely "catalysis by oxido-reduction". This means that no attempt is made to explain all of the phenomena in this field. The so-called peroxide catalysis, i.e. the "oxygen activation" in the more restricted and more or less historical sense of ENGLER-BACH-MANCHOT will deliberately be omitted from this sketch. It will be discussed in detail on p. 118. Secondly, the theory is called "general", because many important points are not yet elucidated though the leading workers have reached an agreement on certain fundamentals.

The theory to be outlined here does not lay claim to absolute correctness. It is to be taken as a working hypothesis; it is subject to changes in accordance with evidence- forthcoming in future work.

The general theory begins with the movement of the hydrogen. It goes to a catalyst representing a reversible redox system; the catalyst, thereby being reduced, is the first hydrogen acceptor in the chain. It is in turn dehydrogenated by a thermodynamically stronger system. If the latter is to be molecular oxygen, the special condition must be satisfied that the reduced form of the catalyst must be autoxidizable.

### a) Dehydrogenation and Hydrogenation as Electron Shifts.

Every oxidation is by necessity coupled with a reduction, every dehydrogenation with a hydrogenation. The fundamental reaction

$$DH_2 + Acc \rightarrow D + AccH_2$$

is independent of the kind of donator or acceptor; the acceptor may, of course, be molecular oxygen. The over-all process is nothing but a redistribution of the electrons within this system. For the purposes of the exposition of the general theory we shall retain, for the present, WIELAND's assumption of simultaneous transfer of two hydrogen atoms. The increasing importance of two-step reduction and oxidation, i.e. of transfer of unpaired electrons, will be discussed on p. 100.

The reader need hardly be reminded that oxidation of a reactant is equivalent to loss of electrons, and reduction to gain of electrons. In the case of "simple" oxidations this statement covers all of the events. If we take the case of oxidation of ferrous to ferric iron

$$Fe^{++}(Cl^-)_2 + \tfrac{1}{2} Cl_2 \rightarrow Fe^{+++}(Cl^-)_3$$

the $Fe^{++}$-ion yields one electron to the neutral chlorine atom, is thereby oxidized to $Fe^{+++}$-ion and can now bind the freshly charged Cl-ion. However, if we are dealing with dehydrogenation processes the electron shift is accompanied by $H^+$-ion shifts. Hydrogenation consists in transfer of entire hydrogen atoms, i.e. of $H^+ + \varepsilon$. In case the H-atom is already bound in ionized form, the preformed ion together with the electron are simply transferred to the neutral atom to be hydrogenated, for example

$$S^=(H^+)_2 + I_2 \rightarrow 2\ I^-H^+ + S.$$

Similarly, in polyphenols the hydrogen is bound in a dissociable manner so that it may easily be given off resulting in quinone formation (See below). Lately this case has acquired interest from the enzymatic standpoint with reference to the activation of molecular hydrogen: STEPHENSON and STICKLAND [18] have discovered an enzyme in bacteria, called Hydrogenase, capable of transferring molecular hydrogen to various acceptors. This enzyme is, then, a true reducase as contrasted to the oxido-reducases. It is responsible for the long known uptake of free hydrogen by micro-organisms and also for the utilization of nascent hydrogen produced by the action of other enzymes, the Hydrolyases. With respect to the mechanism of action of hydrogenase it may be mentioned that hydrogen peroxide which is formed by aerobic organisms (e.g. *Acetobacter peroxydans*) which are devoid of catalase may function as acceptor of this hydrogen (WIELAND 1321)); in anaerobiosis carbon dioxide may be the acceptor. B. coli act towards HD like platinum black insofar as they rearrange it in presence of water to $H_2$ + HDO (CAVANACH et al. 165)).

In ordinary dehydrogenation reactions with organic compounds the hydrogen atoms subject to transfer are bound not in ionized state but as atomic hydrogen. Here, the reaction must therefore begin with the labilization and subsequent ionization of the hydrogen, thus enabling the electrons to combine with the acceptor and charge it negatively. This in turn, renders the acceptor capable of binding the $H^+$-ions. This uptake may result in a polar nature of the hydrogen with respect to the rest of the molecule; or else the hydrogen may be bound in ionized form. The dehydrogenation of succinic acid by a quinoid pigment, here symbolized as quinone itself, may serve as an example:

$$CH_2 \cdot COOH \atop CH_2 \cdot COOH \quad \rightarrow \quad {CH \cdot COOH \atop CH \cdot COOH} \quad + 2\,H^+ + 2\,\varepsilon +$$

Of course, the hydrogen may also come to rest in a more stable type of bond, in truly main valency linkage. In that case the whole hydrogen atom, $H^+ + \varepsilon$, is transferred as a unit. This occurs, for instance, if acetaldehyde is changed into ethyl alcohol by the uptake of 2 H from some kind of donator:

$$DH_2 + CH_3 \cdot CHO \rightarrow D + CH_3 \cdot CH_2OH.$$

Finally, this is also true if molecular oxygen is the acceptor. In this case water is the little dissociating end product, hydrogen peroxide in which the hydrogen is linked in a strongly polar manner being the intermediate. The change from the neutral oxygen molecule to the charged ionized oxygen is the important feature, e.g. in the spontaneous dehydrogenation of leuco dyes by air. The reduced form of the dyestuff may again be represented by hydroquinone:

$$\rightarrow \quad + 2\,H^+ + 2\,\varepsilon + O_2 \rightarrow H^+ \cdot {}^-O - O^- \cdot H^+$$

A second molecule of hydroquinone furnishes another $2\,H^+ + 2\,\varepsilon$:

$$H - O - O - H + H_2 \rightarrow 2\,H_2O.$$

This acquisition of a charge by the oxygen is always preceded by a close approach of the oxygen molecule to the substrate, whether we are dealing with a spontaneous process (autoxidation) or with a catalytic oxidation. In extreme cases this mutual contact may lead to peroxide formation involving chemical bonds to the oxygen molecule. Due to the intimate contact with the substrate the O=O molecule presumably suffers a deformation which lowers its energy of activation. The oxygen is thereby rendered more reactive. This first act might be called "activation of oxygen" if it occurs in a catalysis, i.e. at a surface. In this sense even WIELAND 1301) has no objection to this term. Here we have a primary step common to all theories: The addition of —O—O— takes place in any event, whether the result be a hydrogenation with subsequent splitting off of hydrogen peroxide or whether an organic peroxide is formed (for details see p. 14).

For these fundamental considerations the particular nature of the acceptor does not matter. All that is required is, 1) it should be capable of taking up hydrogen, 2) the over-all reaction should yield free energy, 3) the acceptor should have affinity for the donator-catalyst system (p. 34).

It has been known for a long time that besides the quinoid dyestuffs other substances of various types, e.g. nitrates or chlorates, may act as acceptors. Due to the recent attempts to discover the nature of the enzymes active in bacterial reduction processes this aspect has acquired renewed interest. It has been shown that the various thermodynamically possible acceptors are biologically differentiated in a strange manner. The animal cell is incapable of utilizing a number of acceptors which bacteria use with the greatest of ease. According to BURK [22] the fixation of molecular nitrogen is brought about by an enzyme system called Azotase; an enzymatic component, Nitrogenase, transfers active hydrogen to molecular nitrogen. The reduction of nitric oxide, NO, to nitrous oxide, $N_2O$, by a biological system consisting of the yellow enzyme and hexosephosphate has been described by MEYERHOF [799]. Whether this is a true catalysis or some other mechanism is not yet clear, inasmuch as NO may be reduced readily without a catalyst, e.g. by leuco dyes; the reduction by pyrogallol, however, requires the presence of palladium ([122]). The reaction is strongly exothermal ($\triangle U = -51,000$ cal.). The mechanism of the reduction of organic sulfur compounds to hydrogen sulfide by an enzyme found by TARR [1139] is likewise still obscure. In any event, it is a desmolytic process, probably a reductive scission of the carbon-sulfur linkage by activated hydrogen. An amino group in the substrate appears to be necessary for the reaction. [*] Similarly, hydrogenations of inorganic sulfur compounds have been observed, e.g. that of sulfate at the expense of ethyl alcohol which is dehydrogenated (BAARS [49]). In the course of this process, as brought about by *Vibrio desulfuricans* and *Vibrio Rübentschickii*, hydrogen sulfide is formed. This and similar reactions have been discussed by KLUYVER [577]. The ethyl alcohol is quantitatively transformed into acetic acid. Analogous results have been obtained by STICKLAND [1101] and TARR [1139]. It is possible that the Rhodanese of LANG [695] belongs to this group. This enzyme is present in preparations obtained from animal tissues. It synthesizes rhodanide from hydrogen sulfide and hydrocyanic acid; the former originates from thiosulfate as the donator.

The biological reduction of nitrate by bacteria is catalyzed by a carbon monoxide-sensitive iron system (QUASTEL [944]); the cyanide-inhibition of the enzyme is well-known. The enzyme has not yet been isolated. It is also of interest that $CO_2$ (or rather $H_2CO_3$) may be an acceptor in biochemical reactions. It is reduced by molecular hydrogen to form methane with the aid of Hydrogenase ([1070]). [**] This reaction had previously been observed with living bacteria. Hydrogen sulfide is used by sulfur bacteria as donator for the photochemical reduction of $CO_2$. In this respect it is important that WASSERMANN [1283] has observed a rapid decomposition of $H_2S$ by an iron system plus hydrogen peroxide. A rather complete biological oxidation system has been found by BARKER [71]: methane bacteria form methane from $CO_2$ at the

---

[*]   The existence of an enzyme, called Cysteinase, which forms $H_2S$ from *l*-cysteine, has now been demonstrated in B. coli (DESNUELLE and FROMAGEOT, Enzymologia, **6**, No. 1 (1939)).

[**]   See, however, OPPENHEIMER's "SUPPLEMENT", p. 1699.

oxpense of alcohol which is dehydrogenated to acetic acid. The question is whether free hydrogen is formed as an intermediate. It is true that enzymatic formation of free hydrogen by hydrolyases has been observed, e.g. in the case of B. coli which dehydrogenate formic acid to $CO_2$ and $H_2$ (STICKLAND and STEPHENSON 1070)). Methane formation has been observed also in this instance, probably by reaction of the two reactants in statu nascendi. Otherwise, $H_2$ is released as such, without any acceptor; this is probably due to the negative potential of the coli bacteria. The hydrolyases are not identical, one enzyme of that group forms hydrogen from glucose; the hydrolyase, specific for formic acid, is different from formic acid dehydrogenase which will not act without an acceptor. The hydrolyase action is reversible (WOODS 1344)).

Electron Formulas: An attempt has been made recently, especially by E. MÜLLER 843), to represent the activation of hydrogen, i.e. the labilization of the electron, more in detail on the basis of the octett theory. His formulas are based on the assumption that the substrate exist not only in its normal or stable configuration (A) but also in an electron isomer form (B) in which the hydrogen atoms later subject to dehydrogenation are already present in labile form. Labile is used in the sense that they are linked only through one electron so that they are already polar and almost "ionized". This conception of labile hydrogen is a quite familiar one nowadays; hydrogen peroxide is an example (see p. 22).

This question has been treated in detail by FRANKE (cf. 288)). One of the formulas discussed by him, relating to the dehydrogenation of alcohol, follows:

$$
\begin{array}{ccc}
\text{H  H} & \text{H  H} & \text{H  H} \\
\text{H : C : C : O:} \rightleftarrows \text{H : C : C : O :} + \text{Acc} \rightarrow \text{H : C : C : : O} + & \text{H·}\\
\text{H  H  H} & \text{H  H  H} & \text{H} & \text{Acc} \\
\text{A} & \text{B} & & \text{H··}
\end{array}
$$

For another formulation, based on the number of oxidation steps, see JIRGENSONS 503).

### b) Energetics and Redox Potentials.

The coupling of oxidation-reduction reactions is required from the standpoint of energetics: the over-all reaction dehydrogenation-hydrogenation, like all other chemical reactions, is a spontaneous process if it yields free energy. But in all cases which are of interest here, the dehydrogenation, i.e. the ionization and detachment of hydrogen, is not spontaneous but requires the input of free energy *). The process is possible under biological conditions only because the hydrogenation, i.e. the acquisition of a negative charge by the acceptors and the attachment of hydrogen atoms, yields more free energy than is consumed by the dehydrogenation. Thus the reaction may serve the energy production by the living cell, by utilizing the free energy of hydrogen in a gradual manner. Of course, if the ultimate oxidation of hydrogen is

---

*) An exception is the dehydrogenation of aldehyde hydrates to carboxylic acids which proceeds spontaneously as WIELAND demonstrated in 1912. The (hypothetical) hydrate of acetaldehyde decomposes spontaneously without an acceptor when brought in contact with palladium. The claim of WIELAND that hydroquinone is also dehydrogenated to quinone by palladium black has recently been refuted by GILLESPIE and LIN 403) both on thermodynamical and on experimental grounds. The reduction of $CO_2$ by hydrogenase systems which we have mentioned above is an apparent exception. But this reaction does not occur with the intermediary formation of aldehydes and it is exothermal owing to the accompanying formation of water. The alleged reduction of butyric acid to butyl alcohol during fermentation should likewise be a coupled reaction.

effected by molecular oxygen, the entire energy of the combustion of hydrogen to water is released at once.

This is the content of the postulate of WIELAND's theory (see p. 108) according to which the hydrogen can only combine with a "thermodynamically permissible" acceptor. It implies the statement that the over-all reaction cannot occur spontaneously and cannot, therefore, be catalyzed, if this basic postulate is not fulfilled. A thermodynamically "weaker" acceptor can never bring about the dehydrogenation of a "stronger" donator. VESTERBERG 1198) has calculated for some dehydrogenation processes which acceptors are thermodynamically admissible. OPPENHEIMER 902), following WIELAND, had tabulated previously the possible and the impossible coupled reactions on the basis of the well known caloric values. Today it is possible, in many instances, to perform the calculations using the free energy values which lead to a more correct picture. If the coupling between dehydrogenation and hydrogenation is fully reversible and if true equilibria are attained, the electron shifts or rather the tendency of the electrons to shift may be directly measured in terms of the redox potentials. From the potential values the position of the equilibria and from the latter the affinity, i.e. the free energy of the reaction (see p. 50) may be calculated. Another method for the calculation of free energy values has been made available by the determination of the entropy of formation of the compounds. Thereby a much more secure basis for such considerations is provided than was formerly available by the caloric values. If the potentials or the entropies are known it is possible to foretell exactly the manner in which different redox systems may react with each other, in other words, to predict which dehydrogenation-hydrogenation processes are possible. Whether they will actually occur is no longer a question of thermodynamics but of reaction kinetics.

If we look at the process in the direction of the dehydrogenation of an isolated system which is at equilibrium, dehydrogenation can take place only if the oxidized form of another redox system which is added has a stronger affinity for hydrogen than the first system. In terms of potentials this would mean that the added system, in order to function as acceptor of hydrogen, must have a higher (less negative or more positive) potential than the system to be dehydrogenated. In that case the first reaction step $DH_2 + Acc \rightarrow D + AccH_2$ will take place, provided there exist no kinetic hinderances. The extent to which this reaction will proceed and where equilibrium will be reestablished depends, under otherwise identical conditions, on the difference between the affinities of the two systems as expressed by the difference in redox potential. Complete dehydrogenation can occur only if the difference between the potential values reaches a certain value, in other words, if the maximum of work is performed when the equilibrium of the reaction just given is completely shifted to the right. This will be illustrated by numerical examples on p. 50. If, as is frequently the case in biochemical reactions, the potential difference is small, the dehydrogenation reaction does not go to completion. Equilibria result which may be abolished in various ways. Either the dehydrogenated form of the original donator is freshly reduced by another donator or the reduced form of the original acceptor becomes a donator for a stronger acceptor. In either case the equilibrium is disturbed and the dehydrogenation progresses. If we consider the reversible redox system succinic acid-fumaric acid as an example,

$$\begin{array}{cccc} CH_2 \cdot COOH & & CH \cdot COOH & \\ | & \rightleftarrows & \| & + H_2 \\ CH_2 \cdot COOH & & CH \cdot COOH & \end{array}$$

it is perfectly stable if considered as an isolated system. A given mixture of the two components remains unchanged. Suppose p molecules of succinic acid yield 2 H each to p molecules of fumaric acid. On the other hand this would give rise to p molecules of freshly formed succinic acid, but at the same time p molecules of fumaric acid would be formed by the dehydrogenation of the succinic acid molecules. The system is therefore in dynamic equilibrium. But if a quinoid dye system of similar potential, like methylene blue, is added to this equilibrium mixture, new equilibria are established as follows: If we start with succinic acid (S) and methylene blue (MB) only, and if we designate fumaric acid as F and leuco methylene blue as LMB, the reaction in the system $S + MB \rightleftharpoons F + LMB$ will of course proceed only from the left to the right. The more fumaric acid and leuco methylene blue are formed the stronger the tendency becomes for hydrogenation of fumaric acid by the leuco dye, i.e. for the back reaction. At a certain numerical ratio the reaction reaches dynamic equilibrium between the four molecular species. The position of the equilibrium is thermodynamically fixed by the value of the maximum work; it can be determined by measuring the redox potential of the equilibrium mixture of the four components. If the electrode equations and the normal potentials of both systems are known, the measured value at the equilibrium point permits one to calculate the concentrations of all four reactants and the ratios of oxidized to reduced form in both systems. If a new acceptor in the shape of a system of more positive potential is added to the equilibrium mixture the process will again go on. The acceptor dehydrogenates the leuco methylene blue to methylene blue which, in turn, will be able to dehydrogenate more succinic acid. In this way the reaction may proceed via different redox systems. The potential will rise gradually until eventually the systems with very high potentials, $O_2$ or $H_2O_2$, come into play and the hydrogen transfer is completed by a practically irreversible reaction entailing the loss of the last free energy content of the hydrogen.

## c) Catalysis.

So far we have dealt with the thermodynamical aspects of dehydrogenation. It is a different question whether such a reaction will take place at all and at what time. This question relates to the "friction" in the reacting system and to its diminution by the "lubricants" represented by catalysts. Of course, there are reactions with a very small internal friction so that they can proceed without being catalysed. By catalysis we refer here, in agreement with MITTASCH, exclusively to true increase in the rate of reaction and to control of the direction of a reaction by chemical substances and not to the effects of heat, radiation, ionization, and so on.

Amino acids, for instance, are dehydrogenated by quinone without a catalyst, just as by alloxan at the boiling point (STRECKER's reaction). There are also simple systems which react with molecular oxygen without the mediation of catalysts and which are, therefore, truly autoxidizable. The leuco forms of methylene blue and of similar dyes appear to be such compounds. According to MACRAE 742) they are autoxidizable in weakly alkaline solutions (ph = 8). However, REID 971, 972) finds that the process occurs only on surfaces, particularly on protein surfaces (at ph 7.5), and that narcotica inhibit the oxidation. At acid reaction, the reoxidation of leuco dyes is a cyanide resistant copper catalysis (HARRISON 453)), MACRAE 742)). Hydroquinone and dihydroxymaleinic acid are said to be autoxidizable (WIELAND and FRANKE, see p. 31). According to HILL 484) dialuric acid is autoxidizable at neutral reaction, whereas at acid reaction iron functions as a catalyst (STRECKER's reaction, p. 107).

The existence of systems which will react directly with molecular oxygen is the prerequisite of every catalytic end oxidation. In many instances it is the catalyst itself which is autoxidizable in the reduced form. The most important case of this kind is undoubtedly the autoxidation of complex salts of ferrous iron, e.g. that of ferrous cysteine or of heme derivatives. With the exception of the latter, autoxidizable substances in living matter have little to do with the important reactions. In almost all biologically significant cases neither dehydrogenations by an acceptor nor reactions with molecular oxygen occur without suitable catalysts.

.It may be well to point out at this time that the typical model reaction consisting of dehydrogenation of succinic acid by methylene blue or other quinoid dyes does not take place at an appreciable rate unless tissue extracts, containing succinic dehydrogenase, are added. The catalysts participating in hydrogen transfer during anaerobic metabolism may be enzymatic or non-enzymatic in nature. The quinone of adrenaline, called the "Omega Substance" by KISCH, is an example of the latter.

The general basis of our discussion is that, according to the views of WIELAND, the first step in biological oxidation is the activation of the hydrogen. At the same time, the surface forces of the catalyst activate the acceptor. In case the acceptor is molecular oxygen, this means an activation of oxygen. We repeat that by "activation" we designate the process whereby the reactants are acted upon by the field of surface forces of the catalyst resulting in a deformation of both the donator and the acceptor molecules; both are thereby made more reactive. This will be discussed in detail on p. 14 with respect to the molecular oxygen and on p. 27 with regard to hydrogen.

There exist no qualitative but only quantitative differences between the dehydrogenases and the oxidases. The main difference is not one of catalytic mechanism but of thermodynamics. The enzymes which, by historic custom, are called oxidases, dehydrogenate substances of so high a potential that, practically speaking, only molecular oxygen or hydrogen peroxide, besides some quinones, may act as acceptors. In the case of some oxidases, e.g. of glucose oxidase, the situation is still obscure; perhaps, certain acceptor specificities play a role in addition to the potential range. Both groups of biocatalysts are sub-classes of the Hydrokinases (WIELAND). The oxidases may be further divided into "true" oxidases, containing heavy metal, and the cyanide resistant oxidases which occupy an intermediate position between the true oxidases and the dehydrogenases. (See p. 127).

The general theory of dehydrogenation catalysis including the reaction with molecular oxygen, as outlined above, combines the "antithesis" of WARBURG and of WIELAND.

## Oxidation by Oxygen, Activation of Oxygen:

If we speak of "activation" of oxygen we imply that the ordinary, free oxygen is inert when brought in contact with many substances which are readily oxidized by oxygen in another form. In compounds containing "active" oxygen this element is not bound as the molecule $O = O$ but in main valency linkage, as in permanganate, nitric acid, and peroxides of various kinds, e.g. hydrogen peroxide. It is evident that before molecular oxygen will react it must first be converted into a form corresponding to that existing in compounds with "active" oxygen. The primary step in any "autoxidation", i.e. in any reaction with molecular oxygen, is a loose combination of the substrate (or catalyst) with the oxygen; it may be an adsorption complex or a residual valency compound.

QUASTEL **943**), in agreement with the modern conception of the catalytic activity of surfaces *), assumes that the substrate is bound by the "active centers" in the catalyst surface where the deformation of the $O = O$ molecule is achieved by strong polar forces. The excess energy of the active centers is transferred to the oxygen molecule as energy of activation (ROMAN **997**)) which is required for the subsequent reaction. The valency forces are hereby shifted, as can also be calculated with the aid of quantum mechanics. The molecule is somehow "unfolded". It is assumed that it is combining loosely with the substrate in the form —O—O—, forming an a d d u c t (addition compound). This activation does, therefore, not primarily concern the free oxygen molecule but is inseparably connected with the binding of the molecule by chemical forces **).

This is true for every reaction involving molecular oxygen, whether the other reactant is an organic compound or a simple metal salt or a metal complex. It is valid even in the case of the simplest reactions in the gas phase, e.g. between sodium vapor and oxygen, as has been proven in a strictly kinetic manner by HABER **435**). He could show that previous to the reaction one atom of sodium combines with one oxygen molecule. The same holds for non-catalysed dehydrogenation reactions like the yielding of two hydrogen atoms to oxygen by leuco methylene blue, also for the oxidation of ferrous to ferric iron. This primary reaction in catalytic processes is important if the reaction is concerned with a reversible redox system and if the latter acts as an oxidation-reduction catalyst by virtue of the fact that its oxidation or dehydrogenation is fully reversible.

Following this primary step the process may proceed in two different ways. The loosely bound "unfolded" —O—O— molecule may accept two electrons and thereby be transformed into the oxygen ion $O_2^=$. Simultaneously, 2 $H^+$-ions combine with this ion to form H—O—O—H which is removed from the surface of the substrate ***). This is the general scheme of typical dehydrogenation. The other possibility is that the

first adduct is changing into a better defined peroxide, in the case of iron $Fe^{++} \diagup \!\!\! \diagdown \genfrac{}{}{0pt}{}{O}{O} \Big|$ .

It is difficult to draw a line between an adduct and a peroxide. Both may react in the same manner. The second possibility, therefore, is only important if the peroxide formation cannot be followed by the splitting off of hydrogen peroxide. The mechanism of catalysis may then be a quite different one.

The first mentioned scheme is the normal one for most of the biological catalyses with which we are here concerned. It holds for purely organic and for metallo-organic (heavy metal) systems. The only difference is that in the case of the purely organic redox systems the hydrogen comes directly from the molecule of the substrate, whereas

---

*) For fundamental considerations concerning these questions see FRANKENBURGER **357**), SCHWAB **1039**), PIETSCH et al. **927**).

**) There exists also a true, primary activation of oxygen by light, in presence of fluorescing dyes. In this photosensitized oxidation the oxygen molecule stores additional energy and acquires an activated state which may lead to the formation of free atoms. This mechanism, however, does not seem to be of importance in normal biological oxidative catalysis.

***) In the reaction between atomic hydrogen and molecular oxygen hydrogen peroxide is formed too (see FRANKE **288**), pp. 118, 159). Here $H_2O_2$ is the product of chain reactions ($H + O_2 \rightarrow HO_2$; $HO_2 + H_2 \rightarrow H_2O_2 + H$) (BATES **88**)).

in heavy metal catalysis it is derived only indirectly from the substrate and directly from the water which is formed in the reaction itself (or from HCl, etc.); and that, owing to this hydrogen shift, the valency of the metal is changed. It will be shown later that the hydrogen ions are indeed originating, in the last analysis, from the substrate.

If oxygen is the acceptor and leuco methylene blue the donator, the simple equation holds:

$$1)\ DH_2 + O_2 \xrightarrow{\ 2\,H^+ + 2\,\varepsilon\ } D + H\!-\!O\!-\!O\!-\!H$$

In the other case, if the ferrous iron, bound either in a salt linkage or in complex linkage, is symbolized as $>Fe^{++}$, we have the equation:

$$2)\ 2\ >Fe^{++} + 2\,H_2O + O_2 \xrightarrow{\ 2\,H^+ + 2\,\varepsilon\ } 2\ >Fe^{+++}\!\cdot\!OH^- + H\!-\!O\!-\!O\!-\!H.$$

Hydrogen peroxide, however, cannot be demonstrated to be one of the product molecules because it is further decomposed catalytically to water and oxygen (p. 22).

The catalysis proceeds due to the fact that D takes up fresh hydrogen from another donator, D thereby becoming an acceptor. In heavy metal catalysis, $>Fe^{+++}$ also attracts hydrogen atoms, deprives them of their electron and is thereby reduced to $>Fe^{++}$; the anion $OH^-$ reacts with the proton ($H^+$) to form water (p. 19). The bulk of biological oxidation is a catalysis by heavy metal, notably by the hemin system of WARBURG. To a smaller extent metal-free, purely organic systems may be responsible for oxidative catalysis, e.g. the flavin system (see p. 108). The "methylene blue respiration" is the model reaction for such pure "acceptor respirations".

It is possible that some borderline cases may eventually be shown to be dehydrogenation reactions, e.g. the oxidation of carbon monoxide to carbon dioxide in aqueous solution with palladium as the catalyst. In this case an oxidation by oxygen formed by water decomposition might be assumed (W. TRAUBE 1185)). However, WIELAND 1310) takes the point of view that in the course of this heterogenous catalysis formic acid is first formed by simple hydration and that this acid is dehydrogenated in the usual fashion by $O_2$: $CO + H_2O = H \cdot COOH$; $H \cdot COOH + O_2 = CO_2 + HOOH$. Hydrogen peroxide has been detected both in the CO- and in the $H_2$-flame by DIXON in 1882. (For further reference to the discussion of this process by WIELAND and W. TRAUBE see FRANKE (l.c. 288) p. 160)).

The conception that in every case of interaction with molecular oxygen the $O_2$ molecule must f i r s t  b e  b o u n d and deformed or activated, has been of great value for the unification of the theory. This may be demonstrated with respect to three points.

First, this conception has lead to the important modification of WIELAND's theory, namely, to its supplementation by the postulate of a c c e p t o r  s p e c i f i c i t y, notably by BERTHO (p. 34) but also by WIELAND himself. The acceptor specificity explains the different behaviour of donator-catalyst systems towards molecular oxygen. If the $O_2$ molecule cannot be bound, the system is not autoxidizable and no reaction takes place. It is useful in interpreting the problem of peroxidase specificity. Here, not oxygen but hydrogen peroxide is the acceptor. The peroxidases act as phenol dehydrogenases only because of their ability to bind specifically the acceptor $H_2O_2$. How far there exists an a b s o l u t e specificity for $O_2$, even if other acceptors are thermodynamically permissible, and whether this specificity is the reason for the behaviour of the "oxidases" cannot definitely be decided as yet. It is also possible that such acceptor specificities play a more important rôle in purely anaerobic acceptor catalyses than has so far been suspected. They may explain why certain catalyses fail to occur which are thermodynamically possible (see p. 34).

Second, this concept is the key to the fundamental change which WARBURG has made in his theory and which concerns the v a l e n c y c h a n g e of the iron. Previously he reckoned with some sort of peroxide formation at the iron. The peroxide was supposed to transmit atomic oxygen to the previously "activated" substrate in agreement with the views of ENGLER and BACH. Today, WARBURG assumes that primarily a short-lived adduct is formed between the ferrous iron of the heme of the respiratory enzyme and the molecular oxygen. The model for this adduct is oxyhemoglobin which is readily dissociable and which yields its oxygen in non-activated form. The oxygenation product of the hemin enzyme is unstable and rearranges itself to form the true oxidized form of the enzyme with ferric iron. The change from ferrous to ferric iron involves, of course, the loss of an electron which is taken up by the oxygen. The model for the ferric form of the respiratory enzyme is methemoglobin. Just as hemoglobin-methemoglobin is a redox system (CONANT 179)), so is the system formed by the ferrous and the ferric form of the enzyme.

**Peroxide Formation:** Thirdly, this concept helps to explain the fact that aside from the dehydrogenation catalysis there exists still another form of oxidative catalysis where the older theory of ENGLER-BACH retains its value. Oxygen is capable of combining with metals without valency change to form peroxides, and also with organic biological systems where there is no primary dehydrogenation. We think of the mere addition of oxygen to valency gaps, e.g. to —C = C— double bonds, to form

—C—C—.
 |   |
 O—O

In discussing such organic peroxides two questions suggest themselves. The questions 1) as to how such peroxides are formed and 2) whether they are catalytically active; in other words, do they represent reversible systems of the type $X + O_2 \rightleftarrows X \cdot \cdot O_2$ which can transfer peroxidic, i.e. activated, oxygen? The second question has lost much of its importance. Most of the substances which formerly were looked upon as organic peroxides with catalytic activity have later been shown to be quinoid redox systems. There remains a certain possibility, however, that the catalytic oxidation of unsaturated fatty acids proceeds via such peroxides. According to FRANKE 354), carotenoids are active as catalysts in such reactions, perhaps through intermediary peroxide formation. This and the role which defined peroxides play not as catalysts but as intermediaries in dehydrogenation, e.g. in the case of aldehydes, will be discussed below (p. 118).

The formation of these organic peroxides is of considerable interest. A variety of them are known. They arise partly through simple autoxidation, e.g. from aldehydes, partly through heavy metal catalysis. It has already been mentioned that a loose adduct of molecular oxygen to a substrate may undergo molecular rearrangement resulting in peroxide formation. The pronounced polarity of peroxides, of hydrogen peroxide for instance, indicates that they represent not quite saturated main valency compounds. Peroxide configuration will result instead of mere adduct formation if there are no $H^+$-ions available for the splitting off of $H_2O_2$, i.e. if true dehydrogenation is not (or at least not instantly) possible. The scheme for organic peroxide formation may be illustrated for the example of aldehyde autoxidation in a non-aqueous medium:

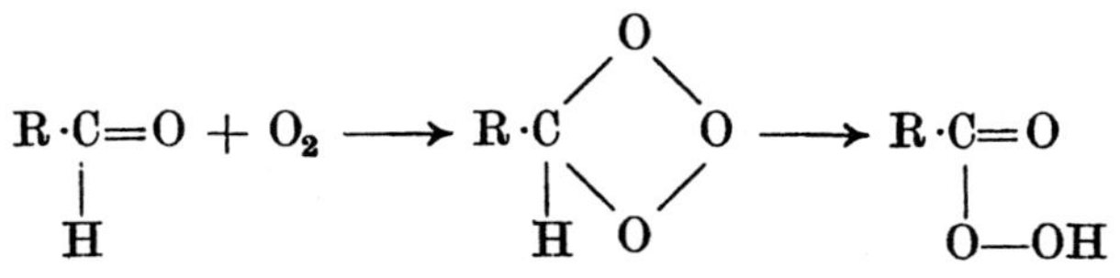

$$R \cdot C = O + O_2 \longrightarrow R \cdot C \overset{O}{\underset{H \quad O}{\diamondsuit}} O \longrightarrow R \cdot C = O$$
$$| \qquad\qquad\qquad | \qquad\qquad\qquad |$$
$$H \qquad\qquad\qquad\quad\quad\quad\quad\quad\quad O-OH$$

**Aldehyde**          **Adduct**          **Peracid**

(ENGLER's Moloxide)

In this case we cannot speak of a hydrogen activation; the residual affinity of the double bond (or "activation of the carbonyl" in WIELAND's terminology) suffices to "unfold" the oxygen molecule and to bind it. With benzaldehyde this takes place even in aqueous solution. It is noteworthy that such reactions, according to WIELAND, are encountered as intermediates in true dehydrogenation. By reacting with a second aldehyde molecule the peracid yields the carboxylic acid:

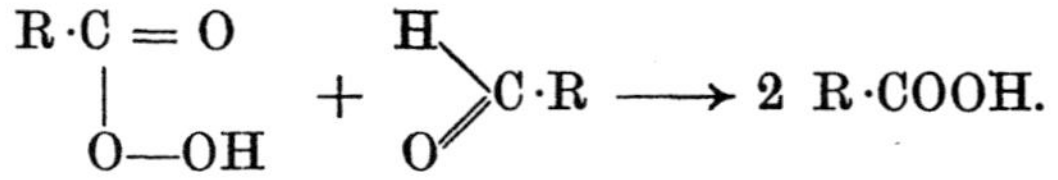

$$R \cdot C = O \qquad H$$
$$| \qquad\qquad + \qquad > C \cdot R \longrightarrow 2\ R \cdot COOH.$$
$$O-OH \qquad\quad O$$

This is the same result as if the aldehyde were dehydrogenated in the usual manner via the aldehyde hydrate.

WIELAND **1315)** points out that such peroxides or adducts are also intermediates in quinone formation; their formula may be written as follows:

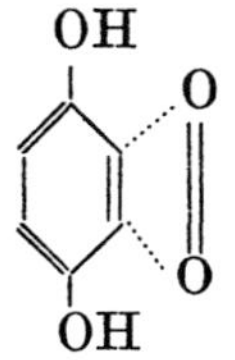

or

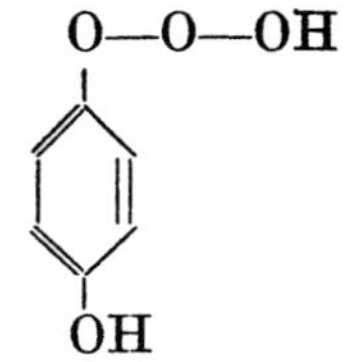

**Adduct**                 **Peroxide**

Therefore these questions may be discussed from the point of view of WIELAND's dehydrogenation theory unless the peroxides become subject to a new type of purely oxidative catalysis (p. 118).

The catalysis of the formation of peroxides by heavy metals involves a problem of fundamental importance. Since the substrate is not dehydrogenated by the heavy metal system, the simple concept of valency change, namely, dehydrogenation by $Fe^{+++}$, and reoxidation of the $Fe^{++}$ thus produced to $Fe^{+++}$, cannot be applied. This refers us back to the older views of ENGLER-BACH-MANCHOT who assume peroxide formation with the iron. The next question is, whether these metal peroxides may act as catalyts and in which manner. This causes one to enquire as to what extent the hypothesis of dehydrogenation catalysis by valency change is valid and whether it is permissible to encompass the peroxidatic and catalatic catalysis by iron from the point of view of valency change. (Discussion of the fundamentals on p. 19 and following).

**Special Hypotheses on Oxygen Activation.**

In this section some theoretical considerations, concerning the primary process of deformation and addition of the oxygen molecule, are briefly discussed. This process is interpreted with the aid of electron models, on the basis of the octett theory.

RAIKOW **955)** goes back to the old concept of "ozone" like substances (SCHÖNBEIN) by assuming that the first product of oxidation in aqueous solution is an adduct of $H_2O$ to the residual valencies of the $O=O$ molecule. This "ozone hydride" (I) rearranges itself into the desmotropic "pseudo ozone hydride" (II). In the latter the $O^*$ is already "activated"; it oxidizes the substrate while the preformed $H_2O_2$ is split off:

$$\text{I. } \diagdown O = O :: O \overset{\cdot\,H}{\underset{\cdot\,H}{\phantom{.}}} \rightleftharpoons \text{II. } > O^* :: \underset{\underset{O-H}{|}}{O-H}$$

The unfolding of the oxygen molecule is thought to occur in this case without a substrate, the charge being supplied by electrons derived from water molecules. In contrast to this view MILAS **833)** thinks that the electrons originate from the substrate. This is a more plausible assumption. A novel feature of his theory is his interpretation of true peroxide formation without instantaneous release of $H_2O_2$: the two substrate electrons are entirely transferred to the oxygen and share the orbits of its other electrons. In organic substrates the two electrons are exposed in a special manner ("molecular valency electrons"). This concept is nothing but an expression, in other words, of the "activation of hydrogen" where two hydrogen atoms are "activated" to $2\,H^+ + 2\,\varepsilon$. These are said to be unpaired electrons with parallel spin. This hypothesis has been attacked by STEPHENS **1067)**.

ZELINSKY **1380)** formulates the process which we designate as peroxide formation via addition to $-C = C-$ linkages with the aid of the octett theory. He makes use of the desmotropic biradical formula of the ethylene linkage and of a special representation of the activated oxygen molecule, containing a single bond and therefore two "lone" electrons (I), according to scheme II.

$$\text{I. } \overset{..\;\;..}{\underset{.\;\;\;.}{:O:O:}} \qquad \text{II. } \overset{H\;\;H}{\underset{.\;\;\;\;.}{R:\overset{..}{C}:\overset{..}{C}:R}} + \overset{..\;\;..}{\underset{.\;\;\;.}{:O:O:}} = \underset{\overset{..\;\;..}{:O:O:}}{\overset{H\;\;H}{R:\overset{..}{C}:\overset{..}{C}:H.}}$$

It appears strange that according to this picture the oxygen should be present in the peroxide as complete octett, i.e. very firmly bound.

On the whole, these attempts do not contribute very much to the understanding of the phenomena. At any rate, we cannot detect any special "activity" of the oxygen beyond that of deformation, unfolding, and charging. According to STEPHENS we are dealing here with special oscillation states and with the formation of a peroxide in statu nascendi. The latter has a higher energy level compared with the ready stabilized one so that it becomes more reactive. Here, too, it is difficult to draw a line between the energy rich "adduct" and the normal "peroxide".

### Heavy Metal Catalysis.

In treating the oxidative catalysis with oxygen it is necessary to begin with the primary stage of the process. Suppose that a lower valency form, e.g. ferrous iron, is the oxidation catalyst. As usual the primary step will consist in the addition of $O=O$ and in its activation. Following this, either $H_2O_2$ is split off taking with it electrons from the iron and leaving the catalyst in the trivalent ferric form (valency change), or a peroxide is formed. The latter contains more oxygen than calculated for ferric iron but does not represent a form with higher valency number. The peroxide still contains ferrous iron; it may be written, for instance, as iron salt of hydrogen peroxide:

$$Fe\diagdown\begin{matrix} O \\ | \\ O \end{matrix}\ .$$

If we take the simplest formula $FeO_2$, the catalysis may proceed as follows. One atom of oxygen ("active oxygen") is given off in accordance with the old scheme of ENGLER:

$$AO_2 + B \longrightarrow AO + BO \text{ (where A the catalyst and B the substrate)}$$

The result is that not only is the substrate (B) oxidized but that also the catalyst (A) is oxidized. For ferrous iron as the catalyst this means that one electron is lost and that ferric iron results. However, this is not a true catalysis but a coupled reaction. The reaction comes to a stand still after the inductor, A, is used up. Ferric iron as such does not act catalytically.

In order to establish a true catalytic mechanism we would have to postulate that in this case the whole oxygen molecule, probably in the form —O—O—, is to be split off. This has been discussed by OPPENHEIMER 902) as applied to platinum catalysis. Following this a scission of —O—O— into single O atoms and a direct oxidation of two acceptor molecules may take place. Theoretically one can also imagine that the unfolded —O—O— molecule forms an adduct with the organic acceptor. Thereby the primary process at the ferrous iron is reversed. Such a cycle would mean that the catalyst permits the autoxidation of inert molecules by transmitting to them activated oxygen molecules. In that case it would be of secundary importance whether this adduct leads to dehydrogenation by the transfer of $H^+ + \varepsilon$ to —O—O— (example: hydroquinone) or whether it leads to genuine peroxide formation at organic valency gaps if no hydrogen atoms are available. The possibility that the peroxide yields all of its oxygen to an acceptor has recently been discussed by MANCHOT (p. 123).

In reality it appears certain that this modus of detachment of —O—O— and of its transfer to the substrate either does not take place at all in heavy metal catalysis with biological substrates or else only in cases where a dehydrogenation is not possible (see for instance the case of direct addition of —O—O— to the $C = C$ double bond, p. 14). Later it will be shown that even the last mentioned example is not yet firmly established. FRANKE, for instance, does not believe that iron participates as catalyst in the formation of peroxides from unsaturated fatty acids (p. 123).

When an organic substrate is dehydrogenated in such a way that $H^+$ions are released, in addition to the electrons, the other mode of heavy metal catalysis is the preferred one and probably the only one. By this other mechanism —O—O— is not detached unchanged and the organic substrate is not dehydrogenated through the primary adduct, but a valency change of the iron takes place. It is the f e r r i c  i r o n which oxidizes the substrate or rather dehydrogenates it. The hydrogen split off of the substrate dissociates into the hydrogen ion and the electron. The former gives rise to water (or HCl) formation while the latter is taken up by the ferric iron. The ferric iron is thereby reduced to ferrous iron and adduct formation with molecular oxygen may again take place. If there are $H^+$ions available (from water or HCl molecules) the combination of $H^+$ with —O—O— takes place much more rapidly than a possible detachment of —O—O—; $Fe^{++}$ loses an electron and once more becomes

$Fe^{+++}$. The cycle leading to restoration of the catalyst (ferric iron) is thereby completed. This concept embodies the new idea that, contrary to the inorganic ferric ion, $Fe^{+++}$ may also be a catalyst in complex linkage, i.e. that it promotes dehydrogenation. This property of complex ferric compounds may be further increased and differentiated by providing a suitable colloidal bearer. The hemin enzyme of WARBURG is such a complex. WARBURG has demonstrated in this case the valency change spectroscopically.

According to this view heavy metal catalysis is due to a cyclic process involving loss and return of electrons. It is important that the catalyst acts while in complex linkage. Two possibilities present themselves. Firstly the metal originally is added to the system as a free salt. In tat case it must be present in the lower valency form and combine with the substrate to form a complex; higher valency forms do not form complexes subject to dehydrogenation. In this complex $Fe^{++}$ is first oxidized to $Fe^{+++}$ by $O_2$, then reduced by the hydrogen of the substrate, thus regenerating the catalyst, $Fe^{++}$.

Secondly there exists already a complex containing the metal of higher valency form ($Fe^{+++}$). The complex, acting as catalyst, attracts the substrate, labilizes its hydrogen and is, in turn, reduced to $Fe^{++}$. Then only $O_2$ enters into play and reoxidizes the complex back to the ferric stage. The catalyst, $Fe^{+++}$, is now restored and ready for new action.

If we consider the action of simple ionized salts first, the process begins with a complex formation between the substrate and the salt, e.g. between cysteine and $Fe^{++}$ salt. By combination with the strongly charged metal ion the stability of all linkages in its vicinity is profoundly altered; it will attract oxygen contained in hydroxyl groups and repulse hydrogen, therefore "labilize" it. Complex formation will take place only with the metal in the lower valency form ($Cu^+$ or $Fe^{++}$). This step is autoxidizable. After the entrance of $O_2$ the hydrogen contained in the organic residue shifts over to the oxygen. The catalyst acts, therefore, as a dehydrogenase. Catalyses of this type certainly have some biological significance. Thiols or ascorbic acid may be oxidized in this manner in living cells. Undoubtedly, some reactions heretofore ascribed to "oxidases" will find their explanation in such mechanisms.

Biologically of greater importance, however, is the other mechanism where the catalyst exists already as an organic heavy metal complex, as for instance in hemin catalysts of the cell. These are active only in colloidal state, namely, in combination with bearer proteins. Such complexes are, by definition, to be classified as enzymes. In these cases the hydrogen of the substrate is not labilized by transfer of the whole organic molecule into a complex with the heavy metal, but the polar forces of the macromolecular catalyst affect the donator by means of residual valency linkage and labilize its hydrogen (WIELAND).

Here, the catalyst starts its activity in the oxidized state ($Fe^{+++}$). It dehydrogenates the substrate in such a manner that the field of force of the heavy metal renders one hydrogen atom labile. A hydrogen ion is split off while the electron goes to the metal and reduces it to the next lower valency stage. As a consequence the metal is forced to release one anion ($OH^-$ or $Cl^-$) which combines with $H^+$ to form $H_2O$ or $HCl$. Regeneration of the catalyst is brought about by reoxidation by a stronger acceptor which deprives the metal of one electron and thereby transforms the catalyst to the original, active form. Whether molecular oxygen may be the stronger acceptor, depends on the structure of the catalyst. Cytochrome c, for instance, is not autoxidizable and is therefore only reoxidized by a specific oxidase which very probably is nothing other than the hemin enzyme of WARBURG (Indophenol oxidase = Cytochrome oxidase).

In order to illustrate the process of dehydrogenation by a simple metal salt through intermediary complex formation, the events leading to oxidation of cysteine to cystine may be depicted according to MICHAELIS (804) p. 140): Cysteine forms a complex with ferrous iron. Ferrous cysteine is autoxidizable, forming ferric cysteine with oxygen. Ferric cysteine undergoes an intramolecular rearrangement, the ferric iron being reduced to ferrous iron by the labile cysteine hydrogen. Ferrous cystine is unstable and decomposes spontaneously into its components. The reader will not find it difficult to visualize the movements of hydrogen and electrons during this cycle.

Similar processes take place if the catalyst is a colloidal heavy metal compound. The most important example of such a system, the respiratory hemin enzyme of WARBURG, begins the catalysis while the iron in the pheohemin group is in the trivalent state (symbol: $> Fe^{III} \cdot OH$). Upon collision with a suitable substrate it attracts $H^+ + \varepsilon$ and is reduced to a derivative of heme with bivalent iron. This no longer possesses the positive charge required to hold the anion. The latter must consequently combine with the hydrogen ion:

$$1) \quad > Fe^{III} \cdot OH + H^+ + \varepsilon \longrightarrow > Fe^{II} + H_2O$$
$$2) \quad 2 > Fe^{II} + O_2 + 2 H_2O \longrightarrow 2 > Fe^{III} \cdot OH + H\text{—}O\text{—}O\text{—}H.$$

The system containing ferrous iron is autoxidizable. If in contact with oxygen, according to equation 2), the iron yields two electrons to the oxygen, the charged molecule $^-O\text{—}O^-$ removes $2 H^+$ from the water, and the remaining two hydroxyls satisfy the positive charges on the $Fe^{+++}$ atoms. The cycle is now completed; the catalyst is regenerated and the hydrogen atoms removed from the substrate are combined with oxygen.

The theory of HABER and WILLSTÄTTER 437), which will be given in detail on p. 44, takes a somewhat different point of view. Here, two fundamental points of that theory are of importance. The theory assumes in accordance with the view developed above that the oxidation catalysts are reversible redox systems and that they "activate" the substrate hydrogen by accepting one electron and consequently reducing the prosthetic group of the enzyme. A stronger acceptor, e.g. molecular oxygen, reoxidizes the "desoxy-enzyme" to the active form. The second point is the postulate that the enzyme is reduced in a monovalent fashion, i.e. that only one electron is accepted and one $H^+$ released at a time. Thus free radicals are formed which carry on the reaction in the form of a reaction chain in the solution without further participation of the catalyst. The formation of hydrogen peroxide is not postulated by this theory. On the contrary, its formation would break the chain due to the recombination of two radicals, $\overset{\wedge}{O}H$, which could otherwise propagate the chain.

### Formation and Removal of Hydrogen Peroxide.

The appearance of hydrogen peroxide in the course of heavy metal catalysis has already been discussed. It is a necessary corollary to the scheme of valency change of the metal. The question of the formation of $H_2O_2$ was one of the main objects of the now historic dispute between WIELAND and WARBURG. According to the former, hydrogen peroxide should always be formed in biological oxidation, according to the latter, never as long as one assumed peroxide formation at the iron and direct oxidation of the substrate by this intermediate.

In retrospect it is difficult to appreciate this dispute, inasmuch as it had been known for a long time that the prototype of heavy metal catalysis, i.e. the oxidation of hydrogen with palladium as catalyst, constitutes a hydrogenation of $O_2$ via $H_2O_2$. Now the dispute is settled. It is generally accepted that $H_2O_2$ is formed during the "valency change" catalysis by heavy metal as well as during the autoxidation of organic substances, and we know that the detection of $H_2O_2$ in heavy metal catalysis is obviated because the peroxide is decomposed by the same metal system as fast as it is formed. The decomposition is either of the "catalatic" type, i.e. a simple breakdown into water and oxygen, or of the "peroxidatic" type, i.e. a dehydrogenation of the substrate by the peroxide oxygen. Undoubtedly, hydrogen peroxide may also arise by other reactions, e.g. by the interaction of peroxides with water. Formerly WIELAND rejected this possibility; today (1302)) he is less definite on this point since peracids, for instance, will yield hydrogen peroxide by reacting with water. The autoxidation of aldehydes may possibly take place in this manner. On the other hand, this mechanism of $H_2O_2$ formation is hardly of great biological significance unless it should be demonstrated that the first product of autoxidation of organic substances is not a simple adduct but a defined organic peroxide as has recently been suspected by KREBS, KEILIN, and others. The essential thing, in our opinion, is the fact that hydrogen peroxide acts in statu nascendi (see below).

If it is granted that $H_2O_2$ arises regularly in the course of biological oxidation, the mechanisms for its removal must be considered. It is not only highly toxic, killing the anaerobic organisms which produce $H_2O_2$ when brought in contact with air (p. 33), but its further hydrogenation to water entails the largest gain in free energy. The first stage, consisting in the reduction of oxygen to hydrogen peroxide, yields only about $1/4$ of the total free energy while the second stage, i.e. the reaction $H_2O_2 + H_2 \longrightarrow 2\,H_2O$, contributes the bulk. The relatively small amount of work made available by the first step is the reason why molecular oxygen, thermodynamically speaking, represents a rather weak acceptor. $O_2$ is just capable of dehydrogenating aromatic hydroxy- and amino compounds, but this reaction leads to equilibria (FRANKE 288)). It is the second step, namely, the renewed dehydrogenation by peroxidatic systems, which is the decisive factor. This is well known for the case of the "oxidase reactions". Besides the peroxidatic mechanism there is the catalatic one which restores a part of the $O_2$, thereby subdividing the release of the hydrogenation energy of oxygen into several increments. The integral effect, however, remains unaltered.

The complete utilization of the energy stored in hydrogen peroxide is achieved by the respiring cells in various ways. The various mechanisms are: 1) The direct, non-catalyzed oxidation of substrates by the primarily formed $H_2O_2$. Examples are the oxidation of pyruvic acid (SEVAG, see below) and of thiols (p. 63). However, this mechanism does not appear to be very important. In the heavy metal-free system of dehydrogenase + hypoxanthine (or aldehyde), addition of alcohol does not cause an oxygen uptake. KEILIN and HARTREE's 546) explanation of this failure is that alcohol is not attacked by the hydrogen peroxide formed by the dehydrogenation of the purine. 2) Most important of all systems removing $H_2O_2$ are undoubtedly heavy metal catalysts acting either as "catalases" or as "peroxidases". They may be enzymes, like the respiratory ferment of WARBURG, peroxidase, catalase, or oxidases of the type of uricase, or else other hemin derivatives present in the cell, e.g. free hematin. In this connection it is advisable to report briefly the main results of recent experiments by KEILIN and HARTREE 546) which, incidentally, may throw new light on the physiological significance of catalase. These investigators find that the cyanide-sensitive uricase as well as the cyanide-resistant d-amino acid dehydrogenase (KREBS) specifically require oxygen as the acceptor and that $H_2O_2$ is formed by their action. They are, therefore, to be termed oxytropic dehydrogenases or, in short, oxidases. While in the case of the prototype of such oxytropic dehydrogenases, the SCHARDINGER enzyme, hydrogen peroxide will only react with added oxidizable substances if peroxidase or an iron system are present (HARRISON and THURLOW 462), HARRISON 456)), these catalysts are not required in the case of uricase or amino acid dehydrogenase. p-Phenylenediamine or alcohol will react with the $H_2O_2$ produced by the enzyme. It is particularly interesting that the hydrogen peroxide does not act in stoichiometrical proportion but that it increases the oxygen consumption of the system about twofold. With p-phenylenediamine, for instance, the figures for oxygen uptake, according to KEILIN 544), are: uricase + uric acid: 362 cmm. $O_2$; uricase + phenylene diamine + peroxidase : 40 cmm.; uricase + uric acid + diamine + peroxidase (complete system): 757 cmm. There is no doubt that in the first place, under the influence of the iron system, the $H_2O_2$ is used up in an oxidation reaction. The increase of oxidation beyond this limit as shown by the oxygen uptake must be interpreted as follows: $H_2O_2$ + peroxidase oxidize the phenylene diamine stoichiometrically; the hydrogen peroxide is, of course, destroyed in this reaction. More "activated" hydrogen is liberated from a fresh amount of the substrate which is uric acid in the present case. It is accepted by freshly added $O_2$. The end effect is, then, that double the amount of substrate (uric acid + diamine) has been oxidized which accounts for the increased oxygen uptake. The uricase system, presumably containing heavy metal, is aided in a secondary fashion in its effect by the peroxidatic activity of the new heavy metal system as displayed towards the diamine. In the case of alcohol the path of this coupled oxidation is such that $H_2O_2$ + "peroxidase" oxidize it to aldehyde. Amino acid dehydrogenase behaves in the same manner. The results obtained with the SCHARDINGER enzyme are the most interesting. Here, alcohol is not oxidized by hydrogen peroxide and no secundary uptake of oxygen occurs. But if a trace of catalase is added to the system [donator + SCHARDINGER enzyme + alcohol] the same additional alcohol oxidation occurs as with the natural system uricase or amino acid dehydrogenase. KEILIN and HARTREE assume that in this case $H_2O_2$ + catalase dehydrogenate the alcohol to aldehyde and that the aldehyde thus produced is dehydrogenated in the usual manner by the SCHARDINGER enzyme. The peroxide + catalase alone have no effect whatsoever on alcohol. Instead, the ordinary catalatic decomposition of $H_2O_2$ into $O_2 + H_2O$ takes place. It is to be concluded that $H_2O_2$ will act as an oxidizer of alcohol only if it is split by catalase under such conditions that it can act in statu nascendi. In fact, catalase + other peroxides (barium, cerium, ethyl peroxide) react with the alcohol. In other words, it is essential that the hydrogen peroxide must be formed by a reaction taking place in the system itself.

These results are interesting from several points of view. Firstly, they seem to suggest, just as has been pointed out by STERN 1073) and HAUROWITZ 468), that the catalatic decomposition of hydrogen peroxide proceeds via highly active groups, perhaps free radicals. Furthermore, they seem to support the suggestion, recently again discussed by HAUROWITZ 468), that catalase and peroxidase are closely related and that the apparent entirely different behaviour is a question of kinetics. If this is true and if KEILIN's systems may be regarded as biological models, the role of catalase in the living cell is open to reinvestigation. This enzyme is commonly considered to be a "safety valve" for the removal of hydrogen peroxide. May it not be, that the enzyme has some significant function as a "peroxidase" in biological oxidation, even if not in the main chain comprising cytochrome and the respiratory ferment? This secondary heavy metal catalysis with hydrogen peroxide and catalase or peroxidase appears to assume an increasingly important physiological significance. It is possible that such reactions may go beyond the stage of simple dehydrogenation and lead to deeply changed products (p. 32). Ascorbic acid, for instance, is not directly oxidized by indophenol oxidase but by hydrogen peroxide formed by autoxidation (SZENT-GYÖRGYI 1118)). It is quite plausible that this mechanism is of importance in the breakdown of fatty acids (p. 254), and perhaps also in the oxidation of the intermediary catalysts ascorbic acid and glutathione (pp. 93, 95).

If we disregard the above possibilities, it may be stated that $H_2O_2$ disappears in every respiring cell due to the activity of catalase and that it cannot be detected for this reason. Where catalase is absent, other iron containing systems will catalyze the decomposition. In systems devoid of catalase and of other iron compounds, hydrogen peroxide is used up more or less completely by direct reaction with certain substrates. In these cases sometimes the peroxide may be detected. Such uncatalyzed reactions, e.g. with thiol compounds, are hardly significant if catalase or other iron systems are available. The only exception is pyruvic acid which will react more quickly with hydrogen peroxide than catalase:

$$CH_3 \cdot CO \cdot COOH + H_2O_2 \longrightarrow CH_3 \cdot COOH + CO_2 + H_2O.$$

This has been demonstrated for pneumococci by SEVAG 1045) and confirmed by FUJITA 383). A catalase resistant intermediate peroxide appears to be formed: $CH_3$—C—COOH (WIELAND). Addition of pyruvic acid will protect these anaerobes

$$\overset{\displaystyle \wedge}{OH \quad O \cdot OH}$$

against hydrogen peroxide poisoning just as efficiently as catalase. Methyl glyoxal acts similarly.

The explanation of the peroxidatic and catalatic activity of iron systems offers theoretical problems of fundamental importance. Are we able to explain the peroxidase function of iron systems also on the basis of a catalysis with a valency change of the metal or rather on the basis of peroxide catalysis? The evidence favoring the former point of view will be discussed on page 70 and that supporting the latter on page 123.

There is no difficulty in explaining the peroxidase function by WIELAND's theory as is done by BERTHO 105), p. 727). From the standpoint of dehydrogenation catalysis and acceptor specificity it does not matter whether $O_2$ is added as —O—O— to the bivalent iron or whether $H_2O_2$ is attracted to it by secondary valencies. In that case two electrons of the metal and two hydrogen ions from water (derived from the substrate) would add on to the hydrogen peroxide and hydrogenate it to water: H—O—O—H + 2 H = 2 $H_2O$, or, written more accurately

$$2 > Fe^{II} + 2 H_2O + H^+ \text{—}^-O\text{—}O^-\text{—}H^+ \xrightarrow{\quad 2H^+ + 2\varepsilon \quad} 2 > Fe^{III} \cdot OH + 2 H^+ — O^= —H^+.$$

In the case of simple iron salts this may be an induced reaction; in the case of iron complexes which are capable of repeated reduction of the ferric stage (especially if the iron is contained in the molecule of hemin ferments like peroxidase*) this process becomes a true catalysis. It is readily seen that in the case of cysteine metal complexes, for instance, the hydrogenation of $H_2O_2$ is a direct continuation of the preceding $O_2$

*) See, however, HAUROWITZ (468), 469)).

catalysis leading to further dehydrogenation of substrate molecules (MICHAELIS 800)). However in this case and especially in induced reactions, the possibility of peroxide formation on the iron and copper, as postulated by MANCHOT and discussed by WIELAND, must not be overlooked (p. 123).

The strong catalatic activity of heavy metals cannot be explained so readily. WIELAND has acknowledged that the apparently simple (monomolecular) decomposition of $H_2O_2$ into $H_2O + O$ is out of the question since no atomic oxygen is formed. He interprets the reaction as consisting of the dehydrogenation of one hydrogen peroxide molecule and hydrogenation of another peroxide molecule:

1)  $H-O-O-H$ (+ Catalyst) $\longrightarrow O_2 + 2 H$
2)  $H-O-O-H + 2 H \longrightarrow 2 H_2O.$

The monomolecular course of the reaction is probably simulated by the fact that only reaction 1) proceeds with measurable velocity while 2) is instantaneous (FRANKE l.c. 288), p. 177)). One might think, therefore, that here also a valency change of the iron takes place so that the ferric iron is reduced by $H_2O_2$ to ferrous iron which in turn transfers the electrons and hydrogen ions to the second $H_2O_2$ molecule. The difficulty is that ferric iron, according to MANCHOT, does not react with $H_2O_2$. For this reason, MANCHOT favors a mechanism with peroxide formation at the metal; however, the regeneration of ferrous iron from $Fe..O_2$ which is to be postulated if the scheme $> Fe + 2 H-O-O-H \longrightarrow Fe..O_2 + 2 H_2O$ is to result in catalysis, is somewhat doubtful. MANCHOT's formulation has been attacked by HABER who interprets the process as a complicated chain reaction (p. 123). WIELAND assumes that a small amount of ferrous iron exists in stationary concentration in the system due to interaction of ferric iron with $H_2O_2$ as donator. The donator function of $H_2O_2$ is especially supported by the decomposition of diethyl peroxide by ferrous salts (1307)) yielding ethyl alcohol and acetaldehyde. This is explained by hydrogen transfer from one peroxide molecule to another (for details see FRANKE (l.c. 288), p. 178)). While the mechanism of the catalysis by simple iron systems is still obscure, for hemin catalysis a possible explanation has recently been put forward by HAUROWITZ 468) based on

the polar structure of $H_2O_2$. He attributes to it the formula $\underset{H}{\overset{H}{>}}O^+\!-\!O^-$, or simply, $\underset{H}{\overset{H}{>}}O \rightarrow O$. This formulation would explain the well known fact that hydrogen peroxide may act both in a reducing and an oxidizing manner. In accordance with such a structure it could, as suggested by WIELAND, serve as a hydrogen donator by hydrogenating the terminal O atom of another molecule so that two molecules would yield two $H_2O$ molecules. BANCROFT 62) on the basis of his potentiometric studies writes the reaction as follows:

$$2 H + O:O + H_2O_2 \longrightarrow 2 H_2O + O_2.$$

He deduces from his measurements that a substance is reduced by $H_2O_2$ if its true electromotoric force as referred to any type of electrode and any type of solution is greater than that of the $H_2O_2$ in the same system, and that it is oxidized by $H_2O_2$ if the reverse is true, provided, of course, that the reduction or oxidation is kinetically possible. Such a reaction is promoted by the hemin catalyst without a valency change. The ferric iron of the hemin holds the $H_2O_2$ (just like HCN, etc., p. 73) at the terminal O in coordinative linkage and effects its hydrogenation by hydrogen from any available source. The source may be electrolytically produced hydrogen at the cathode, a reducing chromogen or $H_2O_2$ itself. The general equation would be:

$$Fe^{+++} \cdots O \leftarrow OH_2 + DH_2 \longrightarrow Fe^{+++} + 2\ H_2O + D,$$

with $H_2O_2$ as donator:

$$Fe^{+++} \cdots O \leftarrow OH_2 + H_2O \rightarrow O \longrightarrow Fe^{+++} + 2\ H_2O + O_2.$$

Such complexes have been discovered by spectroscopy (p. 80). According to Stern 1073), the same scheme holds for the enzyme catalase. $H_2O_2$ is attached to the enzyme molecule. At the surface of the catalyst the peroxide is split into the free radicals $\overset{\wedge}{O}H$ which enter a chain reaction. In the case of catalase Stern (1085), see p. 176) was able to demonstrate complex formation with ethyl hydrogen peroxide but not with hydrogen peroxide. An intermediary valency change of the catalase iron in the course of the catalysis of $H_2O_2$ has been postulated by Keilin and Hartree 549) (see p. 177) although the ferric iron of catalase is remarkably stable and cannot be reduced by hydrosulfite.

A unitary theory of peroxidase, catalase, and also oxidase action had already been attempted by Shibata 1049), but his concept, at least in details, conflicts with the more recent experimental results. Therefore it may suffice to note in passing that Shibata refers back to the old hypothesis of the "decomposition of water". The metal complex is supposed to form a labile aquo compound (oxidases) or a compound with $H_2O_2$ (peroxidase, catalase) without a valency change of the metal. The aquo complex is split in $H \cdots OH$, the hydrogen peroxide complex in $H—O| O—H$. The free radical $\overset{\wedge}{O}H$ oxidizes the substrate, the hydrogen hydrogenates the acceptor (quinone, oxygen). In the catalase reaction two $H_2O_2$ molecules (just as in the theory of Haurowitz) react with the catalyst. One of them is split into $HO| OH$ while the other is split into $H \cdots O—O \cdots H$. The two sets of split products yield $2\ H_2O + O_2$. The theory has been criticized by Zeile (1374) p. 721).

It may be mentioned parenthetically that Remesow 977) reports that cholesterol in colloidal solution shows cyanide resistant catalase and peroxidase activity. However, the presence of traces of heavy metals apparently are not completely excluded, just as in analogous findings with phosphatides (p. 122).

## II. Extension and Reshaping of Previous Theories.

In the preceding general discussion it has been attempted to give the skeleton of a theory of catalytic oxidation. It is now our task to render a more detailed description of the development of these problems and, as far as possible, of the modern unitarian theory. The framework is provided, but many special problems are still in a controversial state *). Inasmuch as the modern concept of biological oxidation places the main emphasis on the hydrogen transfer which will be designated as "activation" for the sake of brevity, the changes which Wieland's theory has undergone will be dealt with first. Following this, the reshaping of Warburg's theory will be described.

### 1) Development of Wieland's Theory.

#### a) On the Nature of the Activation of Hydrogen.

From the very beginning, Wieland has attempted to define the concept of the

---

*) See the comprehensive treatment of the subject by Franke in Euler's book (288)).

,,activation of hydrogen" which is the main foundation of his theory. The activation which he visualizes is certainly, as a rule, not to be confused with an activation consisting in the production of hydrogen atoms. In most cases it means a deformation of the hydrogen molecule by residual valency forces. In any event, as in every catalysis at surfaces, the substrate must be "bound" to the catalyst. The molecule of the donator is thereby distorted and the hydrogen is "dislocated" from its firm seat in the substrate molecule (BÖESEKEN). The assumption underlying this concept is that it is preferably the hydrogen which represents the part of the molecule attached to the

catalyst, as follows: $D\diagdown\genfrac{}{}{0pt}{}{H:}{H:}\;\cdot Cat.$ If free $H_2$ is utilized in the hydrogenation the

bond between the hydrogen atoms is weakened by the attraction $\left.\genfrac{}{}{0pt}{}{H}{H}\right\}$ $\cdots\cdots$ Cat.

Also on palladium surfaces this deformation of the molecule is the essence of hydrogen activation and not its splitting up into atoms. During the hydrogenation of the acceptor which is likewise adsorbed on the catalyst the hydrogen severs its residual valency linkage and goes to the acceptor (WIELAND 1302), QUASTEL 943)).

The concept of a labilization of hydrogen and of its attraction by the active center of the catalyst is especially important, because it leads without difficulty to the modern hypothesis that the hydrogen is completely taken over by the catalyst which is thereby hydrogenated or at least reduced in the case of a heavy metal. In the latter case the liberated hydrogen atom is not attached as an entity to the catalyst but only its electron is taken up while $H^+$ is set free.

Though from the outset one could predict that the complicated and in its premisses improbable theory of oxidation-reduction of BACH would eventually be absorbed by the the more comprehensive theory of dehydrogenation-hydrogenation of WIELAND, BACH has made several attempts to preserve it by trying to establish differences between "oxidation" and "oxido-reduction". BACH and MICHLIN 52) state that in presence of $O_2$ xanthine dehydrogenase does not effect a reduction of oxygen but that uric acid is formed by some kind of dismutation. This has been refuted by WIELAND and ROSENFELD 1325). The milk enzyme dehydrogenates aldehydes (and xanthine) in presence of oxygen and it dismutes aldehydes in absence of oxygen (WIELAND 1319)). This is not a general rule. In the case of acetic acid bacteria there exists a separate "mutase" (WIELAND and PISTOR 1321)). It is true that according to BACH 54) highly purified aldehydrase from milk yields more salicylic acid anaerobically with acceptors than under aerobic conditions. But the reason for this phenomenon is that the hydrogen peroxide formed in the reaction inactivates the purified enzyme which is protected by catalase in the less pure preparations (WIELAND 1319, 1325)). The same fact may possibly explain the claim of REICHEL 966) that aldehydrase (from milk) is anoxytropic. Only in the presence of yellow enzyme is it stated to dehydrogenate aldehyde with simultaneous hydrogenation of the flavin enzyme; otherwise, it is said to act as a mutase (see p. 262). Another claim of BACH 53) that the dehydrogenation of succinic acid in presence of oxygen or of methylene blue as acceptors is brought about by different mechanisms has been refuted by F. G. FISCHER 337). In both cases the same dehydrogenase is effective. The reason for the inferior behaviour of methylene blue is that leuco methylene blue is very slightly soluble and is strongly adsorbed by the muscle fibers present in the preparations. The inferiority of quinone is due to its toxicity towards the enzyme (WIELAND 1311)).

The "hydrolytic" theory of BATTELLI and L. STERN has been refuted by OPPEN-HEIMER 902) and recently by FRANKE (l.c. 288), p. 126).

Dehydrogenation by Metals: Another important step in the development of WIELAND's theory was the demonstration of its applicability to certain heavy metal catalyses occurring in the absence of oxygen. Not only platinum or palladium surfaces but also other heavy metals are capable of transferring hydrogen in the absence of oxygen. From the point of view of "valency change" catalysis it makes no theoretical difference but is only a question of potentials and acceptor affinity whether a heavy metal catalyst in the reduced form is reoxidized by oxygen (or hydrogen peroxide) or by another acceptor.

Even more important are those experiments which concern the role of iron in oxidation by molecular oxygen as well as by hydrogen peroxide and which have led to entirely new points of view.

The previous theory concerning active peroxides (MANCHOT) or peroxide like substances (like the poorly defined higher stages of iron oxides of WARBURG) was sufficient to explain only induced reactions in which ferrous iron via intermediates of about the type $Fe \cdot \cdot O_2$ promotes the oxidation of the substrate by the scheme of ENGLER ($AO_2 + B \longrightarrow AO + BO$) and comes to rest as ferric iron (see OPPEN-HEIMER 902)). For a catalysis, this scheme is insufficient since it cannot explain satisfactorily how the regeneration of the ferrous form is brought about. The possibility that the peroxide yields its entire oxygen to an acceptor and is thus reduced to the ferrous form would not be plausible as a general scheme.

Thus WIELAND was led to develop a different concept based on the assumption that ferrous iron does not activate the oxygen but the hydrogen of the substrate and that this process is brought about by the formation of a complex between ferrous iron and substrate. One of the most important observations suggesting this hypothesis was the "Primärstoss" (initial peak). If an oxidizable substance in weakly acid solution is allowed to react with ferrous salt and hydrogen peroxide an immediate extremely rapid peroxide decomposition is observed which is much greater than that calculated for an induced reaction involving peroxide formation. In the case of some substrates the reaction will stop at this point, in the case of others a catalysis follows which may be due to dehydrogenation of the substrate and a concomitant reduction of ferric to ferrous iron. Tartaric acid, for instance, is oxidized to dihydroxymaleic acid by the "Primärstoss". The subsequent catalysis is carried on via tartaric acid and dihydroxymaleic acid:

$$\begin{array}{ccc} CHOH \cdot COOH & & C \cdot (OH) \cdot COOH \\ | & \xrightarrow{\;-\;2\;H\;} & || \\ CHOH \cdot COOH & & C \cdot (OH) \cdot COOH. \end{array}$$

A further important milestone in this development was the recognition of the significance of complex formation with heavy metals which was promoted not only by WIELAND but also by MICHAELIS and by BÖESEKEN. It is probable that all true catalyses are based on the phenomenon that the decisive valency change does not concern simply ionized metals but metal complexes showing entirely novel features. It appears that simple salts are able to cause only induced reactions leading to the inert ferric stage. The new features depend, of course, on the structure of these complexes. On the one hand, complex formation tends to lower the potential of the $Fe^{+++}/Fe^{++}$ system (FRANKE 352)) which in itself would produce an increase in reaction rate, especially in the ph range of 5—7, on the other hand SMYTHE 1061) points out that it should be more difficult to remove an electron from a nucleus carrying a double positive charge (ferrous ion) than from a non-ionized ferrous complex. Furthermore, the rearrangement of the valencies in the organic structure due to the introduction of the heavy metal causes a labilization of hydrogen. FRANKE 353) has studied the structure and the properties of a number of iron complexes. In the case of the intramolecular oxidation of thiol groups in the cobalt cysteine complex this labilization of hydrogen has been experimentally demonstrated with the polarograph by BRDIČKA.

For these considerations is it not necessary to restrict oneself to complexes in the narrow sense of the term. Salts of simple structure are also catalytically active if the metal has an opportunity to exert, besides its main valency, also residual valency forces. If, for instance, the metal salt of a hydroxyacid is formed, the metal will replace the hydrogen atom of the carboxyl group, but it will also exhibit covalency bonding with the hydroxyl group (or a thiol group, as the case may be). This intails a deformation of the molecule. By attracting the O of an OH group the H is repelled by the metal and thereby mobilized. For the example of ferrous tartrate SMYTHE 106) has developed more detailed formulas. Similar schemes might be applied to the catalytic oxidation of cysteine (p. 22) especially to that phase of the reaction where the ferric cysteine, initially formed by the oxygen, is rearranged intramolecularly to ferrous cystine. Also in carbonyl compounds (aldehydes) the metal may attract the oxygen thus changing the charge on the carbon atom and mobilizing the hydrogen. Finally, WIELAND assumes that even oxygen (or $H_2O_2$) is involved in the complex formation and that the ferrous iron is thus protected for a certain period and is given an opportunity to ,,activate'' the oxygen (see FRANKE (288) p. 199)).

On the basis of this concept, WIELAND has encompassed reactions as instances of hydrogen activation which formerly had been interpreted as peroxide formation with the iron, e.g. the ''autoxidation'' of arsenite to arsenate by ferrous iron which is an induced reaction. In other, similar reactions, e.g. in the system $Fe^{++}$—$O_2$—hypophosphite, WIELAND and FRANKE 1313) found that they are no true induced reactions: The induction factor is much too large, they are incomplete catalyses and stop because the metal eventually is stabilized in an inactive ferric state. The authors consider these processes as cases of hydrogen activation where $Fe^{++}$ transfers activated hydrogen to oxygen until it is gradually oxidized by $O_2$ or $H_2O_2$ to $Fe^{+++}$. These consider become true catalyses if a minute amount of complex forming organic acids (dihydroxymaleic acid, thioglycolic acid, various $\beta$-ketoacids) are added (combined autoxidation systems). The latter act catalytically because their ferric complexes are readily reduced to ferrous complexes. An excess of these ''activators'' acts in an inhibiting manner because the formation of the main complex, e.g. of the hypophosphite complex, is prevented. Various organic substances like thioglycolic acid or pyrocatechol behave like hypophosphite. With simple fatty acids, however, only induced oxidation has been observed. Here, the ferrous iron is able to transfer oxygen only until it is oxidized to ferric iron since no complex formation between the Fe and the substrate takes place. It is only after the addition of complex forming substances, e.g. thioglycolic acid, that the simple fatty acids are catalytically oxidized. It should be mentioned that, according to BOCKENMUELLER 129), in the oxidation of hypophosphite there is first formed a phosphoric acid radical which propagates the further oxidation as a chain reaction (p. 44).

We see that as a result of this trend the theory of primary dehydrogenation was slowly penetrating into the realm of heavy metal catalysis and that WIELAND, in accord with WARBURG, has stressed the significance of the valency change for this type of catalysis. This is coupled with the assumption that the participation of oxygen is only due to the fact that the reduced form of the catalyst is autoxidizable.

### b) The Role of Hydrogen Peroxide.

For some time the question whether in the course of an oxidation by molecular oxygen hydrogen peroxide is formed was considered to be of prime importance in deciding whether activation of hydrogen or of oxygen had taken place. WIELAND assumed that $H_2O_2$ is formed by necessity; WARBURG argued against this view (OPPENHEIMER 902)). This controversy is a thing of the past. Whenever a catalysis involving initial dehydrogenation, with or without heavy metal, occurs, hydrogen peroxide is produced by the hydrogenation of oxygen. Besides this mode of $H_2O_2$ formation there may occur a secundary $H_2O_2$ production by the reaction of peroxides with water which is no longer denied by WIELAND 1302).

It is mainly the merit of WIELAND's school that on the one hand the formation of $H_2O_2$ has been demonstrated both in model systems and in biological experiments, and that on the other hand the cause for its frequent failure to appear has been revealed, namely, that the peroxide is decomposed in the course of the reaction. It need hardly been mentioned that hydrogen peroxide cannot be detected in biological systems if the cells contain catalase (see for instance WIELAND 1299)).

#### Model Experiments.

It has been shown by WIELAND and FRANKE 1313) and by MACRAE 742) that the non-catalyzed HCN-resistant autoxidation of methylene blue, pyrocatechol, pyrogallol, and dihydroxymaleic acid represents a typical hydrogenation of oxygen leading to $H_2O_2$ production. Detection under these circumstances is possible since $H_2O_2$ reacts only slowly with the leuco forms (MACRAE). Hydrogen peroxide is also formed by the autoxidation of methyl glyoxal + HCN, the end products of which are acetic acid and formic acid (SMYTHE 1062)). If methyl glyoxal reacts with HCN under anaerobic conditions pyruvic acid and a polymer of $C_3H_6O_2$ are formed by dismutation (1062)).

If a heavy metal catalysis replaces the simple metal-free autoxidation of such organic donators, hydrogen peroxide disappears almost invariably.

It is especially true for iron catalysis, less for copper: When leuco methylene blue is dehydrogenated with copper as a catalyst hydrogen peroxide may be detected (REID 971)). If pure iron salts are permitted to autoxidize in the presence of $H_2O_2$ the latter disappears while it will not be destroyed during the oxidation of copper salts and only partly by that of complex cobalt salts. The kinetics of this reaction has been studied in detail by WIELAND and FRANKE 1312, 1314, 1315, 1316). Under the same set of conditions $H_2O_2$ will react 10 times faster than oxygen in the presence of copper and 100 times faster in the presence of ferrous iron. The disappearance of $H_2O_2$ in presence of iron is due to two types of reaction, a catalatic and a peroxidatic one. Iron salts will decompose $H_2O_2$ catalytically into water and oxygen. Palladium will do the same. During the dehydrogenation of alcohol, hydrogen peroxide which has been added to the system is destroyed. The demonstration of hydrogen peroxide formation during the hydrogen-oxygen reaction which has been used as an argument against WIELAND's theory (TANAKA 902), p. 1289) is made possible by kinetic circumstances, namely, by the greater H-concentration prevailing at the surface of the metal (WIELAND 1310)). In the case of palladium black, MACRAE 743) has succeeded in fixing and demonstrating $H_2O_2$ formation in the dehydrogenation of methanol and alcohol as in the $H_2 + O_2$ reaction. Secondly, all these heavy metal systems, expecially iron, have a strong peroxidase action. The complexes which they form with the organic substrate are attacked by $H_2O_2$ which, in turn, is thereby destroyed. We are dealing here partly with induced reactions and partly with true catalyses. The second substrate may be a diphenol, p-phenylene diamine, dihydroxymaleic acid, or a thiol compound (for details see p. 70). It may be mentioned here that HARRISON 462) was able to demonstrate these relationships by adding ferrous salt to an autoxidizable, heavy metal-free system, e.g. hypoxanthine plus dehydrogenase. There will occur a dehydrogenation reaction closely resembling peroxidase action. This reaction is gaining more and more in importance: it may well be the normal sequence of reactions and it may have a considerable biological significance (p. 22). The first step consists in the autoxidation reaction proper yielding $H_2O_2$. In the subsequent stage the metal acts on the peroxide in statu nascendi as a peroxidase. This process may extend far beyond that of a reversible oxidation-reduction reaction. Such processes are apparently involved in pigment formation. A few examples may be given:

In the course of the iron catalysis of hydroquinone at first $H_2O_2$ is formed. In conjunction with iron there follows a peroxidatic change of the quinone to humin substances. A similar change occurs with the polyphenol ethers as the substrate; they are stable against $Fe + O_2$ (BACH 50)). No purpurogallin is formed in the oxidation of pyrogallol by $Fe + H_2O_2$, as is the case with peroxidase as the catalyst, but humin substances. The primary product in this case is an inactive pyrogallol-iron complex (WASSERMANN 1285)). Epinephrine behaves similarly (see e.g. MERRIT WELCH 780)). A similar explanation may hold for the observation of BINGOLD 120) that hemin $+ H_2O_2$ will decolorize methylene blue irreversibly. The oxidation by a metal $+$ hydrogen peroxide is a very drastic process and goes much further than ordinary biological dehydrogenations. In the decomposition of fatty acids by copper acting as peroxidase a drastic degradation has recently been observed (SMEDLEY-MACLEAN et al. 1059)). It might be possible that in this instance the model reaction may possess true biological importance. Further details will be found on pp. 70, 254.

The theoretical significance of these observations is to be seen in their classification by WIELAND as an acceptor hydrogenation of $H_2O_2$ by the substrate, thereby postulating complete equality for molecular oxygen and hydrogen peroxide. This is equally important for the development of the view that the heavy metals too first act as dehydrogenases and then in turn, are dehydrogenated themselves, as it is for the interpretation of the biological action of peroxides, i.e. of the peroxidase function. It is in keeping with the assumption of WIELAND that the catalatic breakdown of hydrogen peroxide represents a coupled dehydrogenation and hydrogenation of two $H_2O_2$ molecules.

### Hydrogen Peroxide in Biological Systems.

Some time ago $H_2O_2$ formation was first detected in biological systems. It was found in catalase-free bacteria and also in a cell-free enzyme system, namely, when xanthine is acted upon by the purine dehydrogenase of milk (THURLOW, DIXON). A little later WIELAND and FISCHER 1310) and WHELDALE-ONSLOW 1297) observed

$H_2O_2$ production in the dehydrogenation of polyphenols by a thermostable and cyanide-resistent inorganic salt mixture obtained from mushrooms. These findings, however, did not carry very much weight since the peroxide concentrations were very small owing to the presence of iron containing oxidases. Moreover, the observations made with the salt mixture from Lactarius could not be reproduced. The true oxidizing enzyme present in the same source does not yield $H_2O_2$ (WIELAND and SUTTER 1325)).

Since then, the formation of hydrogen peroxide by the action of the SCHARDINGER-enzyme which dehydrogenates xanthine and aldehydes has been confirmed by WIELAND 1311, 1319, 1320) in extensive experiments. The quantitative method which he employed is based on the following principle:

$H_2O_2$ is stabilized as cerium peroxide by adding ceric salt; the peroxide is decomposed with acid into $H_2O_2$ and ceric salt. The hydrogen peroxide is destroyed by catalase and the ceric ion is determined by titration. If the biological system contains catalase to begin with, the method will fail because the enzyme reacts much more rapidly with $H_2O_2$ than does the cerium salt. In other instances (lactic acid bacteria), it has been possible to determine the $H_2O_2$ by direct titration. KI is added and the iodine liberated is titrated with thiosulfate.

Other examples of $H_2O_2$ formation are reactions catalyzed by cyanide resistent oxidases, e.g. tyramine oxidase (HARE 451)) and amino acid dehydrogenase. The same holds for the cyanide sensitive uricase (KEILIN and HARTREE 546)). One molecule of $O_2$ yields one molecule of $H_2O_2$. In certain coupled reactions, however, the $H_2O_2$ is hydrogenated by the substrate, e.g. alcohol, phenylene diamine.

It has since been possible to detect and to assay quantatively $H_2O_2$ in living cells which are void of catalase. The experiments by PLATT 933) and HEWITT 480) with pneumococci and streptococci confirmed the earlier observations, e.g. of AVERY, but they were only qualitative in nature and therefore not quite decisive. DUBOS 221) showed that the limiting factor in peroxide formation by pneumococci is the donator concentration; addition of sugar will cause a continuation of $H_2O_2$ formation. Eventually BERTHO and GLÜCK 107, 108) succeeded in solving this problem with the aid of catalase-free, facultative anaerobic lactic acid bacteria (B. Delbrücki, acidophilus, Jugurt). They showed that the entire $O_2$ which is absorbed is converted into $H_2O_2$ (R.Q. = 0.5). Finally, the bacteria are poisoned by the peroxide and they die. The same observations have been made by FROMAGEOT 378) with B. bulgaricus.

These cells represent very special cases of enzymatic systems. They contain no hemins (not even cytochrome) and they are accustomed to live anaerobically. The rather unphysiological respiration which is "forced" upon them by bringing them into oxygen is catalyzed by the yellow ferment (p. 42). We are dealing here, then, with an acceptor respiration (p. 108) which may be increased by adding methylene blue up to 300 per cent and which is not affected by cyanide or carbon monoxide. The substrate which is utilized in this respiration is some product of the anoxybiontic carbohydrate breakdown which is symbolized by DAVIS 195) as "lactic acid."

Acetobacter peroxydans is the exceptional case of a strongly respiring bacterium with hemin enzyme systems which is free from catalase and still does not produce hydrogen peroxide. WIELAND's 1321) investigation has revealed that the bacterium compensates for the lack in catalase by a strong peroxidatic activity of its hemin systems; even extraneous $H_2O_2$ is utilized for the dehydrogenation of alcohol or, in the model reaction, of p-phenylenediamine. We are dealing here with a case of special adaptation, since the bacterium is capable of oxidizing molecular hydrogen by hydrogen peroxide only under anaerobic conditions. Upon admission of $O_2$ this reaction ceases, probably because $O_2$ displaces the $H_2O_2$ from the surface of the catalyst by virtue of its greater affinity.

In the case of cells containing catalase but possessing a cyanide resistant respiration, e.g. Chlorella, the attempt to detect $H_2O_2$, even after inhibition of the catalase by HCN, has failed to yield positive results (TANAKA (see OPPENHEIMER 902)). Presumably also in this case the $H_2O_2$ is decomposed in a peroxidatic reaction by the hemin system which may be present but not always be active. It is to be remembered that the respiration of Chlorella occurs via the flavin system only if sugar is lacking; otherwise the respiration is catalyzed in the usual manner by the ferment hemin (p. 42 and 260).

## c) Acceptor Specificity.

The most important development of WIELAND's theory in recent years has been the recognition of specificity of dehydrogenases also with respect to the acceptor. In other words, we have come to realize that the acceptor too exhibits specific affinity for the catalyst, and not only the donator. The term affinity is applied here in the same sense as in the exposition of the general theory of catalysis. It represents a tendency to form complexes which may be interpreted either as adsorption or as residual valency compounds.

Even in his first publications on biological oxidation WIELAND considered the possibility of adsorption of the acceptor on the catalyst. Furthermore, he pointed out that certain compounds of considerable oxidizing power, e.g. persulfates, are not suitable as acceptors in dehydrogenase systems. But WIELAND did not recognize the specific element; for all ordinary cases his theory explicitly postulated that any possible and thermodynamically permissible acceptor could be utilized. The same enzyme, for instance, was supposed to dehydrogenate acetaldehyde, and to transfer the hydrogen to another molecule of aldehyde (dismutation) or to a quinoid dye (SCHARDINGER-reaction) or to $O_2$. The only exception that WIELAND mentioned was that, in view of the specificity of the peroxidase action, $H_2O_2$ cannot act as acceptor for the ordinary dehydrogenases of the cell but only for peroxidases. It has previously been pointed out (OPPENHEIMER 902)) that this justified assumption punctures WIELAND's theory with regard to an essential point. If in one case acceptor specificity had to be admitted there was no reason why the principle of non-specificity should be maintained for the other acceptors or enzymes.

As a matter of fact, the hypothesis of free choice of acceptor has been refuted by subsequent developments. When the acceptor was varied with the same catalyst-donator combination, differences were observed which have led to the rule that in each case the catalyst, and especially the dehydrogenases, are functioning in a specific ternary system and that the acceptor is, at least quantitatively, of decisive importance for the effect.

There are, for instance, true dehydrogenases which, contrary to the case of the xanthine dehydrogenase and aldehydrases which have been particularly stressed by WIELAND, are unable to react with molecular $O_2$ unless another autoxidizable system (heavy metal catalyst) is present. This conclusion is inevitable if it is admitted that poisoning by cyanide in very small concentrations is adequate proof for the participation of an iron system. The first case of this kind was that of succinic acid dehydrogenase (see OPPENHEIMER **902**)). The enzyme action is resistant against cyanide if a dye is used as acceptor, while the hydrogenation of oxygen is cyanide-sensitive. The same is true for citric acid and for lactic acid dehydrogenase. These enzymes have been designated by THUNBERG as **anoxytropic dehydrogenases**. Here it has indeed been possible to prove that they no longer react with $O_2$ in purified state. LEHMANN **710)**, continuing the work of L. STERN and of HAHN, has recently shown that if succinic dehydrogenase is kept at 0° the reaction with $O_2$ is completely abolished. Conversely, almost all "anoxytropic" dehydrogenases may be made to utilize oxygen if a heavy metal catalyst is added. This has been demonstrated for the "anoxytropic" glucose dehydrogenase by HARRISON **458)**. However, his findings could not be confirmed by OGSTON and GREEN **895)**. The weak aerobic action in the system dehydrogenase-glucose was increased sixfold if he added cytochrome c and respiratory ferment (in the form of washed sheep heart muscle). In the absence of the dehydrogenase the reaction did not proceed. A similar finding was made by BARRON and HASTINGS **83)** in the system lactic acid dehydrogenase (from gonococci) $+$ nicotine hemochromogen or cresyl blue.

With respect to these anoxytropic dehydrogenases it appears to be established that they are unable to react with molecular oxygen because they show no affinity for it. FRANKE (l.c. **288)**, p. 240) explains this by the fact that the small oxygen molecule is adsorbed and deformed only with difficulty and that a particular structure is required for combination with oxygen, e.g. complex metal configurations. One might mention here also the affinity of oxygen for polyphenols. As a matter of fact, only few organic substances and only those belonging to a few classes of the system are autoxidizable. The acceptor specificity is, then, proven to satisfaction in these instances.

When testing for purposes of comparison a number of different acceptors, physical-chemical and toxic factors may confuse the issue. Experiments with intact cells, in particular, are not quite suitable for obtaining clear cut results. The situation is much simpler if isolated dehydrogenases are used. It is true, though, that even here affinities may be simulated by inhibition of the enzyme by certain acceptors. Methylene blue and quinone are cases in point (WIELAND **1320)**). Quinone poisons the succinic dehydrogenase of muscle tissue (WIELAND and FRAGE **1311)**), the aldehydrase of yeast (WIELAND and CLAREN **1308, 1309)**) and, more slowly, also the aldehydrase of milk (**1320)**). The isolated alcohol dehydrogenase of yeast reacts with $O_2$ and methylene blue, but not at all with benzoquinone (MÜLLER **842)**). On the other hand, the enzymatic activity may be harmed with $O_2$ as the acceptor if the purification of the dehydrogenase has eliminated the last traces of catalase so that the toxic $H_2O_2$ which is formed in the enzymatic process is no longer destroyed. The aldehydrase of milk is an example (WIELAND and ROSENFELD **1325)**).

The phenomena observed with whole cells are very complex. Besides considerations of acceptor specificity there enters the question of diffusion and permeability in general, furthermore specific inhibition of the enzyme proper (see above) and lastly secondary reactions of the acceptor with other components of the cell, e.g. with the hemin enzymes. There is the further complication that in comparative experiments with acceptors and $O_2$ with respiring cells the reduced form of the acceptor will not react directly with $O_2$ but that it will be reoxidized by the ferric form of the respiratory ferment (WARBURG). Thereby the dehydrase action becomes cyanide sensitive in an indirect manner even if the enzyme as such should be cyanide resistant. In any case these disturbances mask the true kinetics of the enzyme reaction.

In the first place problems of diffusion have to be considered when working with intact cells. BERTHO (**105)**, p. 732) states that besides $O_2$ biological acceptors like acetaldehyde will readily penetrate into the cell. As a rule, this holds also for quinone which is a very strong acceptor. Equal affinity for the enzyme surface assumed, these acceptors will act similarly to $O_2$ even with intact cells. This is not true for the quinoid dyes. The efficiency of methylene blue as an acceptor, for instance, will sometimes be found to be close to that of quinone or oxygen, as with lactic acid bacteria (BERTHO and GLÜCK 108)), in other cases it is found to be much less efficient, e.g. with acetic acid bacteria (WIELAND and BERTHO 1305), REID 973)), with yeast (WIELAND 1308, 1309)) and THUNBERG 1173)), or with minced muscle (WIELAND and FRAGE 1311)).

The participation of permeability factors is brought out by experiments by AMBRUS et. al. 22) with yeast. Methylene blue which has once penetrated into the cell is reduced three times faster than the dye outside the cell. Furthermore, the apparent inferiority of methylene blue compared with $O_2$ and quinone disappears, both in the case of yeast and of acetic acid bacteria, if acetone preparations with destroyed cell structure are employed (WIELAND 1301)). Plasmolyzed yeast reduces methylene blue twice as fast as intact yeast (HARRISON 456)).

That some complications are caused by the cell hemins is indicated by the different behaviour of hemin containing cells. BERTHO 104) reports that those cells which are free from hemin (lactic acid bacteria) react with methylene blue and with quinone as well as with $O_2$ (Ratio $O_2$: Methylene blue: Quinone = 1 : 2.5 : 4.0). The cells reacting abnormally have a true respiration and a complete heminsystem. In acetic acid bacteria, for example, quinone inhibits the oxygen uptake (WIELAND and BERTHO 1305)); methylene blue will do the same in the case of yeast with acetate as the substrate (WIELAND and CLAREN 1308, 1309)). The interpretation of these observations is still in a state of controversy. WIELAND thinks that the inhibitor blocks the catalyst so that oxygen cannot react with it (e.g. in the quinone inhibition of acetic acid fermentation) while WARBURG's school (REID 973)) prefers to think of an interaction between the acceptors and the cell hemins, just as WARBURG has conclusively demonstrated in the case of the blood hemins (p. 82). According to this view the acceptor does not react directly with the substrate, at least not exclusively, but in the first place with the hemin and thereby initiate an acceptor respiration which is only slightly cyanide-sensitive. This fact is interpreted by WIELAND as a replacement of the cyanide from the catalyst surface by the acceptor, while, according to REID, quinone reacts chemically with HCN. It may also be that, conversely, the acceptor dehydrogenates the substrate and hands on the hydrogen to the final oxidation system, either directly to the respiratory ferment or to cytochrome (see also TANAKA 1136)). Experiments with cells the structure of which has been destroyed cannot decide the question which of the two factors, permeability or reaction with cell hemins, is the more important one. The increase in effectiveness of methylene blue under these conditions may be due to facilitated access to the interior of the cell as well as to destruction of the hemin system which is known to be linked up intimately with the cell structure. The methylene blue respiration of such cells is no longer cyanide sensitive (MEYERHOF) (see also p. 108).

Inspite of all these complications significant differences with various acceptors when used in the same donator-catalyst system may be observed. A few examples may illustrate this statement (see also WIELAND 1311) and FRANKE (l.c. 288), p. 210)). Firstly, there exists a general acceptor specificity since certain thermodynamically powerful acceptors *) are completely inactive in dehydrogenase systems, e.g. diethyl

---

*) The striking observation that aliphatic carboxyl groups are never acceptors in spontaneous processes has a thermodynamic reason. The dehydrogenation of aldehyde hydrate (CH(OH)$_2$ →

peroxide and potassium persulfate (BERTHO 103)). On the other hand, the resistance of disulfide groups towards dehydrogenases of acetic acid bacteria is a special case since the S—S group may be utilized as acceptor in other instances. That it is the affinity between acceptor and donator-catalyst system which matters is shown by the fact that the "main acceptor" methylene blue is not utilized at all in the deamination of amino acids though, according to recent energy data, it is thermodynamically permissible. It may be that we are faced here with structure-linked specificity since all preferred acceptors for amino acid hydrogen contain $C = O$ groups (quinones, alloxane, isatin) (see also FRANKE l.c.288), p. 154). The new outlook with regard to acceptor specificity will probably help to reinterpret many biological observations, e.g. the fact that nitrates cannot always serve as acceptors for animal tissues and sulfates not at all, while both substances are utilized in this capacity by plant tissues and bacteria.

It is not possible to speak in a mere qualitative sense of the suitability of an acceptor; we are dealing with quantitative differences. It is to be noted that thermodynamic relationships are not the only ones to be considered. This may be concluded from the different effect of various dyes on the acceptor respiration (p. 108). Of course, thermodynamic questions play an important part. The well-known fact, for instance, that the classical "oxidases" (phenolases etc.) react exclusively with $O_2$, is very probably due to the inability of the usual acceptors to dehydrogenate these substances of very positive potential. That thermodynamics is not the only consideration is demonstrated by the example of the glucose oxidase of MÜLLER which has exactly the same effect as the anoxytropic glucose dehydrogenase of HARRISON (formation of gluconic acid); but that reaction in contrast to the latter, occurs only with $O_2$, quinone and certain dyes of high $E'_0$, but not with methylene blue, nitrate, etc. (W. FRANKE and F. LORENZ, Lieb. Ann., 532, 1 (1937)).

In view of the importance of the subject some special examples of measurements with aldehydrases may be given here. The relationships in this field are particularly intricate. First of all, the question of donator-specificity is still to be answered. It is as yet undecided, for instance, whether purine dehydrogenase and aldehydrase are identical, as is believed by the school of HOPKINS, or whether they are different entities as WIELAND maintains. In order to find out whether the aldehydrase and the xanthine dehydrogenase of milk are identical WIELAND and MITCHELL 1320) have studied the enzyme reaction with methylene blue and quinone as acceptors. The substrate was a mixture of xanthine and aldehyde. With methylene blue as acceptor the entire xanthine was dehydrogenated first and it was not until afterwards that the aldehyde was dehydrogenated. In the case of quinone the opposite is observed: 80 per cent aldehyde and 20 per cent xanthine are attacked. This demonstrates that the combination between enzyme and substrates is markedly affected by the acceptor. In presence of methylene blue the active center of the aldehydrase is blocked by xanthine and becomes available only after the latter is dehydrogenated. This is not the case with quinone. It is probable, however, that these results are vitiated by the methods used.

Another example is the aerobic dehydrogenation of acetaldehyde by yeast previously freed from reserve substrates by aeration. Here, the acceptor function depends on the donator concentration. In concentrated aldehyde solution the chief acceptor is again aldehyde; dismutation is the result. In dilute aldehyde solution acetic acid is formed by the acceptor $O_2$. With acetic acid bacteria, the predominant process is the

---

$(COOH + H_2)$ proceeds with a loss in free energy (p. 11) so that the hydrogenation of COOH to $CH(OH)_2$ would represent an endothermal reaction (see Table on p. 53).

dehydrogenation at any substrate concentration (WIELAND et al. **1305, 1308, 1309)**). As has already been mentioned, the situation is complicated by the fact that not $O_2$ but "hemin oxygen" acts as acceptor in this organism. This same holds for the experiments by REICHEL **968)** who found a decrease in dismutation upon adding indophenol oxidase or quinone. REICHEL claims that aldehydrase will react with $O_2$ only in presence of the yellow enzyme.

In the case of plant enzymes nitrate reduction and dismutation proceed in entirely different fashion (MICHLIN **832)**). Methylene blue is hardly utilized. While the dehydrogenase from potatoes does not act as mutase, the corresponding enzyme in peas acts almost completely as mutase. On the other hand, the potato aldehydrase is so sensitive to oxygen that BERNHEIM **98)** was unable to decide whether it may utilize $O_2$ as acceptor. DIXON **216)** has succeeded in separating the mutase and aldehydrase from each other.

Also in the dehydrogenation of alcohols by bacteria differences are found with respect to various acceptors and donators; for results on acetic acid bacteria the review paper by BERTHO **104)** and with respect to butyric acid bacteria that of WIELAND and SEVAG **1326)** should be consulted; tables are found in FRANKE (**288)**, p. 212)). For instance, ethyl alcohol is not dehydrogenated by quinone, but methylene blue or oxygen may serve as acceptors. The ratio for iso-butyl alcohol (and similarly for iso-propyl alcohol) is, $O_2$/quinone/methylene blue $= 52/112/5000$; for propyl alcohol all three acceptors are equally well suitable. For an account of the strange choice of acceptors by anaerobic vibrios which dehydrogenate ethyl alcohol to acetic acid with the aid of sulfate, see BAARS **49)**.

These differences between the various acceptors led WOOLF **1345)** to develop the conception of the general necessity of substrate-enzyme-acceptor complex formation. BERTHO **103)** has developed these ideas further for the case of the dehydrogenase of acetic acid bacteria and he has applied it to the SCHARDINGER-enzyme. His kinetic considerations show that the conclusion drawn on the basis of free acceptor choice, namely, the linear dependence of turn-over on acceptor concentration, at optimal donator concentration, is not correct, but that there exist much more complicated relationships. The kinetic data favor unquestionably the view of a specific linkage, acceptor and donator and the enzyme thus forming a ternary system.

According to this view molecular oxygen can be hydrogenated only if it is bound by the catalyst.

The theory of the acceptor specificity is not limited to oxygen but has a far more extended range of validity. It may be applied to peroxidase action. WIELAND's view that peroxidases are phenol dehydrogenases with acceptor specificity for $H_2O_2$ has gained a high degree of probability. HABER and WILLSTÄTTER **437)** accept this point of view which previously was not shared by WILLSTÄTTER **1336, 1338)**. MANN **753)** attempts to explain the phenomena of peroxidase action on the basis of acceptor specificity or of a system substrate-peroxidase-$H_2O_2$ respectively.

The concept of invariable acceptor specificity leads directly to the modern idea that the enzymes of preparatory desmolysis, i.e. the dehydrogenases, must be redox systems themselves. In the case of the yellow enzyme or of the pyridine containing enzymes this has already been experimentally verified. The oxidized form of the enzyme represents the acceptor combining with the substrate, the enzyme in turn combines with the following acceptor which may again be an enzyme or a non-enzymatic metabolic intermediate (p. 87).

Above all the recognition of the acceptor specificity makes it possible for WIELAND's school to agree with the main postulate of WARBURG that his enzyme hemin which has the rare property of being truly autoxidizable is essential for the complete combustion of metabolites in respiring cells.

## 2) Development of Warburg's Theory.

In contrast to WIELAND, WARBURG did not develop his theory on the basis of model reactions but on that of his findings in the field of cell respiration. His main postulate has been from the very beginning that molecular oxygen never reacts directly with metabolites in respiring cells but only via the autoxidizable iron nucleus of the respiratory ferment. The inhibition of respiration by minute amounts of hydrocyanic acid ($< 0.0001$ M.) was explained by complex formation with this iron. The question as to how the oxygen is activated by the iron was relegated to the background. WARBURG spoke of a "higher oxidized form of the iron" which is formed from $Fe^{++}$ by oxygen and which after reacting with the substrate is changed back into $Fe^{++}$. Later, he no longer spoke of an "activation" but of a transfer of the oxygen. He is of the opinion that a substance is in active state if it reacts. Only indirectly from the known facts concerning the old peroxide theories could it be surmised that WARBURG did not refer to $Fe^{+++}$ when speaking of higher oxidation stages of iron but to some kind of peroxide in the sense of MANCHOT. After his own experiments had demonstrated that $Fe^{+++}$ is not inactive if it is contained in a certain complex linkage but that it is reduced by organic "activated" molecules did WARBURG identify this higher stage with ferric iron. When viewed in this light, the catalysis does not begin with an activation of oxygen but with a reduction of ferric to ferrous iron. It could no longer be disputed that hydrogen is transferred in this catalysis. The reduction of the ferric iron corresponds to a dehydrogenation of the substrate. Subsequently, WARBURG has discovered the participation of hydrogen transferring enzymes in this catalysis (p. 43).

Since we are dealing with a cyclic reduction and reoxidation of iron, we cannot designate either of them as the "beginning" of the catalysis. Since WARBURG **1261)** found that in respiring cells the rate of oxygen uptake is practically independent of the outside oxygen pressure (down to $10^{-5}$ atm. $O_2$) he assumes that the iron is preponderantly present in the higher oxidation form, as $Fe^{+++}$ which has been formed by the reaction of the $Fe^{++}$ (produced by reduction) with $O_2$ via the short-lived oxygenation product $Fe^{++} \cdots O_2$ (corresponding to oxyhemoglobin). The factor determining the rate in this case is the reduction of the ferric iron by the metabolites. The intermediary existence of a true ferrous peroxide is no longer held probable (REID **974**, p. 341). According to the equation

$$2 > Fe^{II} + 2\ H_2O + O_2 = 2 > Fe^{III}.OH + H\!-\!O\!-\!O\!-\!H$$

the oxidation of the ferrous stage should yield hydrogen peroxide. WARBURG has not particularly stressed this point but he has not denied it. His collaborator REID **972)** has actually identified $H_2O_2$ formed during the copper catalysis of leucomethylene blue oxidation. Thereby another controversial point has been removed from discussion. Only as long as WARBURG assumed that the "higher oxidation stage" of iron yields "active" oxygen, did he have to deny the formation of $H_2O_2$. It has been indicated above that very probably $H_2O_2$ does appear during the catalysis and that the failure to detect it in most cases is due to its rapid removal by iron catalysis.

## The Hemin Ferment of Warburg.

All these modifications of the original theory have been developed by WARBURG himself in the course of his work on the "oxygen transferring enzyme of respiration" or "hemin ferment". The fundamental observation which led to the trail of this enzyme was the discovery that the carbon monoxide inhibition of respiration is light-sensitive **1221)**. This suggested the close relationship of the enzyme to the blood hemins the CO complexes of which for some time were known to dissociate reversibly in strong light (HALDANE and SMITH). The ingenious method employed by WARBURG in the indirect determination of the absorption spectrum of the hemin ferment ("photochemical efficiency spectrum") rests on this observation. He was able to show that by far the largest fraction of cell respiration is carried out via this complex iron system. The great importance of this work is not impaired by WARBURG's more recent discovery of an iron-free respiration which may occur under special conditions, especially when the iron system is lacking. We refer to the yellow enzyme which catalyzes a true acceptor respiration by an organic autoxidizable system without the participation of heavy metal. The descriptive chemistry of the hemin enzyme will be taken up later. Here it may suffice to say that it is a pyrrol-iron compound somewhat similar to blood hemin but containing more oxygen; it is a pheohemin which is also related to chlorophyll. Like all hemins it contains ferric iron which upon reduction is transformed into the heme stage, with $Fe^{++}$. This is the physical explanation for its catalytic action (WARBURG **1233)**).

This particular hemin is contained in a special linkage. It is not simply combined with a colloid bearer protein but it is linked to the structure of the cell. The respiration of the cell is a function of its structure and is abolished upon cell death. This situation prohibits a chemical isolation of the enzyme by ordinary methods. It is to this linkage that the enzyme owes its enormous activity which is estimated to correspond to a turnover of $10^5$ molecules of substrate per second by one enzyme molecule. Oxidase function is found even in the common blood hemin. But whereas the latter, in the absence of a specific bearer, like other complex Fe salts exhibits some peroxidatic and catalatic activity besides oxidase function the oxidase property is found highly specialized and tremendously increased in the respiratory ferment. This "activation" enables the minute traces of ferment hemin in the cell ($10^{-7}$ M) to take care of the respiratory requirements.

The reaction of the enzyme with the well-known respiratory inhibitors permitted WARBURG and his colleagues (**1234, 1277, 1280)**) to demonstrate the valency change of the iron of the ferment with the spectroscope (in acetic acid bacteria). The non-inhibited enzyme will show a characteristic shift of the absorption bands when oxygen admission is restricted (reduction of the enzyme by metabolites) and when oxygen is readmitted (reoxidation). The long-wave absorption band of the ferric form is in the red region (639 m$\mu$) and that of the reduced form is in the yellow region (589 m$\mu$) (For criticisms of these findings see p. 149).

The type of inhibition points to the existence of two valency stages: HCN inhibits the reduction, not the oxidation of the enzyme, while CO inhibits only the oxidation. Both valency stages may be stabilized by these inhibitors.

The unstable intermediate oxygenation stage may also be spectroscopically observed. Upon saturation of a suspension of bakers yeast with oxygen a band is seen in the yellow region; it disappears upon removal of the oxygen by the respiration. If the $O_2$ uptake is inhibited by cyanide poisoning the band in the yellow remains visible.

The same spectroscopic observations yielded the very important result that the ferment hemin, of all cell hemins, is the only autoxidizable system. KEILIN's **540)** cytochrome is not autoxidizable in the physiological ph-range.

Cytochrome, like the respiratory ferment, is a hemin system. It represents a mixture of three components (a, b, and c) one of which, cytochrome c, contains a hemin group closely related to protohemin in combination with a bearer protein more basic than globin (ZEILE, **1372a)**). KEILIN himself has observed that cytochrome c alone is not autoxidizable; its oxidation is accomplished by a special "oxidase" which was originally identified by KEILIN as the well-known, more or less structure linked, indophenol oxidase. Today, both WARBURG and KEILIN agree that the oxidase is identical with the respiratory ferment. If the entire hemin system of the cell is poisoned in reduced state by HCN, upon admittance of oxygen the absorption band of the ferrous form of the respiratory ferment in the yellow is shifted to red while the absorption band of the reduced cytochrome in the green remains unchanged. The oxidation of cytochrome is inhibited by HCN because the ferric form of the respiratory ferment is blocked by HCN and is therefore unable to oxidize cytochrome (WARBURG). For an analogous reason the oxidation of the cytochrome is also inhibited by CO; here, the ferrous form of the respiratory enzyme is blocked by the poison. Cytochrome proper does not react with carbon monoxide within the physiological ph-range; it will do so only in strong alkaline solution, around ph 12 (KEILIN **540)**).

### The New Schema of Cell Respiration.

The interpretation of the relationship between the cytochromes and the ferment hemin by WARBURG and by KEILIN is another step towards a reconciliation with WIELAND's theory. Inasmuch as the respiratory ferment alone is autoxidizable, the cytochromes can represent only carrier systems. They carry the labile hydrogen of the substrate to the respiratory ferment. They deserve equally well the name "hydrogen transporting systems" (THUNBERG); we propose here the term Meso-catalysts". The fact that cytochrome contains iron does not matter.

That iron may participate in anaerobic processes had previously been found in WARBURG's laboratory (see TODA, **1177)**). WARBURG has studied the question of heavy metal catalysis in fermentation (p. 71). The over-all fermentation is sensitive to HCN and $H_2S$ but not to CO. The assumption is made that the metal in these instances transfers "bound", not free, oxygen (**1223)**). It is still an open question as to what extent dehydrogenating heavy metal systems participate in vital anaerobic processes (p. 71). In any case, it is worthy of consideration that a cyanide-resistant hemin system, cytochrome, is not directly but only indirectly concerned with catalytic oxidation.

The new schema of WARBURG, proposed in agreement with KEILIN, as put forth for the case of acetic acid bacteria (**1280)**) is as follows:

$$O_2 \rightarrow \underbrace{Fe^{++} - Fe^{+++}}_{1} \rightarrow \underbrace{Fe^{++} - Fe^{+++}}_{2} \rightarrow \underbrace{Fe^{++} - Fe^{+++}}_{3} \rightarrow \underbrace{Fe^{++} - Fe^{+++}}_{4} \rightarrow$$

Intermediate Enzymes → Alcohol.

The redox potential in this chain is increasingly positive from right to left. The hydrogen will, consequently, go from right to left in accordance with the potential slope. When it has reduced the first cytochrome component (2) it is taken over by the respiratory ferment (1) and finally offered to the oxygen. WARBURG's collaborator HAAS (**430a)** has shown that in the actively respiring yeast cell the entire respiration goes through this chain. The scheme might, therefore, be written just as well in the following form:

Donator → Dehydrogenases as acceptors → Cytochrome I → Cytochrome II → Cytochrome III → Hemin Ferment of Respiration → Oxygen. The order in which the three cytochrome components are linked in the chain is still not definitely settled. Recent experiments by BALL 58b) and others indicate that the order is: $O_2$ ← Respiratory Ferment ← Cytochrome a ← Cytochrome c ← Cytochrome b . . . . .

According to its donator specificity, and in accordance with WIELAND's terminology, the hemin ferment of WARBURG might be called an oxytropic dehydrogenase of cytochrome III. Since there is no longer any differentiation between dehydrogenases and oxidases, the hemin ferment might also be termed cytochrome oxidase (THUNBERG).

### The Yellow Enzyme and the Hydrogen Transferring Enzymes.

The yellow enzyme which has recently been discovered by WARBURG is a metal-free respiratory system capable of reacting with molecular oxygen. It shares the property of being reversibly oxidizable and reducible with a good number of other biological systems; like the respiratory ferment it is autoxidizable. It may therefore act as the first catalyst in a respiratory chain just like other quinoid dyes, e.g. methylene blue or pyocyanine (p. 108).

The way to this discovery has been paved by a number of isolated observations on cyanide resistant respiration systems. We quote as examples the respiration of the alga Chlorella in media lacking sugar (WARBURG), the small respiration of facultative anaerobic bacteria (MEYERHOF and FINKLE) and other, less certain observations on cyanide resistant residual respiration (p. 260). WARBURG 1230) had found such a respiration in LEBEDEW juice from bottom yeast. There are also his observations on red blood corpuscles treated with phenyl hydroxyl amine (1264)). Such cells are capable of burning additional glucose with the aid of the redox system hemoglobin-methemoglobin (p. 82). The most interesting feature of this phenomenon is the fact that glucose is oxidized by methemoglobin only in living cells but not by the isolated pigment. We have to conclude that previous to being attacked by methemoglobin the substrate must be activated by an agent produced by the cell. If the treated blood corpuscles are laked (1237)) glucose is no longer utilized. However, the catalysis may proceed with hexose monophosphate as the substrate.

Still more information was obtained by experiments with untreated red cells. If normal erythrocytes are hemolyzed and if the stromata are centrifuged off, the clear supernatant fluid will not attack glucose. Hexose monophosphate is likewise not oxidized by this system in presence of oxygen. However, if methemoglobin or methylene blue are present this substrate is burned. By adsorption on aluminium hydroxide gel a catalyst is removed from the system thereby rendering it inactive towards hexose monophosphate even in presence of methemoglobin or methylene blue. WARBURG has found that this "activator" consists of an enzyme ("Zwischenferment") and a coferment. Both may be obtained separately from red blood cells. By using a sufficiently high concentration of these two substances hexose monophosphate may be oxidized by molecular oxygen without adding any other acceptors (methemoglobin, methylene blue). The catalysis by this water soluble, hemin-free system is resistant both to HCN and CO. In contrast to the enzyme the coenzyme is comparatively heat stable. Soon afterwards WARBURG discovered, in addition to these two catalysts, a thermolabile substance occurring in blood cells and, in fact, in any cell yet examined which supplements the "zwischenferment"-coferment system to yield a complete, iron-free, respiratory chain capable of reacting with molecular oxygen. The isolation of this new substance from bottom yeast (LEBEDEW juice) led to its chemical characterization (**1239, 1240, 1242, 1243, 1244**)). It has become known under the name "yellow oxidation enzyme"; it is the combination of the phosphoric acid ester of lactoflavin (vitamin $B_2$) with a protein bearer (p. 190). It represents a reversible redox system the oxidized form of which cannot react directly with hexose monophosphate but only with the intermediary participation of the auxiliary ferment system "Zwischenferment-coferment". The latter has since been shown to be a pyridine nucleotide (WARBURG). There exists a specific interaction between the reduced form of this pyridine ferment and the oxidized form of the yellow enzyme which may be classified as oxytropic dehydrogenase (**1158**)). The reoxidation of the reduced yellow enzyme by oxygen yields $H_2O_2$. It does not react with HCN or CO. The reduced enzyme may react not only with $O_2$ but also with isolated cytochrome c (**1159**)) and with methylene blue.

WARBURG **1237**) has emphasized the fact that in actively respiring cells the yellow enzyme cannot compete successfully with the hemin ferment; the latter reacts at a much greater rate with $O_2$. The main function of the yellow enzyme in such cells is very probably concerned with hydrogen transfer in the preparatory, anaerobic phases (p. 87). THEORELL **1158**) has demonstrated that the yellow enzyme, at the partial $O_2$ pressure existing in tissues, can be directly oxidized to only a very small extent; probably only a minute fraction of cell respiration takes place via the flavin enzyme (p. 265).

On the other hand, where the ferment hemin is lacking the yellow enzyme may replace it, e.g. in certain bacteria. This is only possible for short periods since $H_2O_2$ is formed and these bacteria are devoid of catalase. This is the reason why WARBURG has called the oxygen uptake by these facultative anaerobic organisms an "unphysiological respiration". He has demonstrated the presence of the yellow enzyme in these bacteria by spectroscopy (**1244, 1246**)) and he has proved that in BEETHO's experiments with facultative anaerobic lactic acid bacteria the enzyme is responsible for the entire respiration. The failure of BERTHO to detect $H_2O_2$ as a product of yellow enzyme catalysis and to find the theoretical amount of $H_2O_2$ in presence of HCN is explained by WARBURG as due to the presence of catalase (p. 265). WARBURG (**1277**)) also takes the view that the small and cyanide resistant respiration of Chlorella in sugar-free medium is caused by this enzyme. The same appears to hold for the residual respiration observed in mammalian tissues after HCN poisoning and in hemin free LEBEDEW juice (p. 265). The respiration caused by the yellow enzyme is a striking instance of a biological acceptor respiration.

## The "Pyridine" Ferments.

Of late, WARBURG **1234**) has emphasized that the combination hemin ferment-oxygen, does not suffice to react with the substrate but that the substrate must previously be "activated". This activation is achieved by enzyme systems which are identical with WIELAND's dehydrogenases. The "zwischenferment"-coferment system of WARBURG and CHRISTIAN **1239**), for instance, is an anoxytropic hexose monophosphoric acid dehydrogenase. It may be supplemented either by the yellow enzyme or by the hemin ferment. The components of this dehydrogenase system have recently been isolated by these workers (**1255**)) from yeast and from blood cells. The "zwischenferment" is nothing but the protein bearer while the "coferment" is the prosthetic group of the enzyme. The individual components are inactive (NEGELEIN and HAAS **882**)). HARDEN-EULER's Cozymase, according to independent work by WARBURG and by EULER, is chemically closely related to the coferment of the hexose phosphate dehydrogenase. It appears that combinations of varying specificity may be formed by the combination of the same coenzyme with different protein bearers and also by that of different coenzymes with the same protein (p. 212).

We see that WARBURG and WIELAND, on the basis of their own recent experimental contributions, have reached almost complete agreement on the more important features of biological oxidation. Their views and individual theories have merged to yield a unitary theory which is capable of explaining satisfactorily the known facts and which appears equally suited for extension to encompass the discoveries of the future.

## III. Theory of Chain Reactions.

In outlining their theory of oxidation processes, HABER and WILLSTÄTTER 437) have endeavoured to consider not only qualitative but also quantitative problems as relating to energetics and kinetics. The two outstanding features of their theory are the assumption of transfer of unpaired hydrogen atoms (as contrasted to the usual dehydrogenation equations considering only transfer of pairs of H atoms) and the assumption of reaction chains propagated by the free radicals produced by monovalent dehydrogenation. Inasmuch as these chains, once started, will go to completion without further participation of the catalyst this theory would help to explain the misproportion between reaction rate and mass of the catalyst. KUHN 648) has estimated that in the case of the hemin containing enzymes (respiratory ferment, catalase, peroxidase) one enzyme molecule will promote the reaction of $10^5$ molecules of substrate ($H_2O_2$ and $O_2$ respectively) per second.

The point of origin of the theory is the inorganic model reaction involving the "autoxidation" of sulfite by copper catalysis which had been studied by FRANCK and HABER 351a). The chain is started by the reaction:

$$SO_3^= + Cu^{++} + H_2O = \overset{\curvearrowright}{S}O_3H + Cu^+ + OH^- \qquad (1).$$

The free radical $\overset{\curvearrowright}{S}O_3H$ (monothionic acid) sets up a reaction chain:

$$\overset{\curvearrowright}{S}O_3H + O_2 + H_2O + SO_3^= = 2\,SO_4^= + OH + 2\,H^+ \qquad (2).$$

$$\overset{\curvearrowright}{O}H + SO_3^= + H^+ = OH^- + \overset{\curvearrowright}{S}O_3H \qquad (3).$$

The second free radical, produced without the catalyst, is the uncharged $\overset{\curvearrowright}{O}H$ which reacts with fresh sulfite ion to form the first radical, $\overset{\curvearrowright}{S}O_3H$ (equation (3)). The net result of the process is the oxidation of sulfite to sulfate by oxygen. The chain breaking step consists in occasional collisions between two $\overset{\curvearrowright}{S}O_3H$ radicals leading to formation of dithionic acid, $H_2S_2O_6$. The cuprous copper formed in reaction (1) is reoxidized by $O_2$ to cupric ion (regeneration of the catalyst).

The application of the general scheme of this inorganic model to organic reactions and organic catalysts (enzymes) encounters no theoretical difficulties. There is, for instance, the dehydrogenation effected by hemin complexes containing ferric iron. The reaction with aldehyde would be as follows:

$$CH_3 \cdot CHO + Fe^{+++} = CH_3 \cdot \overset{\curvearrowright}{C}O + Fe^{++} + H^+ \qquad (4).$$

The symbols $Fe^{+++}$ and $Fe^{++}$ stand for the enzyme and the monodesoxyenzyme respectively.

It will be seen that, at variance with the original scheme of WIELAND, the monovalent dehydrogenation yielding the free radical does not require a previous hydratation of the donator, in this case $CH_3 \cdot CH(OH)_2$. The $H^+$ combines with the anion lost by the Fe in the reduction. The main links of the ensuing chain reaction with the participation of $O_2$ would be:

$$CH_3 \cdot \overset{\curvearrowright}{C}O + CH_3 \cdot CHO + O_2 + H_2O = 2\ CH_3 \cdot COOH + \overset{\curvearrowright}{O}H + H_2O \qquad (5)$$

$$\overset{\curvearrowright}{O}H + CH_3 \cdot CHO = \overset{\curvearrowright}{C}H_3 \cdot CO + H_2O \qquad (6).$$

There may exist intermediary peroxides in this reaction (BÄCKSTRÖM 56), BOCKEN-MÜLLER 129)).

In anaerobic acceptor dehydrogenations, the place of $O_2$ is taken by the organic acceptor; consequently, instead of $\overset{\curvearrowright}{O}H$ a corresponding organic radical is produced in the chain reaction.

The dismutation of aldehyde to alcohol and acid (CANNIZZARO-reaction) is formulated in the following manner:

$$CH_3 \cdot \overset{\curvearrowright}{C}O + HOH + CH_3 \cdot CHO = CH_3 \cdot COOH + CH_3 \cdot \overset{\curvearrowright}{C}HOH \qquad (7)$$

$$CH_3 \cdot \overset{\curvearrowright}{C}HOH + CH_3 \cdot CHO = CH_3 \cdot CH_2OH + CH_3 \cdot \overset{\curvearrowright}{C}O \qquad (8).$$

The dehydrogenation of the donator by a quinoid dye as acceptor is given as:

$$CH_3 \cdot \overset{\curvearrowright}{C}O + HOH + \text{quinone} = CH_3 \cdot COOH + \text{meri-quinone *}) \qquad (9)$$

$$\text{Meri-quinone} + CH_3 \cdot CHO = CH_3 \cdot \overset{\curvearrowright}{C}O + \text{hydroquinone} \qquad (10).$$

The chain length is of the order of $10^5$ links. A different formulation of the intermediary radicals has been proposed by KENNER 561).

Analogous reactions are assumed to occur in non-enzymatic heavy metal catalysis, e.g. in the so-called autoxidation of benzaldehyde (p. 123) or in its dismutation. In both reactions the same radical, $C_6H_5\overset{\curvearrowright}{C}O$, is supposed to be formed; it reacts subsequently according to equations (5) and (7) respectively. Whether, in fact, these non-enzymatic reactions represent heavy metal catalyses (KUHN 652), HABER 434)) is, according to recent experiments by WIELAND 1323, 1324), not quite decided (p. 123). The catalatic activity of $Fe^{++}$ has also been explained by HABER and WEISS 157) in terms of a chain reaction with the radicals $\overset{\curvearrowright}{O}H$ and $\overset{\curvearrowright}{O} \cdot OH$. A similar interpretation of various other model reactions has been discussed by FRANKE (288), pp. 201, 262.

Quantitatively, these reaction chains are supposed to cause the main bulk of the substrate turnover while the initial step involving the enzyme itself is only important as the chain starting process. In order to be able to start another reaction chain the cata-

---

*) While HABER and WILLSTÄTTER assume that a meri-quinone (hybrid molecular compound between quinoid and benzoid form) is produced in this monovalent hydrogenation of the acceptor, this may be thought to be replaced by a semiquinone (free radical) in accordance with the theory of L. MICHAELIS (p. 100). Such a change would necessitate no modification of the scheme for the chain reaction.

lyst must be regenerated in the oxidized form. In the presence of oxygen this is achieved by autoxidation (transfer of $H^+ + \varepsilon$ to $O_2$, formation of $H_2O_2$). In the case of organic acceptors the desoxy-enzyme is directly dehydrogenated to the fully active enzyme. According to WEISS 1289) the concept of reaction chains is also able to explain the empirically established relationship between redox potential and rate of reaction. The reoxidation of the reduced form of the catalyst by $O_2$ may also involve an induced oxidation of a substrate equivalent (X):

$$Fe^{++} + HOH + X + O_2 = Fe^{+++} + OH^- + \overset{\wedge}{O}H + XO.$$

Peroxidase and Catalase action are formulated by HABER and WILLSTÄTTER in a similar manner. In the former case the enzyme will react first with a polyphenol molecule:

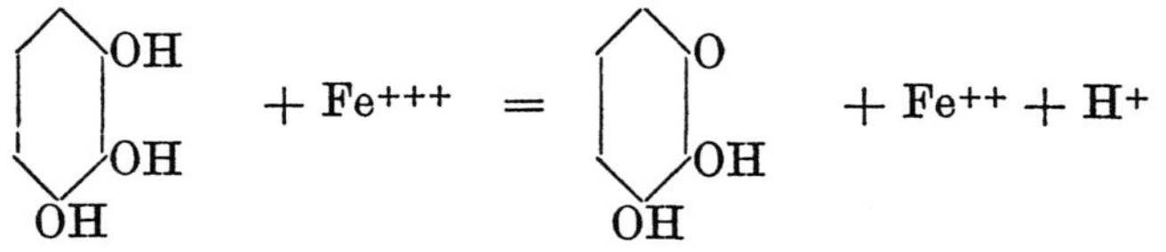

(11)·

The subsequent reaction chain is depicted in equations (12) and 13):

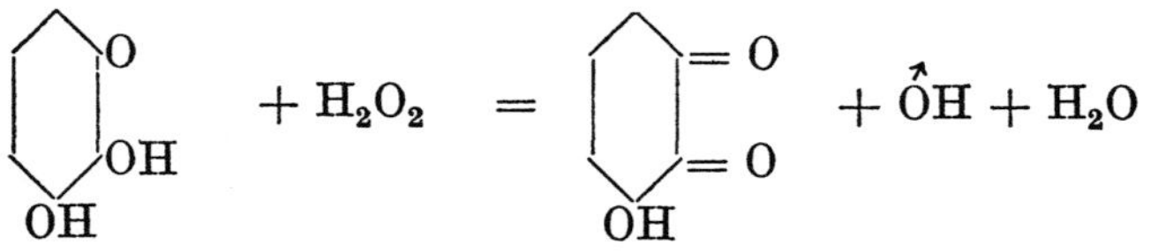

(12)

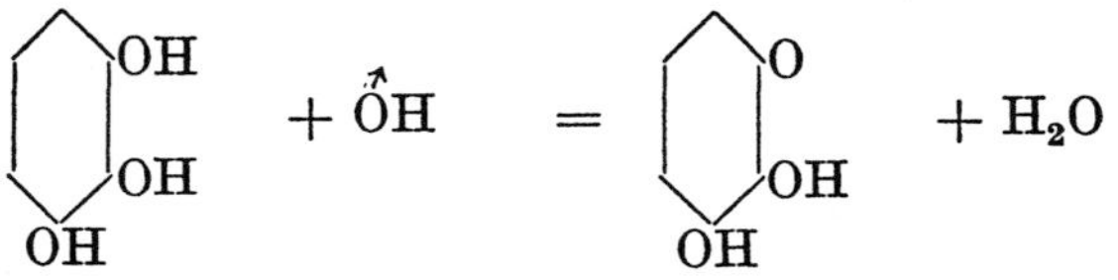

(13).

For the catalatic reaction the following scheme is proposed:

$$H_2O_2 + Fe^{+++} = \overset{\wedge}{O}\cdot OH + Fe^{++} + H^+ \tag{14}$$

$$\overset{\wedge}{O}\cdot OH + H_2O_2 = O_2 + \overset{\wedge}{O}H + H_2O \tag{15}$$

$$\overset{\wedge}{O}H + H_2O_2 = H_2O + \overset{\wedge}{O}\cdot OH \tag{16}.$$

The experimental evidence for the unusual stability of the ferric form of catalase and against an intermediary reduction of this form in the course of the enzyme reaction has led to the proposal of a somewhat different scheme (STERN 1073)):

$$H_2O_2 + Fe^{+++} \rightarrow [Fe^{+++}\cdot H_2O_2] \rightarrow Fe^{+++} + 2\ \overset{\wedge}{O}H \tag{17}$$

$$\overset{\wedge}{O}H + H_2O_2 = H_2O + \overset{\wedge}{O}\cdot OH \tag{16}$$

$$\overset{\wedge}{O}\cdot OH + H_2O_2 = O_2 + \overset{\wedge}{O}H + H_2O \tag{15}.$$

According to this hypothesis, the primary step would consist in the intermediary formation of an enzyme-substrate compound containing ferric iron and of its breakdown into the free enzyme and two uncharged hydroxyl radicals. An intermediate of this type has recently been observed spectroscopically in the catalase-monoethyl hydrogen peroxide catalysis (STERN 1085)). The subsequent alternating chain links would be the same as those postulated by HABER and WILLSTÄTTER, but in reverse order.

There exist some obvious objections against the chain theory which have, in part, already been discussed by HABER and WILLSTÄTTER. The energetics and kinetics of the catalytic primary reaction especially are still rather obscure. After a consideration of the rather uncertain energy values concerned, FRANKE (288), p. 261) concludes that, at standard concentration, the process should actually run in the reverse direction. Only under special conditions of concentration ratios of the reactants the reaction should proceed as postulated. The kinetics of the reactions, too, may be reconciled with the theory only with difficulty. Aside from such important objections against the primary reaction, the concept of the reaction chains propagated by free radicals has likewise been criticized.

RICHTER 981) favors the substitution of "energy chains" of the type first described by CHRISTIANSEN for the radical chains. An energy chain is propagated by energy transfer from activated product molecules to fresh substrate molecules. The objection made by WILLSTÄTTER, namely, that the energy should be dissipated and lost by collisions with the solvent molecules has been met by RICHTER by the reasonable assumption that the energy transfer is rather specific.

It has been pointed out by HALDANE 442), BERTHO 105), and FRANKE 288), p. 200) that the idea of radical chains, at least in the form proposed by HABER and WILLSTÄTTER, conflicts with the observed specificity of the enzyme reactions. If the enzyme drops out of the reaction after the first step and if the substrate turnover is mainly brought about by means of the free radicals which are supposed to be identical in various types of processes ($\overset{\rightharpoonup}{O}H$ figures in the schemes suggested for oxidase, peroxidase, and catalase reactions), one is unable to account for the specificity of the enzymes concerned. Furthermore, HALDANE fails to see how the chain reaction concept may explain the known relationships between enzyme concentration and substrate turnover. If each primary reaction between an enzyme molecule and a substrate molecule initiates a long reaction chain yielding many product molecules, an increase of the catalyst concentration should produce an exponential increase in product molecules rather than a proportional increase (linear relation) as is actually found to be the case within certain limits of enzyme concentration.

SCHWAB and his collaborators 1040) have attempted to obtain experimental evidence for or against the chain character of the catalatic reaction by studying its inhibition by small concentration of certain reagents which are supposed to act as chain breakers due to their reaction with intermediary radicals, especially with $\overset{\prime}{O}H$, and by comparing the extent of inhibition thus produced with that obtained in processes which are generally considered to be chain reactions (e.g. styrol polymerization and benzaldehyde autoxidation). The result of this investigation was inconclusive and the authors leave the question open as to whether the radicals propagating the hypothetical chain may not differ in the reactions studied.

If the radicals formed in the various processes were chemically different the chain theory would not conflict with the facts known concerning the specificity of enzyme reactions. However, the established relationships between enzyme concentration and substrate turnover would still be an objection to the chain mechanism. It should be pointed out, though, that these relationships are still rather obscure in the case of many enzymatic processes.

A further important objection, raised by HALDANE, is that all known chain reactions occur in homogeneous systems while the size of the enzyme molecules renders the enzyme reactions microheterogeneous. This question is related to the problem as to whether the reaction chains, if there are any, assume a considerable individual length or whether they are stopped in an early stage. This is especially pertinent in those instances where two OH radicals might combine to form $H_2O_2$. This type of chain breakage is observed to a large extent in the simple catalytic oxidation of hydrogen by oxygen at surfaces and must therefore also be considered in biological surface reactions with the participation of enzymes. BERTHO 105), (p. 745) stresses the fact that in enzyme reactions all of the oxygen which is used up is frequently converted into hydrogen peroxide. This would mean that the reaction would proceed completely according to equation (5) (p. 45) and that equation (6) would be eliminated due to the disappearance of the OH radical by dimerization so that the chain would be stopped.

On the other hand, a chain might conceivably be stopped by the combination of two organic radicals; this offers interesting possibilities concerning biochemical syntheses. It is known, for instance, that in the course of fermentation and also at other occasions there are formed acetoins from aldehydes in statu nascendi. This has been explained by NEUBERG as being due to a special enzyme, carboligase. DIRSCHERL has thrown doubt on this explanation. The synthesis could be explained without difficulty by the HABER-WILLSTÄTTER theory on the ground of combination of the two radicals postulated by equations (7) and (8) leading to breaking of the chain and acetoin formation:

$$CH_3 \cdot \overset{\shortmid}{C}O + CH_3 \cdot \overset{\shortmid}{C}HOH = CH_3 \cdot CO \cdot CHOH \cdot CH_3.$$

The problem of the reality of the chain reaction mechanism has recently again been attacked by GERENDÁS 397) in the course of investigations by SZENT-GYÖRGYI and his collaborators 65) on the mechanism of dehydrogenative autoxidation. The special case chosen for scrutiny was that of the oxidation of sulfite by heavy metal which has served as the starting point of HABER and WILLSTÄTTER's comprehensive theory (see p. 44). The two alternatives between which GERENDÁS tried to obtain a decision are: 1) the metal catalyzes the oxidation of every molecule of the substrate and 2) the metal starts a reaction chain which continues without further participation of the metal and causes the oxidation of many substrate molecules until it is terminated by a chain-breaking step. If the entire oxidation is mediated by the metal, the rate of alternating oxidation and reduction of the iron or copper ("Wechselzahl") must be comparable to that of the autoxidation of the sulfite. The rate of oxidation of the iron is rapid compared with that of the reduction. The latter may be determined with the aid of α, α'-dipyridyl. The experiment shows that the velocity of the autoxidation of the sulfite and that of the reduction of the iron are of the same order of magnitude. Inasmuch as the total autoxidation may be accounted for by the oxidation-reduction cycle ("Wechselzahl") of the metal, GERENDÁS and SZENT-GYÖRGYI consider the chain hypothesis to be ruled out. Although this may be correct for the cases here reinvestigated, other processes, such as the catalatic reaction, may still be chain reactions.

### Antioxidants and Inhibitors.

The inhibiting action of certain organic substances, especially phenols, on oxidative processes is probably due to their property to break off reaction chains. MOUREU and DUFRAISSE 836—840) who did the pioneer work in this field, have extended their experiments to various other classes of chemical compounds, notably nitrogen containing substances (837)).

The antioxidative efficiency, according to MATILL 759), depends on the number and the position of the phenolic hydroxyls: o- and p-compounds are inhibitors while m-isomers are inactive; $\alpha$-naphthol but not $\beta$-naphthol inhibits. The mechanism of the inhibition of oxidation by such compounds as proposed by MOUREU and DUFRAISSE is essentially based on the old scheme of ENGLER and depicts these processes as induced reactions. This may hold for the case of the various inorganic systems where it has been tested experimentally, but it appears rather unlikely for biological processes. While MOUREU and DUFRAISSE incorporate HCN in their list of antioxidants (837)), there can hardly be any doubt that its specific action is due to a different type of mechanism (p. 73). It is quite possible that the same inhibitor specifically blocks certain oxidation processes in vivo while it does not affect others; it might thus act as a regulator. Epinephrine, for instance, will promote the decomposition of amino acids in its quinoid form (p. 105) while it strongly inhibits the autoxidation of unsaturated fatty acids (FRANKE 354)). Some sulfhydryl compounds (glutathione, cysteine, thioglycolic acid) which ordinarily promote dehydrogenation reactions as intermediary catalysts, will inhibit other oxidations, e.g. the reoxidation of leuco methylene blue with copper as catalyst, by virtue of complex formation between the Cu ions and the SH-groups (SCHÖBERL 1030)). De CARO 163) interprets the inhibition of the oxidation of fatty acids by thyroxin as a complex formation of this type. On the other hand, iodine compounds have been found to be typical chain breakers (FRANKE (l.c. 288), p. 200)).

It may be, then, that the inhibitors stop reaction chains by using up free radicals and thereby becoming oxidized themselves. This aspect is discussed by EULER (l.c. 288), p. 260). According to RICHTER, energy chains might transfer the excess energy to the inhibitor molecules and thus be terminated. In accordance with this view only such substances may function as inhibitors which are unable to transfer the absorbed activation energy to other substrate molecules (BODENDORF 130)). For the theory of the antioxidants the papers by PERRIN 920), by DUPONT 225), and by MILAS 833) should be consulted.

# C. The Phenomena of Oxidative Catalysis.

## I. Energetics and Potentials.

**On the Energetics of Hydrogenation and Dehydrogenation:** Let us begin with some fundamental questions relating to thermodynamics. Until a few years ago affinity relationships in hydrogenation processes had to be discussed on the basis of heats of reaction, Q, as calculated from thermochemical data, because there were no determinations of free energy changes, $\triangle F$, available at that time. Since then substantial advances have been made. With the aid of entropy determinations instituted largely by G. N. Lewis and his school, the free energy change, $\triangle F$, may be calculated from the heat of reaction, $\triangle H$, and the entropy change, $\triangle S$, by means of the equation:

$$\triangle F = \triangle H - T \triangle S.$$

With respect to the theory in detail the book by Lewis and Randall 713) may be consulted. A short introduction has also been provided by Oppenheimer 905). A certain number of entropy values now available for organic substances have been compiled by Franke 356). Recently Borsook and his colleagues have determined the thermal values for some amino acids, purine and thiol compounds; these results, excepting the last paper by Stiehler and Hoffmann 1103), have also been tabulated by Franke 356). Reference is also made to the review by Borsook 135).

The electromotive force set up by a reversible redox system may also serve as a basis for calculating $\triangle F$ since the free energy change during a reversible process corresponds to the maximum useful work. Here, the potential imparted to an inert noble metal electrode by a reversible redox system is measured electrometrically, in volts. The normal potential, $E_o$, of a redox system at ph 0 is defined as the potential difference between a 1 : 1 mixture of the oxidized and the reduced form of the redox system and the normal hydrogen electrode. The free energy change may be obtained by means of the equation:

$$\triangle F = n \cdot F \cdot E_o \text{ volts, or } n \cdot 23,06 \cdot E_o \text{ Calories,}$$

where n is the valency difference between the oxidized and the reduced form ("electron number"), and F is the electrochemical equivalent (Faraday). $E_o$*) is obtained from the electrode equation:

$$E_o = E - \frac{RT}{nF} \ln \frac{[Ox]}{[Red]},$$

where E is the potential difference measured in a system of arbitrarily chosen concentration ratio of the oxidized form [Ox] to the reduced form [Red], at a given ph, and referred to a hydrogen electrode of the same ph and absolute temperature, T (°K); R the gas constant in electrical units. With respect to the theory and experimental methods, the monographs by Michaelis 804), Wurmser 1354), and the review by Thunberg 1168) may be consulted.

---

*) By convention $E_o$ and E represent potential differences and require no $\triangle$ prefix.

Like the hydrogen electrode most biological redox systems show a dependence on the hydrogen ion concentration. The cause of this phenomenon is the property of these systems to dissociate hydrogen ions at certain levels of ph (dissociation constants).

The normal hydrogen electrode which is the reference point of most measurements (and of all measurements reported here) *) is a platinum electrode plated with platinum black, immersed in a solution which is 1 N· with respect to hydrogen ions, and in equilibrium with hydrogen gas of atmospheric pressure. To this electrode the potential zero has been ascribed by convention **). Upon increasing the ph (decreasing the hydrogen ion concentration) the potential of the hydrogen electrode decreases in linear fashion. It is also dependent on temperature; at ph = 1, $E_h$ is —0.0601 v. at 30°, —0.0579 v. at 25°, and — 0.0581 v. at 20°. With an increase in ph of 1 unit the potential decreases approximately 0.06 v.; at ph = 7.0 (the neutrality point, so important in biological considerations) the potential of the hydrogen electrode is $E_h = — 0.421$ v.

With a view to avoiding the change of sign required upon passing the zero point of the redox potential scale SZENT-GYÖRGYI 1120) suggested that the oxygen electrode should be chosen as the reference point ($E_o$ ($O_2$) = 0 v. at ph = 7.0, instead of $E_o = +$ 810 v.) and that all other potentials should be calculated with a positive sign with respect to this electrode. The usual connotation, however, has so deeply penetrated into the literature that such a revision, if adopted, would create considerable confusion.

Many of the systems of interest for our discussion exhibit the same ph-dependence as the hydrogen electrode so that $E_o$ may be calculated from the measured $E_h$ values by extrapolating back to ph = 0 (adding 0.06 v. for each ph unit). It is to be remembered, however, that the slope of the $E_o'$/ph changes at those ph values where a dissociation constant (or the logarithm of the dissociation constant (pK)) is situated. Correct calculation of the $E_o$ of a system requires, therefore, either a) a knowledge of the various dissociation constants of the oxidized and reduced forms of the system, which may be obtained independently by potentiometric acid-base titration, employing a hydrogen or a glass electrode, or b) a charting of the $E_o'$-data for an extensive ph range (say from ph 1 to 12) to reveal the location of the dissociation constants as inflection points on the $E_o'$/ph-curve (see CLARK 169a) and MICHAELIS 804)).

For practical purposes and as an analogy to the term ph, W. M. CLARK 169a) at one time created the term rH as a measure of the oxidation-reduction potential. rH = — $\log_{10}$ P, where P is the hydrogen gas pressure at the electrode in equilibrium with the solution containing the redox system. Since $E_h = — \dfrac{RT}{nF} \ln P — \dfrac{RT}{F}$ PH, rH may be calculated from this equation after the simple transformation:

$$rH = — \ln P = \frac{nF}{RT} E_h + n \, ph,$$

<hr />

*) We summarize: Potential, referred to the hydrogen electrode of identical ph, = $E_h$ ; $E_{oh}$ = potential at ph = 0, as referred to the hydrogen electrode. Since in biological systems $E_{oh}$ is almost always given, this shall be written simply as $E_o$. $E_o$, then, represents the n o r m a l   p o t e n t i a l   at ph = 0 referred to the hydrogen electrode, at a [Ox]/[Red] ratio of unity. Frequently potentials measured at other hydrogen ion concentrations are also designated as "normal potentials", provided that the concentration ratio of the two forms of the redox systems is unity. Such potentials shall be designated here as $E_o'$; if not otherwise stated, this $E_o'$ is given for ph = 7.

**) In actual practice, however, a calomel electrode is used as the half cell operating opposite the redox electrode to complete the electromotive circuit. The values thus obtained are recalculated for the hydrogen electrode as the reference point in the manner described by MICHAELIS (804, p. 133)).

with the aid of the experimentally found values for $E_h$ and ph (for details consult MICHAELIS 804), p. 48), THUNBERG 1168), and NEEDHAM 873).

In the following table, the potentials and the rH values of the most important redox indicators at 50 per cent reduction and at ph = 7 are given; the rH values have been taken from the papers by NEEDHAM 874) and by COHEN et al. 175) whereas the potential values for these and other indicators may be found in the tables compiled by STERN 1081).

It should be emphasized that the calculation of $E_o$ from $E_h$ or $E_o'$ is only permissible if the ph-dependence of the potentials of the system is known. CLARK has warned against an indiscriminate usage of the rH term for similar reasons. However, rH is being widely used in biological publications.

TABLE 1.

| Indicator | E°' in mv. at ph = 7.0 | rH |
|---|---|---|
| (Oxygen Electrode) | + 810 | 41.0 |
| Phenol-m-sulfonate-indo-2,6-dibromophenol | + 273 | 23.1 |
| o-Chlorophenol-indophenol | + 233 | 20.9 |
| 2,6-Dibromophenol-indophenol | + 219 | 20.2 |
| o-Cresol-indophenol | + 194 | 19.7 |
| 2,6-Dichlorophenol-indo-o-cresol | + 181 | 19.3 |
| 1-Naphthol-2-sulfonate-indophenol | + 123 | 17.4 |
| 1-Naphthol-2-sulfonate-indo-2,6-dichlorophenol | + 118 | 17.2 |
| Cresyl Blue | + 31 | — |
| Methylene Blue | + 25 | 14.4 |
| Indigo tetrasulfonate | — 58 | 11.9 |
| Indigo trisulfonate | — 90 | 10.3 |
| Indigo disulfonate | —129 | 9.2 |
| Neutral Blue | —192 | — |
| Janus Green | —256 | — |
| Neutral Red | —340 | — |
| Benzyl Viologen | —359 | — |
| (Hydrogen Electrode) | —421 | 0.0 |
| Phenosafranine | —525 | — 3.5 |

Most of the direct potentiometric measurements of the redox potentials of isolated organic chemical systems, especially of quinoid dyes and of hemin complexes, are due to CLARK, BIJLMANN, MICHAELIS, CONANT, and their collaborators. Some authors, e.g. WURMSER, NEEDHAM, CHAMBERS, and their colleagues, have attempted direct measurements of redox potentials in complex biological systems, e.g. by injecting indicators into living cells with the aid of micromanipulators (see NEEDHAM and NEEDHAM 872) and AUBEL 34)). The interpretation of the data thus obtained is made somewhat difficult because of toxic effects of some of the dyes used and also because the amount of indicator required for visibility of the color of the equilibrium state with the surrounding cytoplasm may cause a shift in the cell potential due to "poising action". The ideal conditions for measurements of intracellular potentials would be provided by the discovery of a colored reversible redox system occurring within the cell in a concentration sufficiently high to serve as an "internal" rH indicator. It is possible that some of the already known biological redox systems, e.g. pyocyanine or lactoflavin, may be suitable for this purpose.

Even so it must be kept in mind that the redox potentials in various parts of a cell may differ greatly.

It has been possible to determine the potentials of certain colorless, reversible but electromotively inactive biological systems by allowing them to come into equilibrium with electromotively active systems which serve as potential indicators. The system succinic acid — fumaric acid is a case in point. In presence of succinic dehydrogenase this system is fully reversible, i.e. the potential is, at any time, determined by the ratio of the concentration of the oxidized form (fumaric acid) to that of the reduced form (succinic acid). However, a noble metal electrode when immersed in this equilibrium mixture will not register any stable potentials. The cause can be assumed to be that the exchange of electrons between the system and the electrode is very sluggish. If an amount of methylene blue is added which is too small to shift the equilibrium to any appreciable extent, the resulting stable potential observed, though actually the potential of the methylene blue/leuco methylene blue system, will also represent the potential of the biological enzyme-substrate mixture providing sufficient time has been allowed for establishment of equilibrium between the two systems. The first observation of this kind was made by THUNBERG. Valuable information along these lines has been obtained by QUASTEL, LEHMANN, BORSOOK, WURMSER, BARRON and other workers (see also FRANKE 355).

**Tables of Redox Systems of Biological Interest:** In the following tables the normal potentials, $E_o$ and $E_o'$ for the given ph values, the heats of reaction, $\triangle H$, and the free energy change, $\triangle F$, are given for important metabolite systems and for reversible biological redox systems including certain intermediary catalysts and co-enzymes. Many of the data have been taken from the table by FRANKE 356).

In that table, FRANKE has critically discussed the reliability of the various values given. The reader is referred to his papers (356, 355)) for this discussion. The table below has been supplemented by more recent data. A large number of the potential values and information pertaining to their significance will be found in the tables by STERN 1081). The values in rectangular brackets have been calculated for the present purpose./ When the $\triangle F$ and $\triangle H$ values carry a negative sign, the energy relationship may be regarded as abnormal, i.e. dehydrogenation is then exothermal and hydrogenation endothermal.

Whereas in several cases (see Table 2, p. 54) the normal potential of substrate systems in the presence of the corresponding enzyme could be determined, we know almost nothing about the potential of isolated oxidation enzymes though it is now widely held that a catalyst which promotes oxidation-reduction processes must in itself represent a reversible redox system (see for instance L. MICHAELIS 804)). The experimental determination of enzyme potentials is difficult not only because very few enzymes have, as yet, been prepared in pure state, but also because the colloidal protein bearer introduces technical difficulties of the type encountered in the potentiometric study of the methemoglobin-hemoglobin system (179)). The only enzyme potential so far directly measured is that of the yellow oxidation enzyme (636)) (see Table 2). It is interesting to note that $E_o'$ at ph 7 was found to be —0.06 v. (and $E_o = + 0.37$ v. (at ph 0)) whereas $E_o'$ of the free prosthetic flavin group is considerably more negative ($E_o' = — 0.28$ v. at ph 7 (822, 636)). It appears to be a general rule that the attachment of a prosthetic group to a protein bearer raises the normal potential; $E_o'$ at ph 7 for free heme is near —0.2 v. (at ph 9 it is —0.23 v.) (see Table 2), while $E_o'$ for methemoglobin-hemoglobin is + 0.15 v.

## TABLE 2.

### Metabolite Systems.

#### 1. Hydrogenation Reactions.

| Reaction | t° | $E_0'$ mv. | ph | $E_0$ mv. | ΔH Cal. | ΔF Cal. | Ref. |
|---|---|---|---|---|---|---|---|
| $H_2 + HCO_3' \rightleftarrows HCOO' + H_2O$ { enzymatic | 25 | — | — | — | 5.407 | 0.171 | } 1844 |
| thermodyn. | 25 | — | — | — | 2.854 | 0.742 | } 1844 |
| Formic acid (liquid) + $H_2 \rightleftarrows$ Formaldehyde (gaseous) + $H_2O$ | 25 | — | 0 | ≶0 | (—1.8) | (—7.6) | (356) |
| Acetic acid (liquid) + $H_2 \rightleftarrows$ Acetaldehyde (gaseous) + $H_2O$. | 25 | — | — | — | (—1.6) | (—6.7) | (356) |
| Acetaldehyde (gaseous) + $H_2 \rightleftarrows$ Ethyl alcohol (liquid)   enzymatic | 30 | —90 | 7.45 | (+356?) | — | (16.4?) | (709) |
| thermodynamic | 25 | — | — | — | 20.8 | (10.7) | (356) |
| Ethyl alcohol (~1 M.) — $H_2 \rightleftarrows$ Acetaldehyde. | 25/35 | —190 | 7.0 ~ | +190 | — | 9.25 | (1857) |
| Acetone + $H_2 \rightleftarrows$ Iso-propyl alcohol. | 25 | — | — | — | 21.7 | 11.0 | (356) |
| Iso-propyl alcohol (~1 M.) — $H_2 \rightleftarrows$ Acetone | 25/35 | —260 | 7.0 | +176 | — | 8.1 ± 0.2 | (1857) |
| Iso-propyl alcohol (gas.) — $H_2 \rightleftarrows$ Acetone (gas.) | 25 | — | — | — | — | 7.0 | (1857) |
| Glucose + $H_2 \rightleftarrows$ Sorbitol | 25 | — | — | — | 14.5 | 7.0 | (356) |
| Reducton $\rightleftarrows$ Oxydoreducton + 2 H· + 2 ε' | 38 | — | 0 | +282 | — | [12.16] | (1365) |
| Ascorbic acid $\rightleftarrows$ Dehydroascorbic acid + $H_2$ *) | 25 | —81 | 7.0 | +332 | — | (15.4) | (379) |
|  | 35.5 | —66 | 7.0 | +405 | — | (18.6) | (139) |
|  | 30 | —54 | 7 | +375 | — | (17.3) | (414) |
|  | 30 | +51 | 7.2 | +390 | — | .. | (58a) |
| Succinic acid + $H_2 \rightleftarrows$ 2 Acetic acid | 25 | —94 | — | — | 9.0 | 9.8 | (356) |
| Maleinate + $H_2 \rightleftarrows$ Succinate | 37 | — | 7 | — | — | 26.65 | (690) |
| Pyruvic acid + $H_2 \rightleftarrows$ Lactic acid | 25 | —180 | — | — | 22.0 | (13.4) | (356) |
| Pyruvate ion + $H_2 \rightleftarrows$ Lactate ion (in presence of enzyme) | 35 | —200 | 7.01 | +248 | 21.64 | 11.44 | (84) |
|  | 37 | — | 0 | +249 | — | 11.49 | (1120) |
|  | 37 | — | 7.0 | +252 | — | 11.63 | (1868) |
|  | 25 | — | — | — | — | 11.86 | (356) |
| Pyruvate ion~ + $NH_2^+$ + $H_2 \rightleftarrows$ Alanine + $H_2O$ | 32 | — | 0 | +316 | (18.4) | 14.572 | (93) |
| Xanthine + $H_2O \rightleftarrows$ Uric acid + $H_2$ | 25 | — | — | +182 | — | 17.2 | (356) |
|  | 35 | — | 0 | +123 | — | — | (1857) |
| (in presence of enzyme) | 38 | — | 0 | — | 5.0 | 5.7 | (335) |
|  | 30 | — | 0 | (+60) | 13.9 | (2.8) | (356) |

*) Laki (689) was unable to find reversibility. Borsook and Keighley (189) explain his results by the fact that the oxidized form is unstable in neutral solution (see p. 94 and p. 209).

| Reaction | t° | E'₀ mv. | ph | E₀ mv. | ΔH Cal. | ΔF Cal. | Ref. |
|---|---|---|---|---|---|---|---|
| Hypoxanthine + $H_2O$ ⇄ Uric acid + $H_2$ | 38 | —410 | 7.31 | +68 | 12.0 | 6.3 | (335) |
| Hypoxanthine + $H_2O$ ⇄ Xanthine + $H_2$ | 35 | — | 0 | +14 | — | — | (1357) |
|  | 30 | — | 0 | (+50) | — | (2.3) | (356) |
| Fumaric acid + $H_2O$ ⇄ Succinic acid | 25 | — | — | — | 30.7 | 22.1 | (356) |
| Fumarate ion + $H_2O$ ⇄ Succinate ion thermodynamic | 25 | — | — | — | 29.8 | 20.46 | (356) |
|  | 45 | — | — | — | — | 19.5 | (951) |
| enzymatic | 25 | — | 0 | +436 | 29.85 | 20.14 | (356) |
|  | 18 | — | 0 | +443 | 29.85 | 20.4 | } (709) |
|  | 37 | — | 0 | +430 | 29.85 | 19.8 |  |
| Cystine + $H_2$ ⇄ 2 Cysteine | 30 | +15 | 6.9 | +433 | — | 19.95 | (1167) |
| l-Cystine + $H_2$ ⇄ 2 l-Cysteine | 25 | — | 0 | (+79)** | — | (3.7) | (356) |
| S-S-Glutathione + $H_2$ ⇄ 2 SH-Glutathione | 25 | (+62)** | 7.0 | — | 3.8 | —1.7 | (413) |

**) The potential measurements in sulfur containing redox systems, according to GREEN (418), are not reliable; if a mercury electrode is used there occurs complex formation between the sulfur and the metal.

2. Other Reactions.

| Reaction | t° | ΔH Cal. | ΔF Cal. | Reference |
|---|---|---|---|---|
| Fumaric acid + $H_2O$ ⇄ Malic acid | 25 | —0.7 | —2.9 | (356) |
| Fumaric acid + $NH_3$ (gas) ⇄ Aspartic acid | 25 | 26.0 | 12.9 | (356) |
| Fumarate ion + $H_2O$ ⇄ Malate ion | 25 | 0.7 | 0.7 | (356) |
| Fumarate ion + $NH_4^+$ ⇄ Aspartate ion enzymatic | 25 | — | 2.8 | (356) |
| thermodynamic | 25 | — | 2.95 | (356) |
| Pyruvic acid ⇄ Acetaldehyde + $CO_2$ (gas.) | 25 | —1.4 | 15.2 | (356) |
| Pyruvate ion + $H^+$ ⇄ Acetaldehyde + $CO_2$ | 25 | —0.1 | 13.6 | (356) |
| Phosphopyruvic acid + $H_2O$ ⇄ Phosphoglyceric acid | 20 | — | 0.55 | (356) |
| Cyanic acid (liquid) + $NH_3$ (liquid) ⇄ Urea (liquid) | 25 | 19.2 | 18.4 | (356) |
| Acetic acid (liquid) ⇄ $CH_4$ + $CO_2$ (gas.) | 25 | —4.5 | 11.5 | (356) |
| Glucose (liquid) ⇄ 2 Lactic acid | 25 | 29.0 | 32.8 | (356) |
| Glucose ⇄ 2 Ethyl alcohol + 2 $CO_2$ (gas.) | 25 | 20.6 | 55.3 | (356) |
| Glucose + $H_2O$ ⇄ Acetone + 3 $CO_2$ + 4 $H_2$ | 25 | —27.5 | 45.1 | (356) |
| Hexose phosphoric acid ⇄ 2 Dioxyacetone phosphoric acid (enzymatic) | 20—40 | 33.5 | — | (793) |

Metal-free Intermediary Catalysts.

| Reaction | $t°$ | $E'_0$ mv. | ph | $E_0$ mv. | $\Delta H$ Cal. | $\Delta F$ Cal. | Reference |
|---|---|---|---|---|---|---|---|
| Porphyrexid + H ⇌ Leuco radical | 18/19 | +725 | 7 | 1028 | — | [24.95] | { 644 |
| Porphyrindin + H ⇌ Leuco radical | 18/19 | +565 | 7 | [+1055]* | — | [24.3] | |
| o-Quinone of adrenalone + H₂ ⇌ Adrenalone | 30 | — | 0 | +909 | — | 41.8 | (356, 60, 61) |
| o-Quinone of nor-homoadrenalone + H₂ ⇌ Nor-homo-adrenalone | 30 | | 0 | ×822 | | 37.8 | (356, 60, 61) |
| o-Quinone of adrenaline + H₂ ⇌ Adrenaline | 30 | | 0 | +809 | | 37.2 | (356, 60, 61) |
| o-Quinone of dihydroxyphenylalanine (Dopa) + H₂ ⇌ Di-hydroxyphenylalanine | 30 | | 0 | +800 | | 36.8 | (356, 60, 61) |
| o-Quinone + H₂ ⇌ Pyrocatechol | 30 | | 0 | +792 | | 36.5 | (356, 60, 61) |
| p-Quinone + H₂ ⇌ Hydroquinone | 18 | | 0 | +704 | 42.8 | 32.5 | { (356) |
| | 25 | | 0 | +699 | 42.8 | 32.2 | |
| | 30 | | 0 | +696 | | 32.0 | (60, 61) |
| p-Quinone of homogentisinic acid + H₂ ⇌ Homogentisinic acid | 20 | +250/60 | 7 | +680 | | 31.8 | (128) |
| | 25 | | 0 | +688 | | 31.7 | (346) |
| 5, 6-Quinone of dihydro-indol-2-carboxylic acid ("Red Body" of tyrosinase action) ⇌ Leuco compound | 17 | +171 | 4.62 | — | | — | (370) |
| Phoenicin + H₂ ⇌ Leuco compound | 20 | + 47 | 7 | — | | — | (369) |
| Juglon + H₂ ⇌ Leuco compound | 20 | + 33 | 7 | +495 | | 20.0 | (373, 368) |
| Pigment from Arion rufus | 20 | + 25 | 7 | — | | — | (373, 367) |
| Hallachrome + H₂ ⇌ Leuco compound | 20 | + 22 | 7 | +440 | | 20.2 | (365) |
| Murexid ⇌ Leuco compound | 20 | + 20 | 7 | — | | — | (651) |
| Alloxan + H₂ ⇌ Dialuric acid + H₂O | 25 | | 0 | +367 | 21.0 | 16.9 | (356) |
| | 30 | | 0 | +364 | | 16.7 | (356) |
| Dimethyl alloxan + H₂ ⇌ Dimethyl dialuric acid + H₂O | 25 | | 0 | +365 | 21.5 | 16.8 | (356) |
| Cyanohermidin + H₂ ⇌ Hermidin | 30 | — 45 | 7.3 | +367 | | 16.9 | (160) |
| Pyocyanine + H₂ ⇌ Leuco compound | 20 | — 33 | 7 | — | | — | (376) |
| Toxoflavin (Bact. bongkrek) ⇌ Leuco compound | 30 | (—52)* | 7.4 | (+390) | | (18.0) | 376, 233 |
| Yellow Ferment (Flavoprotein) ⇌ Leuco compound | 38 | — 49 | 7 | +370 | | — | (1083) |
| Lawson + H₂ ⇌ Leuco compound | 20 | — 60 | 7 | +352 | | — | (686) |
| Chlororaphine + H₂ ⇌ Leuco compound | 30 | —139 | 7 | — | | 16.2 | (373, 368) |
| Alloxazine | 30 | —139 | 7.42 | — | | — | (234) |
| 9-methyl alloxazine | 30 | —170 | 5 | — | | — | (822) |
| | 30 | —188 | 7 | — | | — | (822) |

| Reaction | t° | E'₀ mv. | ph | E₀ mv. | ΔH Cal. | ΔF Cal. | Reference |
|---|---|---|---|---|---|---|---|
| Glucoalloxazine ............ | 30 | −163 | 7 | [+230]** | — | · | (822) |
| Lactoflavin (riboflavin) + H₂ ⇄ Leuco compound ......... | 20 | — | 0 | +197 | — | 9.1 | { (356) |
| Gluco-, arabo-, and riboflavin ...... | ? | — | 0 | +187 | — | 8.6 | } |
| Phtiocol + H₂ ⇄ Leuco compound ...... | 30 | −208 | 7 | +180 | — | — | (822) |
| Lapachol + H₂ ⇄ Leuco compound ......... | 30 | −208 | 7.3 | +299 | — | 18.8 | ( 57) |
| Lomatiol + H₂ ⇄ Leuco compound ......... | 30 | −209 | 7.32 | +300 | — | — | { ( 58) |
| Echinochrome + H₂ ⇄ Leuco compound ......... | 30 | −212 | 7.32 | +300 | — | — | } |
| Lumiflavin + H₂ ⇄ Leuco compound (from milk)....... | 20/30 | −220 | 7.0 | +200 | — | 9.2 | (161, 365) |
| Lumiflavin + H₂ ⇄ Leuco compound ......... | 30 | −223 | 7 | — | — | — | (822) |
| Lumiflavin + H₂ ⇄ Leuco compound (from liver and yeast) | 30 | −227 | 7.0 | (+160) | — | (7.4) | (1080, 356) |

## Hemin Systems and Hemocyanin.

| Reaction | t° | E'₀ mv. | ph | E₀ mv. | ΔH Cal. | ΔF Cal. | Reference |
|---|---|---|---|---|---|---|---|
| Hemocyanin ⇄ Methemocyanin ......... | — | +540 | 7 | — | — | — | (177) |
| Hemoglobin ⇄ Methemoglobin ......... | — | +152 | 7 | — | — | — | (179) |
| Cytochrome-c ⇄ Reduced cytochrome ..... | — | +123 | 7.0 | — | — | — | (415) |
|  | — | +253 | 5—8 | — | — | — | 1858, 1359 |
| Nicotine ferrohemochromogen ⇄ Nicotine ferrihemochromogen ......... | — | + 53 | 9.5 | — | — | — | (77) |
| Pyridine ferrohemochromogen ⇄ Pyridine ferrihemochromogen ......... | — | + 17 | 9.5 | — | — | — | (77) |
| α-Picoline ferrohemochromogen ⇄ α-Picoline ferrihemochromogen ......... | — | − 22 | 9.50 | — | — | — | (77) |
| Histidine ferrohemochromogen ⇄ Histidine ferrihemochromogen ......... | — | −138 | 9.50 | — | — | — | (77) |
| Pilocarpin ferrohemochromogen ⇄ Pilocarpin ferrihemochromogen ......... | — | −156 | 9.50 | — | — | — | (77) |
| Cyanide ferrohemochromogen *) ⇄ Cyanide ferrihemochromogen ......... | 30 | −183 | 9.89 / −18.0 | −183 | — | — | (85) |
| Ferroheme ⇄ Ferriheme ......... | — | −285 | 9.50 | — | — | — | (77) (see also 180) |

*) In presence of an excess of CN'.

It is possible, however, to deduce approximate potential levels for certain enzymes from indirect evidence or from general considerations. Thus, Barron 76) believes that the potential of the α-ketooxidase from gonococci is in the neighbourhood of —0.034 v. because it may be replaced by reversible systems (hemochromogens) of this potential range. Once the place of an oxidation enzyme or catalyst has been determined in the chain of biological oxidation it is possible to predict where its potential will be eventually found because otherwise the catalyst would not be capable of fulfilling its function. Let us consider the important case of the respiratory enzyme of Warburg as an example. We know that it stands at the top of the series of catalysts participating in the main chain of cell respiration. It should, consequently, be less positive than molecular oxygen (or the oxygen electrode) and more positive than the next lower catalyst in the chain which is one of the cytochrome components. The potential of cytochrome c only is accurately known (Wurmser and Filitti 1359)). The order in which the three cytochrome components are lined up is not yet sure, but it is believed, on the basis of the experiment by Haas 430a) in Warburg's laboratory), that the three cytochromes are not connected in parallel but in series in the chain b→c→a (Ball 58)). It would seem, therefore, safe to predict that the potential of the respiratory ferment will be found to lie between + 0.260 and + 0.800 v. and probably much closer to the former value than to the latter. *)

Table 3 contains some of the directly measured potential values for cells, tissues, and cell suspensions. A recalculation of the $E'_o$-values determined at the ph indicated in the table, in terms of $E_o$, has been omitted because of the experimental uncertainty still inherent in such potentials which are only slowly and sluggishly established in most cases, and because the nature of these potentials is still controversial. Some workers believe that the aerobic and the anaerobic potentials (i.e. those measured in the presence and in the absence of air respectively) of cells correspond to the normal potential of individual reversible redox systems which exert maximum poising action in this range, whereas others regard such potentials as the resultant of a multitude of oxidation-reduction processes occurring in the cell which need not necessarily involve strictly reversible systems. In other words, it is still an open question whether these biological potentials correspond to equilibrium states or to steady states of kinetic character.

Cell potentials may be determined by colorimetric or by electrometric methods. In the former, an indicator of suitable potential range is added to the system and its final state of oxidation is considered to be identical with that of the entire system, if sufficient time has been allowed to attain equilibrium between indicator and biological material. The absolute amount of indicator must be kept as small as compatible with the optical requirements in order to avoid appreciable poising action on the part of the indicator. For electrometric measurements a noble metal electrode is placed in the cell suspension and the potentials are followed several hours until a practically steady state has been reached. It is important to employ several electrodes and, if possible, electrodes of different material (gold, platinum )to eliminate the individual properties or pecularities of any one electrode; only if several electrodes register closely agreeing potentials is the inference permissible that the actual potential of the biological system has been measured. Both methods may be combined by adding an indicator and performing potentiometric measurements. The dye is functioning here as an e l e c t r o a c t i v e  m e d i a t o r between the electrode surface and the cell content. The final potential value, in this case, is attained more rapidly than in the absence of an indicator dye.

Along with the $E'_o$ values, the table lists the corresponding rH values. In the use

_____

*)  See the recent measurements by Ball (58b) and Laki (694).

of the latter the reservations mentioned above (p. 59) should be borne in mind.

Further observations on cell potentials of a more qualitative character will be found in the publications by PAVLOW and ISSAKOVA-KEO 919), RAPKINE and WURMSER 962), FABRE and SIMONNET 330), TSUKANO 1188), AUBEL 36), MAYER and PLANTEFOL 761), and ELEMA, KLUYVER and VAN DALFSEN 237). The lowest and highest potential values compatible with the vitality of the cell nucleus have been determined by QUASTEL and WOOLDRIDGE 954) and AUBEL and AUBERTIN 38). The changes of $E_h$ and rH in the course of the development of the chick embryo have been studied by FRIEDHEIM 360).

It has been observed by AUBEL et al. 37) that there develops in buffered solutions of fructose and glucose, without the addition of other substances and under anaerobic conditions, a potential level of approximately —180 mV. (at ph 7.5 and 20°) as registered by blank platinum electrodes (details in (1356) and (1354)). WURMSER and his colleagues assume that an equilibrium of the type Glucide $\rightarrow$ G' $\rightleftarrows$ A $+$ $H_2$ is established where A represents an oxidation product of the active substance G'. They believe that this active compound (Glucide X) is the most important redox buffer of the living cell. It is of interest to note that these anaerobic sugar potentials are in the neighbourhood of the ascorbic acid potentials. GEORGESCU 393) finds the redox potential in anaerobic glucose-buffer solutions of rH = 12.6 $\pm$ 0.5 at 20° (colorimetrically and electrometrically), and for ascorbic acid an rH-value of 15 $\pm$ 0.5. Still closer agreement between ascorbic acid and the "glucide X", later designated "redoxin", was observed by WURMSER and LOUREIRO 1360, 1361), who, therefore, suggest that the two substances may possibly be identical. However, it is quite possible that the reducing substances formed under these conditions are reductones (see p. 210).

### Oxidation-reduction potentials in heterogeneous systems:

In the interpretation of potentials measured in cell suspensions and in the evaluation of data obtained on isolated cell constituents in homogeneous solution for an integration of the events in the living cell, it must be kept in mind that the behaviour of reversible redox systems may be greatly affected by the presence of interfaces as they exist in a heterogeneous system like protoplasm. As CLARK 170) has pointed out, "the heterogeneity of the living cell places restrictions upon the application of data from homogeneous systems".

MICHAELIS (quoted by ABRAMSON and TAYLOR 2)) had observed that methylene blue when adsorbed on filter paper was not reduced when placed in a suspension of platinized asbestos saturated with hydrogen. This shows that the hydrogen is not active at any distance from the surface of the metal catalyst. ABRAMSON and TAYLOR 2) confirmed this observation and were able to show, furthermore, that soluble reducing agents like sodium hyposulfite ($Na_2S_2O_4$), cysteine, or thiourea will readily reduce the adsorbed dye in a reversible manner. If the excess of the reducer is removed by superficial washing, molecular oxygen as well as quinone or ferricyanide will reoxidize the leuco dye to methylene blue. Similar observations were made with litmus paper and with phenosafranine adsorbed on cellulose. ABRAMSON and TAYLOR point out that "a surface having selective adsorption for one form of the constituents of a reversible oxidation-reduction system could shift the oxidation-reduction potential of the system." "In heterogeneous systems the attainment of "equilibrium" in solution does not necessarily indicate the reduction intensity at the surface."

Work in the same direction has recently been carried out by KORR 587) who extended ABRAMSON and TAYLOR's observations. He finds, upon adding adsorbents like

## TABLE 3.
### Oxidation-Reduction Potentials Measured in Cells.

| Material | Conditions | Method | t° | ph | E°' mv. | rH | Refer. |
|---|---|---|---|---|---|---|---|
| Acetobacter rancens, A. pasteurrianum, A. aceti, A. peroxydans | aerobic | Color. | 25 | 5.5 | +350 ? | 23 | (188) |
| Acetobacter melanogenum, A. suboxydans | aerobic | Color. | 25 | 5.5 | +300 ? | 21 | (188) |
| Vitreous body of eye of killed rabbit | | +Electr. | 37 | 7.6 | +200 | 22.1 | (965) |
| Azotobacter vinelandii | | Color. | 23 | 7.0 | +200 | | (157a) |
| Eggs of sea urchin Paracentrotus lividus Lk. | | Color. | | 6.6 | — | 19—22 | (872) |
| Sea urchin eggs | | Electr. | | | | 20.2—21 | (1196) |
| Living tissues (muscle, kidney, testis, liver, brain) of guinea pig, rabbit, dog | aerobic | Color. | 37 | | | ~20 | (41) |
| Caterpillars of Galleria mellonella, living | aerobic | Color. | | | | ~20 | (40) |
| Nyctotherus rodiformis (Ciliate) | aerobic | Color. | | 7.1 | | 19—20 | (874) |
| Salivary gland cells of larvae of Chironomus and Calliphora erythrocephala | | Color. | | 7.2 | | 19—20 | (963) |
| Oocytes of Asterias rubens and Paracentrotus lividus Lk. | | Color. | | 7.2 | | 19—20 | (963) |
| Blood, dog, in vivo | | Electr. | | 7.4 | +156—186 | | (448) |
| Vitreous body (eye of living rabbit) | | Color. } Electr. | 37 | 7.6 | +110 | 19.1 | (965) |
| Blastodermal sack of sea-urchin larva, Paracentrotus lividus Lk. | | Color. | | 7 | +150 | 19,0—19.6 | (959) |
| Liver and muscle of frog (living) | | Electr. | 27/29 | 7 | +150 | | (899) |
| Liver and muscle of rat | | Electr. | 27/29 | | +100 | | (899) |
| Spleen, pancreas, intestine, stomach of rabbit and dog (killed after injecting dye) | | Color. | 37 | 7 | | 16—20 | (361) |
| Amoeba proteus, interior of cell, independently of $O_2$-pressure | | Color. | | 7.5—7.7 | 120—150 | 17—19 | (874) |
| Valonia (marine alga), cell sap | | Color. | | 6.02 | —210 to | 17.9—18.4 | (150) |
| cytoplasm | | Color. | | — | +480 | ~18 | (150) |
| Amoeba proteus and Amoeba dubia | aerobic | Color. | 20—25 | 6.9 | | <18 | (175) |
| Spirogyra, interior of cell | anaerob. | Color. | | 6±0.2 | | 14.4—17.6 | (961, 1358) |
| Vitreous body, eye of dead rabbit | | Color. } Electr. | 37 | 7.6 | +51 | 16.9 | (965) |
| Spirogyra, interior of cell | | Color. | | 6±0.2 | | 14—16 | (963) |
| Muscle, kidney of dog, killed after injection of dye | | Color. | 37 | 7 | | 14—16 | (361) |
| Eggs of Asterias forbesii and Echino arachnius parma (starfish) | aerobic | Color. | | | | ~12 | (176) |

| Material | Conditions | Method | t° | ph | E°' mv. | rH | Reference |
|---|---|---|---|---|---|---|---|
| Living tissues (muscle, kidney, testis, liver, brain) of guinea pig, rabbit, dog | anaerobic | Color. | 37 | 4.6—5.5 | +71 to -166 | 11—12 | (41) |
| Fermenting lactic acid bacteria, in absence of redox indicators | anaerobic | Electr. | 30 | | | 5.05 to 12 2 | (580) |
| Yeast cultures | aerobic | Electr. | 24 | 6.4 | +20 to+35 | 9.5—10.5 | (719) |
| Nyototherus rodiformis (ciliate) | anaerobic | Color. | | 7.1 | | >9.9 | (874) |
| Amoeba proteus and A. dubia | anaerobic | Color. | 20—25 | 6.9 | | | (175) |
| Eggs of Asterias forbesii and Echino arachnius parma | anaerobic | Color. | 20 | 7.0 | +33 to -139 | ~9 | (176) |
| Plant cells | | Electr. | | | | | (368) |
| Fermenting yeast, in presence of redox indicators | anaerobic | Electr. | 30 | 5.2 | -43 | 9 | (580) |
| Brain cortex, liver of dog, rabbit, killed after injecting dye | | Color. | 37 | ~7 | | ~9 | (361) |
| Carcinoma, mouse | | Electr. | | | -156 | 8.8 | (361) |
| Spleen of guinea pig, rabbit, hog | | Electr. | | | -110 to -179 | 7.0—8.7 | (361) |
| Muscle of guinea pig, rabbit, hog | | Electr. | | | -117 to -297 | 3.7—9.7 | (361) |
| Azotobacter vinelandii | anaerobic | Color. | 31 | 7.0 | -150 | | (157a) |
| Fermenting lactic acid bacteria, in presence of redox indicators | anaerobic | Electr. | 30 | 3.9—6.3 | -60 to -214 | 5—6 | (580) |
| Yeast cultures (in presence of indicators) | anaerobic | Electr. | 24 | 6.4 | -90 to -130 | | (719) |
| Fermenting Yeast cultures | anaerobic | Electr. } Color. | | 7 | -160 | 7 | (39) |
| Lymph ganglion, guinea pig, hog | | Electr. | | | -189 to -216 | 6.5 to 7.6 | (361) |
| Extract of chick embryos, 5 days old | anaerobic | Electr. | | 7.07 | -209 | 6.72 | (360) |
| Agrolimax agrestis (snail), living | anaerobic | Color. | | 7.5—7.6 | | 5—6 | (40) |
| Kidney of guinea pig, rabbit, and hog | | Electr. | | | -234 to -324 | 3.1 to 6.4 / 6.4 | (361) |
| Thyroid gland of hog | | Electr. | | | -288 | 4.75 | (361) |
| Brain of guinea pig, rabbit | | Electr. | | | -343 to -335 | 4.0 to 4.3 | (361) |
| Testiculum of guinea pig, hog, beef | | Electr. | | | -269 to -343 | 2.0 to 4.8 | (361) |
| Liver of guinea pig, rabbit, hog | | Electr. | | | -327 to -344 | 2.1 to 2.56 | (361) |
| B. coli cultures | | Electr. } Color. | 7 | | -420 | 0 | 39, 1102 |

cellulose, kaolin, or silica gel, to solutions containing redox indicators at various ph and varying degrees of reduction, that the adsorbent will differentially adsorb the various forms of the dye. This is concluded from the ensuing potential changes in the solutions as well as form color differences between the adsorbate and the solution. In the course of these experiments the interesting observation was made that the semiquinoid radical form of pyocyanine may accumulate at the surface of the adsorbent. It is also shown that a dissolved dye may act as "hydrogen carrier" by being reduced at a catalytic surface, e.g. platinum asbestos, and then reoxidized at the surface of the adsorbent covered with the reduced form of a dye. Such an interaction between dissolved and adsorbed dye molecules may be considered to be a model for the "coupling link" visualized by BORSOOK 135) to exist between various enzyme centers in the living cell. Such a mechanism may be operative in the transfer of energy from one surface, where an oxidation-reduction process releases energy, to another surface where this energy is required, e.g. for the purposes of synthesis.

## II. Catalysis of Oxido-Reduction.

### 1) Heavy Metal Catalysis.

#### a) General Considerations.

It is proposed to describe the phenomena of heavy metal catalysis on the basis of the valency change theory of WARBURG and WIELAND. Certain difficulties will be discussed later in the section on peroxide catalysis (p. 123). There also the recent attempt by HAUROWITZ to explain the peroxidatic and catalatic action of ferric iron without a valency change will be described (p. 79).

It is assumed that the catalysis in the dehydrogenation reactions to be discussed presently proceeds with the aid of complex salts of various heavy metals. Of paramount importance are iron and copper; occasionally manganese and a few other metals are active. The well-known model experiments with noble metals (platinum, palladium) which have been performed by WIELAND, using alcohol, aldehydes, and amino acids as substrates, will receive little further attention in view of their comparatively small physiological interest.

There is some difference with regard to the phenomena and the mechanism between a catalysis by simple salts and by macromolecular complexes. In the former case a complex is formed between the substrate and the ionized metal, while in the latter case the substrate is adsorbed on the surface of the macromolecular catalyst by residual valency forces and subsequently "activated". If the catalyst is an enzyme the substrate combines with a prosthetic group which is linked to a colloidal protein bearer. In some instances the prosthetic group alone (hematin) is already somewhat catalytically active; combination with the bearer protein raises the activity and renders its action highly specific. Whether the metal complex brings about an anaerobic catalysis or one in which molecular oxygen is utilized depends solely on the a u t o x i d i z i b i l i t y of the substrate-catalyst symplex. Of primary importance are those heavy metal catalyses concerned with true oxidation. They may be subdivided into oxidative and peroxidatic catalyses. The catalatic processes where heavy metals split hydrogen peroxide into water and oxygen belong to this class only in an indirect manner.

## Oxidative Catalyses.

The model reactions of greatest importance for heavy metal catalysis in biological systems are those dealing with s u l f h y d r y l  c o m p o u n d s (thiols). For careful studies on the mechanism of such catalyses we are indebted particularly to MICHAELIS 809, 1035) and to CANNAN 162) and their collegues.

The so-called autoxidation of cysteine is actually, at least in the acid range, a heavy metal catalysis; iron, copper and manganese are active (WARBURG). The inhibition of the reaction with oxygen by HCN is due to complex formation with the heavy metal and not with the thiol (HARRISON 455)). Upon adding HCN the violet color of the ferric cysteine complex disappears and the blue color of prussian blue (ferri-ferrocyanide) appears instead (GERWE 399)). The iron catalysis is inhibited by pyrophosphate which reacts specifically with iron but not with copper, manganese, or SH-groups (WARBURG). Carbon monoxide inhibits both the iron and the copper catalysis, but only the orange CO-ferrous cysteine complex is reversibly dissociated by light (CREMER 191)). While in biological systems the cysteine oxidation is mainly catalyzed by iron that of glutathione is mainly effected by copper salts (VOEGTLIN et al. 1205)); pyrophosphate, in this case, does not act as an inhibiting agent but rather as an activator (ELVEHJEM 251, 252)). It is claimed that both the copper and iron catalyses of glutathione are resistant to CO, also the hematin catalysis of SH-compounds (DIXON 210)).

At the same time there appears to take place a true, though very slow, autoxidation which is resistant even to m/100 cyanide (GERWE 399)) *). This demonstrates, once more, that HCN reacts exclusively with the metal and not with the thiol group. On the other hand, HCN will react with S-S-linkages (VOEGTLIN et al. 1205, 1005)), particularly in the absence of iron; the disulfide is reduced to the corresponding thiol (e.g. cystine to cysteine) while the HCN is oxidized to cyanate or transformed into thiocyanate. Contrary to earlier findings, iron-free cystine has no effect on the oxidation of cysteine (GERWE), in contrast to metal-free dithiodiglycolic acid (HARRISON 453)). Instead of molecular oxygen, methylene blue (see p. 108) or indigo disulfonate (DIXON 217)) may be acceptors in this system.

The catalytic oxidation of cysteine proceeds probably beyond the cystine stage, just as the iodine titration method of cysteine will yield correct values only if, by cooling the reaction mixture, further oxidation of the cystine formed is prevented (LEWIS and VIRTUE). According to MEDES 772) cystine sulfoxide $(C_6H_{12}O_6N_2S_2)$ represents a product of intermediary cystine metabolism. In presence of copper total oxidation of cysteine to form $CO_2$, $NH_3$, and $H_2SO_4$ may occur (ROSENTHAL and VOEGTLIN 1006)). This will be discussed further below. In alkaline solution, according to SCHÖBERL and WIESNER 1032), oxalate and thiosulfate are formed. Besides the oxidative scission hydrolysis of disulfides may take place:

$$R—S—S—R + HOH \rightleftarrows R—SH + HO—S—R \ (1031)).$$

This mechanism operates probably preponderantly if disulfide compounds of all types (cystine, SS-glutathione, proteins in the oxidized SS-stage) are brought into strongly alkaline solutions. The inactivation of fully active SS-insulin in alkaline solution is probably due to this cleavage (FREUDENBERG). For a discussion of the chemical reactivity of SS- and SH-compounds the review article by BERSIN 101a) should be consulted.

The oxidation of a thiol is preceded by complex formation between the SH-group and the metal (MICHAELIS 809); PIRIE 928)). However, not all of these complexes are autoxidizable. The course of events following complex formation depends on the con-

*) See, however, ELVEHJEM 252).

stitution of the substrate (cysteine, SH-glutathione, SH-proteins) and the nature of the metal. The most active metal is always copper (see also RONA 999)); the others act less strongly or not at all (see below). For pure metals, BERSIN 101) found the following order: As, Cu, Sb, Zn, Cd, Ag, Fe, Ni. $H_2O_2$ cannot be detected as a reaction product because it it decomposed by the metal and also because it reacts rapidly with the residual substrate; cysteine, for instance, is quickly dehydrogenated to cystine by hydrogen peroxide alone (SCHÖBERL 1029, 1030)). This reaction is the basis of some methods for the preparation of disulfides from the corresponding thiols; it will proceed quantitatively with cysteine, glutathione. and probably also with proteins.

At this place, a few words concerning the sulfur groups of proteins and their oxido-reductive behaviour may be said. It has been known for some time (HEFFTER) that denatured proteins will give a positive test with nitroprusside while the test with the corresponding native protein is negative. The appearance of free SH-groups by denaturation and their disappearance by renaturation has recently been extensively studied by MIRSKY and ANSON 833a). In the native state, most proteins exist entirely in the oxidized SS-form. In some of them, however, "potential" SH-groups may be detected by shifting the ph to the alkaline side; hemoglobin is such a protein. A native SS-protein may be reduced to SH-protein without appreciable denaturation if mild and specific reagents, e.g. cysteine or thioglycolic acid, or HCN, are employed. Upon denaturation additional SH-groups appear. If such a denatured protein is further reduced by the same reagents the number of SH-groups detectable is stated to be the same as that found upon complete hydrolysis of the protein. Reoxidation of these SH-groups is possible with cystine and similar small-molecular disulfides; $H_2O_2$ may also be used for this purpose. Of particular interest for the biologist are those proteins which show a specific physiological activity, the degree of which depends on the SS $\rightleftarrows$ 2 SH-equilibrium in the protein molecule. There appears to be no definite rule governing this relationship: while certain proteinolytic enzymes, e.g. cathepsin and papain, appear to be active in the SH-form only (see, however, ANSON 31a)), insulin conversely is stated to be physiologically active only in the SS-form. Treatment of insulin with cysteine, reduced glutathione, HCN, decreases the activity markedly. It has recently been shown (STERN and WHITE 1097)) that if denaturation of the protein hormone during reduction is avoided, the more or less soluble SH-form retains about half of its original activity. The loss of activity is concomitant with the appearance of only few SH-groups (2 to 3 from a total of 20). It is concluded that a few SS-linkages in insulin play the part of "activating groups" in the sense of LANGENBECK's terminology 697) and that the nature of the "active center" which actually participates in the catalysis is still obscure. It should be mentioned that FREUDENBERG and WEGMANN 359a) succeeded in reactivating reduced insulin to a small extent by treating it with hydrogen peroxide in the presence of a large excess of cysteine as protective substance. The authors realize that the resulting active substance may not be the original hormone but a partly artificial insulin where SH-groups of insulin have been joined in the course of the reoxidation to SH-groups of cysteine molecules. Reduced native insulin of 50 per cent activity may be oxidized by molecular oxygen in presence of certain heavy metals (Cu, Fe, Mn) (WHITE and STERN 1298)). Although the oxygen uptake corresponds closely to the amount required to reoxidize the SH-groups, as also shown by independent SH-analyses, the resulting products, instead of being more active, have lost almost all of their residual activity in the course of the oxidation. It may be that under these conditons still other important reducing groups of the protein are oxidized.

The catalytic oxidation of thiols may be coupled with other reactions. The system Cu + thioglycolic acid, for instance, may oxidize hypophosphorous acid or tartaric acid (WIELAND and FRANKE 1317)). The coupling with biological systems by means of intermediary catalysts will be discussed on p. 95.

Cysteine appears to be only slightly selective with respect to the metal catalyzing its oxidation; besides Cu, Fe and Mn (WARBURG 1231)), Co, but not Ni (MICHAELIS 809)) are effective. In the case of inorganic SH-compounds (alkalisulfides and H$_2$S) Mn and Ni are particularly active catalysts (KREBS 597)). There is a possibility that Pb, at least in insects, is a catalyst (RONA 999)). The potential of hydrogen release in cysteine solutions is lowered by Co and Ni (BRDIČKA 145)); this is explained by dipole formation due to production of the complex, the dipole, in turn, contributing to the labilization of H. The system cystine-cysteine has been studied with the polarograph by RONCATO 1001). According to electrometric measurements of MICHAELIS and BARRON 810) the potential of cysteine solutions is influenced also by metallic mercury, platinum and gold (only as deposit on platinum but not as massive metal). Here, the reaction with oxygen takes place preferably in alkaline solution with a ph-optimum at 12.8. Cyanide inhibits the platinum but not the mercury catalysis. In both metal catalyses, also with Co, methylene blue may replace oxygen as the acceptor (813)). Silver is inactive. Under anaerobic conditions the potential is independent of the nature of the metal forming the electrode (813)). Under aerobic conditions Hg will react with SH to form an electroactive complex (80)). Copper acts most strongly and effects eventually a total oxidation, while iron and manganese effect only a dehydrogenation to cystine (1006)). The Cu-catalysis is fastest in pyrophosphate medium (251)) but, in this case, brings about dehydrogenation only (1006)). ELLIOT 241) finds that the reaction proceeds more rapidly in phosphate-free solution, the rate increasing with the oxygen pressure and also proportionally to the amount of copper in the range of small Cu concentrations. The somewhat more stable cobalt complexes which are more readily isolated and useful as models have been studied by MICHAELIS 809, 810) and by KENDALL 1035) and their collegues. SCHUBERT 1035) was able to isolate the Fe-complexes of thioglycolic acid which had previously been studied in solution by CANNAN and RICHARDSON 162) with respect to the kinetics of their oxidation and to their potentiometric behaviour. The formation of the Cu-cysteine complex possesses an optimum temperature (ELVEHJEM 251)). The oxidation of thioglycolic acid by Fe, Mn, and Cu has recently been reinvestigated by KHARASCH, GERARD et al. 562); Cu was found to be most active, its action being promoted by phosphate while Fe and Mn are thereby inhibited.

The so-called "spontaneous" autoxidation of glutathione is attributed to Cu (DIXON 211)). Iron is much less active (ELLIOTT 241)) or completely inactive in contrast to Pd, Au, Co (VOEGTLIN 1206)). The situation here is not so simple as with cysteine since glutathione forms regular salts (through the COOH-group, e.g. with Cd or Cu) but apparently no complexes (through the SH-group) with heavy metals. Therefore, the presence of a compound capable of such complex formation seems to be necessary. Added cysteine may serve in this capacity (MELDRUM and DIXON 778)) *). With glutathione stored as such cysteinylglycine, formed by decomposition, acts as the intermediate (MASON 757)). Presumably cystine is formed in the course of the reaction and this dehydrogenates the glutathione.

The SH-groups in proteins are dehydrogenated but not fully oxidized by Cu, Mn,

---

*) This observation could not be confirmed by VOEGTLIN 1005).

Fe (**1006**)). After coagulation ovalbumin behaves like free cysteine: Mn dehydrogenates whereas Cu oxidizes completely.

A significant contribution to the problem of the mechanism of autoxidation has recently been made by SZENT-GYÖRGYI and his associates **1126a**). Among other systems, they studied the oxidation of c a t e c h o l by molecular oxygen as catalyzed by heavy metal. The starting point of their experiments was the observation that ferrous iron which is only slowly oxidized by $O_2$ when it exists in simple inorganic salt linkage, is rapidly oxidized to ferric iron when added to catechol. The iron is, therefore, "activated" by complex formation with catechol. At the same time, the autoxidation of the catechol is accelerated by adding a ferrous salt at neutral reaction. Inasmuch as no free $Fe^{III}$ ions may be detected in the system, it follows that the oxidation of the catechol is not brought about by ferric ions but that it is due to an intramolecular electron shift from the ferric iron to the catechol within the complex. This process is analogous to that postulated by WARBURG in the case of the oxidation of cysteine by iron salts (cf. **191**). It was found, furthermore, that the reaction between the ferrous-catechol complex and $O_2$ is strongly inhibited by CO. Inasmuch as it can be shown that there exists a competition between the $O_2$ and the CO for the complex which depends on the relative concentration of the two gases, it is concluded that the first step of the reaction consists in a reversible oxygenation of the iron-catechol complex. There must follow an intramolecular electron shift between the $O_2$ and the central $Fe^{II}$ -atom. The $Fe^{III}$ thus formed accepts an electron from the catechol and is thereby regenerated to the $Fe^{II}$ form. The over-all result of this cycle is the oxidation of the catechol to quinone and the combination of the charged oxygen with the hydrogen ions of the water to form $H_2O_2$. This picture conforms with that developed by WIELAND **1302a**) on the basis of his extensive investigations with FRANKE. An analogous scheme is developed for the oxidation of ascorbic acid by Fe salts $+ O_2$.

Other substances the catalytic oxidation of which has been frequently studied are the s u g a r s. Some time ago WARBURG and MEYERHOF showed that fructose is oxidized by molecular oxygen in the presence of phosphoric acid and traces of Fe or Cu. Of the various sugars tested only fructose and sorbose were rapidly attacked (WIND).

According to MEYERHOF and LOHMANN **792**) the reactivity (placing that of fructose arbitrarily at 100) is as follows: fructose phosphate (NEUBERG-ester) 220, ROBISON-ester (equilibrium mixture between aldose and ketose monophosphate) 46, glucose 14, zymophosphate 0. In the last case complex formation with the heavy metal seems to be impossible due to substitution of both terminal groups. In solutions containing buffer substances other than phosphate fructose is rapidly and other sugars are more slowly oxidized by heavy metal plus $CaCl_2$, at ph 8.5. The catalysis is inhibited by HCN ($10^{-3}$ M.), $H_2S$, or pyrophosphate (KREBS **593**)).

The products of oxidation are acid. Consequently, the oxidation will soon be arrested if the buffer concentration is insufficient to maintain an alkaline ph (NICLOUX **888**)). NICLOUX states **887**) that glucose, fructose, lactose, galactose, maltose, and invert sugar but not sucrose are attacked. No tests were made concerning the metal content of the solutions. Small amounts of $CO_2$ are said to be formed. The production of various decomposition products depends on the hydrogen ion concentration. (W. L. EVANS et al. **328**)).

On the other hand, SPOEHR **1064**) found another complex iron-pyrophosphate system not so highly specific; all sugars were attacked. Another system, consisting of $Na_2HPO_4 +$ ferric phosphate, seems to be capable of oxygen transfer only in the presence of methylene blue.

It would appear that complex formation with pyrophosphoric acid is an essential requisite for catalysis (DEGERING 198), MALKOV 749)), According to KUEN 625), however, the alkaline ph — and therefore the formation of reductone-like substances by alkali — is of greater import than the presence of pyrophosphate. In recent experiments HAURO-WITZ 468) finds oxidation only with fructose as the substrate while glucose is decomposed with $CO_2$ formation, but without $O_2$ uptake. This agrees with the statement by THE-RIAULT et al. 1165) that oxidation of glucose in ferrophosphate is due to contamination with bacteria. This statement has, however, been refuted by GOERNER 308) who confirmed the desmolysis of glucose by SPOEHR's ferriphosphate model system under conditions which excluded the presence of bacteria. It will be remembered that KREBS 593), working in WARBURG's laboratory, observed oxidation of all sugars tested in alkaline solution and in presence of heavy metal, especially of Cu. Pyrophosphate, HCN, and $H_2S$ were found to inhibit the reaction. PALIT and DHAR 914) report induced oxidation of all carbohydrates tried and also of uric acid in the presence of freshly precipitated ferrous hydroxide and of certain non-metallic reducing agents, e.g. sodium sulfite. This oxidation of arsenite etc. by atmospheric oxygen, as "induced by sugar" has been explained by HARNED 452) by the formation of sugar peroxides; the participation of traces of iron is not denied because these reactions are sometimes inhibited by cyanide. In some instances a stimulation by cyanide was found. Upon shaking alkaline sugar solutions in air SHAFFER and HARNED 1047) were unable to detect the formation of organic peroxides; only $H_2O_2$ could be found. The amount of hydrogen peroxide formed was increased by adding allegedly metalfree glass or of methylene blue. These observations suggest that autoxidative reactions take place in the course of these complex reactions. The oxidation of fatty acids, carbohydrates, and proteins in presence of Fe and Ce hydroxide has been interpreted as an induced reaction with the initial formation of higher metal oxides, e.g. $FeO_2$ and $Ce_2O_5$ (PALIT and DHAR 913, 915)).

Trioses are oxidized in similar complexes (15)). Iron is most active here when added in the form of ferrous sulfate while hemin is less effective. The following compounds are oxidized by Cu in phosphate solution: glycerol aldehyde and dihydroxyacetone (rapidly), fructose and methyl glyoxal (slowly). The following substances show no oxygen uptake under the same conditions and exhibit partly inhibiting and partly stimulating effects on the oxidation of the above compounds (773)): glyoxal, glyoxylic acids, glycol, glycolic acid, pyruvic acid, valerianic acid, butyric acid, and propionic acid. A decarboxylation has also been observed in this system, increasing with the content of carbonyl groups. Glycol and glycolic acid do not react. We are probably dealing with coupled reactions. If animal charcoal is used as the catalyst, only sugar breakdown products, but not glucose itself are attacked (380)). Glycerol is dehydrogenated at palladium surfaces (MAZZA and CARERA 768)). As acceptors for the glycerol hydrogen $O_2$ and, less effectively, methylene blue may be employed. HCN but not $H_2S$ or CO inhibit this catalysis.

The "autoxidation" of ascorbic acid is likewise a heavy metal catalysis and as such inhibited by cyanide (at acid reaction up to ph 7.6); Cu is again the most efficient catalyst (265, 257, 760, 554, 79)). The first oxidation product is dehydroascorbic acid (KELLIE 554)). This is the true oxidant of the reversible redox system, the reductant of which is ascorbic acid. Its preserved vitamin activity is probably due to its reconversion to the vitamin proper under physiological conditions. Dehydroascorbic acid is rather unstable and is readily further oxidized to irreversible decomposition products. It is this situation (which is somewhat analogous to that existing in the epinephrine system) which is responsible for the controversy on the reversibility of ascorbic acid oxidation and on the thermodynamic significance of potential measurements in the system. Autoxidation in alkaline solution causes irreversible destruction of ascorbic acid (see p. 93). Hemochromogens containing bases like nicotine may also promote the oxidation of ascorbic acid (BARRON 79)). According to ZILVA, SZENT-GYÖRGYI and others, there exists a specific ascorbic acid oxidase (see also HOPKINS and MORGAN 499)). Its occurrence has been reported in plants like squash and cabbage. Recently it has been claimed that the responsible agent is not an enzyme but heavy metal, probably copper, in a rather simple chemical linkage (KING et al. 1105a)). STRAUB 1107a), however, finds that fresh cucumber extracts have a 5 to 10 times higher ascorbic acid oxidase activity than ash solutions obtained from the same material. He assumes that the copper is activated by a linkage to a specific protein pheron. Details will be found in OPPENHEIMER'S SUPPLEMENT, p. 1587.

The copper catalysis of ascorbic acid oxidation, just like the oxidation of leuco methylene blue by Cu (SCHÖBERL 1029, 1030)), is inhibited by SH-groups (265, 257, 102, 869)) due to their affinity for Cu and resultant complex formation. Once SS-groups have been formed by oxidation of the thiols, the inhibition ceases because of lack of affinity of SS-groups for Cu. According to SCHÖBERL 1029, 1030) SH removes intermediary $H_2O_2$; it is thereby oxidized to SS while the amount of $H_2O_2$ available for the main reaction is diminished. This interpretation is questioned by the recent finding of BARRON 78) that the inhibitory effect of SH-glutathione is only observed when ascorbic acid is oxidized by Cu as the catalyst. The dehydrogenation of ascorbic acid by hemochromogens or plant extracts is only inhibited by HCN but not by glutathione. Tissues contain a protective substance (554)) which prevents the oxidation of ascorbic acid. It is thermostable and dialyzable and does not consist, at least not entirely, of glutathione (760)). Epinephrine (p. 93), proteins, and amino acids act also as protectors in animal tissues (326)).

A b i e t i n i c   a c i d is oxidized by cobalt salts. The change in the absorption spectrum indicates complex formation between the metal and the substrate. The oxidation of o l e i c   a c i d by air as catalyzed by various metal complexes (ferricyanide, molybdicyanide, copper glycine and copper pyridine) and by indophenol had been studied by CHOW and KAMERLING 168). The rate of oxygen uptake depends on the redox potential in the system: ferricyanide ($E_o' + 0.42$ v.) is most effective. Upon adding increasing amounts of ferrocyanide and thereby decreasing the potential the oxygen uptake decreases. The potentials of iron complexes and their ph-dependence varies greatly with the nature of the complex forming anion. The redox potential of the iron cyanide system is $+ 0.42$ v. in the range of ph 4 to 8 whereas the potentials of the acetate, malonate, and pyrophosphate complex will decrease along this range (814)). U r i c   a c i d and its salts are aerobically oxidized by basic zinc salts (388)). Cobaltammines act as catalysts on p y r o c a t e c h o l but not on tyrosine (939)).

The first model reaction studied by WARBURG was the oxidative catalysis of a m i n o  a c i d s   a t   c h a r c o a l   s u r f a c e s. Today the significance of these experiments is more historical than experimental. The importance of the charcoal model for the heavy metal theory of biological oxidation has been lessened by the observation of WARBURG himself that iron-free sugar charcoal is only a little less active than the iron-rich blood charcoal originally used. From a descriptive point of view the catalysis consists in a dehydrogenation and decarboxylation yielding the next lower aldehyde. Thus leucine gives rise to valerianic aldehyde. In charcoal containing iron the active agent is Fe linked to N. The linkage from the iron to the the nitrogen atoms of the pyrrol rings in hemin appears to remain intact upon carbonification though the rings themselves and the methin bridges of the tetrapyrrolic porphyrin structure are, of course, destroyed. In heavy metal-free charcoal the active agent is the carbon itself. In the first instance, but not in the latter, cyanide is an inhibitor. The oxidation of amino acids and also of glucose, pyruvic acid, and several fatty acids by iron-free plant charcoal has been studied by MAYER and WURMSER 762). It was found that the $O_2$ as well as the substrates are adsorbed by the charcoal and that the process represents a typical heterogeneous catalysis on a carbon surface (see p. 22). The formation of $CO_2$ as an end product of catalytic oxidation was also observed to a small extent with manganese dioxide as the solid phase.

HANDOVSKY 446) reported that iron-containing charcoal and iron in powder form show activity only after they have been heated and permitted to cool in a hydrogen but not in a nitrogen atmosphere. He assumed that there was formed as the active agent hydrogen peroxide from the adsorbed hydrogen. His hypothesis has been refuted by WARBURG 1226) who points out that the amount of hydrogen adsorbed would have to be infinitely large in order to explain the permanent catalytic activity of the charcoal surface. Inorganic iron compounds are not capable of catalyzing the oxidation of amino acids; the same holds for ferrous salts after addition of indifferent colloids. The catalytic action of iron powder is stated to be inhibited by HCN (HANDOVSKY 446)). $Fe^{II}$-ions adsorbed on charcoal are ineffective (KUHN and WASSERMANN 678)), only "embedded" iron is active. On the other hand $Fe^{II}$-ions adsorbed on charcoal exhibit a strong catalatic activity towards $H_2O_2$ which is of the same order as that of hemin. $Fe^{\cdot\cdot}$ adsorbed on metatin acid is less effective while $Fe^{\cdot\cdot}$ adsorbed on aluminium hydroxide is inactive. It appears that the "activation" of $O_2$ requires more energy than the decomposition of $H_2O_2$. Synthetically prepared charcoals containing iron are oxidative catalysts apparently only if the iron is linked to nitrogen. According to RIDEAL 982) we have to distinguish between different active centers of varying activity, e.g. Fe—C—N, Fe—C, C—C.

With respect to the specificity of the oxidation on charcoal it has been found that amides, proteins, peptones, heterocyclic N-bases, urea derivatives and sugar are hardly attacked (FÜRTH 380)), whereas peptides and fructose are stated to be oxidized in presence of alanine (1352)). Aliphatic amino acids with a long carbon chain and aromatic amino acids are more rapidly oxidized than alanine (FÜRTH). The reaction never proceeds to completion. In alkaline solution a considerable rate increase is observed together with a complete change with respect to the mechanism. No more amino acid molecules disappear than in acid reaction, but the oxygen uptake and still more the $CO_2$ evolution are greatly increased (RQ > 2) (411)).

It is not always possible to formulate the process as a dehydrogenation since N-substituted amino acids also are oxidized, e.g. dimethylamino isobutyric acid to acetone, dimethylamine, and $CO_2$. With ozone the same oxidation products are obtained. Monomethylaminoisobutyric acid is also decomposed by charcoal, though more slowly. Probably oxygen is attached to the nitrogen to form an oxide or peroxide of pentavalent nitrogen which subsequently breaks down (95)).

The statement that amino acids are hydrolyzed at charcoal surfaces (1351, 1352)) has been refuted by WIELAND 1303). Due to the retention of oxygen by charcoal dehydrogenation always takes place. Iron- containing charcoal dehydrogenates succinic acid to fumaric acid; oxygen as well as methylene blue function as acceptors. The activity of the charcoal increases with its iron content. With oxygen as acceptor quinone and KCN act as inhibitors (1130, 1131)). Oxalic acid is likewise oxidized by charcoal. The process is accelerated by ether, alcohols, and phenylurethane (GOMPEL 409)).

## Peroxidatic Effect of Heavy Metal Systems.

The pronounced oxidizing action of iron salts + hydrogen peroxide has been known since SCHÖNBEIN, and the system was later regarded as a peroxidase model. This appears justified inasmuch as the true peroxidases have been found to contain iron in complex formation with porphyrin (heme) and a specific protein (p. 178). While the true peroxidases are generally considered to act almost specifically on certain phenols, the simple iron systems will attack vigorously a number of biologically important compounds, e.g. sugars, fatty acids, and amino acids (DAKIN and others). This property is made use of for preparatory purposes, e.g. in the FENTON reaction of the breakdown of hexoses, in the preparation of dihydroxymaleic acid from tartaric acid, etc. They act, of course, also on phenols (405)). It has already been mentioned (p. 22) that such reactions probably also occur in cell metabolism where they may be catalyzed by simple iron complexes or by hemins.

These effects appear to be based, at least partly, on the dehydrogenation of complexes which in the case of sugars, especially fructose, have been studied in detail by KÜCHLIN and BÖESEKEN 617, 618, 619, 620). The primary dehydrogenation of the sugar-ferrous iron complex yields osones which, in turn, reduce the ferric iron formed. It is not decided as yet whether actual complex formation or only a certain orientation of the polar sugar molecules by the iron ions is necessary. Phosphates, according to MALKOV 749), inhibit, while KUEN claims that they accelerate the reaction just as in the oxidation by $O_2$ (625)). The intermediary formation of iron peroxides is unlikely (618)).

Specific iron complexes as they have been postulated by BAUDISCH 91) in mineral waters (see also HEUBNER 478)) are not recognized by SIMON 1056, 1057) and by FRESENIUS 359). SIMON is of the opinion that the solution contains ferrous ions which will form a complex only when in contact with the substrate. According to GOLDSCHMIDT et al. 408) such a complex formation is not always necessary since even ethyl alcohol which shows so little inclination for complex formation may be a substrate. Furthermore, the authors were unable to find a complex formation between Fe and glycolic acid as the substrate upon spectrographic examination. In contrast to WIELAND and FRANKE it is assumed by GOLDSCHMIDT that the peroxidatic reaction proceeds via ionogen and not complex bound iron. He believes that the reaction is a chain process involving an intermediate peroxide (p. 123).

The peroxidatic and catalatic activity of iron oxides and other Fe compounds, and of metallic iron in various physical states have been studied by BAUDISCH et al. (90,92)). KUHN and WASSERMANN (678, 679, 680, 1283)) studied the catalatic, oxidatic, and peroxidatic activity of iron in various states of adsorption. The oxidation of $H_2S$ by $H_2O_2$, for instance, is very strongly catalyzed by free $Fe^{II}$-ions and practically not promoted at all by iron complexes (phosphate, simple organic complexes, porphyrins) (1283)), while in other model systems the porphyrin compounds will show very pronounced peroxidatic activity. Concerning the qualitative differences in the action of various Fe-compounds the paper by UCKO 1189) should also be consulted.

The kinetics of the oxidation reactions with diphenols, p-phenylene diamine, dihydroxymaleic acid, and linoleic acid has been carefully investigated by WIELAND and FRANKE (1312, 1314, 1316). With diethylperoxide + Fe·· formic acid, lactic acid, amino acids, pyrogallic acid, etc. may be dehydrogenated (1307, 1306)). Without a donator the peroxide is decomposed by ferrous salt into alcohol and aldehyde (intramolecular hydrogen shift). The peroxidatic reactions which are brought about with the aid of this peroxide are evidence for the dehydrogenative character of the catalyses since it cannot give rise to an iron peroxide of the type depicted by MANCHOT (p. 123). WIELAND assumes that an intermediate reactive complex between $Fe^{··}$, peroxide, and substrate is formed. Upon the breakdown of the complex the iron is oxidized by the excess peroxide. We are dealing here with an induced reaction involving the formation of a "critical" complex in the sense of BRÖNSTED; the process may become a true catalysis if the ferrous iron is regenerated by the donator. It is noteworthy that this Fe-catalysis is cyanide resistant.

Similar experiments with thiols which are dehydrogenated by hydrogen peroxide alone — Fe and Cu accelerate — have been performed by PIRIE 929) and by SCHÖBERL 1029, 1030). The fact that the dehydrogenation of $H_2O_2$ is not inhibited by cyanide 1115) is proof that it may be brought about without the participation of heavy metal. The rate of the reaction is greater at alkaline than at acid ph; potato oxidase and peroxidase do not accelerate and are, conversely, not inhibited by SH-compounds while acting on other substrates (454)). Obviously the direct oxidation of SH by $H_2O_2$ must have a mechanism different from that promoted by heavy metal. Under certain conditions the SS-groups formed without the participation of metal may themselves bring about dehydrogenations (462)).

The effect of various simple iron salts and complex salts on benzidine has been examined by SIMON 1057). Partly they represent induced reactions, partly they are catalytic in nature but involve a desactivation of the catalyst. Hemoglobin is more effective than the other Fe compounds tried (p. 82). Copper but not nickel, zinc, or manganese salts exert a peroxidatic action on fatty acids (succinic acid, etc.) (SMEDLEY-MacLEAN 1059, 89)). Fe acts weakly in this case. Nucleic acid salts are attacked by basic zinc salts (GANASSINI 388)). Other data concerning the catalatic and peroxidatic effects of Fe, Cu, Co, and other heavy metal complexes will be found in SHIBATA's publications (1051, 1052)). As substrates oxyflavins like myrecitin were used. HCN inhibits these reactions (1049)).

The peroxidatic activity of complex iron systems is of considerable theoretical importance: it makes it impossible to detect $H_2O_2$ in autoxidizable iron systems, simply because the peroxide is utilized in a new type of oxidative catalysis as soon as it is formed. In the case of copper the kinetic relationships are different from those of iron. $Cu^+$ reacts faster with $O_2$ than with $H_2O_2$, while $Fe^{++}$ will react about 1000 times faster at ph 7 and about 2—3000 times faster at ph 4—5 with $H_2O_2$ than with $O_2$ (WIELAND and FRANKE 1315)). This explains why $H_2O_2$ may be detected in Cu but not in Fe catalyses.

The catalytic action of simple copper complexes, e.g. $Cu(NH_3)_4$, Cu-pyridine, and of analogeous silver compounds has been studied by EULER (295)). Colloidal carbon, according to SCHWOB 1041), is peroxidatically but not catalatically active.

### Participation of Heavy Metals in Anaerobic Processes.

While it is probable that heavy metals play an important part not only in oxidative but also in anaerobic biological processes, our knowledge has so far not penetrated beyond a few isolated observations and details. These observations are mostly limited to the accelerating effect of metal salts and to the inhibiting effect of substances known to form stable metal complexes on certain model reactions and biological processes.

To begin with model systems, TODA 1178) in WARBURG's laboratory and HARRISON 453) demonstrated the accelerating action of $Fe^{··}$ and $Cu^{··}$ on the dehydrogenation of cysteine and thioglycolic acid by methylene blue. The same has been found in the case of sugars and similar substances; e.g. for fructose in phosphate solution (BLIX 126)) and for dihydroxymaleic acid (WIELAND and FRANKE l. c. 288), p. 198). Some further data will be found in the paper by ANDO 27). Aldehydes, sugars, and other substances are said to reduce methylene blue in presence of $Fe^{..}$. EULER 304) finds an acceleration of the spontaneously slow dehydrogenation of ascorbic acid by methylene blue. Iron-containing charcoal dehydrogenates succinic acid with methylene blue as acceptor. It appears that also in certain anaerobic biologic processes iron is playing the part of a catalyst. However its participation can only be concluded from an inhibition of the over-all process by reagents specific for iron. The same is true for copper as the catalyst. It might be mentioned here that THUNBERG 1172) observed an acceleration of methylene blue reduction by plant extracts in the presence of Zn, Cd, Cu and Hg.

According to WARBURG 1218) the alcoholic fermentation of yeast is inhibited to an extent of 90 per cent by CN' in a concentration of $10^{-2}$ M. This is a high concentration when compared with that effective in respiration ($10^{-4}$ M.) but it is an effect 200 times stronger than would be predicted by the adsorption constant. In other words, this inhibition cannot be due to an unspecific surface action. 30 times the amount of HCN in $5.10^{-3}$ M. solution is required for inhibition of fermentation as compared with the aerobic dehydrogenation of alcohol (WIELAND 1322)). It is claimed (PATTERSON 918)) that KCN in concentrations greater than $5 \cdot 10^{-3}$ M. affects only the rate, but not the total extent of fermentation by prolonging the induction period in intact yeast and by producing an induction period in yeast juice. KCN, then, appears to influence the phosphorylation rather than other phases of fermentation. $H_2S$ acts like CN' (875)).

Carbon monoxide affects neither alcoholic nor lactic acid fermentation. KEMPNER 555) made the interesting observation that the butyric acid fermentation of *Clostridium butyricum* is very sensitive towards CN' and CO. While at first he was unable to find an effect of light on the CO inhibition, in analogy to the respiratory ferment, later a diminution of the inhibition by very strong illumination could be demonstrated (KEMPNER and KUBOWITZ 557)). By employing monochromatic light sources these workers were able to determine the photochemical absorption coefficient of the enzyme-CO compound for a small number of wavelengths.

Subsequent work by KUBOWITZ 612) showed that the CO inhibition is not concerned with the first phases of sugar cleavage to $C_3$-compounds, but with later stages leading to the normal end products butyric acid, hydrogen and carbon dioxide. In the CO-poisoned bacterium lactic acid is the end product. The inhibition appears to affect an enzyme containing heavy metal which splits pyruvic acid into $CO_2$, acetic acid and $H_2$. A carbon monoxide inhibition of formate scission by *Bacterium coli* has previously been observed by STEPHENSON et al. 1070). TAMIYA 1133) found an inhibition of methylene blue reduction by acetic acid bacteria in presence of CO. He interprets this effect as one of displacement by adsorption rather than of specific poisoning of an enzyme.

It is noteworthy that nitric oxide (NO) will inhibit fermentation in a reversible manner, an effect which has been explained by a reaction with heavy metal (WARBURG 1223)). According to WIELAND 1322) sodium fulminate affects respiration to the same extent as fermentation.

The participation of copper in glycolysis, particularly in that of tumors, has been inferred from its inhibition by substances (amino acids, pyrocatechol disulfonate, etc.) which will form complexes with copper and which are able to protect animals against lethal doses of copper salts (232, 472, 589)). The inhibition of formic dehydrogenase of bacterial origin by aminonaphthol sulfonic acid has also been attributed to the presence of copper in the enzyme (COOK, HALDANE and MAPSON 181)). On the other hand, the importance of loosely bound ferrous iron for glycolysis has been postulated by ZUCKER-KANDL et al. (1383)) in view of his finding that $\alpha, \alpha'$-phenantroline which forms stable complexes with Fe·· inhibits both alcoholic and lactic acid fermentation; $\alpha, \beta'$-phenantroline which forms no such complexes is ineffective. USTVEDT 1191) was unable to confirm these observations in the case of muscle glycolysis.

## Inhibition of Metal Catalysis.

In many instances the inhibition technique has been and still is the only method available in the quest for the elucidation of the nature of biological catalysts. Almost everything that is known about the respiratory ferment, for instance, is derived from spectroscopic studies of its inhibition by carbon monoxide. It should not be overlooked, however, that the applicability of this tool is limited. The same heavy metal, iron for example, will exhibit a different behaviour towards supposedly specific reagents depending on its particular type of linkage in the catalyst molecule. This may even go to the extreme of non-reactivity: $\alpha, \alpha'$-dipyridyl which is a highly specific reagent for ferrous iron and with which it forms a stable red complex (tridipyridyl ferrous sulfate) will not react with the ferrous iron in heme complexes. It is obvious that the deciding factor here is the ratio of the affinities of the complex forming reagents to the metal. WARBURG has stressed the point that the absence of heavy metal in a given catalysis must not be inferred from negative results with metal inhibitors. On the other hand it must not be overlooked that in spite of complex formation between the reagent and the heavy metal catalyst, the process may proceed unchecked or even more rapidly as before, simply because the newly created complex is catalytically active. A case in point is the formation of cyanhematin which promotes the oxygen uptake of unsaturated fatty acids. Conversely, the inhibition of an enzyme by cyanide is not sufficient proof that a heavy metal is at work. Xanthine oxidase is inactivated by cyanide in a relatively slow reaction which bears the marks of cyanhydrin formation rather than of ferric cyanide complex production (p. 90). In other cases cyanide has been found to activate enzymes, e.g. papain, cathepsin, by reducing the inactive —S—S— linkage to the active —SH— form.

This section will be devoted mainly to a discussion of the interaction of inhibitors and metalcontaining model systems. The effect of these inhibitors on enzymes containing heavy metal, particularly a hemin grouping, will later be treated in greater detail.

The two reagents for heavy metals which interest us most in this respect are c a r b o n m o n o x i d e and the c y a n i d e ion. While an inhibition by CO indicates unambiguously the presence of heavy metal, the action of cyanide, as has been pointed out, is not decisive proof of this type, especially if only higher CN' concentrations are found to be effective. In the case of iron the mode of action of these two important inhibitors is fundamentally different: CO combines with f e r r o u s i r o n and prevents its oxidation; the iron carbonyl complexes exhibit varying degrees of photodissociation or rather light sensitivity. The cyanide ion has a greater affinity for f e r r i c than for ferrous iron. It prevents the reduction of the iron (WARBURG (p. 76), KREBS 596)). In the case of the respiratory ferment this can be shown by virtue of the fact that the degree of cyanide inhibition is independent of the $O_2$-pressure. This concept has been further supported by direct spectroscopic observations in microorganisms (WARBURG 1280)). The affinity of ferric iron for cyanide is exhibited not only by heme derivatives but also by simpler iron complexes and salts.

Carbon monoxide: A variety of heavy metals are capable of forming complexes with carbon monoxide (carbonyl complexes). Iron pentacarbonyl ($Fe(CO)_5$), iron tetracarbonyl, diferro nonacarbonyl ($Fe_2(CO)_9$), nickel carbonyl are examples of the simplest type of these compounds. Complex heavy metal salts may also form CO compounds provided that they are not coordinatively saturated. Thus, potassium ferrocyanide where the covalence forces of the metal are satisfied will not absorb CO while the less saturated trisodium ferropentacyanammin ($Na_3FeCy_5NH_3$, 6 $H_2O$) will form complexes not only with CO but also with nitric oxide (NO) and oxygen. The CO replaces the $NH_3$ molecule in the complex (MANCHOT). MANCHOT has also determined the composition of certain copper and mercury CO complexes: $CuCl(CO)$ (2 $H_2O$), $Hg(OC_2H_5)(CO)$ Cl. The most important CO complex from a biological point of view is, of course, carbon monoxide hemoglobin. It is not only interesting for its own sake but also as a convenient model for similar complexes, e.g. the respiratory ferment-CO compound, which have thus far not been isolated and which at best are only present in infinitesimal concentrations in living systems. Another interesting model is the ferrous cysteine CO complex which was discovered by CREMER (191) in WARBURG's laboratory.

The outstanding property of the iron carbonyl complexes is their dissociation by light. MOND and LANGER, in 1891, discovered the reversible photodissociation of $Fe(CO)_5$. Soon afterwards, HALDANE and SMITH accidentally observed the splitting of CO-hemoglobin by sunlight. The quantum yields obtained in various model systems and in the respiratory ferment-CO system have been measured by WARBURG (p. 136). The affinity between the CO and the metal compound and the light sensitivity are important individual constants depending on the kind of metal, the nature of the other constituents, and on the type of linkage existing in the molecule. Nickel carbonyl complexes are less light sensitive than the corresponding iron compounds, while copper carbonyl complexes are stated to be light-stable.

Compared with hemoglobin the respiratory ferment has a higher affinity for oxygen and a lower affinity for carbon monoxide. This may be demonstrated by the fact that in small cells where diffusion plays no large role (*Micrococcus candicans*) respiration will go on undiminished, i.e. the respiratory ferment will be saturated, down to oxygen concentrations of the order of $10^{-5}$ M. where oxyhemoglobin is completely dissociated. Conversely, hemoglobin will be fully saturated with CO in presence of $O_2$ at a $O_2/CO$ ratio of 1/4, whereas the respiratory ferment is still not fully combined with CO at a ratio of 1/40 (WARBURG (p. 142). CO-hemoglobin is considerably less light sensitive than the respiratory ferment-CO compound or CO-pyridinehemochromogen. The light sensitivity of the hemoglobin complex may be greatly increased by methyl carbylamine (WARBURG 1278)). The theory of photodissociation of these CO compounds will be treated later (p. 136).

Ferric iron forms no reversible CO complexes. But certain hemins (e.g. phaeohemins) are capable of oxidizing CO to $CO_2$:

$$2 > FeOH + 3 \ CO = 2 > FeCO + CO_2 + H_2O \ (NEGELEIN \ 876)).$$

Whether biological copper catalyses are inhibited by CO is still controversial. REID's 971) finding that CO inhibits the copper catalysis of leuco methylene blue has not been confirmed by MACRAE 742).

According to BARRON et al. 79) the copper catalysis of ascorbic acid oxidation by molecular oxygen is inhibited by CO. Illumination fails to alleviate the CO-poisoning of the catalyst. These workers assume that the metal, in the course of the reaction, undergoes a cyclic change from the cupric to the cuprous form and that it is the latter which combines with the CO. The polyphenol oxidase from potatoes which has been demonstrated to be a copper-proteid complex (KUBOWITZ 613)) is likewise inhibited by carbon monoxide. In conjunction with a phosphopyridine nucleotid-proteid the copper-proteid represents an "alcohol dehydrogenase" system which oxidizes alcohol to aldehyde and which is inhibited both by HCN and CO. On the other hand, the "alcohol dehydrogenase" composed of alloxazine-proteid (yellow ferment) and the same phosphopyridine nucleotid-proteid is affected by neither inhibitor.

The question of the formation of complexes between CO and heavy m e t a l - t h i o l c o m p o u n d s has a rather involved history. CREMER 191) found that the catalytic cysteine oxidation by iron salts is inhibited by CO. Carbon monoxide will form an orange colored complex with ferrous cysteine where 2 CO molecules are absorbed by one atom of iron. This complex is dissociated by light in a reversible manner. DIXON 210) could confirm the formation of this complex but not the inhibition of the catalysis. He was likewise unable to confirm the CO-inhibition of the cysteine oxidation by hematin which had been reported by KREBS 595, 596). DIXON explained the discrepancy by the observation that hematin decomposes in solution and that the breakdown products will combine with CO but not with cyanide. In the case of glutathione it was reported at first (HARTMANN 463)) that the nickel but not the iron catalysis was inhibited by CO. A reinvestigation (KUBOWITZ 612a)) had the result that not only nickel but also iron and cobalt will form complexes with glutathione (also with thioglycolic and thiolactic acid) which in turn will combine with CO. These phenomena are observed only if an excess of glutathione is employed with respect to the iron. The Fe-glutathione-CO complex but not the others are dissociated by light.

Animals carrying hemoglobin as the protein for oxygen transport purposes are not suitable for the demonstration of inhibition of respiration by carbon monoxide because anoxemia due to CO-hemoglobin formation will cause death before cell respiration is stopped owing to CO inhibition of the respiratory ferment. The specific interaction between CO and the respiratory ferment has been demonstrated in unicellular organisms, e.g. in yeast (WARBURG 1270)), in leucocytes (FUJITA 381)), and in insects carrying hemocyanin (*Galleria*) (HALDANE 441)). While revival by illumination of whole animals poisoned with CO had previously been tried with negative results owing to their opacity to light, FLEISCHMANN et al. (348a)) have recently succeeded in this respect with translucent young *Tenebrio* larvae. They became immobilized by treatment with a mixture of 80 per cent CO and 20 per cent air in the dark; mobility was restored by illumination with strong light. The experiment could be repeated up to six times with the same animals. The heart rate of fish larvae (*Fundulus*) is slowed down by treatment with $CO/O_2$ mixtures: illumination will restore to some extent the original frequency (FISHER and IRVING 345)).

YAMAGUTCHI, working in the laboratory of SHIBATA, has recently reported (1367)) that an indophenol oxidase preparation in solution is CO resistant. The conclusions drawn by the Japanese authors from this observation concerning the role of WARBURG's respiratory ferment in the cell are not convincing until the identity of the two enzymes is demonstrated beyond doubt. WARBURG has expressed the view that the respiratory ferment is linked to the structure of the cell in such a manner that it cannot be separated from it without a change in stability and specificity (change of the protein bearer ?).

According to observations from WARBURG's laboratory (REID 973)); WARBURG et al. 1280)) CO will not react with the cytochrome system. KEILIN however believes that the rather labile cytochrome-a component is blocked by CO; cytochrome-c in solution will combine with CO only outside the physiological ph range, namely, beyond ph 12. (See, however, ALTSCHUL and HOGNESS (20 a)).

Peroxidase and Catalase represent hematin protein complexes in a stabilized ferric state. Consequently, neither of them is inhibited by CO (ELLIOTT and SUTTER 248); K. G. STERN 1089)). There exist also reports to the contrary (KUHN et al. 649); CALIFANO 159)). KEILIN and HARTREE 548), after poisoning catalase with hydroxylamine or sodium azide and adding hydrogen peroxide, observed a change in the absorption spectrum in presence of CO which they interpret as due to the formation of a ferrocatalase-CO complex. The same authors (549)) observed an inhibition of catalase by CO in some instances.

Other important biological processes which are affected by carbon monoxide are the nitrate reduction by bacteria (QUASTEL 944)) and photosynthesis (WARBURG 1225a)). WARBURG was inclined to put the inhibition of the thermal dark reaction phase of photosynthesis (BLACKMAN reaction) by cyanide and CO in parallel to the cyanide inhibition of catalase. According to him iron was the active agent in both reactions. These observations have lately been reinvestigated and reinterpreted by GAFFRON 386a).

Cyanide: The inhibition of heavy metal catalysts by cyanide is probably always caused by the formation of a complex between the catalyst and the inhibitor leading to a blocking of the active group of the former and thereby preventing the acceptance of an electron, i.e. the reduction. In the catalysis by simple iron salts the complex formation with cyanide may prevent the production of the autoxidizable complex with the substrate, e.g. that of ferrous cysteine. It has been known for a long time that hemin derivatives with ferric iron, e.g. methemoglobin, have a high affinity for HCN. While WARBURG seems to picture this compound as a simple complex salt of ferric iron, of the type $> Fe^{III}(CN)$, HAUROWITZ 464, 467) assumes a coordinative linkage where the OH-group (or a water molecule) is replaced by the HCN molecule. It is true that cyanide may also combine, under certain conditions, with ferrous iron. Hemochromogens are combinations between ferroheme or ferriheme *) and a nitrogen containing base, e.g. pyridine, nicotine, hydrazine. Cyan may also function in this capacity. Usually two molecules of the nitrogen base are combined with one iron atom (MIRSKY and ANSON 833b, 833c)). The cyanide hemochromogen and its dissociation constants have recently been carefully studied by spectroscopic methods by BARRON et al. 489a). It is important that denatured proteins also form hemochromogens with heme. Of the native proteins apparently only globin may combine with heme. When hemoglobin is denatured the spectral type changes to that of globin hemochromogen. The corresponding ferrihemochromogen is cathemoglobin which may be obtained by denaturation of methemoglobin. Native ferrohemoglobin does not combine appreciably with cyanide because the bonding between the ferrous iron and the globin is stronger than between $Fe^{II}$ and cyanide. However, there exists some evidence that cyanide may form a loose complex with hemoglobin of the type of oxyhemoglobin. The light sensitivity of carbon monoxide hemoglobin is increased by adding cyanide just as it is by methyl isocyanide (methyl carbylamine) (WARBURG et al. 1278)). The light sensitivity is defined as the ratio between photodissociation and dark dissociation. It is obvious that this ratio may be increased either by increasing the photodissociation or by decreasing the dark dissociation. The latter may be achieved, for instance, by lowering the temperature which will slow down

---

*) In this discussion the recently proposed terminology for heme derivatives by PAULING and MIRSKY l.c. 918a) is employed. It appears preferable to the one used by KEILIN who calls the combinations between oxidized heme (hematin) and bases *parahematins* and only those between reduced heme (heme) and bases *hemochromogens*.

the thermal dark reaction but not the photic reaction. Cyanide and isocyanides have a similar effect; WARBURG concludes that they must combine with hemoglobin and CO, forming a ternary compound in order to be able to affect the rate of dark dissociation which is a characteristic constant for each individual heme complex.

Whether a given heavy metal catalysis is inhibited by cyanide depends on the nature of the metal and on the type of its linkage to other constituents in the catalyst molecule. Predictions cannot be made; cytochrome c for instance, which is a combination of hematins with proteins, will combine with CN' neither in the ferrous nor in the ferric state. That in cyanide poisoned cells, e.g. in bakers yeast, the spectrum of reduced cytochrome is observed even under aerobic conditions is due to the block of cytochrome oxidation by the combination of cyanide with the ferric form of the respiratory ferment which represents a specific cytochrome dehydrogenase (or cytochrome oxidase). The inhibition of cell respiration by cyanide is apparently solely caused by this inhibition of the oxygen transferring enzyme. It is true that the respiration of certain cells, e.g. of the algae Chlorella 123a) is cyanide resistant under certain conditions. It is believed that in such cells respiration does not proceed via the phaeohemin enzyme of WARBURG but via metal-free catalysts of the type of the yellow oxidation enzyme. On the other hand, it is not possible to deny that HCN might act as inhibitor in metal-free catalyses. In the case of chain reactions, for example, HCN may be an active "antioxidant" like phenol or hydroquinone by causing a breaking of the chain (p. 48). WIELAND's theory of cyanide inhibition as due to an unspecific adsorption on the surface of the catalyst is certainly not valid for the range of low cyanide concentrations and for the case of hemin derivatives in general.

Not only the ferric form of the respiratory ferment but also c a t a l a s e and p e r o-x i d a s e both of which contain ferric iron are strongly inhibited by HCN. ZEILE and HELLSTRÖM 1377, 1371) have described the formation of a spectroscopically well defined complex between catalase and HCN. The dissociation constant as derived from inhibition experiments agrees in the order of magnitude with that computed from rough optical measurements. One molecule of the enzyme will combine with one molecule HCN. The assumption of ZEILE that the inhibitor will combine with a group in the enzyme different from that reacting with the substrate appears improbable and unnecessary. The cyanide inhibition of the respiratory enzyme as well as of catalase is fully reversible. It is claimed that the cyanide inhibition of peroxidase is irreversible (1328, 400)); this finding requires critical reinvestigation since KEILIN and MANN 551) have shown that peroxidase is also a ferric compound containing very probably protohematin as the prosthetic group. The combination of protohematin derivatives with HCN has been found to be reversible in all other instances investigated. A number of other enzymes, e.g. uricase, carbonic anhydrase, and succinic dehydrogenase have also been stated to be cyanide sensitive (see however (711)). Table 4, showing the inhibition of enzymes by cyanide and carbon monoxide has been taken from the review article by HAND 444).

Copper catalyses are less cyanide sensitive than iron catalyses, at least as far as model reactions with thiols as substrates are concerned. Polyphenol oxidase from potatoes which contains copper in the prosthetic group and a protein bearer (KUBOWITZ 613)) is inhibited by HCN. Manganese as catalyst in the cysteine oxidation is completely cyanide insensitive. Not all hemin catalyses are inhibited by cyanide: neither the oxidation of unsaturated fatty acids (KUHN and MEYER 653)) nor that of cysteine (WRIGHT and ALSTYNE 1349)) is poisoned by HCN.

N i t r i l s: Though nitrils form no complexes with heavy metals they behave like

TABLE 4.

Affinities of Oxidases and Simple Iron-Porphyrins.

| Specificity (Terminology as by RAPER 957)) | Affinity for HCN | Reference | $CO/O_2$ at 50 per cent inactivation | Reference | Probable Valency of Iron |
|---|---|---|---|---|---|
| Catalase . . . . . | 50 per cent active at $M = 8.10^7$ | (1377) | No affinity for CO | (1299) | $Fe^{III}$ |
| Peroxidase . . . . | 50 per cent active at $M = 5.10^{-6}$ | (1328) | No affinity for CO | (1328) | $Fe^{III}$ |
| Methemoglobin . . | Stable Compound | (1381) | No affinity for CO | | $Fe^{III}$ |
| Cytochrome oxidase. | Poisoned by $M = 10^{-3}$ to $10^{-4}$ | (1217) | 7 to 30 | (1261) | $Fe^{II} \rightleftarrows Fe^{III}$ |
| Indophenol oxidase. | Poisoned by $M = 10^{-3}$ | (538) | 5.6 to 9.8 | (538) | — |
| Dopa oxidase . . . | Poisoned by $M = 2.10^{-3}$ | (957) | — | | — |
| Polyphenol oxidase. | Sensitive to HCN | (538) | 1.5 | (538) | — |
| Laccase . . . . . . | Sensitive to HCN | (443) | — | (443) | — |
| Tyrosinase . . . . | Sensitive to HCN | (443) | — | (443) | — |
| Hematin (reduced) . | 50 per cent bound at $M = 10^{-3}$ | (596) | large affinity | (596) | $Fe^{II}$ |
| Hemoglobin . . . . | No affinity for HCN | (485) | 0.005 | (70) | $Fe^{II}$ |
| Cytochrome c . . . | No affinity for HCN | (486) | No affinity for CO | (486) | $Fe^{II} \rightleftarrows Fe^{III}$ |
| Xanthine oxidase . | No affinity for HCN | (538) | No affinity for CO | (538) | — |
| Succinooxidase . . | Probably no affinity for HCN | (538) | No affinity for CO | (538) | — |

CN' in certain aerobic processes, e.g. in the dehydrogenation of succinic acid (WIELAND and FRAGE 1311), SEN 1042)) and of lactic acid (BARRON 83)). It has been shown in WARBURG's laboratory (TODA 1177)) that certain nitrils, e.g. valeronitril, have a narcotic action which is in agreement with their adsorption constant.

Isonitrils: The ester of hydrocyanic acid (carbylamines) show a selective inhibiting power. In model experiments they inhibit the oxidation of fructose and of cysteine. They have no effect on respiration, fermentation and $CO_2$-assimilation (WARBURG 1220)) and none on catalase (TODA 1177)). They inhibit the PASTEUR reaction in yeast and tumor tissue as evidenced by the increase in aerobic fermentation in their presence. Carbylamine will combine with hemoglobin, thereby increasing the light sensitivity of CO-hemoglobin (WARBURG 1275)), but not with methemoglobin (WARBURG 1220)). It is to be concluded that carbylamine shows a high affinity for ferrous iron and little or none for ferric iron.

Cyanate has no effect on tissue respiration (ROSENTHAL and VOEGTLIN 1005)).

Hydrogen sulfide (or $Na_2S$) inhibits the aerobic dehydrogenation of alcohol (WIELAND 1322)), it inhibits peroxidase (1328)), catalase (1377, 1073)), indophenol oxidase (1295)), and the oxidative hemin catalyses with linseed oil as substrate (598, 875)). The reaction $H_2S + H_2O_2$ is catalyzed by Fe and by hemin but not by peroxidase (1282, 1283)). The cyanide-sensitive uricase is $H_2S$-resistant (546)). The copper containing potato oxidase is inhibited by $H_2S$ (613)).

Sodium azide: This reagent has recently been introduced as a heavy metal inhibitor by KEILIN. It inhibits catalase reversibly (123)), furthermore it inhibits peroxidase, indophenol oxidase, phenolase. Cell respiration is poisoned only below ph 6.7 but no longer at ph 7.5 (KEILIN 543)). It also interferes with alcohol oxidation by yeast (WIELAND 1322)).

Hydroxylamine: The inhibition of catalase by hydroxylamine has been known for a long time (JACOBSOHN). It has recently been confirmed by BLASCHKO 124) who tried to employ it as tool in deciding the question as to the importance of the enzyme for cell respiration. The result was, unfortunately, not clear cut: while tissue catalase is poisoned in every case, the respiration of kidney slices is strongly inhibited by hydroxylamine, that of testis hardly at all. The existence of a combination between catalase and the reagent has also been made probable by spectroscopic experiments (KEILIN and HARTREE 548)).

Pyrophosphate: The action of pyrophosphate as a heavy metal inhibitor varies with the type of metal and of the substrate. In cysteine oxidation pyrophosphate inhibits the action of iron and manganese but not that of copper (WARBURG 1222); ELVEHJEM 251)); in the case of sugar oxidation it is the copper catalysis which is affected (KREBS 593)). According to VOEGTLIN and ROSENTHAL 1201) only the total oxidation of cysteine but not the dehydrogenation to cystine by Fe or Mn is inhibited by pyrophosphate. The inhibition of cell respiration by this reagent, according to LELOIR and DIXON 711), is due to a poisoning of succinic dehydrogenase; in cells where this enzyme is not essential for the oxygen uptake, e.g. in bakers yeast, pyrophosphate has no effect on respiration.

## b) Hemin Catalyses.

The term "hemin catalysis" encompasses all catalyses brought about by derivatives of the porphin skeleton in combination with a heavy metal nucleus. A subdivision may be made depending on whether chemically well defined heme compounds or colloida systems consisting of a prosthetic heme group and a bearer protein are at play. The latter include the blood pigments, e.g. hemoglobin and methemoglobin, as well as the hemin complexes of the cell (respiratory ferment, cytochrome, etc.).

### Action of hemins in model systems:

Inasmuch as hemin (ferriheme) is reduced by a variety of substances to heme (ferroheme) and inasmuch as the latter compound is readily autoxidizable (CREMER 190)), the conditions for catalysis are met whenever the reversible redox system Ferriheme $+ H_2 \rightleftarrows$ Ferroheme is formed; this system transfers hydrogen from suitable donators to molecular oxygen. The autoxidation of ferro heme is inhibited by carbon monoxide. The iron in heme is coordinatively tetravalent. Upon combining with bases like pyridine or with proteins it undergoes a change to coordinatively hexavalent iron. This is not dependent on whether the metal exists in the ferric or in the ferrous state. One molecule of CO is absorbed by one ferro heme complex molecule, i.e. by one iron atom. This holds even for the simplest compound of this type, namely, CO-heme (HILL 484b)). The sixth place in the coordination shell of the iron is probably occupied by $H_2O$ in this instance. The potential of the ferro/ferri heme system is more negative than that of any hemochromogen (CONANT 180); BARRON 77)).

Free hematin has been observed spectroscopically in living cells (KEILIN 538)). KEILIN has speculated on the possible significance of this heme for cell respiration as the precursors of cytochrome (538)). According to the same author (540)) these cell hemins are also responsible for the thermostable peroxidase activity of hemoglobin- and peroxidase-free cells (bacteria, yeast, etc.)

As in any type of heavy metal catalysis the question must be raised as to whether the v a l e n c y  c h a n g e  h y p o t h e s i s is capable of explaining all phenomena observed in hemin catalyses or whether other mechanisms, e.g. that of intermediary peroxide formation, must also be taken into consideration. In order to render the valency change mechanism plausible for any given case it is first necessary to ascertain whether reversible oxidation-reduction of the hemin derivative may occur under the experimental conditions. The action of inhibitors with special affinity for one valency stage only or spectroscopic observations are the most prominent tools in such investigations. Some hemin derivatives like peroxidase and catalase contain iron in a stabilized ferric state (ZEILE and HELL-STRÖM 1377); STERN 1089)). In such cases an attempt has been made to explain the catalysis without the assumption of an intermediary valency change (p. 46). Like catalase (STERN 1085)) other ferriheme derivatives form spectroscopically defined complexes with peroxides while in the ferric state: methemoglobin with hydrogen peroxide (KOBERT 584), HAUROWITZ 467)) and with ethyl hydrogen peroxide (KEILIN 545), STERN 1082)); peroxidase with $H_2O_2$ (KEILIN and MANN 551)); hemin with $H_2O_2$ (EULER 297), HAURO-WITZ 469)). HAUROWITZ assumes that the heme-$H_2O_2$ compound contains an activated $H_2O_2$ molecule in coordinative linkage to the iron atom. In this form the heme may catalyze the kathodic reduction of $H_2O_2$ (146)) as well as the transfer of donator hydrogen to chromogens and the hydrogenation of the bound peroxide by another $H_2O_2$ molecule acting as the H-donor. It is interesting to note that other hemins but no other metal-porphyrins function in an analogous manner. Their magnetic moment appears to have no bearing on this question. According to BERGEL and BOLZ 95) the catalytic oxidation of dialkyl aminoacids by hemin (natural amino acids are not attacked) occurs possibly through p e r o x i d e  f o r m a t i o n. The same possibility exists for the hemin catalysis of the oxidation of unsaturated fatty acid which is cyanide resistant (ROBINSON).

The dehydrogenation of t h i o l s to disulfides by hemin was first observed by HARRISON 462). The oxidation of cysteine by free hemin and by hemin combined with cyclic nitrogeneous bases was subsequently studied by KREBS 596). Calculated for 1 mg iron, free hemin will transfer 8300 mm³ $O_2$ to cysteine per hour. The corresponding figure for pyridine hemochromogen is 92,000 mm³ and for nicotine hemochromogen 232,000. Since traces of $FeSO_4$ were found to exert a strong stimulating effect, it appears likely that ferrous cysteine is more rapidly oxidized catalytically than cysteine alone. The catalysis is inhibited by HCN and CO; in fact, it has been used as a model for the catalysis of cell respiration by the respiratory ferment. The point in question was whether the absorption spectrum of a compound as determined by direct spectroscopy is identical in every respect with the photochemical absorption spectrum as determined by the indirect "CO-illumination" method. At the time at which this test was performed the absorption spectrum of the respiratory ferment had not yet been observed directly in living cells. It was, therefore, necessary to choose a compound available in sufficient amounts and of a constitution and properties as similar as possible to the respiratory ferment. Pyridine hemochromogen is such a compound; the substrate was cysteine. The result was that there exists close agreement between the spectra measured by the two independent methods. The subsequent discovery of the long-wave absorption band of the respiratory ferment in yeast and acetic acid bacteria by direct spectroscopy by WARBURG and his associates at the position predicted by the photochemical measurements (p. 136) is certainly as spectacular and significant in biochemistry as a similar event in the field of general chemistry, namely the discovery of certain chemical elements with properties predicted decades previously by MENDELEJEFF and MEYER on the basis of the periodic system. The long-wave absorption band of the respiratory ferment in a z o t o b a c t e r, according to NEGELEIN and GERISCHER 880), is situated in the r e d region instead of in the y e l l o w as in yeast and acetobacter. Unfortunately no photochemical measurements exist with respect to the respiratory ferment in azotobacter. For a crucial test of WARBURG's theory and for a decision of the controversy concerning the direct visibility of the respiratory ferment (p. 148) it appears highly desirable to secure information on the photochemical absorption spectrum of the ferment in azotobacter.

Other substrates employed for the study of the oxidative efficiency of hemin and hemochromogens are ascorbic acid (BARRON et al. 79)) and pyruvic acid (MEYER 783)). Both catalyses are inhibited by HCN. The catalytic oxidation of ascorbic acid yields dehydroascorbic acid, and that of pyruvic acid leads to oxalic acid. In a recent investigation of the oxidation of glutathione by hemin derivatives, LYMAN and BARRON 741a) believe to have observed complex formation between heme and the SS-form of the tripeptide. From the description of the procedure employed for the preparation of this complex, however, it is to be inferred that the green compound formed is not a complex of the type postulated by these workers but a v e r d o h e m o c h r o m o g e n, i.e. the iron complex salt of biliverdin. This compound has been obtained by several earlier workers and has previously been considered to be a "green" (i.e. a chlorophyll-type) hemin. Actually, it is protohemin where one methine bridge between the pyrrol nuclei has been oxidized, thus causing a rupture of the porphin skeleton and the formation of an open-chain bile pigment structure, held together in the original porphyrin form by the complex binding forces of the central iron atom (LEMBERG 711a)).

Heme, besides being an oxidative catalyst, is also peroxidatically and catalatically active. The first systematic investigations of these functions of heme and of its derivatives were carried out by KUHN, BRANN and MEYER 142, 638, 639, 653). They detected the catalatic acticity of these substances. Their observations were extended by ZEILE 1370), EULER 263), HAUROWITZ 466), LANGENBECK et al. 702) and STERN 1075, 1076a). All hemins containing iron are active towards $H_2O_2$; the hemin esters and copper-porphyrins are inactive 1370). Protohemin, deuterohemin, mesohemin, pyratin-SCHUMM were found to be active (263)). The ferrihemochromogens and ferrohemochromogens have an activity of the same order of magnitude, viz. $10^{-1}$ molecules of $H_2O_2$ are destroyed by one molecule of the catalyst per second at 0° (635, 702, 1075, 1076a)). The rather low activity of these simple heme derivatives as compared with the enzyme catalase ($10^5$ molecules $H_2O_2$ per catalase molecule per second) is due to the greater stability and consequent slow decomposition of the intermediary hemin-$H_2O_2$ complex (EULER and JOSEPHSON 297); HAUROWITZ 469)). Imidazolhemochromogens, e.g. histamin hemochromogen, are among the most active of these synthetic hemin catalases (LANGENBECK 702); STERN 1076a)). HCN inhibits only in comparatively high concentrations and at ph values above 8 (263)). Each catalyst appears to have its individual ph-optimum.

Similar observations have been made with respect to the peroxidatic activity of heme and of its derivatives (638, 702, 979)). It is of interest to note that the coupling of heme with the same base may activate the three catalytic functions (peroxidatic, catalatic, oxidatic) in a different manner. Chlorophyll hemins (green hemins) are less active than red (blood) hemins. Some amino acids increase the peroxidatic activity, tryptophane decreases it throughout (979)).

The catalytic activity of heme and of its derivatives is greatly affected by adsorption on various bearers. With charcoal, for instance, the catalatic activity is 200 per cent increased while the oxidatic efficiency is diminished; with metastannic acid both functions are little changed, while both are weakened by alumina (KUHN). Catalysts prepared by adsorbing pyridine- or nicotine hemochromogen on starch or charcoal were found to be highly active oxidative catalysts with cysteine as the substrate (Brit. Pat. E. P. 304731); with respect to the decomposition of hydrogen peroxide, the activity of adsorbed hemochromogens (nicotine, pyridine, histamine) is of the same order as that of the same compounds in solution (STERN 1076a)).

Among the most interesting oxidative functions of ferrihemes is their ability to oxidize hemoglobin to methemoglobin (WARBURG and KUBOWITZ 1262)) and that of certain hemins, notably of chlorophyll hemins and of phaeohemins, to oxidize CO to $CO_2$ (NEGELEIN 876)). In the first case catalytically active methemoglobin is formed on the surface of red blood cells which in turn may burn hexosemonophosphate in presence of the corresponding dehydrogenase system; the second fact explains why in certain manometric experiments, devised to test the inhibitory action of CO on certain hemin catalyses, no inhibition was noted but the gas disappeared instead.

### Catalyses by Hemoglobin and Related Compounds.

Hemoglobin itself is not an oxidative catalyst as is to be expected from a substance developed for oxygen transport to the tissues; oxyhemoglobin is a loose and fully reversible addition compound containing ferrous iron. On the other hand, hemoglobin as well as oxyhemoglobin exhibit, in vitro, a weak catalatic and peroxidatic activity. The latter is somewhat stronger but of the same order as that shown by hemin alone and synthetic hemochromogens; the ph-optimum is at 4.6 (638)) (51)). The catalatic function has been studied, among other workers, by HAUROWITZ 446). A number of qualitative tests for the blood pigment, e.g. the guaiac and the benzidine reaction, are based on these "incidental" properties of hemoglobin. Methemoglobin, the true oxidation product of hemoglobin with ferric iron, is an oxidation catalyst under certain conditions. Though methemoglobin is readily reduced to hemoglobin by various biological systems, e.g. cysteine, glutathione, lactic acid dehydrogenase + lactate, the ferrous iron of the latter is so well protected by the globin component against oxidation that molecular oxygen will ordinarily oxygenate it to oxyhemoglobin and not reoxidize it to methemoglobin. In general, only oxidizing agents of the type of ferricyanide or methylene blue will convert hemoglobin into methemoglobin. By the use of such agents WARBURG and his associates (1262, 1263, 1264)) have succeeded in producing an oxygen transfer with red blood corpuscles via the redox system hemoglobin $\rightleftharpoons$ methemoglobin. Substrates oxidized by this artificially produced system are glucose, lactic acid, hexosemonophosphate, provided that the necessary "activating" enzymes (dehydrogenases) are present. It is somewhat perplexing that the behaviour of such blood cells containing methemoglobin toward oxygen will depend on the nature of the reagent employed for methemoglobin production. Thus, methemoglobin produced by amyl nitrite is catalytically inactive: The hemoglobin formed by the reducing substrate-dehydrogenase system will react with $O_2$ to yield oxyhemoglobin and not methemoglobin as would be required to complete the catalytic cycle. The same result was obtained independently by WENDEL 1292). The reaction comes to a standstill after pyruvate has been formed at the expense of lactate and hemoglobin at the expense of methemoglobin. Phenylhydroxylamine, on the other hand, gives rise to a true catalysis with the redox system hemoglobin $\rightleftharpoons$ methemoglobin, but only if the reduction of the methemoglobin formed by the reagent is effected immediately by sugar in air. Under these conditions the hemoglobin will react in statu nascendi with molecular oxygen to give methemoglobin. If the reduction of the methemoglobin is carried out in an indifferent gas, argon for instance, and if air is subsequently admitted, oxyhemoglobin is produced instead of methemoglobin and no catalysis takes place, just as in the case of amyl nitrite. This shows that the system

$$DH_2 + Methb \rightarrow D + Hb \xrightarrow{\;+\,O_2\;} D + Methb + H_2O_2$$

can only be set up under exceptional circumstances, namely, if the hemoglobin reacts in statu nascendi (see also p. 242). It is however possible to bring about an interesting type of methemoglobin catalysis by introducing a second redox system, e.g. a quinoid dye or hemin. The experiments by WARBURG and his associates dealing with this phenomenon took their origin from the observation of BARRON and HARROP 82) that methylene blue when added to enucleated red blood cells causes the appearance of an artificial respiration. According to WARBURG 1263, 1287) this is not due to an acceptor respiration via methylene blue where the metabolites are dehydrogenated with the aid of dehydrogenases + the yellow enzyme and where the leuco methylene blue reacts directly with molecular oxygen. It is stated that a chain of two redox systems is at work here: First, the methylene blue oxidizes the hemoglobin to methemoglobin; then, the latter oxidizes the metabolites. The hemoglobin thus formed is reoxidized to methemoglobin by the dye *). The catalysis consists, therefore, of an anaerobic (a) and of an aerobic (b) phase:

(a) . . . . $DH_2 + Methb \rightarrow D + Hb$;  $Hb + MB \rightarrow Methb + LMB$

(b) . . . . $LMB + O_2 \rightarrow MB + H_2O_2$

(MB = Methylene blue, LMB = Leuco methylene blue).

---

*) SCHÜLER 1037), in WARBURG's laboratory, finds that the protein component, globin, is oxidized by ferricyanide and therefore probably also by methylene blue

It is not impossible that the experiments by EULER 261) with liver extracts are to be interpreted similarly. U. S. v. EULER 326) suspects that the increase in respiration of muscle provoked by dinitro-α-naphthol is partly due to interaction with muscle hemoglobin (myoglobin) (p. 112).

MICHAELIS and SALOMON 819) have employed a series of redox indicators in an attempt to correlate the catalytic action of the dyes with their redox potential. They found that the stimulation of respiration of erythrocytes increased with the potential of the dyes up to a certain limit where no further increase was observed. They ascertained that the rate of methemoglobin formation was likewise increasing with a rise in potential. It is generally conceded, however, even by WARBURG (1238, p. 225), that it is more or less a matter of concentration ratios to which extent the catalysis will proceed via the dye alone or via the dye plus methemoglobin. This state of things also provides an explanation for the observation of WENDEL 1292) that the catalysis is cyanide-resistant: the pure acceptor catalysis with dyes is not affected by CN'.

It is interesting that a similar coupling of two redox systems for promoting oxygen transfer may be effected by two cell constituents. WARBURG 1262) had found that hemin when added to red blood cells will form methemoglobin. Later he showed that phenylhydrazine destroys hemoglobin by transforming it into free hemin and denatured globin 1263). The hemin oxidizes fresh hemoglobin to methemoglobin and is thereby reduced to ferroheme. The methemoglobin, in turn, is reduced back to hemoglobin by the metabolite-dehydrogenase system. Again we have an anaerobic phase (a) and an aerobic stage (b):

$$(a) \ldots\ldots DH_2 + Methb \rightarrow D + Hb; \ Hb + Hemin \rightarrow Methb + Heme$$

$$(b) \ldots\ldots Heme + O_2 \rightarrow Hemin + H_2O_2.$$

This combination of a non-autoxidizable and an autoxidizable redox system, both of which contain porphyrin-bound iron, represents a model for the catalytic chain of cell respiration. Cytochrome is the non-autoxidizable and the respiratory ferment is the autoxidizable catalyst.

It should be mentioned here that this effect of phenylhydrazine occurs only if blood cells are treated with this substance in vitro. It is known that phenyl hydrazine poisoning in vivo also produces strongly respiring erythrocytes (MORAWITZ cells). Their respiration, however, is not caused by the coupling of methemoglobin with free hemin but by the ordinary respiratory ferment. The MORAWITZ cells are young blood cells, i.e. they respire due to the presence of nuclear material.

### Heavy Metal-Containing Intermediary Catalysts (Mesocatalysts).

A detailed definition of a 'mesocatalyst' system will be given later (p. 87). Here it may suffice to point out that they represent catalysts which can perform their particular function only in conjunction with other enzymatic or non-enzymatic catalysts. The only system of this kind containing heavy metal which we know is cytochrome. Of the three cytochrome components, a, b and c, only the latter has been isolated in pure form and has been found to be non-autoxidizable in the physiological ph-range. According to KEILIN 540) and to ROCHE 990) cytochrome a is autoxidizable or may be readily transformed into derivatives capable of reaction with $O_2$. For the c component, however, the necessity for providing a specific oxidizing system contained in the chain of cell respiratory catalysts is generally accepted. This oxidizing agent has been identified with the well-known indophenol-oxidase by KEILIN and with the respiratory ferment by WARBURG. Now both authors take the view that these two oxidases are identical *).

In model experiments with cytochrome c the necessary "cytochrome-oxidase" must be provided in the form of washed heart muscle (preferably from sheep) in order to make the reoxidation of this carrier possible. Chains of the type: Oxidase + Cytochrome c + cysteine (KEILIN 540)) and Oxidase + Cytochrome c + Glucose-glucose dehydrogenase (HARRISON 458)) have been successfully constructed in vitro to operate with molecular oxygen. Other reducing systems which could be substituted for glucose-glucose dehydrogenase are lactate-lactic dehydrogenase and succinic acid-succinic dehydrogenase (OGSTON and GREEN 895)).

It is very probable that in actual cell respiration a number of iron porphyrin systems is linked up in series. WARBURG has discussed chains consisting of the respiratory ferment and the three cytochrome components and even a chain made up of two respiratory ferments and three cytochrome components 1234).

Cytochrome c is well adapted to a carrier function: It represents a readily reversible redox system with a normal potential intermediary between that of oxygen (and the respiratory ferment) on the one hand and that of the substrate-dehydrogenase systems on the other hand (see p. 160). In addition it has a relatively small molecule (M. Wt. 16.500) considering its constitution as a heme derivative combined with a protein bearer; it is therefore more readily diffusible than hemoglobin (M. Wt. 68000).

## 2) Metal-Free Catalysis (Acceptor Catalysis).

### a) General Considerations.

It is safer to substitute the name "acceptor catalysis" for the more common term "metal-free catalysis". While it is established that metal catalysts of known character and especially hemin systems play no active part in these processes, the possibility that other metal catalysts in subanalytical traces may participate can not be excluded as yet.

The acceptor catalyses are primarily anaerobic reactions, i.e. they proceed without molecular $O_2$. However, in the event that an acceptor is autoxidizable in its reduced form, they may proceed to the ultimate stage of oxidation by oxygen and thus pass over into an aerobic phase. If the reduced form of the last acceptor is non-autoxidizable the coupling link with oxygen may be provided by a suitable heavy metal system. This is the general pattern of cell respiration. Here the ultimate oxidation is effected by the respiratory phaeohemin ferment of WARBURG.

Even if the reduced form of the acceptor should be somewhat autoxidizable this faculty may remain unexploited in the normally respiring cell. WARBURG has pointed out that in the competition for oxygen affinity becomes the decisive factor: the rate of reaction of the reducing systems with the oxygen transferring enzyme is much greater than that with molecular oxygen. This explains why autoxidizable substances like the leuco forms of the yellow enzyme or of pyocyanine, within the living cell, are reoxidized not by $O_2$ but by the ferric iron of the respiratory enzyme.

---

*) It is possible, that these oxidase is nothing else than the autoxidisable cytochrome a itself, and the terminal catalysis acts in the chain: cyt. b $\rightarrow$ c $\rightarrow$ a $\rightarrow$ $O_2$; v. OPPENHEIMER, Suppl. p. 1664.

In principle, regeneration of the oxidized form of an acceptor, e.g. quinone, may be accomplished by any of the following reactions: true autoxidation, reaction with hydrogen peroxide, oxidation by a higher valency form of a heavy metal, reaction with an oxidase, peroxidatic reaction with $H_2O_2$ + iron or $H_2O_2$ + peroxidase.

It is obvious that throughout the field of anaerobic acceptor catalysis the type of reversible reaction occuring is that of WIELAND's dehydrogenation-hydrogenation, i.e. of hydrogen transfer. Whereas this transfer has been usually conceived as a simultaneous shift of pairs of hydrogen atoms or electrons, recent work makes it appear likely that the transfer occurs in two steps comprising o n e H-atom or electron at a time. Intermediary radicals of the semiquinone type have been observed not only in the case of strictly quinoid pigments, e.g. pyocyanine, but also in the case of the yellow enzyme, of cozymase and of codehydrase.

### Autoxidizable and Non-Autoxidizable Systems.

Autoxidizable acceptors or carriers will reduce oxygen to the hydrogen peroxide stage. When they are active within the cell or when they are added to living cells, an "iron-free respiration" results which is an unphysiological process even if no carrier has been added. It is unphysiological because the hydrogen peroxide formed by the reaction leads to the death of the cell. Facultative anaerobic lactic acid bacteria lack hemin components; they contain a considerable amount of yellow enzyme (WARBURG and CHRISTIAN **1244)**). When brought into air they will change from their ordinary type of metabolism, i.e. from lactic acid fermentation, to respiration via the yellow enzyme. They possess no catalase or other iron systems to destroy the $H_2O_2$ formed, hence they perish when a sufficient amount of the peroxide has accumulated. An example of iron-free respiration through an added carrier system is the so-called methylene blue respiration ( MEYER-HOF, BARRON, WARBURG). This type of respiration is not inhibited by CN′ or CO. The cyanide-resistant "residual respiration" of living cells, i.e. the fraction remaining after poisoning of the "main respiration", is very probably such an acceptor respiration, established with the aid of the yellow enzyme or similar carriers. As will be shown later (p. 260) the extent of this residual respiration varies from that of a quantitatively negligeable fraction (as in yeast for instance) to that of a quite appreciable fraction of the total respiration (as in retina for example). Whether it plays any important role in the cell under normal conditions is a matter of conjecture. It may be an "emergency by-path" or it may be a quantitatively insignificant but qualitatively important constituent of the energy-yielding mechanism of living matter.

Let us consider a simple case of acceptor respiration. The system lactic dehydrogenase-lactate reacts with methylene blue to give leuco methylene blue and pyruvate. The process is cyanide-resistant (STEPHENSON **471)**). If this reaction is carried out in an evacuated THUNBERG tube it will stop after all of the acceptor has been converted into the reduced form, provided that an excess of lactate has been used. But if air is admitted the leuco dye will undergo reoxidation by $O_2$ and the reaction will continue. If, on the other hand, no dyestuff is added and air is present from the start no lactate dehydrogenation will occur; lactic dehydrogenase is an "anoxytropic" dehydrogenase, i.e. incapable of utilizing $O_2$ as an acceptor for the hydrogen of the substrate. Under aerobic conditions a very small amount of methylene blue will suffice to bring the reaction to completion by virtue of its cyclic reduction and reoxidation. The dye acts as a c a r r i e r or i n t e r m e d i a r y c a t a l y s t in this set-up, whereas in the THUNBERG tube it merely represents the second substrate or the acceptor. In the living cell, the dehydrogenase-substrate system is linked up in the chain terminating in a iron containing (hemin) system. Lactate oxidation by living cells is, therefore, cyanide-sensitive.

Carrier systems may be classified as e n z y m i c and n o n - e n z y m i c, depending on whether these low-molecular reversible oxidizable groups are linked or not linked to colloidal bearers (proteins). A u t o x i d i z a b l e carriers are, for instance, the yellow enzyme, the "respiratory pigments" of PALLADIN (quinoid plant constituents), oxytropic dehydrogenases, e.g. aldehydrase (?), xanthine oxidase, amino acid dehydrogenases, the glucose oxidase of MÜLLER, pyocyanine, chlororaphin, epinephrine, quinone etc. N o n - a u t o x i d i z a b l e carriers are thiols, particularly glutathione, ascorbic acid, codehydrogenases (cozymases), diaphorase, most of the dehydrogenases, intermediary catalysts derived by desmolysis of food stuffs, e.g. succinic acid, malic acid.

### Enzymes and Intermediary Catalysts (Mesocatalysts).

It becomes almost a necessity to recognize in principle that every reversible redox system found in the living cell represents a catalyst for biological oxidation-reduction reactions. Most of these systems have probably a very restricted catalytic function. They cannot act independently but are active o n l y  i n  c o n j u n c t i o n with the colloidal biocatalysts, i.e. with the enzymes. More specifically, they can only transfer hydrogen which has been labilized ("a c t i v a t e d") by specific enzymes (dehydrogenases). The transfer proper seems to occur on a surface on which both the acceptor and the donator are adsorbed; the surface is provided by the protein bearer of the dehydrogenase. The function of the surface is probably that to effect orientation of the molecules and to lower the activation energy, thereby facilitating the reaction.

The greatest advance in our knowledge of anerobic desmolysis has sprung from the recent work of WARBURG and of EULER and their associates. They have shown that the so-called c o-e n z y m e s are in reality not activators but the actual prosthetic groups which combine with the proteins, formerly thought to be the enzymes, to form the catalytically active complex. The chemistry of these biocatalysts will be dealt with in chapter D (p. 212). Certain coenzymes have the structure of phospho-pyridine-nucleotides. Two different co-enzymes (codehydrase I (cozymase) and codehydrase II), differing in the number of the phosphoric acid radicals, have so far been described. The most important feature of these discoveries is that these co-enzymes may combine with different protein bearers to form enzymes of different specificity. Triose phosphate dehydrogenase, alcohol dehydrogenase, lactic dehydrogenase, hexose monophosphate dehydrogenase are enzymes built according to this schema. Carboxylase is likewise composed of a protein component (hitherto called carboxylase) and a co-enzyme, co-carboxylase. The latter is identical with vitamin-$B_1$-pyrophosphate (LOHMANN) and has actually been synthesized from vitamin $B_1$ in vitro (STERN, EULER). The same compound may be the prosthetic group of pyruvic dehydrogenase (LIPMANN) (p. 205).

The link between the co-enzyme and the protein bearer is dissociable; dilution will establish an equilibrium between free and bound co-enzyme. One protein molecule may thus serve as the active surface for a number of co-enzyme molecules (NEGELEIN et al. 883)). The co-enzymes of these dehydrogenases are reversible oxidation-reduction systems. In the course of the catalysis the pyridine moiety undergoes a cyclic hydrogenation — dehydrogenation with an exchange of two hydrogen atoms. The oxidized form represents a quarternary pyridinium base, while the reduced form is a substituted dihydropyridine derivative (WARBURG, KARRER).

The sequence of events in the dehydrogenation of hexose monophosphate (ROBISON-ester) in vitro has been elucidated by WARBURG and his colleagues (l.c. 1235). It may be given here as an example from which generalizations may be made with respect to similar processes. The hexose monophosphate dehydrogenase consists of a specific protein bearer and of codehydrase II, a triphosphopyridine nucleotide. In the absence of the protein, no reaction between the substrate and the co-enzyme will take place. They will react in stoichiometric proportions, yielding phosphohexonic acid and dihydropyridine nucleotide, only upon adding the protein bearer of the dehydrogenase. The dihydropyridine nucleotide thus formed is non-autoxidizable. It is specifically dehydrogenated by the yellow oxidation enzyme the presence of which is therefore required for the catalysis. The prosthetic group of the latter, riboflavin phosphoric acid, is in turn reduced. The leuco form of this enzyme is autoxidizable. In vivo, however, its reoxidation is not effected by molecular oxygen but by a hemin system or by an intermediary catalyst of the type of oxaloacetic acid (SZENT-GYÖRGYI).*) It has recently been shown that if the yellow enzyme is reduced by hydrosulfite in presence of an excess of co-enzyme (cozymase) at neutral reaction, a red intermediate is formed showing the absorption spectrum of the cationic semiquinoid form of the flavin enzyme (HAAS **430b**)). The interesting point is that the free semiquinone, i.e. the monohydroform, of the yellow enzyme in neutral solution is green and not red (MICHAELIS et al. 822)). The red form observed by HAAS must be regarded either as the semiquinone of a yellow enzyme-codehydrase complex or as the cationic radical of the free yellow enzyme which is somehow stabilized in the neutral medium by the influence of the co-enzyme. Inasmuch as a semiquinoid stage has also been observed (KARRER, HELLSTRÖM, p. 216)) in the reduction of cozymase, a diphosphopyridine nucleotide, it is probable that the hydrogenation-dehydrogenation reactions just described proceed in steps involving single H-atoms or electrons rather than pairs of electrons at a time.

The chain indicated above is also of significance for general enzyme chemistry: One e n z y m e, the hexose phosphate dehydrogenase, acts as the s u b s t r a t e for another enzyme, the yellow ferment. The latter functions here as a d i h y d r o p y r i d i n e  n u c l e o-t i d e  d e h y d r o g e n a s e. The specific oxidation of reduced cytochrome c by the respiratory ferment represents an analogous case. In order to fulfill these functions the oxidizing enzymes must not only have a more positive redox potential than the enzyme system which they dehydrogenate, but, in addition, they must exhibit a chemical affinity for the latter. It is this affinity which ensures the proper interaction of the links in the chain of respiratory catalysts.

### Effect of Metal Reagents on Acceptor Catalyses.

While it is obvious that none of the better known metal complexes, particularly hemins, participate in acceptor catalyses, the absence of any metals in such processes still remains to be proven. It is not known, for instance, whether pigments of the type of methylene blue or pyocyanine may completely dispense with metal traces for the process of their reoxidation or whether they behave like thiols and ascorbic acid. In the case of methylene blue the oxidation of the leuco base in acid solution is a copper catalysis while in neutral or alkaline solution it is catalyzed by "surfaces", e.g. proteins (REID 972)). It is claimed that pyocyanine does not require the aid of a metal. WIELAND himself (1302, p. 13)) suspects, however, that at least subanalytical traces of metals participate also in the so-called "metal-free" reactions.

The only criterion at our disposal when dealing with complex biological systems is

---

*) The reoxidation of dihydropyridine in vivo is probably largely accomplished by the chemically related enzyme d i a p h o r a s e (see p. 226).

still the effect of heavy metal reagents, especially of cyanide and carbon monoxide. While it is already very difficult to remove completely heavy metals even from simple systems, e.g. from methylene blue or ascorbic acid, this is practically impossible in the case of biological preparations, e.g. tissue extracts. Specific inhibition remains, therefore, the only tool; it is unfortunate that this criterion is far from being unambiguous. All that we know about it is that CN' will react with the respiratory ferment, some hemins and most ferric complexes. Other catalysts, even if they contain Fe, are not affected by this inhibitor, copper systems only a little and manganese complexes not at all. A similar situation exists with regard to CO; $H_2S$ and pyrophosphate show again another behaviour. If, therefore, no inhibition is produced by these reagents in biological systems, we are only entitled to conclude that they contain no enzymatic hemin in active state. But they may contain any number of cyanide-resistant heavy metal catalysts. If, on the other hand, an inhibition of an anaerobic process is caused by such reagents it may be due to reaction with a heavy metal, or else it may be due to an entirely different mechanism not involving metal at all. It will be remembered that WIELAND contended that CN' inhibition does not involve iron but the dehydrogenase systems. Though it appears that he has withdrawn in its generality the original assumption that CN' poisons catalase and that the $H_2O_2$ thus conserved harms the dehydrogenases, WIELAND seems to cling to the idea that the mechanism of CN'-inhibition is analogeous to that by quinone or similar agents: It is assumed that HCN has a marked affinity for the catalyst, that it occupies the active centers on the surface and thereby prevents the access of the oxygen just as that of other acceptors. The observation that the "quinone fermentation" (i.e. the fermentation with quinone as acceptor) of acetic acid bacteria is inhibited only by comparatively high concentrations of CN' or CO is explained by the greater affinity of quinone for the enzymatic surface as compared with that of the poisons (BERTHO 106), p. 204). This concept may be found in all of the work of WIELAND concerning itself both with anaerobic (cf. WIELAND and CLAREN 1308)) and with aerobic processes, e.g. with the dehydrogenation of succinic acid by muscle where the dehydrogenase is said to be even more sensitive towards CN' than catalase (WIELAND and LAWSON 1318)). If there are present simultaneously acceptors, oxygen and cyanide, very complex relationships will result (see FRANKE 288), p. 230)). For the majority of aerobic dehydrogenations WIELAND's point of view appears no longer tenable: Inasmuch as cells endowed with a typical respiration contain the respiratory pheohemin ferment, the reaction of which with CN' has been established beyond reasonable doubt, the cyanide inhibition is due here to the ultimate autoxidizable hemin system. On the other hand, we know that a very considerable fraction of the anaerobic model reactions as well as of pure dehydrogenase processes and of the dehydrogenations effected by oxygen in cells devoid of hemin components (e.g. the facultative anaerobic lactic acid bacteria) are cyanide-resistant (see for example LELOIR and DIXON 711)) while methylene blue and quinone in excess may act as competitive inhibitors though, at lower concentration, they represent acceptors. There remain, then, the comparatively few indications of cyanide-sensitivity in anaerobic acceptor catalyses, considering anoxybiosis as a whole, and a few other observations concerning the inhibition of some dehydrogenases by larger CN' concentrations. Recently, OGSTON and GREEN 895) found that a number of dehydrogenases are sensitive to $6.10^{-3}$ M. KCN; among them are enzymes which do not react with cytochrome (glucose and alcohol dehydrogenase). But even the well established cyanide inhibitions are not indicative of the participation of iron since the type of reaction between enzyme and inhibitor appears to be different.

It is true that the SCHARDINGER enzyme is directly poisoned by $5.10^{-3}$ M. KCN while CO has no effect (DIXON 208)), LELOIR and DIXON 711)). BIGWOOD 116) also finds a reversible inhibition of xanthine dehydrogenase by 0.01 M. CN' which, even anaerobically, is said to be detectable only in the presence of traces of oxygen. More recent observations by DIXON and KEILIN 215) suggest that xanthine dehydrogenase (or aldehydrase) represents a special case: when in purified state and when no substrate is present the enzyme is slowly and irreversibly poisoned by HCN; neither oxygen nor methylene blue may function as acceptors afterwards. In presence of purines however, the enzyme is protected against the poison by the substrate; CO and $H_2S$ have no effect under either condition. It is possible that the cyanide effect, in this case, involves a specific reaction with the active group, perhaps of the type of cyanhydrin formation. In this respect, the SCHARDINGER enzyme seems to occupy a unique position among the dehydrogenases, just as it is claimed that pyrophosphate will inhibit selectively succinic dehydrogenase 711). The claim of QUASTEL 946) that malonate is an equally specific inhibitor for this enzyme which has acquired a special significance in view of SZENT-GYÖRGYI's recent theory of tissue respiration (p. 268) is being questioned by WEIL-MALHERBE 1287). Alcohol dehydrogenase may be inhibited by heavy metal and subsequently be reactivated by SH-glutathione (EULER and ADLER 274)). Each dehydrogenase, therefore, appears to behave differently.

At present we have no definite information whether some or all dehydrogenases contain heavy metal. The situation becomes even more involved due to the fact that the same substrate may be attacked by different catalytic systems which may or may not contain a metal constituent. We have already mentioned above the example of alcohol dehydrogenation which, in vitro, may be brought about by a metal-containing (copper proteid- phosphopyridine proteid) or by a metal free (alloxazine proteid-phosphopyridine proteid) system. Only the former is CN'- and CO-sensitive. COOK, HALDANE and MAPSON 181) believe that formic dehydrogenase of B. coli contains copper because it is inhibited by reagents forming stable complexes with Cu (e.g. amino naphthol sulfonic acid). This, of course, is only circumstantial evidence, while the proof offered by KUBOWITZ 613) that the potato polyphenol oxidase contains copper (see p. 181) is unambiguous: the activity is strictly paralleled by the Cu content of the enzyme preparations throughout the purification procedures.

### Affinity and Rate of Reaction.

The existence of a great number of donators and acceptors which may conceivably take part in acceptor catalyses leads to the question as to the principle which governs the sequence of these reactions in biological systems. It is less difficult to understand the terminal position occupied in cell respiration by WARBURG's respiratory hemin ferment. The ferro form is readily autoxidizable and the ferric form exhibits a special reactivity towards reduced cytochrome. One molecule of the $Fe^{II}$-form may "activate" $10^5$ molecules of $O_2$ per second while the $F^{III}$-form will ordinarily react with only $10^2$ substrate molecules (i.e. ferrocytochrome) per second (KUHN 626)). The latter process is, therefore, in vivo the limiting factor of cell respiration. There can be but little doubt that the respiratory ferment represents the most positive redox system in the chain of respiratory catalysts. Beginning with oxygen the chain goes as follows:

$$O_2 \rightarrow \underbrace{Fe^{\cdot\cdot} \rightleftarrows Fe^{\cdots}}_{\substack{\text{Respiratory} \\ \text{Ferment}}} \rightarrow \underbrace{Fe^{\cdot\cdot} \rightleftarrows Fe^{\cdots}}_{\substack{\text{Cytochrome} \\ (1)}} \rightarrow \underbrace{Fe^{\cdot\cdot} \rightleftarrows Fe^{\cdots}}_{\substack{\text{Cytochrome} \\ (2)}} \rightarrow \underbrace{Fe^{\cdot\cdot} \rightleftarrows Fe^{\cdots}}_{\substack{\text{Cytochrome} \\ (3)}}$$

The cytochrome components are arranged in series and not parallel to each other (HAAS **430a)**). The order in which the a, b, and c component are lined up is not yet sure. *)

The last cytochrome component establishes contact with the series of "metal-free" catalysts, i.e. acceptor-enzyme-donator systems, which follow. Among these the number of possible combinations is great. We are still essentially unaware of the principles which cause the donator-dehydrogenase systems to find the right acceptors. In this connection it is important that certain substances which ordinarily are chemically inert show a high reactivity in statu nascendi, in other words, if they occur with higher than average energy content. In that event they become links in reaction chains propagated by energy-rich particles (p. 44).

It is a necessary prerequisite for an oxido-reductive catalysis that the dehydrogenating acceptor must have a higher potential than the donator under the conditions of the experiment. However, this is not sufficient inasmuch as the phenomenon of catalysis is linked up with reaction rates and not with affinity as expressed by redox potentials. Now it is important to realize that, with respect to the potential, the normal potential as calculated for ph 7 ($E'_o$) cannot be always the decisive factor for the actual course of step-wise reactions. This normal potential is calculated for the special case of equal concentrations of the oxidized and the reduced form and also for further "normal conditions", e.g. for a concentration of one mole per liter. These conditions are probably never fulfilled in biological systems. The normal potential represents only one point on the titration curve of a redox system; depending on the electron number, mixtures of the two forms may show potentials differing as much as $\pm$ 30 to 60 millivolts from the value $E'_o$, depending on the ratio Ox/Red. The normal potentials of the very important equilibria Lactic acid $\rightleftarrows$ Pyruvic acid and also Acetaldehyde $\rightleftarrows$ Alcohol, as catalyzed by the corresponding enzymes, are situated close to —200 millivolt at ph 7 (p. 54). If it were only a matter of the value of the normal potential it would be difficult to visualize an equilibrium system of a normal potential sufficiently negative to hydrogenate pyruvic acid or acetaldehyde. However, the occurrence of these reactions in biological systems may be understood if the relative concentrations of the various forms are taken into consideration (cf. FRANKE **355)** and GERSHINOWITZ **398)**). If pyruvic acid, the oxidized form of the pyruvate-lactate equilibrium, is present in excess relative to lactate and if the equilibrium in the donator system is shifted towards the reduced form, a hydrogen transfer from the latter to pyruvic acid should be possible even if the normal potentials of the two redox systems are equal and, indeed, even if $E'_o$ of the donator system happens to be slightly higher than $E'_o$ of the pyruvate-lactate system.

There seems to exist a law in biological desmolysis preventing a hydrogen transfer to the terminal hemin systems in abrupt jumps bridging big potential differences. On the contrary, the hydrogen is passed along over rather small potential increments to increasingly positive systems. Such a concept is held, among other workers, by SZENT-GYÖRGYI **1122)**. He assumes that terminal oxidation in vigorously respiring cells begins mainly with the succinic-fumaric acid system which itself has already a rather high potential (close to 0 millivolt at ph 7). Now there are many systems of a much more negative potential present in the cell and the question is, why is there no i m m e d i a t e reaction between the components of extreme potential, and therefore of greatest difference in level of free energy, omitting the intermediate systems. Since thermodynamics cannot yield the required information we must look to kinetics for help. MICHAELIS

*) See, however, BALL (58)) and OPPENHEIMER'S „SUPPLEMENT", p. 1656: the ordre seems to be B $\longrightarrow$ C $\longrightarrow$ A, cf. p. 85.

(804, p. 146) has pointed out that the efficiency of a redox catalyst is optimal only if it is rapidly reduced by the donator and rapidly oxidized by the acceptor. This condition is fulfilled only in a rather limited region of oxidation-reduction potential where the catalyst exists in a suitable ratio of the oxidized to the reduced form; otherwise one of the two reactions, hydrogenation or dehydrogenation, will suffer. The rate of the over-all process is of course determined by that of the slowest reaction step. MICHAELIS finds, for instance, that only those dyes will efficiently replace methylene blue in acceptor respiration which have a similar $E'_o$ (methylene blue, $+ 11$ mv., pyocyanine, $—34$ mv., hallachrome, $+ 22$ mv.). Indicators with potentials more negative than that of indigodisulfonate and positive indicators of the type of the indophenols are inactive.

It is ordinarily assumed in thermodynamics that there exists no fundamental relationship between the velocity and the free energy of a reaction. However, there is some experimental evidence available in favor of a connection between oxidation-reduction potential of a system, in other words of its free energy, and its rate of reaction with stronger oxidants or reductants. Thus, CONANT 176b) in his studies of the "apparent reduction potential" of certain irreversible systems (nitro compounds, azo dyes, unsaturated 1,4-diketones) observed that their reduction by equimolecular mixtures of the reduced and oxidized forms of a reversible dye was a function of the potential of the latter. Similarly, LA MER and TEMPLE 779a) find that the rate of catalytic oxidation of hydroquinone by manganous salts is a linear function of the available free energy of the system. VOEGTLIN, JOHNSON and DYER 1204b) observed that the time required for reduction of various indicator dyes decreases with an increase in the normal potential of the indicator; they find that the reduction time is approximately a logarithmic function of the electrode potential. In their study of the rate of autoxidation of a series of complex iron salts MICHAELIS and SMYTHE 825) found an increase of the reaction rate with a decrease in the potential. The tridipyridyl ferrous sulfate with a potential higher than $+ 1$ v. is not appreciably autoxidizable. BARRON 75) reports that at constant ph and in the absence of catalysts there exists a linear relationship between the $E'_o$ of a number of indicator dyes and the logarithm of the time necessary to effect an oxidation of the leuco dye from 2 to 50 per cent. Exceptions are 1-naphthol-2-sulfonate-indo-2,6-dichlorophenol and hydroquinone, the former being oxidized at a higher and the latter at a lower speed than predicted by the rule.

Since the publication of BARRON's paper assuring considerations have been forthcoming from physical-chemical quarters. HAMMETT 443a) states: "It is certainly true that there is no universal and unique relation between the rate and the equilibrium of a reaction; it is equally true that there frequently is a relation between the rates and the equilibrium constants of a group of closely related reactions." His paper should be consulted for a review of the known examples of this type of relationship. It may also be mentioned that CHOW 168) has found that the catalytic effect of such redox systems as ferriferrocyanide upon the rate of linseed oil oxidation is determined by the potential of the catalyst. A theoretical basis for these observations has recently been supplied by GERSHINOWITZ 398). This author refers to the theories of the absolute rates of chemical reactions, as developed by EYRING, RICE, STEARN and GERSHINOWITZ, which relate kinetic data with thermodynamic quantities. These authors have shown that the rate of a chemical reaction is given by the concentration of the reacting system in a certain fraction of the total phase space that is available to the systems, multiplied by the velocity with which the systems are passing through this region. The configuration of the reacting molecules which corresponds to this region of phase space is called the a c t i v a t e d  c o m p l e x. The free energy of formation of the activated state is, in general, not identical with the free energy change of the total reaction. Inasmuch as the free energy change for a given reaction is equal to the difference of the activation energies of the forward and reverse reactions, the free energy of formation of the activated state will, therefore, only be equal to the free energy of the total reaction if the energy of activation of the reverse reaction is zero. It is this fact which has made obscure the relation between free energy and the rate of reaction. GERSHINOWITZ develops an equation relating the rate of reaction to the normal potentials of redox systems reacting with each other which differs from an earlier equation from CONANT and PRATT in that it contains no arbitrary assumptions as to the reaction mechanism.

### b) Model Systems and Intermediary Catalysts. (Mesocatalysts) *)

#### General.

For the time being it appears preferable to treat the intermediary catalysts separately from the enzymes, though it is possible that at a later date the mode of classification may require certain changes. We shall discuss here those properties of the intermediary catalysts which are of importance for their participation in enzymatic processes while their chemistry will be discussed following that of the desmolytic enzymes (pp. 208, 228); their rôle in cell respiration will be dealt with in chapter F (p. 261).

The main function of the non-autoxidizable intermediary catalysts in the anoxybiontic breakdown of carbohydrates appears to be related less to the very first stages of activation and scission of the hexose molecule than to the preparation of the metabolites for the ultimate combustion by oxygen. The cell seems to interpose these carriers between the negative metabolite-dehydrase systems and the positive cytochrome-respiratory ferment chain. This is suggested by the potential range of these carriers which is intermediate between the two extreme groups of enzymatic catalysts just mentioned. We have to admit however that, with a few exceptions, the actual function of the otherwise well-known substances of this group, e.g. glutathione, ascorbic acid, and certain quinoid compounds, is still largely a matter of speculation.

#### Metabolite Systems.
#### Ascorbic acid (Vitamin C) and Reductones:

The relationship of the reduced form of the reversible redox system, ascorbic acid, to the oxidized form, dehydroascorbic acid, is that of a dienol to a diketone. Ascorbic acid may be readily oxidized and may thus reduce a variety of substances with the probable exception of molecular oxygen in the absence of special catalysts (p. 209). If the oxidation consists only in dehydrogenation with a loss of two H-atoms, the dehydroascorbic acid may readily be reduced back to ascorbic acid by $H_2S$, e.g. after oxidation by $O_2$ + Cu at acid reaction; after oxidation in neutral solution only a small fraction of the ascorbic acid may be recovered by hydrogenation, while autoxidation at a ph above 7.6 leads to irreversible destruction (BARRON 79)). Apparently the living cell is also capable of reversibly reducing dehydroascorbic acid, since the latter has the same vitamin C activity as the reduced form (cf. ZILVA 1382)).

Ascorbic acid reacts with a number of suitable acceptors: quinone will dehydrogenate it rapidly, methylene blue only slowly in spite of its sufficiently positive potential; Cu catalyzes this reaction (p. 67). The rate goes through a minimum between ph 5 and 7 (EULER et al. 265)). Reductones reduce methylene blue rapidly in comparison with ascorbic

*) See also B. KISCH 571).

acid. The behaviour towards glutathione suggests the establishment of equilibria (see p. 98). It is still a controversial question whether there exists a specific ascorbic acid oxidase the occurrence of which in plants has been reported by SZENT-GYÖRGYI 1119), ZILVA 1382), TAUBER and KLEINER 1144a). On the one hand it is difficult to differentiate clearly between an enzymatic process and a less specific catalysis by Cu (p. 67) or by quinoid substances (ZILVA). On the other hand the possibility exists that, at least in animal tissues where there is no oxidase of this type, pyridine containing enzymes or the yellow ferment may intervene. For details see OPPENHEIMER's "SUPPLEMENT", p. 1587, and TAUBER's review (1144a).

Ascorbic acid is credited with a variety of catalytic or inductive functions. JORISSEN 508) finds an oxidation of lactic acid. HOLTZ 493, 491) reports that ascorbic acid and the reducing substances produced by irradiation of sugar solutions with ultraviolet light or X-rays increase the oxygen uptake of unsaturated fatty acids and cause an induced oxidation of sugar in phosphate buffer. The observation of HARRISON 460) was of particular interest. He observed that the addition in vitro of ascorbic acid to liver and muscle tissue of scorbutic animals provoked an increase in respiration. This was considered to be the first demonstration of an in vitro-effect produced by a vitamin. According to QUASTEL and WHEATLEY 949) this effect is concerned with fatty acid oxidation since the addition of ascorbic acid to liver slices will enhance the conversion of butyric acid and other fatty acids into acetoacetic acid. Upon reinvestigating the results obtained by HARRISON, KING and his associates (1105b)) conclude that the oxygen consumption of tissues caused by ascorbic acid equals the sum of the original respiratory oxygen plus the $O_2$-volume required for the oxidation of the ascorbic acid. If this is accepted, HARRISON's results can no longer be interpreted as an effect of vitamin C on tissue respiration.

Ascorbic acid, as a reducing substance, is stated to protect thiols, epinephrine, and dihydroxyphenylalanine (Dopa) against oxidation (1, 1034)). Conversely, epinephrine has a protective action on ascorbic acid (1369)); this may have some relation to the protective action of tissues on the vitamin (760, 554)).

l-Histidine is converted by ascorbic acid to histamine (HOLTZ 495)). While this indicates that the vitamin is a model for the enzyme Histidase, EDLBACHER et al. 228) have found recently that ascorbic acid in the presence of iron or hemin will deaminate histidine as well as histamine; accordingly, ascorbic acid would be a Histaminase model, showing a pronounced ph-optimum at 7.2 (at 38°). After having reinvestigated the problem, HOLTZ 494) still would prefer to designate ascorbic acid as a histidase rather than a histaminase model in view of the fact that the ammonia formed from histidine and from histamine under the influence of ascorbic acid or histidase is derived from ring nitrogen while the ammonia produced by histaminase originates from the side chain of the substrate molecule.

The activation of papain and cathepsin by ascorbic acid or by its iron complex (KARRER, PURR, MASCHMANN) is probably due to a reduction of the inactive disulfide form of these enzymes to the active thiol form (see BERSIN 101a)).

We know very little about the rôle played by ascorbic acid in cell metabolism. The consideration of the various possibilities hinges upon the question of the reversibility of ascorbic acid as a redox system and on the value of its potential. The experiments of KARRER et al 530a), GREEN 414) and LAKI 689) answered the question as to the reversibility in the negative. The evidence adduced by GEORGESCU 393), WURMSER and LOUREIRO 1360) and BORSOOK and KEIGHLEY 139) spoke in favor of reversibility. Recent work by FRUTON 379) and BALL 58a) justifies the conclusion that the system ascorbic/dehydroascorbic acid is reversible between ph 5 and 7. The sluggish electrode behaviour of the system is due to secondary reactions leading to irreversible changes of the primary oxidation product. By using small amounts of suitable dyes, e.g. thionine, as electroactive mediators between the ascorbic acid system and the noble metal electrodes, BALL was able to determine the equilibrium between the vitamin and the primary dehydro-product. The normal potential of the system at ph 7.2 is $+ 51$ mV. which classifies ascorbic acid as a system of a reducing intensity similar to methylene blue. The fact that it readily reduces TILLMANN's indicator — 2,6-dichlorophenol-indophenol — which is a very positive redox system ($E'_o = + 0. 217$ V. at ph 7) has often been cited, without justification, as evidence for the strongly reducing properties of ascorbic acid. It is, in fact, a weakly reducing system.

The redox potential of pure r e d u c t o n e (enol-tartronaldehyde) has also recently been measured. According to WURMSER et al. 1365) the normal potential at ph 0 equals $+ 0.282$ V.; the system is measurable only up to ph 6.

At present there is hardly any indication for a participation of ascorbic acid in a n a e r o b i c fermentation reactions. It would be easier to assume that its place is in the first stages of oxybiosis, perhaps somewhere in the neighbourhood of the yellow ferment, especially if it is believed that ascorbic acid, similar to the yellow enzyme, may transfer hydrogen d i r e c t l y t o t h e t e r m i n a l h e m i n s y s t e m or to other iron compounds. SZENT-GYÖRGYI, some time ago, visualized a direct oxidation catalysis via ascorbic acid: after dehydrogenation of the metabolites the hydrogen is transferred (through the copper complex and then a specific ascorbic acid oxidase) to molecular oxygen. It is, indeed, not impossible that ascorbic acid and similar carriers cooperate with heavy metal catalysts in ultimate oxidation reactions. One could think of a direct dehydrogenation of ascorbic acid by cytochrome in the course of normal cell respiration. McFARLANE 869), on the other hand, assumes on the basis of model experiments with ionized ferric salts, ferric lactate, iron-protein complexes, that ascorbic acid, in vivo, is able to maintain the entire tissue iron (not the respiratory ferment) in the ferrous state by reduction. Thus, an induced oxidation, by Fe (2), of fatty acids, for instance, might conceivably be coupled with a stoichiometrical regeneration of Fe (2) from Fe (3) by ascorbic acid; since the latter is continually produced in or absorbed by the cell, the over-all aspect of the process would be that of a catalysis. Furthermore, ascorbic acid might be peroxidatically oxidized by the iron systems of the tissues in conjunction with primarily formed $H_2O_2$ (pp. 70, 123). Such a drastic decomposition is, however, unlikely in view of the faculty of living cells to reduce dehydroascorbic acid.

### Thiol Systems, Glutathione:

The most important sulfhydryl compound in living cells is undoubtedly g l u t a t h i o n e. There are hardly any indications that free cysteine or similar simple compounds are of significance for desmolysis, with the possible exception that they may act as auxiliaries in the oxidation of glutathione. Thus, there remain besides this tripeptide of glutamic acid, glycine, and cysteine only the SH-groups contained in proteins and higher polypeptides. It is certain that these are capable of dehydrogenation and reduction in the cell. Cathepsin, the intracellular proteinase in animal tissues, and papain, the corresponding plant enzyme, appear to be proteins and are apparently only able to hydrolyse peptide bonds of their substrates when they exist in the reduced (SH-) state; in the S-S-form they may exert a synthesizing action. This is concluded from VOEGTLIN's 1204a) experiments: proteolysis in tissue suspensions proceeds most rapidly in a nitrogen atmosphere, while in oxygen protein synthesis is favored. *) It may be argued, of course, that this effect is due to the state of oxidation-reduction of glutathione rather than of the enzymes themselves. It is true that both the "zookinase" (WALDSCHMIDT-LEITZ et al.) and the "phytokinase" (GRASSMANN et al.) of cell proteinases have been shown to be identical with SH-glutathione. But the effect of this thiol like that of other activators (cysteine, $H_2S$, HCN) is to be ascribed to their reducing action on the disulfide groups of the enzymes. Therefore, in the last analysis, the effect of the oxygen pressure on proteolysis may be due to the shift in the SS $\rightleftarrows$ 2 SH-equilibrium in the enzymatic proteins with glutathione acting merely as a go-between. However, there exist certain observations which militate against this view. SS-glutathione which, according to this hypothesis, should have the same effect as oxygen on the proteinases is apparently quite inert.

According to ANSON 31a) highly purified cathepsin is activated neither by cysteine nor by HCN. The effects observed with crude enzyme preparations may have been due to the removal of inhibitors by these "activators". It will be remembered that such an explanation was advanced several years ago by KREBS 598a) who found that the activity of papain is increased by the removal of traces of heavy metals from the substrate (gelatine) by complex forming substances, e.g. HCN, prior to addition of the enzyme. However, the incomplete removal of the HCN after the treatment of the substrate might possibly explain the activating effect observed by KREBS. Moreover, his theory cannot explain the activating effect of certain very stable heavy metal complexes, e.g. tridipyridyl ferrous sulfate, on cathepsin (MICHAELIS and STERN 827a). A thorough reinvestigation of the problem with highly purified cathepsin and crystalline papain as they are now available should be able to elucidate the mechanism of the various effects previously observed. The promotion of protein synthesis by increased oxygen tension may possibly find an altogether different explanation: the energy required for the endothermal back reaction might be furnished by simultaneously occurring oxidation reactions (energetic coupling). It should be born in mind that VOEGTLIN's experimental mixtures contained a large variety of enzymes and substrates besides the proteolytic system.

In certain instances the function of the sulfhydryl group appears to consist, indeed, in its combination with inhibiting traces of heavy metal. An example is the "activation" of alcohol dehydrogenase by glutathione, with methylene blue as acceptor (WAGNER-JAUREGG and MÖLLER 1211)); see also SCHÖBERL 1029, 1030)). Another hypothesis is that SH-glutathione removes $H_2O_2$ or "active oxygen" (GOSH 402)).

The main function of glutathione is generally considered to be that of a reversible redox system which operates in conjunction with the other oxidation catalysts of the cell. Here we have to admit that the story of glutathione has been rather disappointing and is still far from being completely written. When HOPKINS announced the discovery of glutathione in 1921 he ascribed to it an important role in intermediary hydrogen transfer processes. Such a role would obviously have to depend on full reversibility of the redox system formed by the SH- and the SS-form under physiological conditions. All attempts to demonstrate such a behaviour in vitro have more or less failed (cf. WURMSER 1354), p. 157), MICHAELIS 804), p. 152), and BUMM 154)). GREEN 413) also reports that glutathione, in vitro, behaves like an irreversible system. This means that mixtures of the reduced and oxidized form fail to impart a stable and constant potential to noble metal electrodes; as in the case of cysteine the potential level appears to be solely determined by the reduced form and, moreover, the values recorded by individual electrodes differ considerably. In the case of cysteine FRUTON has been able to reach an equilibrium

*) For Details see OPPENHEIMER's "SUPPLEMENT", p. 917, 919.

condition or at least a steady state by permitting it to react with the oxidized form of a suitable redox indicator (indigodisulfonate) over a relatively long period. The final degree of oxidation of the dye as derived from the color intensity was assumed to correspond to the potential established in the SS-SH-system. The reversibility was deduced from the fact that the equilibrium state could also be approached by adding cystine to the leuco form of the indicator. It is quite possible that glutathione would show a similar behaviour under these conditions. However, the slowness of this process would be a prohibitive factor in a consideration of the physiological requirements where biocatalysts change their state of oxidation-reduction many hundred times a second. There exists the possibility that in the cell the change-over from one form of glutathione to the other is greatly accelerated by a catalysis. BIERICH et al. 112) are inclined to deny the functioning of glutathione as a redox system in cell metabolism because they were unable to demonstrate the presence of the disulfide form in fresh tissues. Perhaps it is not the disulfide form which forms the redox system with the sulfhydryl form but an intermediary form as KENDALL and NORD 891, 559) suspect.

It would appear that neither the h y d r o g e n a t i o n of glutathione by a donator nor the d e h y d r o g e n a t i o n by an acceptor are as simple as in the case of other carriers. The first attempts to demonstrate a reduction with the aid of dehydrogenase systems (cf. BUMM 154)) were unsuccessful; consequently BERTHO 105) was inclined to ascribe to the S-S grouping a lack in affinity just as some other groupings are unable to act as acceptors. Later the reduction of glutathione could be brought about by experiments with tissue suspensions. This is achieved at the expense of an unknown donator and with the cooperation of a protein containing free SH-groups. The biological materials were muscle (HOPKINS and ELLIOTT 498)), powdered tissues suspended in dilute serum (WURMSER 1355)), sea-urchin eggs (RAPKINE 960)), and erythrocytes (MELDRUM 777)); yeast, according to the last author, was ineffective. RAPKINE found a further catalysis with the participation both of glutathione and methylene blue. The mechanism of these reactions is still obscure. It may be that the role of the donator is played by the strongly reducing sugar derivatives (including ascorbic acid) which, according to WURMSER, are present in the cell. If this were true glutathione would exert its redox function at a rather early stage of desmolysis. One might assume, for instance, that it may take over the hydrogen from the reduced form of the phosphopyridinenucleotide (cozymase). Inasmuch as this oxidation is ordinarily performed by the yellow enzyme or by diaphorase, we would have to assume either a competition between the two redox systems for the dihydropyridine compound or else a reaction between SS-glutathione and the reduced form of the flavoprotein. At present this problem is still entirely in the realm of speculation.

The general significance of sugars for the reduction of glutathione has been demonstrated in model systems as well as with intact cells. Both MANN 754) and SEN 1043) were able to reduce glutathione by glucose plus a dehydrogenase preparation from liver. Neither succinic acid nor lactic acid are suitable as donators (ELLIOTT 240)). In the case of intact cells the part played by sugars is indicated by indirect evidence only: The depleted liver of fasting animals is incapable of reducing glutathione. On the other hand, the addition of glucose, fructose and other sugars to pneumococci (DUBOS 221)) and intact red blood corpuscles (MELDRUM 777)) causes a considerable increase in free SH-groups, presumably attached to glutathione. KÜHNAU 622) and TSUKANO 1188) think of special sugar derivatives, e.g. phosphates, as donators. It has actually been shown by MELDRUM and TARR 779) that hexosephosphate plus pyridine-enzyme will reduce glutathione. The same holds for phosphohexonic acid and zymophosphate but not for phosphoglyceric acid or hexoses. Probably $C_3$-compounds are the active donators as in all experiments dealing with the pyridine-enzyme. How this ties up with the only well-established function of SH-glutathione, namely, to act as the coenzyme of ketonealdehyde mutase (glyoxalase) in the transformation of methyl glyoxal to lactic acid (LOHMANN 731)) is not yet understood. The effect is related to the mechanism of the PASTEUR-MEYERHOF-reaction by BUMM and APPEL (155) and also to the breakdown of carbohydrates without participation of phosphoric acid radicles in certain cell types (p. 244). It is, of course, possible that a part of the "sugar derivatives" is identical with ascorbic acid the behaviour of which towards glutathione is amenable to experimentation. It is claimed that ascorbic acid is partly responsible for the accumulation of glutathione in the SH-form (112, 1005)) and that it maintains the potential required for proteolysis (PURR 940)). Conversely, SH-glutathione protects ascorbic acid against oxidation by air (164)). According to HOPKINS 499) this is to be explained by complex formation between glutathione and catalytically active traces of copper. The protection against oxidation by plant extracts, on the other hand, is explained by the fact that the tripeptide is oxidized more readily by the oxidase than is ascorbic acid. The latter is therefore only attacked after all of the glutathione has been oxidized. However, PFANKUCH 926) observed a rapid reduction of dehydroascorbic acid by cysteine in plant extracts. In the case of liver slices the ascorbic acid is protected against oxidation only if the glutathione is present in large excess. According to BARRON 78) the same is true in the plant extract systems studied by HOPKINS.

Ascorbic acid retards the so-called autoxidation of cysteine (HOLTZ 494)) while, conversely, SH-groups inhibit the Cu catalysis of ascorbic acid. These observations favor the view that the two redox systems may enter an equilibrium reaction. BORSOOK and JEFFRIES 137) have actually claimed that SH-glutathione may reduce dehydroascorbic acid under physiological conditions. This had not been observed by PFANKUCH 926) or BERSIN et al. 102). The latter author, however, states that at ph 6.48 as well as 7.38 SH-glutathione will prevent the "autoxidation" of ascorbic acid. In any event, mixtures of the two compounds are not susceptible to autoxidation.

Still less is known concerning the dehydrogenation of glutathione by defined biological systems. The catalytic oxygen transfer is very small in the case of 11 dehydrogenase systems tested (OGSTON and GREEN 895)); the dehydrogenation and not the reduction is the limiting factor. But it has been shown, at last, that there exists in principle an oxygen transfer via glutathione (and cysteine). The sugar phosphate-pyridine enzyme system of MELDRUM and TARR 779) may utilize oxygen as the acceptor, though less efficiently than flavin or methylene blue (confirmed by OGSTON and GREEN 895)). The reoxidation of glutathione may be effected by a thermostable, cyanide sensitive metal system in the cell, probably identical with Fe-cysteine (HOPKINS 498)). In kidney slices SH-glutathione is dehydrogenated by acetoacetic acid (QUASTEL 950)). In model experiments glutathione (and also cysteine) will transfer oxygen also to lactic and $\beta$-hydroxybutyric acid (HARRISON 462)). The dehydrogenation of glutathione by alloxane is probably an unphysiological model reaction, just as the oxidation by dyes as investigated by KENDALL and NORD 559).

Though it is likely we do not know as yet whether the catalytic oxidation of glutathione in the living cell is accomplished by cytochrome. According to BIGWOOD and THOMAS 118) purified cytochrome c is readily reduced by glutathione. Glutathione may transfer hydrogen to the hemin systems in this manner. It has to be borne in mind, however, that SS-glutathione could not be detected in tissues 112). PIRIE 930) finds that

SH-glutathione, after hydrolysis, cysteine, and also methionine are oxidized to sulfate by liver and kidney slices in vitro. Copper may play the role of a catalyst in this process. The reaction apparently does not proceed via the disulfide stage since cystine is not oxidized in this system. It is also possible, just as in the case of ascorbic acid, that SH-glutathione reduces some iron compounds in the cell, e.g. cytochrome, and that the hydrogen peroxide formed upon reoxidation to the SS-form attacks the disulfide in a peroxidatic reaction and causes the formation of irreversible end products.

Little useful information has been forthcoming from experiments on the effect of glutathione when added to living cells. According to ROSENTHAL and VOEGTLIN 1005) glutathione has no effect on the respiration of normal tissues and yeast. But SH-gluta-thione will decrease the glycolysis in muscle with glycogen as the substrate; in other tissues, where glucose is the substrate, the anaerobic glycolysis is decreased to the level of aerobic glycolysis. GEIGER 389) considers the oxidation of glutathione as the primary process underlying the PASTEUR-MEYERHOF effect. BUMM and APPEL 155), in the contrary, observed no effect of SS-glutathione on the anaerobic glycolysis of tumors; they report a significant rise in the aerobic glycolysis, in other words an inhibition of the PASTEUR reaction, by SH-glutathione. This observation is related to the "co-enzyme" function of glutathione in the glyoxalase system (LOHMANN 731)) which is assumed to dismutate the methylglyoxal formed in tumors by the "phosphorus-free" schema of BUMM and thus to inhibit the resynthesis of glycogen. For details see OPPENHEIMER's „SUPPLEMENT", p. 1292.

\* \* \*

Other possible intermediary catalysts are methylglyoxal (NEUBERG and KOBEL 885); KISCH 568)), phosphohexonic acid (WARBURG 1255)), acetoacetic acid (QUASTEL 512, 950)), oxaloacetic acid (KREBS (p. 274)), citric acid (KREBS and JOHNSON 611)), fumaric acid (SZENT-GYÖRGYI 1124, 1123); QUASTEL 945, 946); BORSOOK and KEIGHLEY 138); WIELAND et al. 1304)), trimethylamine oxide (ACKERMANN 3)). The rôle of the $C_4$-acids will be discussed in the chapter dealing with cell respiration (p. 268).

### Quinone Catalyses.

The general mechanism of this group of processes is as follows: with the aid of a specific enzyme, the oxidized form of a reversible system, mostly a quinoid dye, dehydrogenates a metabolite and is thereby transformed into the reduced form, the leuco dye. The latter is either truly autoxidizable and regenerates the oxidant by reaction with molecular oxygen or else the system contains a heavy metal forming an autoxidizable complex with the leuco compound. It does not matter for the purposes of the present discussion whether the metal renders the reoxidation possible or whether it merely speeds up the process to useful reaction rates. From the descriptive point of view we may divide the phenomena in this field in two categories, viz. the catalytic oxidative deamination of amino acids by quinones and the acceptor respiration proceeding with the aid of exogeneous or endogeneous quinoid dyes.

It has become almost a commonplace that these quinones may oxidize the reduced forms of such redox systems only which possess a more negative potential, thus yielding free energy upon reaction. The occasional neglect of this principle has caused some confusion, inasmuch as the effect of such "oxidizing" systems has occasionally been investigated with the aid of thermodynamically impossible donators. It has been attempted, for instance, to dehydrogenate the strongly positive pyrocatechol by the much more negative methylene blue.

### Semiquinones as Intermediate Steps of Oxidation-Reduction Systems. *)

Reversibly reducible organic dyes have a quinoid structure while the corresponding leuco-dyes have a benzenoid structure. The over-all change between the two forms involves a difference of two hydrogen atoms or two hydrogen equivalents (electrons). Until recently it was held that this bivalent change takes place in a single step:

$$\text{Dye} + 2\ \text{H} \rightleftharpoons \text{Leuco-dye.}$$

This assumption has now been shown to be erroneous. In a certain number of instances it has been demonstrated that this bivalent process actually occurs in two successive univalent steps involving a half-reduced (or half-oxidized) intermediate. This intermediate must be a molecule containing an unpaired electron and it must therefore have the character of an organic radical. In all instances where such an intermediate has been detected, the chromophoric group which is usually identical with the group undergoing reversible reduction consists of two atoms of the same kind, e.g. 2 N-atoms in the phenazine group and 2 O-atoms in the quinones. Compounds like the thiazines, oxazines, or acridines, the chromophoric group of which consists of two different atoms, show no detectable intermediate formation.

Let us consider, as an example, the oxidation-reduction of tetramethyl-p-phenylene diamine:

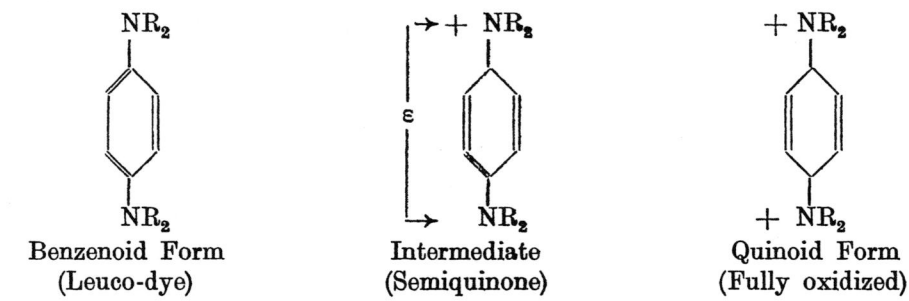

R = CH$_3$; ε is the oscillating electron.

The intermediate form is a half-reduced and half-oxidized radical, a semiquinone in the terminology of MICHAELIS. Its relative stability is explained by assuming that the formula shown is not statically rigid but that the radical has a free electric positive charge oscillating between the two symmetrical halves of the molecule, with the intermediation of the carbon skeleton. This state of affairs is known as a "resonance" phenomenon. The resonance energy is particularly large if the two atoms linked by the odd electron are identical. The greater the resonance energy involved the greater the stability of the radical which explains the observation of such semiquinoid forms only in the case of symmetrically built molecules.

It is obvious that such a radical will show a tendency to change into the fully reduced or fully oxidized form of the dye if a suitable acceptor or donator is provided. The facilitation of an oxidation or reduction process by the intermediate is therefore equivalent to a catalytic effect if the system is compared with one incapable of semiquinone formation. It is highly significant that in all instances of biologically occurring quinoid redox systems the existance of a semiquinoid radical has either been directly demonstrated or else made probable by poten-

---

*) This section is largely based on the recent review of the semiquinone problem by L. MICHAELIS 805).

tiometric or optical methods. In order to name specific examples, semiquinone formation has been proven for pyocyanine, chlororaphin, and the flavin group. It has been made probable for the case of hallachrome, toxoflavin, and pyridine nucleotides. It is very suggestive, therefore, to assume that these biological redox systems exert their particular physiological function by virtue of their ability to form measurable amounts of semiquinone under the conditions of their natural environment.

Intermediates appearing in the reduction of quinones, called quinhydrones or meriquinones, have been known for a long time. The best-known quinhydrone is that of benzoquinone because of its wide use for ph-determinations. Compounds of this type have usually been regarded as molecular addition compounds between the quinoid and benzenoid form of the redox system. This view was based on the belief that a free organic radical would be unstable or would readily polymerize like triphenyl methyl. The first one to question the formulation of the meriquinones as static or dynamic molecular compounds was HANTZSCH 449). He suggested the possibility of a radical-like structure on the basis of spectroscopic observations made on the green, half-reduced form of N-methylphenazonium iodide. Quinhydrones in aqueous solution are usually in a measurable state of equilibrium with the quinoid and benzenoid forms. In other words, a pure, crystalline quinhydrone, upon being dissolved will suffer a dismutation, according to the schema:

2 quinhydrones ⇄ 1 quinone + 1 hydroquinone.

The position of this equilibrium may vary depending on the compound involved and on the ph. Because of this spontaneous dismutation it is practically impossible to prepare pure solutions of meriquinones. This renders difficult the application of the ordinary methods of determination of molecular weight to the solution of this problem. Although WEITZ et al. 1290) claimed to have obtained evidence for the radical nature of several meriquinones, e.g. dipyridilium complexes, by an ebullioscopic method, MICHAELIS 805) contests the validity of his technique and of his conclusions based on his results. Unambiguous evidence for the radical nature of these intermediates was offered for the first time independently by FRIEDHEIM and MICHAELIS 376) and by ELEMA 233) on the basis of potentiometric findings on pyocyanine. They observed that upon reductive titration of this bacterial pigment in weakly acid solution the color change is not from the red, acid quinoid form to the colorless benzenoid form, but that an intermediate green compound is formed. The titration curve shows two well separated steps. The green intermediate differs from the red oxidized and the colorless fully reduced form by one hydrogen equivalent while the molecular size is the same. This finding does not permit of any other formulation of the intermediate than that of a free, cationic radical.

Both MICHAELIS and ELEMA developed a mathematical theory on the subject. Furthermore, they reported on a number of other organic redox systems where semiquinone formation may be demonstrated.

**Mathematical Theory:** If an organic redox system incapable of semiquinone formation is subjected to potentiometric titration, the midpoint of the titration curve is called the normal potential at the hydrogen ion concentration chosen. This potential, as already been explained (p. 50), is that of a mixture of equal amounts of the oxidized and reduced forms which differ by 2 hydrogen equivalents from each other. The slope of the titration curve is, by definition, that of a two-electron system (n = 2). A measure of the slope is the index potential, i.e. the difference in millivolts between the mid-point of the curve (50% reduction or oxidation) and the point corresponding to 25 or to 75 % reduction. The index potential,

$E_i$, for a two-electron system is 14 mV. For a one-electron system, e.g. an iron complex salt, it is 28 mV. If the reduction of the system under study occurs in two individual steps each involving one hydrogen equivalent, the mid-point of the titration curve is no longer the true normal potential, $E_o$, but only of statistical significance. It is now called $E_m$ or the "mean normal potential". In equal distances from this point there are now located the mid-points or normal potentials of the two individual reaction steps: $E_1$ is the normal potential of the system formed by the fully reduced and the half-reduced (semiquinoid) forms while $E_2$ is that of an equimolecular mixture of the half-reduced and the fully oxidized forms. If there is no overlapping between the two individual steps, the slope of each "half-curve" corresponds to that of a one-electron system ($E_i = 28$ mV.), as shown in Fig. 1.

Such a perfect separation, however, is only found under special conditions, namely in a ph-range where the stability of the particular semiquinone involved is high. In some instances this condition is met in the acid ph-range (e.g. Pyocyanine, flavins), in others complete separation of the two steps is only encountered in the alkaline ph-range (indigosulfonates (SHAFFER), quinones). At other hydrogen ion concentrations the two steps will be found to overlap to a smaller or greater extent. Still further, the two curves relating $E_1$ and $E_2$ to ph will intersect and from there on the relationship between the two values is inverted. In other words, the normal potential of the semiquinone-fully reduced form-system is now somewhat more positive than that of the fully oxidized form-semiquinone system. This fact cannot be observed directly, i.e. by optical methods, although in some instances somewhat mixed colors have been noted. The proof for this anomalous state of affairs is based on potentiometric evidence only, i.e. on the value of the index potential in this range. Unfortunately, small differences in index potential correspond here to rather large differences in the position of $E_1$ and $E_2$, and the determination of the index potential is subject to certain experimental errors of an order of magnitude similar to that of the effects to be expected from the anomalous situation of the two normal potentials. In any event, it is important to note that this inverted relationship is encountered in the physiological ph-range. Its significance for biological oxidative catalysis will be discussed further below.

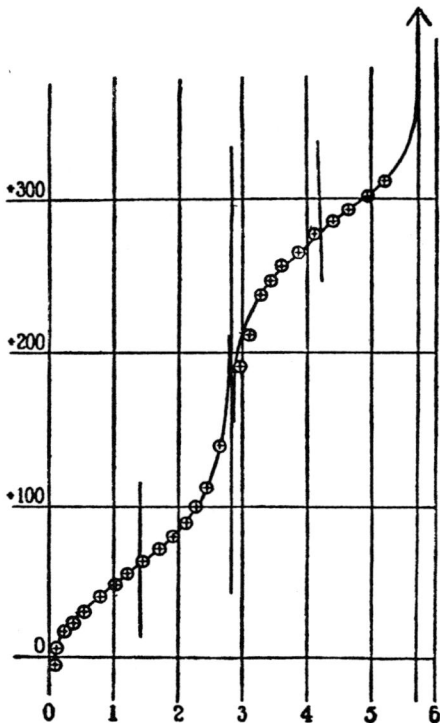

Fig. 1. Potentiometric oxidative titration curve of α-hydroxyphenazine (MICHAELIS 805)). The leuco-dye was titrated with benzoquinone as oxidizing agent at ph 1.00. Abscissa: benzoquinone added; Ordinate: potential as referred to the hydrogen electrode.

It has already been mentioned that the semiquinoid form (S) may reversibly dismutate into equal amounts of the fully reduced (R) and fully oxidized (T) form: $2 S \rightleftarrows R + T$. The equilibrium constant of this dismutation is

$$k = \frac{(R) \cdot (T)}{(S)^2}.$$

The reciprocal value of this dismutation constant, $K = \frac{1}{k}$, is called the formation constant of the semiquinone. The formation constant may be calculated with the aid of the experimentally determined index potential, $E_i$. MICHAELIS (cf. 805) has given a table permitting the interpolation of K for any degree of reduction of the dye ($\alpha$). Later, ELEMA 235) developed the complete equation

$$E = E_m \cdot \frac{RT}{F} \ln \frac{4\alpha}{(1-1\alpha)\sqrt{K} + \sqrt{K(1-2\alpha)^2 + 16\alpha(1-\alpha)}}$$

where F = 1 Faraday or electrochemical equivalent, R = gas constant in the electrical system, and T = absolute Temperature (KELVIN). This equation yields

$$E_i = \frac{RT}{F} \log \frac{1}{2} (\sqrt{K} + \sqrt{K + 12})$$

and

$$\sqrt{K} = 10 \; E_i \frac{F}{RT} - 3.10^{-E_i} \cdot \frac{F}{RT}.$$

Or, if

$$a = 10 \frac{E_i}{0.0601} \quad \cdots \cdots \cdots \quad \text{(ELEMA 236)),}$$

$$K = \left(a - \frac{3}{a}\right)^2$$

It is also possible to determine K by direct graphical interpolation of the experimental titration curve without using $E_i$ (ELEMA 236)). To this end, ELEMA uses the difference, D, (called S by ELEMA) between $E_m$ and the point of intersection of the tangent to the curve through this point with the ordinate. One obtaines thus $D = RT/F (2 + \sqrt{K})$ and, resolved for K and for 30° : $K = (76.63 \, D - 2)^2$.

The formation constant, K, of a semiquinone may assume widely varying values for the same dye depending on the ph and the state of dissociation of the dye-salt. In the case of a dyestuff cation K will increase with decreasing ph-values, and it will increase with ph in the case of a dyestuff anion (see for instance MICHAELIS and HILL 815)). For the case K = 0, $E_1$ becomes equal to $E_2$ and also to $E_m$; the oxidation takes place in a single bivalent step. *)

Recently, the theory has been further developed by MICHAELIS to cover the phenomenon of dimerization of the semiquinoid radicals. Such a dimerization has long been known to occur with organic radicals of the type of triphenylmethyl and diphenylnitride. MICHAELIS and FETCHER 812) have been able to demonstrate the existence of this phenomenon in the instance of a reversible redox system, namely, phenanthrenequinone-3-sulfonate. There exists an equilibrium between the four forms of the dye: the oxidized, the reduced, and the two intermediate forms (monomer and dimer of the semiquinone). In accordance with this equilibrium the formation of the dimer is favored by increasing the total concentration of the dye. They could also show that not only the semiquinone formation constant but also that of the dimer and the dimerization constant depend on the hydrogen ion concentration. MICHAELIS and SCHWARZENBACH 824) have now also

*) The values for individual systems have been compiled by W. ROMAN (Tab. Biol. **16**, 110 (1938)).

reexamined the case of riboflavin from this point of view. The technique consists mainly in carrying out potentiometric titrations at various absolute concentrations of the dye. Inasmuch as the dimerization is a bimolecular process it should be favored by an increase in concentration and vice versa. They find that at low concentrations, including the concentration range occurring under physiological conditions, the intermediate form is entirely present as the monomer radical. In higher concentrations of the dye a partial dimerization of the radical takes place.

MICHAELIS et al. 811, 818) were able, furthermore, to confirm the potentiometric results obtained with phenanthrenequinone-3-sulfonate by measurements of the changes of the magnetic susceptibility in the course of the reduction of the dye.

Perhaps the most complex relationships so far encountered in reversible redox systems are those obtaining in the pentacyano-aquo-ferriate and -ferroate system. MICHAELIS and SMYTHE 827) interpret their potentiometric results by assuming that each of the two iron complexes is present in solution as a quadrimolecular aggregate. Between the ferri and the ferro form there exist three intermediate forms in which a part of the four Fe atoms is in the ferri and the rest in the ferro state. The molecular aggregation is tentatively explained on the basis of hydrogen bonds between the coordinatively bound water molecules of the complexes leading to an eight-membered ring. It is quite possible that these results may some day acquire considerable biological importance in the event that it should be possible to discover such partly reduced and partly oxidized intermediates in the case of hemin proteins. There are four hemin groups contained in one molecule of hemoglobin and it is not impossible that there exist intermediate stages between the fully reduced ferrohemoglobin and the fully oxidized ferrihemoglobin (methemoglobin) which would be formally analogous to the intermediates in the pentacyano-aquo-iron system. In this connection it might be mentioned that it has so far not been possible to secure evidence for the existence of partly oxygenated and partly reduced hemoglobin intermediates (CONANT, ROUGHTON).

### Significance of Two-step Oxidation for the Oxido-Reductive Catalysis.

Let us quote from a recent lecture of MICHAELIS 805) on the semiquinone problem: "Were it not for the existence of intermediate radicals, we might say that oxidations in organic chemistry are of necessity always bivalent. Whenever this bivalent oxidation can be brought about by two successive univalent oxidations, then the kinetics of such a reaction will be greatly enhanced. The thermodynamic possibility of the univalent oxidation, and the existence of the radical as an intermediate step of the complete bivalent oxidation, will depend on the dismutation constant. If this constant is very large, it is equivalent to saying: The formation of the radical involves a very high step in energy. Only if, due to resonance, the formation of the radical requires relatively little energy, will the bivalent oxidation run smoothly. The reason why the oxidation of organic compounds is frequently very sluggish, even when an oxidant of thermodynamically sufficient oxidative power is applied, is that the oxidation has probably, as a rule, to go through two univalent steps; and to go through the intermediate step means, in general, climbing over a large energy hill, except in those cases described above in which the semiquinone formation constant is relatively large. It is the task of all catalysts and enzymes concerned with oxidation-reduction processes to ease the climb over this energy hill, or to convert the substance to be oxidized into some form, or into some compound, in which the intermediate radical will have a stronger resonance and so a greater stability than it has in its original form."

### Oxidative Deamination of Amino Acids by Quinones.

Amino acids are oxidatively deaminated by strong acceptors of the type of benzo-quinone; the first step consists in the transformation of the amino acid into the corresponding imino acid $R \cdot C \underset{\diagdown COOH}{\overset{\diagup NH}{}}$ . This dehydrogenation reaction proceeds in the absence of any enzyme as a homogeneous reaction (see also RAPER and PUGH 958)). Naturally occurring quinoid substances, e.g. chlorogenic acid, are also able to deaminate amino acids. Reactions of this kind play probably an important role in plant fluids where there exists always an opportunity for the formation of quinones by the reaction of poly-phenols with oxidases. In the animal cell, too, they may occur with a d r e n o c h r o m e, the quinone of epinephrine. This process differs from the action of the specific dehydro-genases with regard to its kinetics and specificity. While, for instance, the efficiency of the quinones decreases with an increasing length of the carbon atom chain in the amino acid that of the dehydrogenases will increase (KISCH 571)).

The reaction has been systematically studied by KISCH 573). He succeeded in chan-ging it from a stoichiometric process into a complete catalysis by making provision for a subsequent dehydrogenation of the reduced form of the quinone by $O_2$. The problem of the catalytic dehydrogenation of amino acids in the typical form of primary attack of the amino group has always presented difficulties. The reaction occurs readily with some acceptors (alloxan, isatin, quinone, iron on charcoal) while methylene blue cannot be utilized; this holds for palladium as well as for muscle enzymes as catalysts. On the other hand, bacterical enzymes are stated to utilize methylene blue as acceptor for the attack on amino acids (BERNHEIM et al. 99)). In the dehydrogenation of amino acids by isatin (LANGENBECK 700, 696)) methylene blue may act as acceptor in an indirect manner, i.e. by oxidizing the reduced form of the catalyst (isatyd). It is likely that there exist certain structural chemical requirements with respect to the acceptor, probably the existence of a—C=O grouping (FRANKE (288), pp. 154, 172). If this is provided the reaction will proceed without any special catalyst if there is an opportunity for the reoxidation of the reduced form of the quinone. Dehydrogenation by biologically oc-curring quinones has been demonstrated for chlorogenic acid by OPARIN 901) and for epinephrine simultaneously by EDLBACHER 226) and BLIX 127). Epinephrine represents, according to KENDALL and WITZEMANN 560) and BALL and CHEN 60), a reversible redox system consisting probably of the o-diphenol and the o-quinone form:

It is, of course, possible that there exists a semiquinoid intermediate. The autoxidation of the reduced form is catalyzed by metals and, in that event, will proceed to an irrever-sible stage by a peroxidatic reaction. The primary oxidation product of epinephrine is quite unstable; its life time is less than a second (BALL and CHEN 60)). The equilibrium reaction could therefore be measured only by a modification of the flow-method of HARTRIDGE and ROUGHTON where the reaction mixture of the diphenol and the oxidant after rapid mixing passes a number of electrodes. From the potential gradient established at these electrodes during steady flow it is possible to extrapolate back to zero time and thus to determine not only the normal potential but also the half life time of the oxidized form.

The following table contains some of the results obtained in this way:

TABLE 5.

Normal Potentials and Half Life Times of the Oxidants of Epinephrine and Related Compounds: (Acc. to BALL and CHEN (60)).

| Structure and Name of Reductant | Normal Potential $E_o'$ at ph 7.66 (30° C) | Half Life of Oxidant at ph 7.66 (30°) |
|---|---|---|
| HO<br>HO⟨◯⟩, Catechol . . . . . . . . . . | Volts<br>+ 0.333 | Seconds<br>9.0 |
| R—$CH_2 \cdot CH \cdot NH_2 \cdot COOH$, Dihydroxyphenyl-alanine . . . . . . . . . . . . . . . . | + 0.326 | ·0.14 |
| R—$CH_2 \cdot CH_2 \cdot NH \cdot CH_3$, Epinine . . . . . . . | + 0.306 | 0.05 |
| R—$CHOH \cdot CH_2 \cdot NH \cdot CH_3$, Epinephrine. . . . | + 0.345 | 0.06 |

R = 3,4-dihydroxyphenyl.

Originally EDLBACHER could find an aerobic oxidation by adrenaline only in the case of glycine. He and BLIX observed the activity of other polyphenols in this reaction, among them of pyrocatechol and homogentisinic acid. While adrenaline acts best in acid solution the other polyphenols will react only in alkaline solution with an optimum at ph 10. BLIX found that not only glycine but also alanine, leucine, and valine are attacked; in the latter case acetaldehyde was shown to be one end product. Glycine is also dehydrogenated by adrenalone (aminoacetyl pyrocatechol) with the intermediate formation of glyoxylic acid (BARRENSCHEEN and DANZER 72)). In accordance with these facts it has been observed that amino acids (like cysteine) and ascorbic acid act as stabilizers for adrenaline (WILTSHIRE 1339); MERRIT-WELCH 780)): The oxidation of the hormone is retarded in the tissues because it is kept mainly in the reduced form. However, it is ultimately slowly oxidized by the iron systems of the cell to yield irreversibly dark pigments with an oxygen content much above that of the quinoid form (see for example MEIROWSKY 776)).

The extensive studies of KISCH and his associates (574)) seem to indicate that adrenaline itself is not the active catalyst but rather its mother substance ("c a t a-l y s a g e n"). The true catalyst appears to be the fully quinoid form ("o m e g a substance" of KISCH, a d r e n o c h r o m e of GREEN and RICHTER) which in the course of the reaction with the substrate is not reduced in a bivalent manner to adrenaline, the fully reduced stage, but by a univalent reduction to a semiquinoid form. For details v. OPPENHEIMER, Supplement, p. 1623.

The catalytic deamination of amino acids by quinones is possible not only with molecular oxygen as the acceptor but also under anaerobic conditions with suitable acceptors like m-dinitrobenzene and nitro-anthraquinone; methylene blue is, of course, unsuitable because of the position of its normal potential. The reaction yields up to 85 per cent of the theoretical yield of ammonia and shows a certain degree of specificity. While glycine, serine and cystine are attacked, alanine is not. With dipeptides as the substrate only the free amino group is split off. There is also a dependence of specificity on the hydrogen ion concentration: between ph 8 and 9 hydroxyhydroquinone decomposes specifically glycyltyrosine while at ph 11 only glycine is deaminated. In general, between ph 7 and 8 the dipeptides are more rapidly attacked than their component amino acids. Only in the case of simple amino acids and in the presence of $O_2$ is deamination accompanied by decarboxylation. The different quinones exhibit varying degrees of activity. The most active is the quinone of hydroxyhydroquinone while that of pyrogallol is completely inactive; carboxyl groups in the catalyst molecule, e.g. that in protocatechu acid, exert an inhibitor effect. Thus, not only the value of the normal potential but also chemical structural effects determine the efficiency of the catalyst. This is particularly borne out by the action of metal salts on these catalysts which vary with the quinone tested and also with the buffer used (SCHUWIRTH 1038)). These activators are effective only when oxygen serves as the acceptor.

In the well-known STRECKER Reaction primary amines, notably amino acids, are oxidized by alloxan. The problem of the reversibility of the latter as a redox system has been investigated by RICHARDSON and CANNAN 979a) and more recently by E. S. HILL 484). The reduced form of the system is dialuric acid. The normal potential at ph 7 is approximately $+ 0.06$ volt; this value is extrapolated from the value actually measured in $10^{-3}$ M. alloxantin solutions at ph below 6 since the system is unstable above ph 6. Upon adding small amounts of ferrous salts the potential is stabilized due to complex formation (HILL and MICHAELIS 484a)). Dialuric acid is autoxidizable with a rate maximum at ph 7. Inasmuch as HCN does not inhibit the oxidation at this ph, a true autoxidation is indicated; in acid solution, however, the process is an iron catalysis. The alloxan formed in this reaction will dehydrogenate various amino acids to a small extent (yield less than 5 per cent).

With respect to a different, hypothetical formulation of the reaction with an intermediary coupling between the $NH_2$-group and alloxan to yield uramil which in turn will again react with alloxan,

the review by FRANKE (288), p. 173) should be consulted. Alloxan will also dehydrogenate thiols (LABES 688)).

In the course of the catalytic oxidation of amino acids by isatin there is formed as a probably autoxidizable reductant isatyd:

### Acceptor Respiration.

The following section covers mainly model systems without a full consideration of biological correlations. In describing these models it makes no difference whether the redox systems tried actually participate in the physiological mechanism of cell respiration or whether the entire process is independent of the presense of metals. According to BARRON et al. 775) the catalytic efficiency of the various dyes will depend (1) on the rate with which the dye is reduced by the cell and (2) on the rate with which it is reoxidized by air. These two properties are, other conditions being equal, a function of the redox potential of the dye (p. 90). It need not be mentioned that a dye, in order to affect intact cells, must be able to permeate.

Respiration, evoked by synthetic dyes: When MEYERHOF 784a) discovered the phenomenon of "methylene blue respiration", i.e. the oxygen uptake by dead bacteria and acetone dried yeast cells upon the addition of methylene blue, hopes ran high that this discovery would lead to new and deeper insight into the mechanism of cell respiration. In the retrospective we have to admit that these hopes have been realized to only a small extent. The reason is that these model reactions are frequently too far removed from physiological conditions and that most of the acceptors employed show a more or less pronounced toxicity which tends to cloud the issue. The most important conclusion is that the artificial acceptor respiration will not become appreciably manifest in the presence of a fully active terminal respiratory system; this has been established particularly by BARRON and HARROP 73, 74, 82, 86) who studied the acceptor respiration in non-respiring or weakly respiring cells showing an aerobic glycolysis. The results obtained with enucleated erythrocytes are ambiguous because the addition of the dye will produce the additional phenomenon of methemoglobin catalysis. The observations made with echinoderm eggs, tumor cells, and other cells with an aerobic glycolysis, e.g. retina, are more readily interpreted.

The effect of methylene blue on the small respiration of anaerobic lactic acid bacteria has been studied by BERTHO 108).

One would expect to observe an acceptor respiration of fully respiring cells only if the respiratory ferment is poisoned by HCN. Inasmuch as the terminal respiratory system in such cells, under normal circumstances, is capable of oxidizing practically any amount of metabolic hydrogen, the acceptor respiration should find no place since no acceptor can efficiently compete with the respiratory ferment. But, contrary to expectation, acceptor respiration has also been found in normally respiring cells. This may perhaps find its explanation by two special mechanisms. On the one hand the dye may react directly with hemin systems in the cell, even with the respiratory ferment; it may keep the hemin enzyme in the oxidized form just as methylene blue will produce methemoglobin from hemoglobin. Or the dyes may affect the rate of some of the anaerobic stages of cell metabolism in such a manner that more metabolic hydrogen is placed at the disposal of the terminal oxidation system. Both mechanisms may thus cause an increase in over-all respiration.

As donators for this type of respiration the same metabolites are utilized as for normal respiration, i.e. sugar decomposition products. Sugars themselves are not attacked (BARRON) unless phosphopyridine nucleotide (co-dehydrogenase) is added; the process is accompanied by an esterification of inorganic phosphate, at least in the case of hemolyzed erythrocytes (RUNNSTRÖM 1013), see also 1014)). This helps one to understand why the action of the dyes runs about parallel to the intensity of the anaerobic metabolism of the cells studied (BARRON). Another factor determining the effectiveness of the dye is its redox potential.

The methylene blue respiration of cyanide poisoned tissues has been examined by TORRES 1184) in MEYERHOF's laboratory. She found increases in respiration with liver and spleen of 200 to 400 per cent, with kidney of 100 to 150 per cent. Tissues poisoned by CO behave like those treated with CN' (775)). Embryonic tissue, however, gave different results (BODINE 132)). The "acceptor specificity" found by L. STERN 1100) and FLEISCHMANN 348) in the case of cyanide poisoned tissue as expressed by the different effect of dyes on the oxygen uptake is partly to be explained by the level of the normal redox potential and partly by toxic effects on the dehydrogenases. With succinic acid as donator there was no difference between methylene blue, janus green, cresyl blue, or neutral red. Thionine caused an increase of only 40- to 50 per cent as compared with the dyes just mentioned.

In erythrocytes the respiration, evoked by dyes, is frequently linked up with the formation of the catalytically active, intracellular redox system hemoglobin $\leftrightarrows$ methemoglobin (p. 82). In other words, the dye will not only transfer oxygen to the dehydrogenases but it will also, if sufficiently positive, oxidize some hemoglobin to methemoglobin. In the presence of the dye the reduced hemoglobin will regenerate methemoglobin upon reoxidation by $O_2$ instead of oxyhemoglobin. Whether acceptor respiration in red cells proceeds directly or with the participation of methemoglobin as a secondary oxidant of the metabolites depends essentially on the potential of the dye (MICHAELIS and SALOMON 819)); with respect to the same dye, e.g. methylene blue, the relative concentrations will determine the extent to which the two mechanisms operate (WARBURG). It is interesting to note that dyestuffs of a potential more negative than that of indigo disulfonate are without effect on respiration though they are still able to form some methemoglobin. Free flavins have a potential which is too negative to be effective here (1214, 1215)). All indicators with a positive potential will stimulate red cell respiration provided they are able to permeate and that they are not toxic. Methemoglobin production is not a necessary corrolary of this phenomenon. Nile blue and cresyl blue are stated to stimulate respiration without forming methemoglobin (775)). Another exception is lawson, a naturally occurring pigment of the naphthoquinone type: it has a relatively negative potential, forms no methemoglobin, but it stimulates red cell oxygen uptake (FRIEDHEIM 368, 373)). Pyocyanine, on the other hand, has been shown to form methemoglobin when brought in contact with erythrocytes (STERN 1077a)). In cells free of hemoglobin no other heme compounds appear to act as intermediary catalysts since their acceptor respiration is fully cyanide-resistant (BARRON and HAMBURGER 81)).

The respiration of t u m o r  t i s s u e is also increased by methylene blue, appreciably without KCN but much more strongly in presence of KCN (BARRON), where the main respiration is almost completely suppressed. However, the relationship between dyes, glycolysis and respiration is still quite obscure (ELLIOTT). Almost every dye seems to act differently; according to DICKENS 203), methylene blue increases both respiration and aerobic glycolysis while the latter is decreased by toluylene blue and pyocyanine. The effect of dyes on the respiration of tumors depends on their redox potential. Thionine and brillant cresylblue have the strongest effect. When glucose is added it is oxidized. In the case of retina which, like tumor tissue, will show an aerobic glycolysis, FLEISCH-MANN and KANN 349) find a respiratory increase up to 150 % in the absence and more than 1000 % in the presence of CN'. In sea-urchin eggs not only respiration but also anaerobic and aerobic glycolysis are stimulated by suitable dyes (477)).

Attempts to elicit an additional acceptor respiration in n o r m a l l y  r e s p i r i n g c e l l s have been partly successful; however, they are difficult to interpret. The fact that the extent of this accessory respiration depends, among other things, on the redox potential of the dye added may be taken to suggest an interaction between dye and respiratory enzyme. In any event, the toxicity of most dyes becomes a disturbing factor already in the range of low concentrations; it dominates the picture if high concentrations of dyes are used.

In a concentration of $10^{-3}$ M. which is to be considered a high concentration for an indicator dye and which is often close to its solubility limit, the dyes will inhibit tissue respiration in most instances; in the case of liver and tumor slices and in the presence of glucose a stimulation of respiration is observed (ELLIOTT and BAKER 245)).

A number of dyes will increase the anaerobic and aerobic glycolysis (395, 202, 203)). The latter phenomenon represents an inhibition of the PASTEUR effect, i.e. of the check exerted on fermentation by oxygen (or by the respiration). In other instances, e.g. with pyocyanin, respiration is stimulated and glycolysis is decreased (FRIEDHEIM 372), DICKENS 203)); for details see OPPENHEIMER's "SUPPLEMENT" p. 1292.

The following details may be of interest: upon vital staining of the vacuoles of algae the rate of respiration is increased; the effect is more pronounced with methylene blue than with neutral red which has a more negative potential (GENEVOIS 391)). Similar observations were made with *Elodea* (ALBACH 17)). Bakers yeast which has a strong normal respiration ($Qo_2$ = about 200 cmm. oxygen per mg. dry weight per hour) reacts differently: at acid ph there is only a small initial increase upon adding neutral red, falling below the normal level after about one hour; at ph above 7 the rate of oxygen uptake is diminished from the beginning (GEIGER-HUBER 390)). On the other hand, p-phenylene diamine (HARRISON 456)) and minute concentrations of quinone (SOKOLOFF 1063)) were found to enhance the respiration of bakers yeast. In *Sarcina lutea* methylene blue causes at first a strong increase and later a decrease in respiration (GERARD 396)). When methylene blue ($3 \cdot 10^{-5}$ M.) is added to locust embryos (*Melanoplus*) during the normal diapause which is characterized by a strong decline of respiration, the normal respiratory level is restored; the same holds for cyanide poisoned embryos. Carbon monoxide inhibits the acceptor respiration both when added during normal development or during the diapause. Strong light reverses the inhibition, thus indicating that an iron system takes part in the acceptor respiration (BODINE and BOELL 132)).

The respiration of muscle is stimulated by methylene blue (U. S. v. EULER 326)). According to DICKENS 203) thionine and brilliant cresyl blue raise the rate of oxygen uptake of rat kidney and brain up to 90 per cent. A number of dyes are known to increase the respiration of normal rat tissues (471)); the subsequent decrease suggests that the dyes are somewhat toxic and that the process does not represent a pure acceptor respiration. The same conclusion is to be drawn from the experiments of AXMACHER 48); in certain instances he did not observe a stimulation in presence of CN'. Lactoflavin has been reported to increase the respiration and to decrease the aerobic glycolysis of connective tissue cultures (LASER 705)); the whole flavoprotein (yellow enzyme) showed no effect either because of its more positive redox potential or to its inability to penetrate into the cells (Mol. Wt. 80,000). If liver tissue is carefully freed from the flavin system by washing, pyrrol will act as accessory respiratory catalyst like methylene blue; lactate is a suitable donator (BERNHEIM 100)). It is suggested that a red pigment, present in liver, which is not identical with hemoglobin, is required for the catalysis.

In his experiments on the mechanism of the PASTEUR effect LIPMANN 606) added redox dyes to cell-free yeast extracts. He observed an acceptor respiration with and without an inhibition of fermentation. Dyes with a positive redox potential will inhibit not only yeast fermentation but also the glycolysis in muscle extracts (717)). Pyocyanine, though of a slightly more negative potential than methylene blue, shows a stronger effect (712)). These experiments have been repeated and extended by MICHAELIS and SMYTHE 826). These authors employed a far greater series of dyes than LIPMANN. They conclude that neither the potential of the indicator nor the potential measured in fermentation systems determines the degree of inhibition of fermentation. The dyes studied by these investigators fall in three groups with regard to their effects: The first group causes the yeast extracts to respire without a decrease in the fermentative activity (Gallocyanine, phenosafranine, neutral red); the second group inhibits fermentation by destroying certain enzymes, particularly carboxylase (methylene blue, pyocyanine, and others); the third group, finally, inhibits fermentation by suppressing the formation of the first degradation products of the sugar as evidenced by the fact that only addition of hexose diphosphate is able to revive the fermentation. The acceptor respiration caused by the dyes runs somewhat parallel to their normal potential whether or not the fermentation is affected. It is concluded that respiration and fermentation are uncorrelated processes under these conditions, in contrast to the observations of BARRON made with intact cells.

A special rôle as "respiratory catalyst" has been played by p-phenylenediamine; it is dehydrogenated by "activated" oxygen to the quinone diimine which may be detected by condensation with naphthol (Na-di reagent). BATTELLI and L. STERN and later SZENT-GYÖRGYI employed it extensively as a reagent for "main respiration" via an "oxydone". Building on the ideas of PALLADIN, SZENT-GYÖRGYI revived the concept of the "respiratory pigments" and attributed to p-phenylene diamine the function of an intermediary catalyst. The original experiment of SZENT-GYÖRGYI 609) consisted in the demonstration that washed muscle will oxidize the diamine and that the addition of succinic acid or lactic acid, under anaerobic conditions, will hydrogenate the diamine formed. In the case of lactic acid the system is supplemented by cytoflav, a yellow pigment described by SZENT-GYÖRGYI in his reports on the purification of the co-enzyme of lactic acid dehydrogenation from heart muscle (1125)). Cytoflav is now known to represent riboflavin phosphoric acid ester. Since these early experiments the relationships existing under these circumstances have been partly clarified. Today we know that the so-called indophenol oxidase which appeared to be specific for diamine oxidation is identical with the respiratory ferment of WARBURG and with the cytochrome oxidase of KEILIN. In other words, the catalysis caused by p-phenylenediamine in respiring cells proceeds via the respiratory ferment; the process is cyanide-sensitive (609a)). A complicating factor arises out of the observation that the Na-di reaction is also exhibited by very simple metal systems (WERTHEIMER 1295), WIELAND 1312)). This makes it possible to construct models without the respiratory enzyme. HARRISON 456) has done so by adding $H_2O_2$ + peroxidase to the SCHARDINGER enzyme + p-phenylenediamine. In his experiments with yeast, however, and in those with other cells p-phenylenediamine acted undoubtedly as go-between by transferring hydrogen from dehydrogenase-substrate systems to the respiratory ferment. This explains why RUNNSTRÖM 1011) and ÖRSTRÖM 893) found this accessory respiration in fertilized and unfertilized sea-urchin eggs to be just as sensitive against CN' and CO as has been observed by WARBURG in normally respiring yeast cells; the same holds for the eggs of *Melanoplus* (134)).

**Nitro Respiration:** Not only quinoid dyes but also certain organic nitro compounds are capable of causing an increase in cell respiration which may or may not be due to their functioning as reversible redox systems. The fact that all of these substances are appreciably toxic even in the low concentrations employed for respiration experiments makes it more difficult than in the case of quinones to ascertain where acceptor respiration ceases and where an oxygen uptake due to a disturbance of the physiological relationships by the toxicity of the acceptor begins.

We shall refrain from discussing here the early experiments by LIPSCHITZ with o-dinitrobenzene or his later ones with nitroanthraquinone. The irreversible hydrogenation of these nitro compounds by cell constituents bears no relation to an acceptor catalysis; it may at best serve as an arbitrary measure of the entire donator-dehydrogenase system of the cell. In the case of o-dinitrobenzene, o-nitrophenylhydroxylamine is formed by the reduction. The latter has recently been prepared in pure state by KUHN and WEYGAND 687).

The question of the reversibility of certain dinitro phenols, on the other hand, is still open. Of those the action of which has been studied in some detail, we mention dinitro α-naphthol, 1,2,4-dinitrophenol (Thermol), and 2,4-dinitro-o-cresol. MAGNE et al. 746) found that 1, 2, 4-dinitrophenol causes an enormous increase in the rate of oxidations and hyperthermy in warm-blooded animals. The same effect is found with plant tissues and moulds, to a lesser degree with yeast (PLANTEFOL 932)). Working with dinitro α-naphthol in concentrations of about $10^{-6}$ M., U. S. v. EULER 326) observed an increase in respiration of normal muscle tissue of about 100 per cent; higher concentrations of the reagent will inhibit respiration. Similar findings with various tissues, particularly after adding donators, have been reported by ALWALL 21). The effect disappears slowly. The residual respiration of cyanide- or bromoacetic acid-poisoned muscle is likewise increased by the nitro compound, provided that donators are also added. The stimulation by low concentrations and the inhibition by slightly higher concentrations appears to be a general feature of the dinitro derivatives. It has been observed in the case of 4,6-dinitro-o-cresol and of 2,4-dinitrophenol in experiments with yeast (FIELD 334)). With rat liver and kidney tissue, the optimum concentration of 2,4-dinitrophenol is $10^{-7}$; a concentration of $2.10^{-7}$ M. will already inhibit (McCORD 668)). With yeast, the most effective reagent was found to be 1-hydroxy-2-cyclopentyl-4,6-dinitrobenzene (447).

Dinitrophenol will not only increase the respiration but also the glycolysis, both the anaerobic glycolysis of normal tissues (EHRENFEST 230)) and the aerobic glycolysis of tumor cells (245)); the decrease in respiration in the latter indicates an inhibition of the PASTEUR effect by the reagent. It would appear that the main effect of these compounds consists in a s t i m u l a t i o n  o f  a n a e r o b i c  s u g a r  b r e a k d o w n; this is also true for alcoholic yeast fermentation (KRAHL and CLOWES 591)). In the light of this hypothesis the respiratory increase would merely represent an additional combustion of lactic acid as is also assumed for other dyes. DODDS and GREVILLE 218) have found an increase in respiration of normal rat tissues, including kidney which is one of the tissues devoid of aerobic glycolysis and hence not responsive to methylene blue catalysis (BARRON). In the case of tumor tissue dinitrophenol stimulates respiration as well as glycolysis (219)). Here, as in frog muscle (1002)), the increase in lactic acid production dominates the picture. MUNTWYLER 844) was unable to find an increase of oxygen uptake with rat kidney, but he did find it when working with rat liver and frog kidney. Extensive investigations on the effect of dinitro compounds on yeasts have been carried out by FIELD et al. 334, 332,) GENEVOIS 392), and CRÉAC'H 189).

The question of the reversibility of dinitrophenols as oxidation-reduction systems has been studied by GREVILLE and STERN 421). The product of reduction of dinitrophenol by tissues is 4-nitro-2-aminophenol. This represents a difference of 6 hydrogen equivalents from the dinitro stage. Obviously, then, this compound cannot be the reductant of a reversible system the oxidant of which is the dinitro compound. The reduced forms of all reversible systems differ either in one or at most two H-equivalents or electrons from the oxidant. However, attempts to demonstrate the existence of a reversible intermediate stage met with no success. The potential at which dinitrocresol begins to be reduced is in the neighborhood of — 0.200 V., i.e. in the range of the anaerobic reduction potential of many living cells, e.g. liver. It was possible to reduce dinitrophenol by certain dehydrogenases and their substrates. While xanthine dehydrogenase plus hypoxanthine will reduce the dinitro compound directly, formic and lactic dehydrogenase will do so only in the presence of a suitable redox indicator. It could be shown that the mechanism here was an indirect one: first, the dye is reduced by the enzyme-substrate system to the leuco dye; subsequently, the leuco dye reduces the dinitrophenol. This is another case where a coupling link is required for the interaction of two systems which is thermodynamically possible but which does not take place spontaneously due to a "lack in chemical affinity". The dye may be looked upon as a carrier or as a catalyst. In any event, it is the ease with which electrons are transferred by the two forms of the indicator, its electroactivity, which enable it to play the role of mediator between the two sluggish reactants.

Cyanide poisoning of respiration prevents the dinitrophenols from exerting their effect on the oxygen uptake (332, 590, 133, 624a, 626a)). The same holds for CO (133)) and malonate (419)) inhibition of respiration. Malonate, on the other hand, does not interfere with the true acceptor respiration via brilliant cresyl blue (419)).

Taken all together, it is not very probable that the phenomenon of dinitro respiration represents a true acceptor respiration. Besides the failure to demonstrate reversibility of these compounds in vitro, the main argument is that brought forward by De Meio and Barron 774), namely, the inability of these compounds to produce oxygen uptake in cyanide poisoned cells. It might be argued however that the reductant formed by the cell may not be autoxidizable and that cyanide inhibits its reoxidation. A number of hypotheses have been offered in an attempt to explain the action of dinitrophenols. None of them appears convincing or even sufficiently plausible to merit a detailed discussion at the present time. We refer the reader to the publications by Krahl and Clowes 590, Handovsky et al. 447), Dixon and Holmes 207), and De Meio and Barron 774). Any theory of the mechanism of the phenomenon will have to take into consideration the findings of Clowes and Krahl 174) concerning the block of cell division in sea-urchin eggs accompanying the stimulation of respiration and also their observations on the effect of halogen phenols. The fact that the latter affect living cell in a manner very similar to that of nitrophenols is an indication that the "nitro respiration" is not an acceptor respiration but rather a "pharmacological" effect on the regulatory mechanism of cell respiration; v. a. Oppenheimer, Suppl. p. 1144.

**Natural Pigments** *): One of the most important natural pigments is undoubtedly riboflavin (lactoflavin, p. 184). In the cell it is present mainly in bound form, i.e. as the riboflavin phosphoric acid ester-proteid (yellow enzyme). This is not only indicated by the inability of the pigment to pass cellophane or collodion membranes when ground liver tissue is subjected to dialysis at low temperature (Stern (unpublished)), but also by the shift of the long-wave absorption band of the pigment from 440 to about 460 millimicrons when intact lactic acid bacteria are examined (Warburg and Christian 1244)). It has already been mentioned that the free riboflavin has a relatively

---

*) See also Frei (358)).

negative redox potential ($E_o'$ — 0.2 V. at ph 7) and that its coupling with the bearer protein raises the potential close to that of methylene blue ($E_o'$ — 0.07 at ph 7) (p. 195). We are faced with the situation that addition of the alloxazine proteid to media containing intact cells is likely to be ineffective due to its inability to penetrate (Mol. wt. 80,000 **(553))** while the addition of the free riboflavin is to be considered as unphysiological unless there is an opportunity for the dye to find its bearer protein in the cell under experimentation. It is not surprising, therefore, that free riboflavin does not stimulate the respiration of enucleated red cells: its potential is too negative to make it suitable as an acceptor or to cause methemoglobin formation **(1214))**. In the case of connective tissue cultures, LASER **705)** observed an increase of respiration and a decrease of the aerobic glycolysis; the whole yellow enzyme was ineffective. Flavin increases the respiration of normally anaerobic lactic acid bacteria **(6))**.

The respiration of yeast extracts is increased by non-phosphorylated flavins to only a small extent; the effect does not run parallel to the redox potential. Riboflavin has a very small effect, methyl and gluco-alloxazine have a more pronounced action (MICHAELIS et al. **822))**. It has been demonstrated by EULER and ADLER **(5a))** that riboflavin is a necessary component of several dehydrogenase systems. However, it is unable, except in the case of malico dehydrogenase, to effect the reaction of isolated dehydrogenase systems with molecular oxygen **(895))**. Furthermore, riboflavin is unsuitable as a coupling link between two dehydrogenase systems in contrast to pyocyanine which may be reduced by the more negative system and reoxidized by the positive system **(895))**.

One of the most thoroughly studied natural pigments is pyocyanine, the blue pigment of B. pyocyaneus (See also p. 228). It is both a ph and a redox indicator; it is blue in neutral and alkaline solution and red in acid solution, while the leuco form (dihydropyocyanine) is colorless. Its constitution is that of $\alpha$-oxy-N-methyl-phenazine (WREDE). The normal potential at ph 7 is —0.033 V. (FRIEDHEIM and MICHAELIS **376))**. In acid solution the reduction occurs in two distinct steps each involving the uptake of one electron. The monohydroform is a cationic free radical (semiquinone) showing a characteristic green color **(376))**. We owe careful potentiometric studies of this interesting substance to FRIEDHEIM and MICHAELIS **376)** and to ELEMA **233)**. Although the physiological function of the pigment is naturally restricted to B. pyocyaneus, it has served as a very useful model substance for those reversible and electroactive systems of other living cells which heretofore could not be obtained in pure form. One of the most important features of the pigment is that the monohydro stage (semiquinone) is also formed to some extent upon reduction in neutral solution where its concentration may reach 10 per cent of that of the fully oxidized and fully reduced forms **(822))**.

According to FRIEDHEIM **363)** pyocyanine strongly stimulates the respiration of pigment-free strains of B. pyocyaneus as well as of mammalian red blood cells. He found that the pigment shows an effect preferably on such cells which contain hemin systems; their participation in the catalysis is indicated by the inhibitory effect of cyanide. The reduced form of pyocyanine is autoxidizable. In those cases where the regeneration of the oxidized form is brought about by direct reaction with molecular oxygen, i.e. without iron, cyanide will exert no effect. Examples are the pyocyanine respiration of anaerobic bacteria (Tetanus, (FREI **358)**)) and unfertilized sea-urchin eggs (RUNNSTRÖM **1012)**). In the latter instance HCN will actually increase the pyocyanine respiration while the pigment effect on fertilized eggs, equipped with a hemin system, is abolished by HCN. When pyocyanine acts as the acceptor in the respiration of bottom yeast with hexosephosphate as donator, HCN causes an inhibition of 31 per cent (OGSTON and GREEN 895)). In bacteria the respiratory enzyme system of which has been damaged by acetone treatment the respiration is only little sensitive against HCN and CO (231)). The respiration of Staphylococcus aureus is very strongly stimulated by pyocyanine; the effect on acetone dried bacteria of a B. pyocyaneus strain was found to be very small.

The respiration of mammalian tissues is also increased by the pigment in certain cases. Thus, the oxygen uptake of liver, testis, and tumor cells is raised about 50 per cent. Pyocyanine here acts as a carrier, probably between reducing systems and the hemin system. Consequently, its action is inhibited by cyanide inspite of the autoxidizability of its leuco form. Furthermore it shows no effect on cells with a perfect respiration, i.e. exhibiting no aerobic glycolysis whatsoever (kidney) (FRIEDHEIM 372)). The substrates burned with the aid of pyocyanine are carbohydrates or their breakdown products as judged by the high R Q and also by the stimulation of the pyocyanine effect produced by the addition of such donators (895)).

However, pyocyanine may also act as acceptor in anaerobic THUNBERG tests with succinic acid and amino acids as donators (EHRISMANN 231)).

Pyocyanine possesses certain properties which are not found in other quinoid dyes, e.g. in methylene blue, and which bring it into an intimate contact with the complex reactions occurring during the first stages of carbohydrate breakdown. Pyocyanine not only is frequently more efficient than methylene blue in experiments with cells or in model systems but it behaves qualitatively in a different manner towards the coupled reactions involving phosphorylations and oxidations. In particular, it appears able to replace, to a certain extent, other natural redox systems, e.g. pyridine derivatives (RUNNSTRÖM and MICHAELIS 1014), LENNERSTRAND and RUNNSTRÖM 712)). This property is linked by these workers to the phenomenon of two-step reduction shown by pyocyanine. The oxidation of the pigment by pure oxygen is not more rapid than that of methylene blue while the reduction by mixtures of carbohydrate breakdown products is faster. LENNERSTRAND and RUNNSTRÖM assume that pyocyanine is reduced in two single steps by another natural system, perhaps by the pyridine ferment, which in turn is also dehydrogenated in single steps. *)

Phthiocol, the yellow pigment of Bac. tuberculosis (2-methyl-3-hydroxy-1,4-naphthoquinone, ANDERSON 25)), is a relatively negative redox system; $E_o'$ at ph 7.3 : —0.208 V. (BALL 57)). In the bacteria it is present almost completely in oxidized form. No respiration experiments have as yet been reported. The violet pigment Violacein of Bac. violaceus increases the oxygen uptake of bacteria freed from their own pigment by washing. The reduced form is non-autoxidizable, therefore the dye must act as an intermediate hydrogen carrier (FRIEDHEIM 364)). According to the chemical investigations of WREDE 1346) and TOBIE 1176) the pigment is a pyrrol derivative. Chlororaphin (α-hydroxyphenazine amide, KÖGL 586a)) the green pigment of Bac. chlororaphis, represents a fully reversible redox system sharing many features with pyocyanine (ELEMA 234)). The normal potential, at ph 7.42 is $E_o'$—0.139 V.; the reduced form is stated to be non-autoxidizable. Below ph 4 the reduction is effected in two more or less overlapping steps with intermediate semiquinone formation. The effects of the dye on cells have not yet been determined. They ought to be similar to those of the chemically related pyocyanine though less pronounced because of the more negative potential of chlororaphin.

---

*) Compare p. 104 concerning the fundamental significance of semiquinone catalysis.

Toxoflavin, the prosthetic group of a highly toxic yellow pigment formed by Bacterium bongkrek (van Veen and Mertens 1195)) represents the oxidant of a redox system which is fully reversible and electroactive between ph 4 and 8 (Stern 1083)). The normal potential, $E_o'$, as referred to the normal hydrogen electrode, at ph 7.0 is —0.049 V. The slope of the individual titration curves is atypical throughout the ph range studied and shows an increasing steepness, indicative of semiquinone formation, towards the alkaline range. However no color change suggesting semiquinone production was observed, and the matter is further complicated by the instability of the pigment at ph > 8 which makes potentiometric measurements in alkaline solutions impractical. Although the chemical structure of this pigment is as yet obscure, the evidence available militates against a close chemical relationship to the class of flavin pigments studied by Warburg, Kuhn, Ellinger, Euler, Karrer and Stern. According to manometric experiments by Greville (1083)) toxoflavin will stimulate the respiration of non-nucleated mammalian red blood corpuscles to about the same extent as thionine ($E_o'$ + 0.062 at ph 7). The pigment reacts with oxyhemoglobin to form methemoglobin. Toxoflavin stimulates the respiration of Jensen rat sarcoma in glucose containing media (in one case a 140 per cent increase in $O_2$-uptake was caused by a $2.10^{-5}$ M. solution). The effect however decreases rapidly, probably because of the toxicity of the pigment which is brought out even more clearly by brain slices where a concentration of $10^{-5}$ M. already inhibits strongly the respiration.

Penicillium phoeniceum has been found to contain a reversible system, called Phoenicein, with a potential of —0.037 V. at ph 7.95 (Friedheim 369)). The reduced form is autoxidizable. The mould respires fully without any addition (RQ 1.0); it contains no cytochrome. The respiration here is possibly iron-free. The pigment will increase the oxygen uptake of Bac. pyocyaneus 3 to 4 times. Another accessory respiratory catalyst, according to Friedheim, is the pigment Nigrosin in Aspergillus niger. Its chemical composition is unknown.

Echinochrome (Cannan 161), Friedheim 365)), a pigment or group of pigments occurring in sea-urchins e.g. Sphärechinus granularis, Arbacia punctulata, and Echinus esculentus appears to stimulate the oxygen uptake only of cells having a hemin system (sea-urchin eggs, red blood cells). The normal potential, $E_o'$, at ph 7 is —0.220 V., i.e. as negative as that of free riboflavin (Cannan 161)). The leuco form is not autoxidizable. Recently, pigments of this group have been obtained in crystalline form (Ball; Stern (unpublished)). The empirical formula of the crystalline Echinus esculentus pigment is $C_{10}H_6O_8$. The constitution is not yet elucidated. It is possible that there exist close chemical relationships between this pigment and pigments from Arbacia aequituberculata Bl. and Strongylocentrotus lividus recently isolated by Lederer and Glaser 707). The physiological function of echinochrome is not known. It has been observed that the pigment undergoes some changes in its distribution in the cells during mitosis.

Another interesting invertebrate pigment is Hallachrome, present in the marine worm Halla parthenopea (Friedheim 371)). This reversible system ($E_o'$ at ph 7 + 0.022 V.) was the first example of any redox system to be found to show a tendency for semiquinone formation in the alkaline ph-range; all the previously studied systems like pyocyanine will form appreciable amounts of semiquinone in the acid range. The identity of hallachrome with the "red body" (p. 118) formed as an intermediary in the tyrosine-tyrosinase reaction which has been postulated by Mazza and Stolfi 770) and Raper 958) is questionable. The pigment stimulates the respiration of Ascaris worm eggs, sea-urchin eggs, and red blood cells. The leuco form which is preponderantly present in asphyxiated animals is apparently non-autoxidizable; the pigment, then, would act only in cooperation with hemin or other metal systems.

In this connection one might also mention the pigment of the nudibranch *Chromodoris zebra* which, according to PREISLER **936)**, is a reversible redox system. The reduction of the blue oxidized form to the yellow reduced form appears to involve only one hydrogen equivalent. The leuco form is autoxidizable. The pigment of *Arion rufus* has a normal potential, $E_o'$ of —0.027 V. at ph 7 (FRIEDHEIM **367)**). It is readily reduced, e.g. by cysteine or hydrogen activated by palladium.

Whether the blue Asterinic acid, found by EULER et al. **266)** in certain marine crustacea and echinoderms, is a reversible redox system has not yet been ascertained. It seems to be closely related to Astacin, the carotenoid pigment present as a conjugated protein in the shell, hypodermis and the eggs of the lobster. The latter pigment, termed Ovoverdin, shows a remarkable resistance to reducing agents; it is dissociated reversibly by heat into the carotenoid and the protein moiety (STERN and SALOMON **1096a)**). The constitution of astacin itself, i.e. of the prosthetic group, is that of a tetraketocarotene (KARRER and LOEWE **523a)**).

Certain higher plants contain reversible redox systems in addition to the probably ubiquiteous riboflavin (KUHN, KARRER, STERN) and carotins and xanthophyll, the reversibility of which is questionable. Juglon (5-hydroxy-naphthoquinone), the pigment of walnuts (*Juglans regia*), has an $E_o'$ at ph 7 of + 0.033 V. (FRIEDHEIM **368, 373)**). It exists in the nut mainly in the reduced form. Lawson (2-hydroxynaphthoquinone) is the well-known red henna pigment from the leaves of *Lawsonia inermis*. In accordance with its potential ($E_o'$ at ph 7 : —0.139 V.) this pigment is found mainly in the oxidized form. In experiments with mammalian blood cells both pigments increase the respiration about 5 to 6 fold. The negative potential of Lawson excludes the possibility that the catalysis here proceeds via intermediary methemoglobin formation; the potential of Juglon, on the other hand, would permit of such a coupled catalysis. FRIEDHEIM has used this observation as an argument against the hypothesis that such dyes exert their stimulation on red cell respiration only through methemoglobin formation.

Other naphthoquinones occurring in plants are Lomatiol and Lapachol which are isomer alkylated 3-hydroxy-naphthoquinones. Their potentials have been measured by BALL **58)** (see p. 238). Droseron from *Drosera* **739)**, according to DIETERLE and KRUTA **205)**, is identical with Plumbagin from *Plumbago* (2-methyl-5-hydroxy-naphthoquinone). Its potential has not yet been measured but only been computed.

Hermidin, the chromogen of *Mercurialis*, is oxidized in two stages. The leuco form yields at first an unstable blue pigment, cyanohermidin, finally a brown one, chrysohermidin. Only the former serves as hydrogen acceptor (HAAS and HILL **433)**). CANNAN **160)** found the system to be reversible; he does not comment on the autoxidability. The potential happens to be almost identical with that of the chemically very different pyocyanine ($E_o'$ —0.034 V. at ph 7).

Anthocyans are stated to act as hydrogen acceptors in THUNBERG experiments with purified liver aldehydrase (REICHEL **967)**). They are subsequently reoxidized by air. The potentials and their biological significance remain to be determined.

We may as well append some remarks on certain redox systems present in higher animals, though nothing is known about their actual function. The action of the adrenalin quinone (Omega substance of KISCH, Adrenochrome of GREEN and RICHTER, Biochem. Jl. **31**, 596 (1937)) has already been mentioned on p. 106. KISCH maintains that not adrenaline itself but the quinone acts as a respiratory stimulant. U. v. EULER **325)** however has seen an increase in respiration in muscle tissue when working with adrenalin concentrations of $10^{-8}$ to $10^{-14}$ and at suboptimal oxygen tensions. Only hexose phosphate and glycerophosphate but not lactic acid would act as acceptors. There was no effect to be found with leucocytes. This phenomenon is hardly a pure acceptor respiration.

The system Tyrosine $\leftrightarrows$ "Red Body" (quinone of dihydroxy-indolcarboxylic acid) was found to be reversible (FRIEDHEIM **366, 370, 642, 642a)**. It is thought to function in the chain of reactions leading to melanine formation from tyrosine by tyrosinase (RAPER **958)**. However, the quinone was also found to dehydrogenate anaerobically purines, succinic acid, and cysteine; it may therefore be a biological hydrogen transporter (**374)**). In the presence of heavy metal the oxidation, initiated by tyrosinase, does not arrest itself at the "red body" stage but proceeds irreversibly to melanines.

The system Homogentisic acid $\leftrightarrows$ Quinone acetic acid is also reversible (BLIX **128)**, FISHBERG and DOLIN **346)**). It has the rather positive normal potential of $+ 0.250$ at ph 7. Its function, though undetermined, might conceivably be concerned with the deamination of amino acids like the Adrenochrome.

The chromogen Tyrin of wide-spread distribution which is not identical with tyrosine and which is oxidized by benzoquinone to a red pigment (SZENT-GYÖRGYI **1114)**) is stated to be an unspecific mixture of amino acids (PLATT and WORMALL **934)**).

## III. Oxidative Catalysis via Peroxides.

### 1) Organic Peroxides.

There are instances where an organic compound is catalytically oxidized in the classical sense, i.e. by the uptake of oxygen into the molecule, without the possibility of interpreting this process as a dehydrogenation. Here the oxygen attaches itself to v a l e n c y g a p s, in other words, to double bonds, to form a t r u e p e r o x i d e. Insofar as such a peroxide might conceivably oxidize other molecules by giving off atomic oxygen and thus bring about a true activation of oxygen, a remainder of the old ENGLER-BACH theory of biological oxidation may be preserved. We have to distinguish between two different problems in this connection: (1) formation of well defined peroxides as intermediates and (2) catalytic capability of such peroxides, i.e. their power to transmit the oxygen in active form according to ENGLER's schema $AO_2 + B \rightarrow AO + BO$ or $AO_2 + 2 B \rightarrow A + 2 BO$. It must be kept in mind that such a peroxide, instead of acting as an oxidation catalyst, may also break down without oxygen transfer and thus represent only an intermediate stage in desmolysis.

Organic peroxides may arise through the action of $H_2O_2$ as well as through autoxidation: $H_2O_2$ attaches itself to valency gaps with resulting peroxide formation, e.g. from aldehydes (WIELAND). While this is an interesting possibility from the point of view of model systems, it has no great biological significance since, in general, hydrogen peroxide in respiring cells is decomposed by iron catalysts (catalatic or peroxidatic) at a rate too great to permit a direct interaction with substrates. The direct oxidation of pyruvic acid by $H_2O_2$ in anaerobes must be listed as an exception.

**Bach's Oxygenases:** It will be remembered that BACH considered the oxidases to be complex systems made up of organic peroxides ("Oxygenases") and of peroxidases; the latter, by acting on the former, release active oxygen. This concept has been abandoned in general and in particular with respect to the oxidation of chromogens for which it was developed. Bach's oxygenases are actually quinones insofar as they are not simple heavy metal systems. Of late, the term has been revived by english authors, e.g. J. B. S. Haldane (cf. 181)), to designate the respiratory ferment of Warburg. It has been shown by Szent-Györgyi 1116) that o-quinone will oxidize guaiac to a blue pigment without the aid of a catalyst. Subsequently he has criticized the entire peroxide theory of Bach including the modification proposed by Wheldale-Onslow. The latter author had recognized the relation of the purely organic component of Bach's oxidases to pyrocatechol; however, the peroxide nature of this component was still stressed. It is now established that the pigments of biological interest will not form well-defined peroxides, not even in the photo-oxidation by dyes (Gaffron 386)). The only known peroxides of this type are formed by rubene.

Later Bach has attempted 55) to explain the specific case of the reaction of p-quinone with pyrogallol by peroxide formation ($HO \cdot C_6H_4 \cdot O \cdot OH$). By referring to the fact that no oxidation takes place here in a water-free system he assumes that the quinone splits water and is thus transformed into the peroxide hydrate. It may be, however, that the water is required for dissociation of the diphenol, the dehydrogenation in non-aqueous solution being inhibited by the stability of the hydrogen atoms under these conditions.

It is obvious now that the original theory of Bach has mixed up two entirely different things: the substances thought to be organic peroxides ("oxygenases") are actually quinones and have no direct relationship to Bach's "peroxidases". The latter may be enzymes as well as biological iron complexes of similar action, like hemins, or even simple iron systems, in short every compound able to utilize $H_2O_2$ in reactions of the peroxidatic type. The secondary action of $H_2O_2$ will have to be taken into account whenever autoxidation takes place in the presence of iron. (See also Bach 50), Pugh 938), and Wieland 1328)).

**True Organic Peroxides:** Even though it is to be admitted that the part of Bach's theory dealing with "oxygenases" is untenable in the light of newer knowledge, this does not mean that t r u e organic peroxides may not intervene in oxidation catalyses.

It cannot be denied that certain peroxides arise through simple autoxidation and in such a manner that even Wieland does not interpret the reaction as a dehydrogenation. As has been found some time ago by Wieland and later confirmed by Jorissen and van der Beek 507), aldehydes, when oxidized in the absence of water, will yield peracids. Wieland and Richter 1323, 1324) have observed recently that benzaldehyde forms benzoperacid even in aqueous solution and since quinone or methylene blue are not reduced, not even in presence of ferrous iron, the reaction cannot be interpreted as a dehydrogenation of a benzaldehyde hydrate in a manner analogous to the dehydrogenation of acetaldehyde.

In this instance, then, the primary addition of the "unfolded" oxygen molecule (—O—O—) does not cause the loss of $H^+$ by the donor but rather the transformation of the primary moloxide (I) into a true peroxide, which here is a peracid (II):

$$R \cdot CH \underset{\diagdown O \diagup}{\overset{\diagup O \diagdown}{O}} \quad \rightarrow \quad \begin{matrix} R \cdot C = O \\ | \\ O - OH \end{matrix}$$

I.                                II.

The perbenzoic acid in turn oxidizes a second molecule of benzaldehyde to benzoic acid in the absence of water, while in aqueous solution the aldehyde hydrate is dehydrogenated (see below). WIELAND assumes that in this instance the carbonyl group and not the hydrogen is "activated" which would mean that the oxygen molecule is deformed by the residual valency forces of the double bond in the same manner as is usually accomplished by polyphenols, leuco bases or ferrous iron. According to KUHN (p. 125), though, the so-called autoxidation of benzaldehyde is in reality a heavy metal catalysis; a view held to be too radical by WIELAND **1323, 1324)**. He admits that the reaction, in most cases, is indeed a metal catalysis, but he maintains that even perfectly pure benzaldehyde is oxidized. The autoxidation of benzaldehyde has been treated by HABER and WILLSTÄTTER **437)** as a chain reaction, involving unpaired radicals and initiated by heavy metal. In agreement with this view the process may be checked by substances known to function as chain-breakers (JEU and ALYA **502a)**, SCHWAB et al. **1040)**).

Similar peroxides are produced by the reaction of aldehydes with $H_2O_2$ (WIELAND). Here the $H^+$ migrates to the carbonyl group, yielding for instance dihydroxymethyl

peroxide,    $\overset{\displaystyle H_2}{\underset{\displaystyle OH}{{>}C}}-O-O-\overset{\displaystyle H_2}{\underset{\displaystyle OH}{C{<}}}$ , from formaldehyde. Ethyl alcohol may thus give a

peroxide of the constitution $CH_3\cdot CH\overset{\displaystyle OH}{\underset{\displaystyle O-O-H}{{<}}}$    (WIELAND **1300)**). RIECHE **983)** has

isolated similar compounds. The breakdown of these peroxides presents itself as a normal dehydrogenation within the molecule itself or between two molecules. Thus the formation of acetic acid from peracetic acid in aqueous solution is the result of the dehydrogenation of acetaldehyde hydrate by the peracid:

$$CH_3\cdot CH(OH)_2 + O\cdot OH \quad \rightarrow \quad 2\ CH_3\cdot COOH + H_2O.$$
$$\underset{\displaystyle O}{{>}}C\cdot CH_3$$

In this way peroxides may function as intermediates in dehydrogenations (WIELAND **1300)**). In the course of enzymatic dehydrogenation of aldehydes, however, no peroxides have been detected. It appears that here the oxidation proceeds from the very beginning as a pure dehydrogenation via the aldehyde hydrates.

The oxidation of pyruvic acid by hydrogen peroxide,
$CH_3\cdot CO\cdot COOH + H_2O_2 \rightarrow CH_3\cdot COOH + CO_2 + H_2O$ (SEVAG **1045)**), may proceed through a peroxide stage which has been described previously by WIELAND and WINGLER **1330)**. WIELAND **1300)** assigns the structure $CH_3-\overset{\displaystyle \frown}{\underset{\displaystyle OH\ O\cdot OH}{C}}-COOH$ to this compound. Peroxide

formation is also postulated in the course of oxidative deamination of amino acids. BERGEL et al. **95)** have shown that not only ordinary amino acids but also those carrying substituents at the nitrogen atom are oxidized with animal charcoal as the catalyst. Inasmuch as a primary dehydrogenation is excluded in this case, the authors assume intermediary peroxide formation involving the nitrogen atom:

$$R \cdot CH - N \underset{\diagdown O}{\overset{\diagup O}{<}} \overset{|}{\underset{|}{\big|}} .$$
$$\underset{COOH}{|} \quad (CH_3)_2$$

KREBS 604) hypothecates an analogous schema for non-substituted amino acids, but only for the unnatural d-amino acids, obtaining during the oxidation by amino acid dehydrogenase. The peroxide configuration which he has in mind, however, corresponds to the usual "adduct" rather than to a well-defined peroxide; to postulate the latter would be superfluous in view of the smooth and rapid pure dehydrogenation by acceptors. This point of view is supported by the fact that $H_2O_2$ is produced as a result of dehydrogenase action (KEILIN 546)) and that dehydrogenases will not attack amino acids devoid of H at the N, though this should not interfere with peroxide formation in the sense of BERGEL.

When unsaturated f a t t y  a c i d s undergo autoxidation, a peroxide of the probable configuration R—CH—CH—R·COOH is formed. As an alternative structure the

$$\underset{O - O}{\overset{| \quad |}{}}$$

formula R—C=C—R is discussed by ELLIS 249). The latter formulation, however,

$$\underset{OH \; OH}{\overset{| \; |}{}}$$

may be considered to represent an equilibrium form, existing in the further course of the breakdown of the substrate, rather than a substitute for the peroxide. ELLIS has isolated an "oxidoelaidinic acid" from elaidinic acid subjected to autoxidation; he was also able to synthesize it. The compound is relatively stable. This peroxide formation by addition of —O—O— may be a spontaneous process as indicated by the term autoxidation. But it is possible to accelerate the reaction by suitable catalysts, e.g. hemins (p. 124), other iron compounds, and a variety of organic compounds. The same applies to the case of the c a r o t e n o i d s. Because of their many double bonds they are much more rapidly attacked by oxygen than the fatty acids. The carotenoids are able to transfer oxygen (900)); at the same time they suffer deep-seated oxidative decomposition (FRANKE 354)).

The acceleration of autoxidation by catalysts has been extensively studied by FRANKE 354). Very different substances are active: bases, like aniline, certain amino acids (proline, histidine, arginine, lysine, tryptophane, and leucine), certain sugar like substances, particularly methylglyoxal, ascorbic acid (HOLTZ 492)); and also sterols, bile acids, cysteine, the carotenoids, and vitamin A, SH-glutathione (DIXON 212)). The action of carotenoids and vitamin A has been found by v. EULER 327, 281); MONAGHAN 834) claims that they are only active in the oxidized state and that otherwise they will inhibit the oxidation. PAGE 912) found amino acids without effect on the oxidation of phosphatides. It is not possible to develop a satisfactory theory encompassing all of these activators.

In the case of the bases FRANKE assumes that the entire molecule is made more labile by the salt formation which would thus increase the affinity to the $O_2$-molecule. FRANKE and also RONA 1000) would interpret the catalyses in general as chain reactions. This is not improbable since the only substances found to inhibit the autoxidation are readily oxidizable phenols, the strongest negative action being exhibited by adrenaline (FRANKE). These inhibitors act as "antioxygens" by breaking off the chain. RONA pictures their action as causing a dissociation of the primary addition product of oxygen and substrate into the components. The activating effect exerted by pyridine and nicotine is interpreted in terms of stabilization of the primary adduct. A detailed study of the action of various polyphenols as antioxygens is due to MATTILL 758). According to this author and to his collaborator OLCOTT 900) a prerequisite for this action is the attachment of two OH groups directly to the benzene nucleus with the exception of the naphthols.

It is of interest to inquire whether the promotion of peroxide formation at the double bonds of unsaturated fatty acids by readily oxidizable activators is due to a primary peroxide formation at the latter. This question cannot yet be answered. In the case of catalysis by thiols, for instance, the possibility of a peroxide configuration at the sulfur has been considered (SZENT-GYÖRGYI), whereas HARRISON 455) suggested the formula $R \cdot S \cdot OH$.

The acceleration brought about by carotenoids is of biological interest. They appear to promote not only the oxidation of unsaturated fatty acids, but also that of other systems (v. EULER 258)), e.g. ascorbic acid + glutathione, especially in the presence of adrenaline. JOYET-LAVERGNE 514) reports that vitamin A transfers oxygen to glutathione. His histochemical experiments were carried out with chondriosomes. Complex oxidation systems, e.g. $O_2 \rightarrow$ carotenoids $\rightarrow$ glutathione $\rightarrow$ fatty acids, are another possibility.

Catalytic action in oxidation processes has been attributed to a number of lipoids, e.g. to a water-soluble phosphatide of plant origin (GUTSTEIN 425)), colloidal lecithin (MAGAT 745)), colloidal cholesterol (REMESOW et al. 977)). Although these effects might find their ultimate explanation by a peroxide mechanism, traces of heavy metals cannot be excluded as yet as causative agents. The same applies with even greater force to the "lipoxidases" of ANDRÉ and HOU 28) present in soya bean milk and capable of oxidizing oils and of converting guaiac into a blue dye.

If we concede the possibility that carotenoids and sterols (about ergosterol see p. 125) may form true peroxides by attaching oxygen to their double bonds, we would also have to admit that these peroxides may conceivably promote or catalyze the oxidation of unsaturated fatty acids. They may do so either by transmitting their entire oxygen to the fatty acid and thus converting the latter into an unstable peroxide or by an induced oxidation of the type visualized by ENGLER. The latter would, of course, be no longer a true catalysis, since the carotenoid peroxide, while transmitting some of its oxygen to unsaturated fatty acids, would itself undergo oxidation. FRANKE claims that this is the case. The double function of the carotenoid peroxides, then, would be that of an intermediate in the desmolysis of the carotenoids and of an agent inducing oxidations of other metabolites. The fatty acid peroxides, on the other hand, might perhaps best be regarded mainly as intermediates in fatty acid decomposition, inasmuch as we have no evidence for secondary oxidations induced by them. Just as peracetic acid represents an intermediate which may break down into acetic acid + $H_2O$ as well as into acetaldehyde + $H_2O_2$ (WIELAND 1300)), fatty acid peroxides may further change to hydroxy-keto-acids, $R \cdot CO \cdot CHOH \cdot R' \cdot COOH$, which may either suffer further decomposition in the course of intermediate metabolism or which may polymerize to form resinous products, an example being the drying of oils (GOLDSCHMIDT and FREUDENBERG 407)). This latter process represents probably a chain reaction involving heavy metal (see next section). While studying the autoxidation of linolic and linoleic acid HINSBERG 489) found strongly acid substances of reducing character. It is doubtful how far such mechanisms correspond to physiological happenings and to which extent they play a role in the principal train of metabolism. Most workers consider the so-called $\beta$-oxidation of fatty acids as the preponderant mechanism; here the primary dehydrogenation occurs at a place in the molecule quite distant from the double bonds, namely, between the first and second C atom counted from the carboxyl group (p. 256). Only the quantitatively unimportant oxidation of methyl groups, the so-called $\omega$-oxidation, is to be considered as a true oxidation according to KUHN and KÖHLER 650).

## 2) Heavy Metal Catalysis via Peroxides.

The question as to whether catalytically active metals form true peroxides as intermediates in the transfer of active oxygen, is not new. It still remains to be conclusively answered.

For purposes of discussion we differentiate between oxidative and peroxidative action.

When considering o x i d a t i v e c a t a l y s i s we shall eliminate as a matter of principle all those instances where the metal complex (mostly iron) is a redox system and functions by virtue of a reversible change between two states of valency (e.g. Ferrous $\leftrightarrows$ Ferric form). There remain then only those cases where no primary dehydrogenation can take place. It must be remembered that mere peroxide formation at the metal is not sufficient to establish a truly catalytic schema. If we adopt the schema of ENGLER, the reaction of the peroxide according to $Fe^{II}\cdots(OH)_2 + Acc \rightarrow Fe^{III}OH + Acc.OH$, at least at neutral reaction, is no catalysis (MADELUNG); $Fe^{III}OH$ is changed no further. The experiments of MANCHOT on the oxidation of arsenite via an iron peroxide, $FeO_2$, suggest merely an induced reaction. In their experiments on the oxidation of glycolic acid GOLDSCHMIDT et al. 406), too, arrive at the conclusion that this oxidation by iron salts represents an induced reaction with an intermediary formation of $Fe^{II}$-peroxides. The postulate of WIELAND and FRANKE 131), that the $Fe^{II}$ stage must be regenerated if a c a t a l y s i s is to ensue, is not fulfilled here. This means that the entire oxygen originally taken up must be transferred to other molecules. In contrast to his earlier views MANCHOT is now inclined to believe that the peroxide may indeed yield all of its oxygen to an acceptor.

In general it is often difficult to draw a line between oxidative and dehydrogenation catalysis. Certain phenomena are explained by WIELAND as dehydrogenations which MANCHOT prefers to interpret as true oxidations. The dualism is evident in the catalytic oxidation of arsenite and of sulfite by iron in the form of ferropyrophosphate. The former is explained by SMITH and SPOEHR 1060) as a reaction involving a "moloxide", while in the latter case both mechanism are held possible but the moloxide mechanism considered more probable. WIELAND 1323, 1324) believes that the oxidation of benzaldehyde as catalyzed by metal in aqueous solution, involves a peroxide of the aldehyde, cf. FRANKE (l.c. 288, p. 198)).

A similar situation exists with regard to the p e r o x i d a t i c a c t i o n of heavy metals (p. 70). In most instances the correct explanation will be found in assuming a catalysis involving a change in valence of the metal and a reduction of the $H_2O_2$. While this takes care of the simple metal compounds and derivatives, there remains a residue of non-dehydrogenating catalyses which may perhaps be explainable in terms of peroxide formation at the metal. The hemin catalyses present a particularly perplexing situation inasmuch as they afford evidence neither for a valency change nor for peroxide formation. It is assumed that unchanged ferric iron functions as catalyst both in the peroxidatic and catalatic decomposition of $H_2O_2$ (HAUROWITZ 467)). Methemoglobin, for example, forms a spectroscopically well defined complex with hydrogen peroxide in a molecular ratio of about 1 : 1

(KOBERT 584), HAUROWITZ 467), KEILIN and HARTREE 545). The behaviour of this intermediate towards various reagents suggests that the iron is still in the trivalent state. The complex decomposes spontaneously to form molecular oxygen (and presumably water). Recent observations by KEILIN and HARTREE 549), on the other hand, lend themselves to the interpretation that the enzyme catalase which like methemoglobin possesses proto-ferriheme IX as prosthetic group, when acting upon hydrogen peroxide, is reversibly reduced to the ferro form which subsequently is reoxidized. Hydrogen peroxide is claimed to be a specific reductant of catalase. WEISS 1289a), in a recent theoretical review of the reaction mechanism of catalase and peroxidase in the light of the theory of chain reactions, concurs with KEILIN in this view. He enlarges upon the physical chemical implications as follows: "The rôle of the iron ions in the porphin ring — such as in the respiration ferments and catalase — is to make possible a very quick electron transfer, i.e., the reversible change between the divalent and trivalent state." He adds, in the form of a footnote, "This is probably because the change of valency of the iron in the porphin ring system is taking place without appreciable movement of heavy particles and not as it could be in the case of ferrous and ferric ions, when the water dipoles in the hydration shell undergo a rearrangement when the charge of the central ion is changed. On the other hand, the system of conjugated double bonds around the iron in the hematin group permits — through their loosely bound n-electrons — a rapid "conduction" of the inner electron."

We have as yet no satisfactory explanation for the mechanism of the oxidation of —C = C— bonds by hemins + $H_2O_2$. If the intermediary production of a peroxide at the heme iron is postulated one has to explain how this peroxide reverts to the $Fe^{II}$-stage which is necessary for the establishment of catalysis. WIELAND 1312, 1316) has tried to identify the process as a dehydrogenation catalysis (see also BERTHO (l.c. 105, p. 727)), whereas MANCHOT 750, 751) defends the peroxide hypothesis. This author suggests that not the usual type of a peroxide, $Fe\cdot\cdot O_2$, but a compound of the schematic formula $Fe_2O_5$ or $Fe\cdot\cdot\cdot(OH)_3$ is formed by transformation of an adduct $FeSO_4\cdot\cdot H_2O_2$; the latter is considered as a salt of $H_2O_2$, namely $Fe(OH)_2\cdot\cdot O\cdot OH$, by HABER and WEISS 436). Similar ideas have been expressed by SHAFFER 1046) for the course of the oxidation of ferrous salts and by DHAR 200) for the induced oxidation of formic acid by Fe and Ce ions. The formation of the adduct is considered to be the primary reaction in any case by MANCHOT and PFLAUM 751). Thereby MANCHOT modifies to a certain degree his previous postulate of the formation of well-defined peroxides (see for instance MANCHOT and LEHMANN 750) where a reduction of $Fe_2O_5$ by an excess of $H_2O_2$ was assumed) and he approaches the more general idea of reactive complexes in the sense of BRÖNSTEDT which would be able to explain the mechanism of the catalysis (see also WIELAND 1306), p. 70). A regeneration of the ferrous form by suitable concentrations of $H_2O_2$ and thus a true catalysis is held likely to occur only to a restricted extent, whereas in general $Fe^{III}$ is assumed to be formed. It should be remembered that MANCHOT's entire considerations (see also 752)) refer less to a catalytic process than to induced reactions.

The equation for the over-all reaction, as proposed by MANCHOT,

$$2\ FeSO_4 + 3\ H_2O_2 + 2\ H_2O = 2\ Fe(OH)_3 + H_2SO_4 + O_2,$$

is not accepted by HABER and WEISS 436). These authors assume that the process is a chain reaction. The chains are relatively short and they terminate whenever ferric iron is formed (GOLDSCHMIDT 406)). The effect of peroxides of metals other than iron has been studied by GALLAGHER 387) and COOK 183).

It would appear then that in iron compounds other than hemins, i.e. ions or simple complexes, only the ferrous and not the ferric form is active as "peroxidase"; this has been shown directly by Simon and Reetz 1057) and indirectly by Kuhn 678) who reports that reduction of ferric iron at graphite surfaces will increase its catalytic activity by a factor of $10^5$. If we assume that the ferrous forms act by virtue of peroxide formation, we have to assume that these are changed back to the original ferrous form during the catalysis. But how this happens is not known.

There follow some data concerning oxidations, involving a true uptake of oxygen, which cannot be explained on the basis of dehydrogenation.

Wieland 1323, 1324) has shown that benzaldehyde is autoxidizable without dehydrogenation, perbenzoic acid being an intermediate. In spite of this the ordinary autoxidation of benzaldehyde is a heavy metal catalysis which outweighs by far the weak or negligible autoxidation, e.g. in aqueous solution (652, 964, 782)). According to Raymond 696) radiation is also effective. Kuhn and Meyer 652) find that the oxidation, when catalyzed by traces of Fe, Cu, Ni, Mn, or by pyridine hemin, is cyanide sensitive. The oxidation of aldehydes by $MnO_2$ via peroxides has been studied by v. Braun and Keller 143). Wieland has confirmed the catalytic activity of iron; but his careful investigation has failed to elucidate the mechanism. An orange colored complex which is formed rapidly with ferrous iron and only slowly with ferric iron and which Kuhn suspected to be the catalyst proper has no significance according to Wieland; it is a ferric complex salt of perbenzoic acid. Kuhn and Wieland find that $Fe^{II}$ acts rapidly and that the very low activity of $Fe^{III}$ increases with time. There must exist therefore a mechanism whereby $Fe^{III}$ is reduced back to $Fe^{II}$ after it has arisen in the induced reaction via the peroxide $Fe^{II}O_2$. A typical dehydrogenation does not occur since the aldehyde hydrates which have no other mobile hydrogen, e.g. chloral hydrate, are resistant against catalysis by iron; nor does quinone act instead of oxygen in the system aldehyde-$Fe^{II}$. Perhaps a chain mechanism operates in this case as it does in that of the unsaturated fatty acids (Wright et al. 1350); see below).

From a biological point of view as well as a model the marked acceleration of autoxidation of unsaturated fatty acids by metals assumes importance. This reaction proceeds via peroxide stages; it may be catalyzed both by simple complex salts, e.g. copper ascorbinate (Holtz 492)), as well as by very stable complexes like tridipyridyl ferrous salts or hemins. The latter might well have a significance in metabolism. The catalysis is cyanide resistant. It might possibly account for a part of the so-called iron-free respiration at the expense of fatty acids.

In the oxidation of oleic acid by ferricyanide a complex situation was encountered by Wright et al. 1350). A change in valence cannot be demonstrated; oleic acid does not reduce ferricyanide, and ferrocyanide is not autoxidizable. The process has been studied from the point of view of a chain mechanism by Chow and Kamerling 168). Oleic acid will reduce $Fe^{III}$ only in presence of $O_2$. $Fe^{II}$ alone is completely inactive. Otherwise the rate of oxidation depends on the redox potential $Fe^{II}/Fe^{III}$; the more positive the latter the greater will be the rate of reaction.

Ergosterol has been found to be oxidized by free hemin (Kuhn); this catalysis is cyanide-resistant. According to Mayer 784) a peroxide is produced as an intermediate.

With phosphatides as substrates iron is most active as catalyst while copper only slightly active. Kephalin absorbs more oxygen than other phosphatides; it represents the only phosphatide which is oxidized with cobalt as catalyst (Page 912)).

In view of the fact that the catalysis involves neither a change in valency nor the production of peroxides at the non-autoxidizable highly complex ferrous compounds or at the difficultly reducible ferric compounds, FRANKE (288, p. 195) favors the opinion that the iron catalysis is not at all concerned with the primary formation of the peroxide configuration, —CH—CH—, but    only    with    its    further    change    via

$$\begin{matrix} | & | \\ O & —O \end{matrix}$$

—CO—CHOH— and its oxidative breakdown. He assumes that the iron system reacts peroxidatically with the organic peroxide in such a manner that it is further transformed by dehydrogenation-hydrogenation reactions. It may be that the oxidation of fat in situ by Cu and Fe, as observed by ROSENTHAL and VOEGTLIN 1006), is to be explained in this manner.

# Special Part.

## D. The Enzyme System.

### I. Theoretical Considerations.

#### 1) The Enzymes of the Main Path; Hydrokinases.

The differentiation between Dehydrogenases and Oxidases, though open to considerable criticisms, is retained here for purely practical reasons. Both groups belong to the great class of Desmolases which catalyze oxido-reductive processes of an essential dehydrogenating character. They might also be called Oxidoreducases, but it appears preferable to use instead the term Hydrokinases. The latter was originally introduced by WIELAND as an alternative for "dehydrogenases"; however, this term expresses beautifully the current concept underlying the action of all the enzymes of this class according to which they "cause hydrogen to move" (From the Greek ὕδως, hydrogen, κινεῖν, move). The dehydrogenases and oxidases, as subgroups, are no longer distinguished with respect to fundamental divergence in theory but rather by the superficial form of their action and by their chemical constitution.

The dehydrogenases are defined, in accordance with WIELAND, as enzymes endowed with a rather pronounced donator specificity and a limited acceptor specificity. We have to distinguish between those dehydrogenases which are able to react directly with molecular oxygen (aerobic dehydrogenases (DIXON 212)) or better oxytropic dehydrogenases (THUNBERG 1169)), and those which are unable to do so (anaerobic or anoxytropic dehydrogenases). The latter may utilize oxygen only with the aid of a separate, autoxidizable system which may contain heavy metal or quinoid groups. The anoxytropic dehydrogenases function as important links in the intricate reactions leading the metabolic hydrogen of the primary anaerobic processes to the terminal respiratory system. One of them is cytochrome c, one of the very few non-autoxidizable hemin derivatives. The principle according to which the main chain is constructed is undoubtedly to provide only for one component capable of direct reaction with oxygen, thus forcing the other catalysts to fall in line. The donator closest to the cytochrome may be the fumaric system, or the yellow enzyme (flavoproteid), or diaphorase (see p. 226).

The oxytropic dehydrogenases may fulfill two different functions. On the one hand we have the aldehydrases and alcohol dehydrogenases as important catalysts in anoxybiontic fermentation processes. It must be mentioned, however, that recent work makes their capability to react directly with oxygen rather questionable. Thus EULER assumes that the alcohol dehydrogenase is one of the pyridine enzymes and that it will react with $O_2$ only through the participation of the yellow enzyme. REICHEL claims the same for the aldehydrase. On the other hand, certain oxytropic dehydrogenases appear to catalyze special aerobic reactions occurring apart from the main chain of carbohydrate desmolysis, e.g. the oxidation of purines (xanthine dehydrogenase). One might be tempted to classify the yellow enzyme with the oxytropic dehydrogenases. In the fully respiring cell its action is entirely or at least predominantly anoxytropic, while in special cases this enzyme may establish a true respiration; this takes place if no hemin system is present or if the latter is damaged.

For the sake of systematizing the group of oxidases might be placed at the side of the dehydrogenases. These systems are characterized by two features; they react exclusively with oxygen as acceptor and they are heavy metal systems which may be poisoned by HCN as well as in many instances by CO. The "classical" oxidases, i.e. the chromo-oxidases attacking aromatic chromogens, may be divided into three subgroups, namely, the "indophenoloxidase" which attacks preferably para compounds, the polyphenolase (catechol oxidase) proper which oxidizes ortho compounds, and the monophenolase (previously called tyrosinase). The most important oxidase is the cytochrome oxidase. In addition we encounter forms of intermediate properties, e.g. enzymes which will react exclusively with oxygen as acceptor but which are cyanide resistant. An example are the amino acid oxidases. These and similar enzymes represent alloxazine proteids (see pp. 189 and 197); they may be oxytropic dehydrogenases. The alanine oxidase is such a catalyst. For the time being they may be grouped together as oxhydrases so as to indicate that they are dehydrogenases which will exclusively reduce oxygen. It is as yet not possible to name the reason why certain enzymes show an absolute specificity for oxygen. The cause may be of thermodynamical character, related to the normal oxidation-reduction potential of these catalysts, or it may be related to the chemical structure of these enzymes. In contrast to their rigid acceptor specificity the chromo-oxidases show a less pronounced donator specificity. Some of the classical oxidases may not be enzymes at all but just simple metal salt systems capable of forming autoxidizable complexes with chromogens. Theoretically perhaps the most intricate type of enzymes are the peroxidases which will only use hydrogen peroxide or other peroxides as acceptors. It should be remembered that their action, the reduction of $H_2O_2$, is not only the most important one from the point of view of energy gain (the primary reduction of oxygen to $H_2O_2$ yields much less energy), but also that they effect the ultimate oxidation of the chromogens while the first step, as catalyzed by the oxidases, leads to the establishment of equilibria (see FRANKE (288, p. 152, 284)).

## 2)  Auxiliary Enzymes of Desmolysis.

Desmolysis, in order to proceed smoothly, requires a series of auxiliary enzymes besides the enzymes of the main path, the hydrokinases, which transfer hydrogen from the first donator in anaerobiosis over other redox systems to the molecular oxygen. At various stages of anoxybiosis and oxybiosis there arise substances which cannot be dehydrogenated in the normal way and which have to be transformed prior to further dehydrogenation. Furthermore, one of the most important final stages of intermediary metabolism, the production of $CO_2$, calls for a special type of catalysis. Other auxiliary enzymes serve the purpose of decomposition of the last intermediary product of oxidoreductive catalysis, of $H_2O_2$; it is the ultimate reduction of oxygen to water which yields the last and large amount of free energy stored in the metabolic hydrogen. Then there are certain enzymes of doubtful classification which have been found mainly in bacteria and which perform certain special tasks. We might, then, group the auxiliary enzymes loosely into auxiliary enzymes of anoxybiontic metabolism, of terminal oxidation, and the last mentioned group of special enzymes.

The **auxiliary enzymes of anoxybiosis** may be divided again into three or perhaps four subgroups. The most conspicuous of these is that of the c a r b o x y l a s e s. These catalysts are responsible for the formation of $CO_2$, which together with $H_2O$ represents the ultimate product of any type of metabolism, fermentative or respiratory in nature. While decarboxylation in plants, particularly in yeast, consists of the decomposition of an $\alpha$-ketocarboxylic acid into $CO_2$ and the next lower aldehyde, decarboxylation in animal tissues appears to be of the oxidative type (p. 250). Another subgroup is that of the h y d r a t a s e s. At various stages of desmolysis we encounter intermediates which cannot be directly dehydrogenated. The simplest case is that of the aldehydes which may only be dehydrogenated in the hydrate form $R \cdot CH(OH)_2$ (WIELAND). However, these hydrates arise spontaneously in aqueous solution as can be shown spectroscopically (SCHOU 1033), FROMAGEOT 377)); a special hydratase is therefore unnecessary in this case. The only instance where an enzymatic catalysis of hydratation has been established beyond doubt is the transformation of fumaric acid into malic acid. The equilibrium fumaric acid $+ H_2O \rightleftarrows$ malic acid is catalyzed in a fully reversible manner by the enzyme f u m a r a s e, also called fumaric hydratase (JACOBSOHN). Malic acid in turn may be further dehydrogenated to oxaloacetic acid which may be successively decarboxylated to form pyruvic acid and then acetaldehyde. It is as yet undecided whether the m u t a s e s belong to the group of hydratases. The evidence available indicates that the dismutation of aldehydes to form alcohol $+$ carboxylic acid is not, or at least not always, catalyzed by the a l d e h y d r a s e s proper. DIXON and LUTWAK-MANN 216) have obtained a preparation from milk exhibiting pure aldehydrase activity without any mutase action and a liver preparation with pure mutase action and without any dehydrogenating activity. If we accept their existence, mutases may either be dehydrogenases, i.e. aldehydrases utilizing a second aldehyde molecule as acceptor (Schema I) or they may be hydratases reacting according to the usual scheme of dismutation (II):

$$\text{I.} \quad CH_3 \cdot CH(OH)_2 + CH_3 \cdot CHO \rightarrow CH_3 \cdot COOH + CH_3 \cdot CH_2OH$$

$$\text{II.} \quad CH_3 \cdot CHO + H_2O + CH_3 \cdot CHO \rightarrow CH_3 \cdot COOH - CH_3 \cdot CH_2OH$$

For a better understanding of this problem the following observation of DIXON and LUTWAK-MANN 216) may be significant: enzymatically pure aldehyde mutase will also attack methyl glyoxal and disproportionate it to form pyruvic acid and acetol:

$$\left.\begin{array}{l} CH_3 \cdot CO \cdot CHO \\ \\ CH_3 \cdot CO \cdot CHO \end{array}\right\} \xrightarrow{\phantom{xx} + H_2O \phantom{xx}} \left\{\begin{array}{l} CH_3 \cdot CO \cdot COOH \\ \\ CH_3 \cdot CO \cdot CH_2OH \end{array}\right.$$

As is well known, glyoxalase (ketonealdehyde mutase), on the other hand, will react with methyl glyoxal in such a manner that one molecule of the substrate undergoes an "intramolecular CANNIZZARO reaction" (NEUBERG) yielding lactic acid:

$$CH_3 \cdot CO \cdot CHO + H_2O \rightarrow CH_3 \cdot CHOH \cdot COOH.$$

Thus the two enzymes change the same substrate in a fundamentally different manner, a clear documentation of the principle that a catalyst may not only affect the rate but also the course of a reaction by selectively enhancing one of several possibilities. It appears warranted to consider glyoxalase as a hydratase and the aldehyde mutase as an anoxytropic dehydrogenase with a unique acceptor specificity. The existence of such mutases is also indicated by the recent findings of KREBS (p. 274) concerning the enzymatic dismutation of ketocarboxylic acids, where one serves as the acceptor, being reduced to the hydroxyacid, and the other as the donator, being probably dehydrogenated in the ortho form $R \cdot CO \cdot C(OH)_3$; in the course of this reaction the terminal $CO_2$ looses its foothold as it were and is split off. There is evidence that the agon of these special dehydrogenases is identical with that of carboxylase (vitamin $B_1$-pyrophosphate).

A third group of auxiliary enzymes plays its role in the very early stages of the breakdown of sugars; they are contained in the complex system which is commonly called "enzymes of the first attack". We do not know how many individual enzymes take part here. Even the most recent schemata of sugar decomposition (EMBDEN-MEYERHOF) do not yet embrace the very first stages and begin with fructose diphosphate; glucose is only considered in these schemes at a stage where it is drawn into the process in a secondary manner, namely, in a coupled reaction.

Our knowledge of this "first attack" is limited to three general points: (1) there must take place an intramolecular rearrangement of the pyranoid hexose to a "reaction form of the sugar" the furanoid structure of which is indicated by the fact that fructose diphosphate itself is furanoid. (2) esterification with phosphoric acid must occur, and (3), the phosphorylated hexose is broken down into two molecules of triose phosphoric acid. Reaction (1) is catalyzed probably by MEYERHOF's hexokinase which is closely related to or identical with EULER's heterophosphatese which he now calls p h o s p h o r y-
l a s e *). In the case of glycogen or starch it is highly probable that there takes place a direct phosphorylytic cleavage into hexosemonophosphates (PARNAS). Recent experiments by CORI 185) show that the hexose monophosphate thus formed is hexose-1-phosphate which is subsequently rearranged by an enzyme into hexose-6-phosphate.

Whether ROBISON's p h o s p h o h e x o k i n a s e which converts fructose monophosphoric acid into glucose monophosphoric acid is an enzyme is not yet established. Reaction (2) is most certainly catalyzed by phosphatases. The important step (3) is catalyzed by the a l d o l a s e of MEYERHOF and LOHMANN 794) which is responsible for the following equilibrium reaction

$$\text{Dihydroxyacetone phosphoric acid} \rightleftarrows \text{glycerinaldehyde phosphoric acid}$$
$$\rightleftarrows \text{hexose diphosphoric acid.}$$

This enzyme undoubtedly represents the most important catalyst of the "first attack" and is therefore the most important auxiliary enzyme of the anaerobic sugar breakdown in general. A further enzyme, e n o l a s e, promotes the transition of phosphoglyceric into phosphopyruvic acid **). The enzymatic nature of c a r b o l i g a s e, on the other hand, which allegedly catalyzes the condensation of aldehydes to acetoine, has become doubtful. This process is more likely a non-catalyzed reaction between free radicals where the acetoine formations represent a chain breaking process, or a spontaneous reaction between activated molecules.

Other anaerobic enzymes with special functions are those which operate with molecular hydrogen; this gas is liberated by the action of the h y d r o l y a s e s and fixed by the h y d r o g e n a s e s which are the only true reducases. Other special enzymes, like BURK's a z o t a s e which is responsible for the assimilation of molecular nitrogen by certain bacteria or the n i t r a t a s e of GREEN which promotes the reduction of nitrates, are hardly known as chemical individuals; their presence is inferred from the occurrence

---

*) This form of the "first attack" concerns, naturally, only the f r e e hexoses. Doubts have been expressed lately whether such a phosphorylating attack on free hexoses occurs at all; perhaps the typical breakdown of carbohydrate in a l l cells begins with the direct phosphorylation of poly-saccharides only.

**) For a fuller discussion of these enzyme systems the reader is referred to the recent review article by MEYERHOF in "Ergebnisse der Physiologie" (788)).

of these unique reactions and from the generalization that all life processes are brought about by the action of enzymes.

**Auxiliary Enzymes of Terminal Desmolysis:** These enzymes fulfill a function quite different from those just mentioned. They do not catalyze certain stages of the chemical breakdown of the substrates but they insure the utilization of oxygen to the largest extent possible while they protect the cell against toxic agents. The common biological function of these catalysts appears to be the elimination of hydrogen peroxide which arises as a product of primary dehydrogenation reactions. We refer to the peroxidases and catalases. The chemical nature of these enzymes will be discussed further below (p. 172). While there is no doubt that it is possible to prepare catalase preparations free from peroxidase and vice versa, there appears to be a certain overlapping with respect to specificity; according to recent findings by KEILIN and HARTREE and of HAUROWITZ catalase may take part in peroxidatic reactions, especially under conditions which favor the action of catalase upon peroxides in statu nascendi. The situation is furthermore somewhat complicated by the fact that a number of non-enzymatic systems, e.g. simple heme derivatives and iron salts, show a certain peroxidatic activity.

The action of catalase, acc. to WIELAND, consists in a dismutation of hydrogen peroxide: One $H_2O_2$ molecule hydrogenates a second $H_2O_2$ molecule in such a manner that the over-all reaction

$$2\ H_2O_2 \rightarrow O_2 + 2\ H_2O$$

is the result (see, however, p. 177). The peroxidases, on the other hand, are oxidizing enzymes. This is true because hydrogen from the substrate will reduce the $H_2O_2$:

$$H_2O_2 + 2\ H \rightarrow 2\ H_2O.$$

These reactions do not necessitate a valency change of the heme iron of the catalysts (HAUROWITZ).

If one wishes to assign a place in the general system to the peroxidases they may be designated as phenol dehydrogenases with a pronounced donator specificity towards certain non-autoxidizable polyphenols and amines (guaiacol, pyrogallol, benzidine, triphenylmethane dyes, etc.) and with an almost absolute acceptor specificity towards $H_2O_2$. Other organic peroxides which have been tested were either not acceptors at all or much less effective (WIELAND and SUTTER **1328**)).

### 3) Survey of System of Desmolases.

The following scheme may be taken as a preliminary attempt to bring order in the chaos; as such it has mainly formal interest.

#### A. Enzymes of the Main Path: Hydrokinases.

#### I. Dehydrogenases.

Definition: Dehydrogenating, substrate-specific enzymes capable of reducing chemical acceptors.

##### a) Anoxytropic Dehydrogenases.

Definition: Enzymes reacting only with chemical acceptors and not with molecular oxygen; their complexes with the substrate are therefore not autoxidizable.

### α) Heavy metal-free Systems.

1) Acidodehydrogenases: Acetodehydrogenase, dehydrogenases for higher fatty acids, succinic dehydrogenase, glutamic acid dehydrogenase.
2) Hydroxyacidodehydrogenases: Lactic, malic, β-hydroxybutyric, citric dehydrogenase, etc.
3) Pyruvic dehydrogenase
4) Formic dehydrogenase
5) Oxalic dehydrogenase
6) Group of pyridine enzymes: Agon, pyridine nucleotides; natural acceptor either the yellow enzyme or diaphorase (EULER).
    Alcohol dehydrogenase (Agon, cozymase), triose phosphate dehydrogenase (Agon, cozymase), glucose monophosphate dehydrogenase (Agon, WARBURG's coferment).
7) Other dehydrogenases of the sugar group. Relationship to group 6) not yet established.
    Glucose dehydrogenase (HARRISON), triose dehydrogenase (CLIFT), glycerophosphate dehydrogenase, amylodehydrogenase (THUNBERG).
8) Diaphorase: Agon, an alloxazin nucleotid. Donator Dihydrocozymase, acceptor probably cytochrome b.
    Appendix: Acceptor-free dehydrogenation of organic substrates, mainly formic acid, with formation of molecular hydrogen (Hydrolyases (STEPHENSON)).

### β) Hemin Systems.

Cytochromes. Natural substrates probably flavoprotein (yellow enzyme), diaphorase, succinic acid, malic acid, perhaps glutathione. Natural acceptor exclusively the respiratory ferment (cytochrome oxidase).

### b) Oxytropic Dehydrogenases.

Definition: Enzymes reacting with chemical acceptors as well as with molecular oxygen.
1) SCHARDINGER Enzyme. Agon, an alloxazine nucleotide. Substrates: Aldehydes. hypoxanthine, xanthine. Closely related to this enzyme: Nucleic acid dehydrogenase.
2) Flavoprotein (Yellow Enzyme). Substrate very probably the pyridine enzymes of WARBURG. Acceptors not definitely established, perhaps directly cytochrome b or certain dehydrogenases.

### II. Oxhydrases.

Transition to Oxidases. Definition: Substrate specific dehydrogenating enzymes, apparently metal-free, preferential or exclusive acceptor-specificity for molecular oxygen; even thermodynamically permissible chemical acceptors are not or not appreciably hydrogenated. Action is resistent against heavy metal poisons.
1) Glucose oxidase (D. MÜLLER), Ascorbic acid oxidase (SZENT-GYÖRGYI, ZILVA).
2) Amino acid dehydrogenases (KREBS, KISCH); Agon, an alloxazine nucleotide.
3) Oxhydrases of amines: Diamin-oxhydrase (Histaminase) Monoamin-oxhydrase (Tyraminase.)
    Related enzymes: Proline oxidase, Thiosulfate oxhydrase (PIRIE).

### III. Oxidases.

Definition: Cyanide sensitive heavy metal systems with absolute affinity to oxygen as acceptor.

### a) Soluble Oxidases.

1) Uricase. Related Enzyme: Allantoinase.
2) Chromo-oxidases: Monophenolase (Tyrosinase), Orthophenolases (Cu-Proteins) (Poly-

phenol-oxidase, Catechol-oxidase, Laccase); Paraphenolases (Phenylenediamine oxidase, indophenol oxidase (closely related to the respiratory ferment or perhaps derived from it by partial degradation of the bearer protein)).

3) Luciferase: Donator Luciferin, structure unknown.

### b) "Insoluble" Respiratory Ferment. (Cytochrome oxidase).

Pheohemin on a cell protein as bearer which is stated to be linked structurally to the cell but may be a macromolecular protein. Donator: One cytochrome component (probably cytochrome c). Related to indophenol oxidase and to cytochrome a.

### B. Auxiliary Enzymes of Desmolysis.

### I. Hydratases and Related Enzymes.

1) Fumarase (Fumaric hydratase), Crotonase.
2) Aldehyde Mutases (Mutase of Dismutation, Ketone aldehyde Mutase (= Glyoxalase)).
3) Mutases of Ketocarboxylic Acids (KREBS), perhaps identical with Pyruvic Dehydrogenase.
4) Aspartase (= Fumaric Aminase (JACOBSOHN)).

### II. Carboxylases.
### III. Carbonic Anhydrase.
### IV. Enzymes of the „First Attack" on Sugars.

Aldolase, Enolase, Hexokinase (Heterophosphatese, Phosphorylase (EULER)). Carboligase (Existence doubtful).

### V. Peroxidases.
### VI. Catalases.
### VII. Reducases.

Donator: Molecular Hydrogen.

1) Azotases (Nitrogenases). Acceptor: Molecular Nitrogen (BURK).
2) Hydrogenases. Acceptor: probably dehydrogenase systems, thiols, etc.

## II. Descriptive Chemistry of Enzyme System.

### Introduction.

In order to appreciate the rate at which our knowledge of enzymes has progressed during the past decade it may be well to remember that WILLSTÄTTER 1336a), in 1926, upon reviewing the facts then available, concluded that the enzymes are neither proteins, carbohydrates, nor members of any other well-known group of organic compounds. He introduced the important concept that an enzyme consists of a c o l l o i d a l  b e a r e r and a p r o s t h e t i c  g r o u p. Today we know the chemical structure of the prosthetic groups of a number of enzymes in detail; some of them have even been prepared synthetically. The knowledge of the structure of the colloidal bearers is by necessity limited to the degree to which the structure of proteins in general is known. At the present time, therefore, the structural chemistry of enzymes cannot as yet embrace the protein part of these catalysts.

Several favorable circumstances have contributed to these recent advances in enzyme chemistry. One of them was the amenability of some of the prosthetic groups to spectroscopic study, another the photodissociation of the carbon monoxide compound of the respiratory ferment. In other instances preparative methods have yielded the prosthetic groups in pure and crystalline form and in amounts sufficient for chemical study. A great difficulty in the field of desmolases has been and still is the fact that these enzymes, to an extent far greater than in the instance of the hydrolases, appear more or less linked to the structure of the living cell. Those enzymes which are said to be intimately bound to the vital architecture of protoplasm, eg. the respiratory ferment or the principle associated with the PASTEUR-MEYERHOF reaction and that which is responsible for nitrogen fixation perish with the death of the cell.

A further fortunate factor was the discovery that certain vitamines, namely, Vitamin $B_1$, Vitamin $B_2$(G), and the antipellagra principle, represent components of enzymes or, more specifically, of the prosthetic groups of enzymes. The demonstration of the formation of an enzyme by reversible union of the prosthetic group with the bearer protein in the case of the yellow enzyme (THEORELL, KUHN) has led to a revision of the concept of coenzymes in general. Until then coenzymes, and cozymase in particular, had been generally considered to be a c t i v a t o r s of an enzyme complex ("apozymase"); no fundamental distinction was made between cozymase, for instance, and salts or other agents the presence of which is important for the activity of an enzyme. The new coenzyme concept visualizes the coenzyme as the prosthetic group of the enzyme; what was formerly considered to be the enzyme proper is now recognized as a part of the enzyme, namely, as the bearer protein. (See, however, OPPENHEIMER's „SUPPLEMENT", p. 1502).

This change of concept has, unfortunately, created considerable confusion in the literature as to terminology. Should activators like glutathione which appears to be important for glyoxalase action (LOHMANN) still be termed coenzymes? Should the heme portion of certain enzymes which do not readily reversibly dissociate into their components also be designated as coenzyme? In view of this situation it may perhaps be better to a b a n d o n  t h e  t e r m  c o e n z y m e altogether, although names like cozymase or cocarboxylase which designate specific chemical entities might still be retained. The prosthetic group of an enzyme could more appropriately be design ated as the "a g o n" and the bearer protein as the "p h e r o n", the whole compound as a "s y m p l e x" in accordance with suggestions made by WILLSTÄTTER and by KRAUT.

## 1)  The Hemin Systems.

### a)  General.

Four important biocatalysts have thus far been recognized as hemin derivatives, WARBURG's respiratory enzyme, KEILIN's cytochromes, peroxidase and catalase. Inasmuch as our present knowledge of these substances is mainly due to spectroscopic measurements, either indirect or direct in nature, it is proposed to deal first with the absorption spectra of these catalysts and with their behaviour towards specific inhibitors and other agents. The spectroscopic observations refer mainly to the visible region of the spectrum, i.e. to the range from 400 to 700 millimicron; the absorption in the ultraviolet region is primarily due to the protein components of these enzymes and thus far has not yielded information beyond the fact that the typical absorption band of proteins around 260 to 280 millimicron is shown by preparations of these enzymes.

#### Absorption Spectra:

The first absorption spectrum of any enzyme to be obtained was that of the respiratory ferment. It was measured by an indirect method specially devised for this purpose by O. WARBURG. The principle underlying this method and its theory appear of such paramount importance for the problem of biological oxidation in general that it shall be treated in some detail. Besides the original papers of WARBURG, the review articles by REID 974) and by ZEILE 1372a) shall be drawn upon for this exposition.

Originally, WARBURG 1216a) conceived the "Respiratory Ferment" as the sum of all catalytically iron compounds in the cell. In the course of subsequent work he was able to characterize it as a well-defined enzyme the active group of which is anchored to the cell structure; it contains ferric iron which oxidizes the substrates through the mediation of a number of intermediary catalysts. In accordance with this modification of concept WARBURG proposed to call the respiratory ferment henceforth "oxygen transferring enzyme of respiration". The term respiratory ferment, however, has been so firmly entrenched in the literature that we shall continue to use it here, partly also for the sake of brevity.

The type of linkage of the iron in the respiratory ferment was established by WARBURG's fundamental observation (1221)) that the respiration of yeast, as measured in terms of oxygen uptake, is diminished by adding carbon monoxide to the atmosphere and that this inhibition is partly relieved by illumination; upon returning the cells to the dark the inhibition is restored. This can only be explained by assuming that the respiratory catalyst forms a complex with CO which is catalytically inactive and which is reversibly dissociated by light. Precisely this behaviour had previously been observed with carbon monoxide hemoglobin by J. B. HALDANE and SMITH and with iron pentacarbonyl by DEWAR and JONES. Nickel carbonyl, on the other hand, was known not to be subject to photodissociation. Later, WARBURG and his associates observed the reversible photo-decomposition of CO-ferrous cysteine (CREMER 191)) and of CO-hemochromogens (KREBS 596)). The fundamental principle of photochemistry is that only that part of the radiation which is absorbed by the system can exert any photochemical effect. If the absorption spectrum of a compound is known it can therefore be predicted that that wave-length of light which will have the greatest photochemical effect on the substance will coincide with the absorption maximum as determined by spectrography. If, conversely, the direct absorption spectrum of a substance is unknown, the relative photochemical efficiency of monochromatic radiation of varying wave-length should permit one to obtain the "indirect" or photochemical absorption spectrum of the compound. This is of particular importance where the concentration of the unknown substance is so low or where it is present in such complex mixtures with other light absorbing substances that direct absorption measurements would not be feasible. Exactly this situation existed in the case of the respiratory catalyst at the time when WARBURG carried out his classical photochemical experiments. The sensitivity of the indirect method could be rendered far greater than of the direct method because the effect of radiation on the catalytic activity of the enzyme was measured rather than on the chemical composition of the system. The determination of the relative efficiency of various wave-lengths of light on the CO-inhibition of cell respiration yields the relative photochemical absorption spectrum. Such a spectrum represents the true pattern of the absorption curve but it yields no information as to the absolute value of the exctinction coefficients of the substance at any wave-length. It is obvious that the greater the extinction at any given point of the spectrum of the substance the more rapidly will the photochemical action of the radiation at that wave-length manifest itself. The rate of increase in respiration upon illuminating CO-inhibited cells may thus be utilized to obtain the absolute absorption spectrum. Other catalytically inactive pigments which may be present cannot interfere with the measurements since they do not participate in the photochemical process.

Theory of Photochemical Absorption Spectrum:

The fact that the respiratory ferment transfers "bound oxygen" implies that it may exist in two forms, an oxidized (or oxygenated) form and a reduced form. The bivalent iron of the enzyme (Fe) reacts with molecular oxygen in accordance with equation (1),

$$Fe + O_2 = FeO_2 \quad \ldots \ldots \ldots \ldots \ldots \quad (1),$$

where it does not matter for the theory if $FeO_2$ as a labile intermediate is rapidly transformed into the final ferric form as long as the stability of the two forms ($FeO_2$ and $Fe\cdots$) differs appreciably. Furthermore, it is important for the theory should be that in normally respiring cells the rate of respiration independent of the oxygen tension down to very low values. This has been demonstrated experimentally for the case of micrococcus candicans and red blood cells by WARBURG and his coworkers (1261)). Lately, however, the general validity of this rule has been challenged by workers who experimented with other types of cells. Thus KEMPNER 555a) finds that the respiration of various isolated cells, e.g. leucocytes, erythroblasts (but not non-nucleated erythrocytes), micrococcus candicans, pneumococcus, when undamaged and examined in their physiological environment, decreases with lowered oxygen tension. This variation of respiration with $O_2$-tension is found in cells sensitive and insensitive to carbon monoxide and cyanide. For earlier work concerning this important problem the paper by KEMPNER should be consulted. LASER 705a) has recently measured the respiration of various tissues, e.g. retina, liver, CROCKER mouse sarcoma, chorion, under low (5—20 %) and high (100 %) $O_2$-tension. Except for liver he observed no significant change in the magnitude of respiration under these conditions. The respiratory quotient, though, is lowered under low oxygen tension. In discussing this important point it should be pointed out that the material with which WARBURG experimented (Micrococcus candicans) was particularly favorable; the cells are very small and spherical in shape, thereby facilitating the condition of saturation with $O_2$ even at low partial $O_2$-pressures. The question of diffusion of oxygen into cells assumes a particular significance when the material does not consist of single cells but of tissue slices. For this special case WAR-

BURG has calculated, by means of the formula $d' = \sqrt{8c_o \dfrac{D}{A}}$ *), the limiting layer of

thickness where, for any given $O_2$-tension, saturation of all cells with oxygen is guaranteed. If this limit is exceeded (in the case of liver slices, for instance, 0.2 mm.) quite erroneous results may be obtained. As FRIEDHEIM 372) has pointed out any diffusible substance which acts as an oxygen carrier will increase the oxygen uptake of the tissue slice by enabling the deeper layers to participate actively in the process.

However, it can hardly be questioned that the condition of independence of respiration of oxygen tension was fulfilled in those cells (yeast, acetobacter) which served for the photochemical experiments of WARBURG and his associates. Here, then, practically all of the respiratory ferment may be taken to be present in the $FeO_2$ (or in the equivalent $Fe\cdots$) form; the limiting factor in the respiratory rate is here the reduction of $FeO_2$ by the substrate. The inhibition of respiration by CO is explained by the compe-

---

*) Explanation of symbols: $c_o$, concentration of oxygen immediately outside the tissue slice (in atmospheres); A, rate of respiration of the tissue (in cc. per minute per cc. of tissue); D, diffusion constant of oxygen in the tissue substance (in cc. of $O_2$ under standard conditions per sq.cm. per minute when the pressure gradient is 1 atmosphere per cm.); and $d'$, the thickness of the tissue slice (in cm.) at which the oxygen concentration at the centre layer of the slice is just zero. According to KROGH, D equals $1.4 \times 10^{-5}$ at 38°.

tition of the CO with $O_2$ for the iron of the ferment and the formation of a complex FeCO. The equilibrium between $FeO_2$ and FeCO in the dark may be expressed by equations (2) to (4) in accordance with the law of mass action:

$$\frac{FeO_2}{Fe \cdot O_2} = \frac{B}{Z} = k_{O2}, \quad\ldots\ldots\ldots\ldots\ldots \quad (2)$$

$$\frac{FeCO}{Fe \cdot CO} = \frac{b}{z_d} = k_{CO}, \quad\ldots\ldots\ldots\ldots\ldots \quad (3)$$

where $FeO_2$, FeCO, etc. are the concentrations of the reactions, B and Z and b and $z_d$ are the formation- and decomposition constants of $FeO_2$ and FeCO respectively, $k_{o2}$ and $k_{CO}$ the affinity constants. Equations (2) and (3) may be combined to yield equation (4):

$$\frac{FeO_2 \cdot CO}{FeCO \cdot O_2} = \frac{B}{Z} \cdot \frac{z_d}{b} = \frac{k_{o2}}{k_{CO}} = K'_d \quad\ldots\ldots\ldots\ldots \quad (4)$$

Under conditions where the system respires at an appreciable rate the decomposition of $FeO_2$ is accelerated beyond the value of the "spontaneous" or "dark" decomposition constant, Z; this requires the addition of the term Z', corresponding to the "respiratory" disappearance of $FeO_2$:

$$\frac{FeO_2}{FeCO} \cdot \frac{CO}{O_2} = \frac{B}{Z+Z'} \cdot \frac{z_d}{b} = \frac{k'_{o2}}{k_{CO}} = K_d, \quad (K_d \leqq K'_d). \quad\ldots\ldots \quad (5)$$

The relation to magnitudes which may be experimentally determined is established by the following considerations. If the respiration is partly inhibited by carbon monoxide only a fraction of the enzyme will exist in the form $FeO_2$ and the remainder in the form FeCO; the concentration of the reduced form, Fe, may be neglected, particularly in the presence of CO. Let us assume that the respiration in CO is decreased from $A_o$ to $A_d$; the residual respiration may be defined as

$$\frac{A_d}{A_o} = n_d \quad\ldots\ldots\ldots\ldots\ldots\ldots \quad (6)$$

and the inhibition of respiration as

$$\frac{A_o - A_d}{A_o} = 1 - n_d \quad\ldots\ldots\ldots\ldots\ldots \quad (7)$$

The relation of the residual respiration, $n_d$, to the inhibition of respiration, $1-n_d$, corresponds to the ratio of the concentrations of the active form $FeO_2$ to the inactive form FeCO:

$$\frac{FeO_2}{FeCO} = \frac{n_d}{1-n_d} . \quad\ldots\ldots\ldots\ldots\ldots \quad (8)$$

Consequently, the equation of distribution (5) assumes the form

$$\frac{n_d}{1-n_d} \cdot \frac{CO}{O_2} = K_d \quad\ldots\ldots\ldots\ldots\ldots \quad (9)$$

All terms on the left hand side of equation (9) may be experimentally determined: n is found by the manometric measurements of respiration and the concentrations of the gases are determined by gas analysis. In the event that the cells are not saturated with substrate which may be recognized by a comparison with the value of the maximal respiratory rate which may be attained under optimum conditions, equation (9) is to be replaced by the more general relation

$$\frac{\varepsilon \cdot n}{1 - \varepsilon \cdot n} \cdot \frac{CO}{O_2} = K_d, \quad \ldots \ldots \ldots \ldots \quad (9a)$$

where $\varepsilon$ is the degree of saturation of the system with substrate. If it is ascertained that the oxygen tension is not decreased below the limit required for the maintenance of the full rate of normal respiration and that the cells are saturated with substrate, it is found, in agreement with equation (9), that the degree of inhibition depends solely on the ratio of oxygen to carbon monoxide and is independent of the absolute concentrations of $O_2$ and CO; upon diluting a given $CO/O_2$ gas mixture with an inert gas, e.g. nitrogen or argon, the degree of inhibition remains unaltered.

The above relationships hold for the CO inhibtion in the dark only. Upon illumination the CO-compound of the ferment, but not the $O_2$-compound, suffers a reversible photochemical dissociation which augments to the extent $z'$ (photodissociation constant) the dissociation due to spontaneous dissociation, $z_d$ (dark dissociation constant). This leads experimentally to an increased oxygen consumption in light and theoretically to the "light-distribution equation" (10):

$$\frac{B}{Z + Z'} \cdot \frac{z_d + z'}{b} = K_h \quad \ldots \ldots \ldots \ldots \ldots \quad (10)$$

In analogy to equation (9) we obtain for the "light-state":

$$\frac{n_h}{1 - n_h} \cdot \frac{CO}{O_2} = K_h \quad \ldots \ldots \ldots \ldots \ldots \quad (11)$$

If by "light-sensitivity", L, is understood the increment by which the dark-dissociation constant, $K_d$, is changed per unit light intensity, i, absorbed by the system, we obtain

$$L = \frac{\triangle K}{K_d \cdot i} = \frac{K_h - K_d}{K_d \cdot i} \quad \ldots \ldots \ldots \ldots \quad (12)$$

From equations (10) and (5) there follows

$$L = \frac{z'}{z_d \cdot i} \cdot \quad \ldots \ldots \ldots \ldots \ldots \quad (13)$$

The dependence of the photochemical dissociation constant on the amount of light energy absorbed is furnished by the following considerations. It is obvious that the amount of light energy absorbed, under monochromatic conditions of illumination, is a function of the absorption coefficient, $\beta$, of the absorbing substance for the wave-length of light employed. The complete absorption spectrum of a substance is defined by the sum of the $\beta$-values for all wavelengths.

The amount of light energy in cal. absorbed by a certain amount of the carbon monoxide ferment complex per time unit (min.), when irradiated with monchromatic light of the intensity i under conditions of small total absorption, i.e. if i is constant throughout the solution, is

$$i \cdot q \cdot \beta \cdot c \cdot d \text{ (cal/min)} \quad \ldots \ldots \ldots \ldots \quad (14)$$

where q represents the cross section, d the layer of thickness, and c the concentration of the compound. In order to express the light energy absorbed in terms of quanta, the product (14) is divided by $h \cdot \nu$ (h, PLANCK's quantum constant, $\nu$, the frequency of the radiation); assuming that one quantum of light absorbed will dissociate one molecule FeCO, and introducing AVOGADRO's number, the number of photochemically decomposed molecules is

$$\frac{i \cdot q \cdot \beta \cdot c \cdot d}{N_o \cdot h \cdot \nu} \quad \cdots \cdots \cdots \cdots \cdots \cdots \quad (15)$$

The fraction of the total FeCO ($= q \cdot d \cdot c$) dissociating in unit time, $z'$, is obtained by dividing expression (15) through $q \cdot d \cdot c$:

$$z' = \frac{i \cdot \beta}{N_o \cdot h \cdot \nu} \cdots \cdots \cdots \cdots \cdots \cdots \quad (16)$$

If this is substituted in equation (13), taking equation (12) into account, we obtain

$$L = \frac{\triangle K}{K_d \cdot i} = \frac{\beta}{N_o \cdot h \cdot \nu \cdot z_d} \cdot \quad \cdots \cdots \cdots \cdots \quad (17)$$

There follows for constant $\nu$:

$$\frac{\triangle K_1}{K_d \cdot i_1} = \frac{\triangle K_2}{K_d \cdot i_2} \, ; \quad \frac{\triangle K_1}{\triangle K_2} = \frac{i_1}{i_2} \quad \cdots \cdots \cdots \cdots \quad (18)$$

Conversely, we obtain for constant i:

$$\frac{L_1}{L_2} = \frac{\beta_1 \cdot \nu_2}{\beta_2 \cdot \nu_1} = \frac{\beta_1 \cdot \lambda_1}{\beta_2 \cdot \lambda_2} \quad \cdots \cdots \cdots \cdots \cdots \quad (19)$$

Equation (18) is suitable for testing the theory and equation (19) is suitable for the determination of the relative absorption spectrum.

If the dark dissociation constant, $z_d$, of FeCO is known, the absorption coefficients, $\beta$, may be calculated with the aid of equation (17). Since $z_d$ is related to the r a t e of change in respiration upon illumination and darkening, the determination of that rate provides a measure for $z_d$, $z'$, and, furthermore, for $\beta$. Inasmuch as the experimental measurement of the acceleration of respiration upon changing over from dark to light is difficult to perform, the system is exposed instead to intermittent illumination; the ensuing stationary state is not identical with the mean of the respiration in the dark and in light because the decrease in respiration in the dark occurs more slowly than the acceleration of respiration in light. In this procedure the dark and light periods are of the order of one minute. If the oxygen uptake is followed under these conditions, a curve of roughly the shape shown in Fig. 2 (**1268**)) is obtained.

The amount of $O_2$ absorbed during a certain time interval is represented as the integral by that area which is enclosed by the time ordinates, the abscissa, and by the curve drawn in the figure. If the light and dark periods, t, are permitted to become very large so that the horizontal parts of the curve become very long, $A_i$, the average respiration during light plus dark period, can be obtained with the aid of equation (6) by expression (20):

$$\frac{A_1}{A_o} = \frac{n_d + n_h}{2} \cdots \quad (20)$$

If, on the other hand, light and

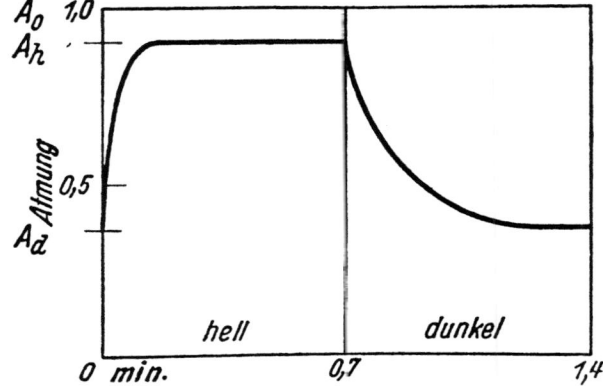

Fig. 2. Course of CO-inhibited respiration with intermittent illumination (WARBURG and NEGELEIN **1268**)).

dark periods follow each other at a suitably faster rate, a state is established which is defined by the equation

$$\frac{A_i}{A_o} \frac{n_d + n_h}{2} + \frac{(n_h - n_d)^2}{\rho} \cdot \frac{1 - e^{\frac{-\epsilon \cdot t}{1 - n_d}}}{2t} \qquad \ldots \ldots \ldots (21)$$

The value for $\rho$ appearing in the bracketed additional term is found by trial and error with the help of equation (21) where all the other magnitudes may be experimentally determined. For $z_d$ a relation containing $\rho$ may be deduced:

$$z_d = \left(\frac{n_d}{1 - n_d}\right)\rho \ldots \ldots \ldots \ldots \ldots (22)$$

If the value is substituted in equation (17), we obtain

$$\beta = L \cdot \frac{n_d}{1 - n_d} \cdot \rho \cdot N_o \cdot h \cdot \nu, \ldots \ldots \ldots \ldots (23)$$

whereby $\beta$ becomes amenable to experimental determination.

### Experimental Method:

Only a bare outline of the technique involved in the determination of the photochemical absorption spectrum can be given here; for details the original publications (616, 1271, 1268, 1270)) should be consulted. In principle it is always the disappearance (oxygen uptake) or the production ($CO_2$, $CO$) of gas which is measured by the sensitive differential manometric method. In accordance with the mathematical development given above, the following determinations are required for obtaining the relative absorption spectrum. The respiration of the cells under investigation, e.g. yeast or acetic acid bacteria, is measured in air as well as in a suitable $CO/O_2$ mixture in the dark; the composition is so chosen as to obtain an inhibition of respiration of about 80 %. The values thus obtained permit the calculation of the dark-dissociation constant, $n_d$. The system is then exposed to strong monochromatic radiation (intensity of the order of $10^{-2}$ cal. per min.) of varying wave-length. The rate of respiration, measured under these conditions, is an index of the light-dissociation constant, $n_h$. When the relative spectrum has been measured, the absolute absorption spectrum may be obtained by determining the absorption coefficient for only one wave-length; $\lambda = 436$ millimicron was chosen because the effect of the radiation is very great at that wave-length. The determination of the light-sensitivity for this point is carried out as indicated above. In order to obtain the value for $\rho$ according to equation (21) the system is first continuously irradiated with white light and subsequently with intermittent light of the same intensity. The dark and light periods last about 1 min. each. The $n'_h$-value as obtained from the respiratory intensities observed under these conditions is different and independent of $n_h$; it serves for the calculation of $\rho$. In concluding the experimental series the respiration is again measured in $CO_2/O_2$ in the dark and in air in order to ascertain that the cells have not been damaged in the course of the procedures. Throughout the experiments the temperature is kept as low as possible ($0-10°$) so as to increase the light-sensitivity of the system. This dependence of light-sensitivity on temperature is to be expected on the basis of the theory given above: The photochemical decomposition of the carbonyl complex, just like other photochemical reactions, is independent of temperature, whereas the "dark" or spontaneous decomposition of FeCO, like ordinary chemical reactions, exhibits the usual dependence on temperature. Consequently, with decreasing $z_d$-values upon lowering the temperature, the value of the ratio representing the light-sensitivity (equation 13) increases as the temperature is decreased. The magnitude of this effect is illustrated by thé fact that the light-sensitivity of the CO-respiratory ferment complex at 0° is 4.5 times greater than at 10°; the specific photochemical dissociation of CO-protoferrohemochromogen at 40° is 52 times smaller than at 0°. It is obvious therefore that a decrease of the experimental temperature is equivalent to an increase of the intensity of the light-sources employed. The limiting factor here is, of course, the effect of low temperatures on living cells: with wild yeast or acetic acid bacteria (616)) as the object, a temperature of 10° could be maintained, while with retina the temperature had to be raised to 20° (1272)).

### Tests of the Theory:

Inasmuch as the main value of the procedure just outlined rests on the ability to obtain the absorption curve of a substance which cannot be measured by direct methods, evidence was required to show that the photochemical absorption spectrum is identical with the directly determined absorption spectrum. This proof was furnished by the study of other iron-carbonyl compounds. The oxidation of cysteine by molecular oxygen is catalyzed by pyridine and nicotine hemochromogen (KREBS 596)). This catalysis is reversibly inhibited by CO which combines with the ferrohemochromogen to form an inactive complex (596)). In this case it is possible to determine both the photochemical absorption spectrum by the method outlined above and the direct absorption spectrum by the conventional methods (photoelectric spectrophotometry, spectrography, etc.). These two spectra were found to be identical.

Another important test of the theory has been performed with the aid of the CO compound of ferrous cysteine. Since it is possible to prepare sufficient amounts of this complex, the actual pressure developed by CO upon photodissociation of the molecule could be measured instead of the more indirect effect of radiation on its catalytic activity. It was found that one quantum of light will dissociate one molecule of the CO complex (191)). In the case of CO-ferrous cysteine this leads to the liberation of two CO molecules. CO-pyridine ferrohemochromogen will yield only one molecule of CO:

$$Fe\text{-cysteine}(CO)_2 + h \cdot \nu = Fe\text{-cysteine} + 2 \ CO$$
$$Fe\text{-pyridine hemochromogen}(CO) + h \cdot \nu = Fe\text{-pyridine hemochromogen} + CO.$$

Thus the validity of EINSTEIN's equivalence law is established for this reaction type. The spectrum of CO-ferrous cysteine as determined by direct measurements coincides with the photochemical absorption spectrum of the compound as determined by plotting the amount of CO split off against the wave-length of light employed (191)).

### Comparison of Light-Sensitivity of CO-Compounds:

The light-sensitivity of iron carbonyl compounds is somewhat related to the affinity of the various iron derivatives to CO. Several methods exist by which the affinity may be determined. One of them consists in determining the concentration of CO at which other components like $O_2$ are displaced from the iron molecule surface, another in increasing the CO-concentration until a reaction catalyzed by the iron compound is markedly or completely inhibited. Thus it was found that the cysteine oxidation as catalyzed by protoferriheme (hematin) is inhibited by CO to about the same extent as the cell respiration which is catalyzed by the respiratory ferment. The oxidation of cysteine when catalyzed by pyridine or nicotine ferrohemochromogen, on the other hand, is appreciably more sensitive to CO. The affinity of ferrocysteine to CO is so small that its catalytic effect on the oxidation of cysteine is not affected even when the CO concentration reaches 95 per cent. The affinities of the respiratory ferment and of hemoglobin to CO and $O_2$ respectively are shown in the following table:

TABLE 6.

$CO/O_2$-Affinity Ratios acccording to WARBURG:

| Respiratory Ferment (Yeast) | Hemoglobin |
|---|---|
| $\dfrac{FeO_2}{FeCO} \cdot \dfrac{CO}{O_2} = 9$ | $\dfrac{HbO_2}{HbCO} \cdot \dfrac{CO}{O_2} \backsim 0.01$ |
| $\dfrac{FeO_2}{Fe\,Po_2} \gtreqqless 2500$ | $\dfrac{HbO_2}{Hb\,Po_2} \backsim 50$ |

The light-sensitivity of the CO-complexes, too, varies greatly. The CO-ferrohemochromogens show about the same light-sensitivity as the CO-compound of the respiratory ferment; CO-hemoglobin and CO-heme have a low light-sensitivity while the carbylamine compound of CO-hemoglobin is highly sensitive (**1278**)). According to equation (17) a great light-sensitivity expresses itself by a small dark-dissociation constant $z_d$. This constant has the value of 0.029 for the case of carbylamine CO-hemoglobin (**1278**)) and 0.5 for the respiratory ferment at 10° (**616**)). In the former instance the large light-sensitivity could be ascertained by two independent methods, namely, (1) directly by the experimental determination of the magnitudes on the left hand side of equation (17) and (2) with the aid of $\beta$ and of the dark-dissociation constant $z_d$, on the basis of the quantum relationship mentioned above. The validity of the mathematical theory is supported by the fact that both methods yield values for the light-sensitivity agreeing with each other.

### Direct Spectroscopy of Hematin-Containing Enzymes:

The first hematin catalysts which were found amenable to direct spectroscopy were the histohematines of MCMUNN, later rediscovered and renamed by D. KEILIN as cytochromes a, b, and c. Whether or not all or some of these cytochromes are to be classified as true enzymes or rather as non-enzymatic catalysts or carriers is still open to debate. The first undisputed hematin enzyme to be studied by the spectroscope was catalase (ZEILE et al. in 1930)). In 1933 and 1934, WARBURG and his associates reported that the long-wave absorption bands of the respiratory ferment may be observed directly with the spectroscope in certain micro-organisms under suitable conditions. As will be shown below this claim has been and still is questioned by KEILIN. The latest enzyme to be obtained in sufficient concentrations to become measurable by spectroscopy is peroxidase (KEILIN and MANN (in 1937)).

Position of Absorption Bands in the Visible Region:

Once it has been established that the position of an absorption band of a hematin derivative is independent of the method by which it has been determined, we may freely compare bands of the various substances without further consideration of the technique employed.

The chief interest attaches itself to the position of the bands rather than to their absolute height, i.e. to the extinction at the maximum. The reason for this is that the position of the absorption bands is a characteristic constant for a heme derivative and that it allows one to draw certain inferences as to the structure of the heme and of the nitrogen base (or protein) with which it is combined. The main absorption band of hematin biocatalysts is situated in the far violet region just as the main band of hemoglobin (SORET band) and other hemochromogens. The position of the main bands of the enzymes and of certain non-enzymatic heme compounds are listed in Table 7.

## TABLE 7.

### Position of Main Absorption Band (γ-Band) of Enzymatic and Certain Non-enzymatic Heme Proteins:

| Substance | State of oxidation | Addition | Method | Main Band m μ | Literature |
|---|---|---|---|---|---|
| MacMunn's histohematins . (Cytochromes a, b, and c) | reduced | None | Direct | 414—420 430—435 446—452 | Warburg and Negelein 1273, 1274) |
| Cytochrome-c ............ | oxidized reduced | None None | Direct Direct | 407.5 415 | Keilin et al. 214) Theorell 1156, 1157) |
| Respiratory Ferment ...... (Yeast, Acetobacter) | reduced | CO | Photochem. | 433 430 | Warburg 1233) Warburg and Negelein 1268,1270,1271) |
| (Bac. pasteurianum) ..... | reduced | CO | Direct | 430 | Kubowitz and Haas 616) |
| (Torula utilis + 10°) .... | reduced | CO | Photochem. | 427—436 | |
| ( „ „ + 0.2°) ... | reduced | CO | Photochem. | 430 | |
| Spirographishemoglobin .... (artificial complex) | reduced | CO | Direct | 434 | |
| Spirographishemoglobin .... | reduced | None | Direct | 445 | |
| Chlorocruorin ............ | reduced | CO | Direct | 439 | Warburg and Negelein 1275) |
| Chlorocruorin ............ | reduced | None | Direct | 449 | |
| Pheohemoglobin-6 ........ (artificial complex) | reduced | CO | Direct | 435 | |
| Pheohemoglobin-b ........ | reduced | None | Direct | 445 | |
| Hemoglobin .............. | reduced | CO | Direct | 420 | Kubowitz and Haas 616) |
| Pheophorbid-b-hemoglobin . | reduced | CO | Direct | 442 | Warburg and Negelein 1275) |
| Peroxidase................ | reduced | None | Direct | 415—420 | Kuhn, Hand and Florkin 649) |
| Catalase ................ | oxidized | None | Direct | 409 | Itoh 501, Stern 1088) |
| Free Protoferroheme ...... | reduced | CO | Direct | ∽410 | Warburg and Negelein 1271) |

An inspection of the table shows that the main absorption band of the CO compound of the respiratory ferment of yeast (Torula utilis) and of acetic acid bacteria (B. pasteurianum) is located at 430 m$\mu$. Both Table 7. and Figure 3. demonstrate that the position of this band is shifted towards the red region as compared with the Soret Band of the CO complexes of the red hemins and of their hemochromogens (e.g. protoferroheme and ferrohemoglobin). The main bands of the CO compounds of green hemins and their derivatives, on the other hand, are situated even further towards the red range than the band of the enzyme. Although the general pattern of the spectrum of the ferment at once established its heme nature, it was also evident that the heme contained in it is very likely neither a red nor a green hemin. Casting about for heme derivatives of a more similar absorption spectrum WARBURG noticed the striking similarity with chlorocruorin, the respiratory pigment of the spirographis worm. This pigment contains a prosthetic group, spirographis hemin, in combination with a protein. While, in principle, it is modeled after the hemoglobin pattern, both the heme and the protein are different from the corresponding components of the vertebrate blood pigment. Not long before the charting of the spectrum of the respiratory ferment by WARBURG, the properties, among them the absorption spectrum, of chlorocruorin had been studied by MUNROE FOX 351). The most interesting property of this pigment is the change in color upon varying the layer of thickness or the concentration of its solutions: in transmitted light the pigment appears green in thick layers (or concentrated solution) and red in thin layers (or dilute solutions). This dichroism led to the designation of spirographis hemin by WARBURG as mixed-colored (green-red) hemin. The spectroscopic similarity between the enzyme and chlorocruorin prompted WARBURG to undertake a more detailed study of the spirographis pigment. In collaboration with NEGELEIN 1275) it was found that the prosthetic group, spirographis hemin, with globin yields a synthetic product with an absorption spectrum still more closely related to the enzyme spectrum (Table 7).

The difference in the spectrum of the artificial spirographis hemoglobin and of the natural pigment, chlorocruorin, bears out the statement of ROCHE 467a) that the bearer protein of the latter is different from globin. The results of the chemical study of the structure of spirographis hemin by WARBURG and by FISCHER will be discussed further below (p. 167). The subsequent discovery of other mixed-colored hemins and the examination of the spectra of their combinations with globin by WARBURG and NEGELEIN 1275) has brought about the recognition of these hemins as a group distinct from both the red and the green. They are now called pheohemins; undoubtedly the prosthetic group of the respiratory ferment is a member of this group.

The main absorption band of cytochrome (also called histohematin of MacMUNN) was first observed by the visual spectroscopy of bee's wing muscles by WARBURG and NEGELEIN 1273). After screening off the red and green part of the spectrum by suitable filters they observed three absorption bands in the blue-violet region with maxima at approximately 450, 430 and 415 m$\mu$ corresponding probably to the a, b, and c-components of cytochrome respectively. A little later the same workers were able to photograph these bands with the aid of a glass spectrograph of small dispersion (1274)). The band at 415 m$\mu$ was later observed by DIXON, HILL and KEILIN 214) in solutions of purified cytochrome-c in the reduced form. The corresponding band of the oxidized (ferric) form is situated still further to the shortwave region. As may be seen from Table 7. the position of the band of reduced cytochrome-c approaches closely that of protoferroheme. Subsequent chemical work by ZEILE and REUTER 1379) has proven the intimate relationship of the cytochrome-c porphyrin to protoporphyrin IX.

The fact that the main absorption band not only of protoheme and of cytochrome-c but also that of peroxidase after reduction with $Na_2S_2O_4$ (KUHN et al. **649**) and of catalase (ITOH **501**), STERN **1088**)) are located near 410 m $\mu$, already suggests that all of these compounds possess the same or a closely similar prosthetic group, viz. protoheme. It will be shown below that this is indeed the case.

While the position of the **main** or **short-wave** absorption bands of the four hemin enzyme systems appears well established and generally accepted, the interpretation of

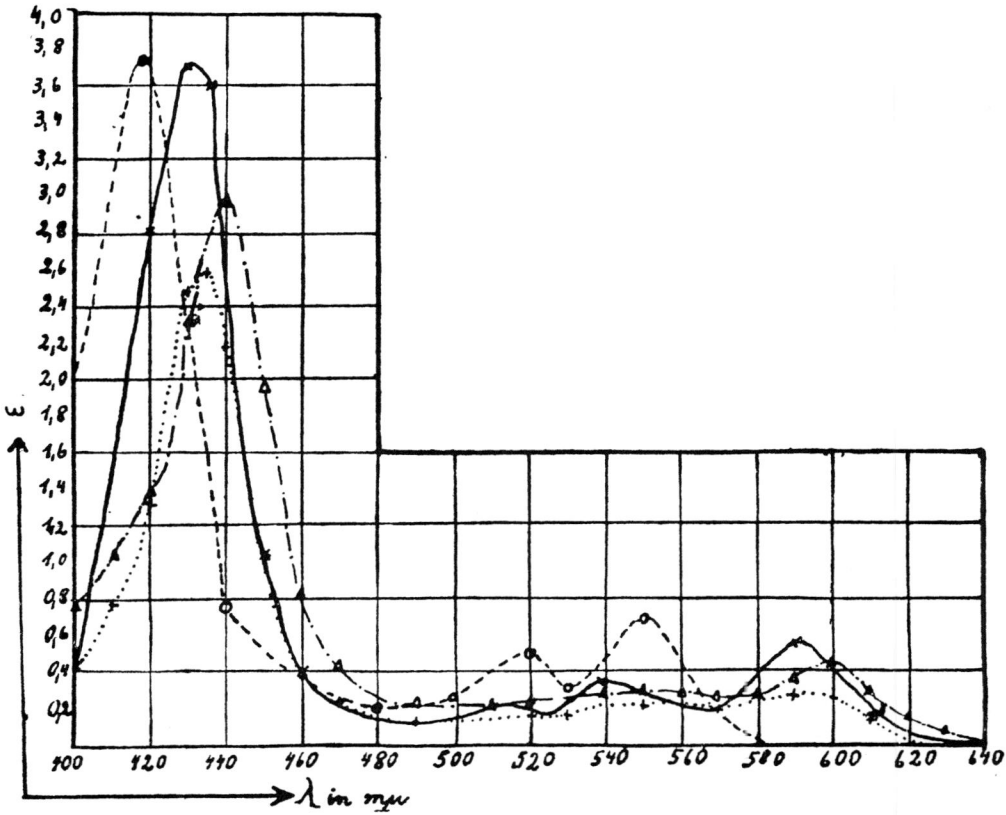

Fig. 3.   Absorption Spectra of the Respiratory Enzyme Spirographishemin proteins and Cytochrome c in the Visible (400 to 640 m $\mu$). Abscissa: Wave-length; Ordinate: Absorption coefficient. Curves:

    ×———×    CO-Respiratory Enzyme (WARBURG et al.)
    +......+    CO-Spirographis hemoglobin.
    △—.—△    CO-Chlorocruorin
    o - - - - o    Cytochrome c (KEILIN et al.)

the numerous **long-wave absorption bands** (in the red, yellow, and green region) which have been observed in solutions of these compounds and in various tissues and micro-organisms is still somewhat controversial. A number of such observations has been collected in Table 8. In some instances the effect of certain reagents and of temperature on these bands has been included. Slight differences in the position of these bands may be due to the subjective factor involved in visual observations and to variations in the biological material used (strains, age, conditions of culturing, etc.). Hence no significance may be attached to them. The symbol "—" signifies "not determined", the symbol "∅", "Band disappears".

TABLE 8.

| Substance (Source) | State of Oxidation | Effect of Reagents or Temperature | β and β¹ | Cytochrome c | α-Band b | α-Band a₁ | α-Band a | α-Band a₂ | Literature |
|---|---|---|---|---|---|---|---|---|---|
| Cytochrome-c ........ | red. | — | 520 | 550 | — | — | — | — | Keilin 214) |
| | ox. | — | 530 | 563 | — | — | — | — | Theorell 1157) |
| | ox. | — | 515 | 575 | — | — | — | 645, 675 | Bigwood 114) |
| Cytochrome-b ........ | ox.? | >ph 13 (Glucose) | 525—545 | 570—580 | 565 | — | — | — | Theorell 1157) |
| | +red. | — | 535—545 | — | — | — | — | — | Urban 1190) |
| Cytochrome a, b, c ...... | red. | — | 518—531 | 550 | 566 | — | 604 | — | Keilin 535) |
| Cytochrome-a (?) B. proteus. Acetobacter) | ox. | — | ~528 | ~566 | ~566 | 590 | — | 630, 636 | Keilin 541) |
| | red. | — | — | — | — | ø | — | 645 | |
| | ox. | — | — | — | — | 590 | — | 630 | |
| | red. | — | — | — | — | 590 | — | 635 | |
| Cytochr.-a (B. pasteur) ..... (Baker's yeast) | red. | Acid + CO | 512—532 | 549 | 564 | ø, 596 | 604 | — | Fink 886) |
| -a, b, c (Bottom yeast) ... | red. | — | 514—534 | 544 | 568 (556) | 588/590 | ø | — | |
| | red. | — | 526 | — | 557/563 | ø | 604 | — | v. Euler and Hellström 298) |
| | ox. | 80° | ø | — | ø | 586 | — | — | |
| | red. | — | 523 | 550 | 557/565 | 590 | — | — | |
| -a, b, c (Baker's yeast) ... | red. | — | 525/530 | 551 | 565 | 586 | 603 | 628 | Keilin 540) |
| -a, b, c (Acetobacter) .... | red. | — | 523 | 551 | 565 | 588 | 608 | 628 | |
| (B. megatherium) .... | red. | — | 526 | ø | 563 | — | — | 634 | |
| (B. coli and proteus).... | red. | CO | 531 | — | 560 | — | — | ø | |
| (B. coli) .... | red. | HCN | 530 | — | 560 | 590 | — | 645 | |
| | ox. | — | 530 | — | 560 | 590 | — | — | |
| (Azotobacter) .... | ox. | CO, HCN | 530 | 550 | 560 | 590 | — | — | Negelein and Gerischer 880) |
| | red. | — | 535 | 550 | 566 | ø | — | — | |
| | red. | — | — | — | 563 | — | — | — | |
| (Horse heart muscle) .... | red. | — | 524,0 | 555,6 | — | 585,2 | — | — | Roche and Bénévent 992) |
| (Sheep heart muscle) .... | red. | — | 525,0 | 555,8 | — | 586,0 | — | — | |
| (Baker's yeast) .... | red. | — | 523,5 | 556,3 | — | 586,2 | — | — | |
| (Bottom yeast) .... | red. | — | 522,5 | 555,9 | — | 587,0 | — | — | |
| (Echalote) .... | red. | — | 522,0 | 556,7 | — | 585,0 | — | — | |
| (Actinia muscle) .... | red. | — | 524,0 | 554,5 | — | 585,0 | — | — | |
| (B. coli) .... | red. | — | 522,5 | 555,9 | — | 585,5 | — | — | |
| (Acetic acid bact.) .... | red. | — | 523,5 | 555,8 | — | 586,0 | — | — | |
| Respiratory Ferment (Torula utilis) | red. | CO | — | — | — | 590 | — | 632 | Warburg 1233) |
| " (Azotobacter) ..... | red. | CO | — | — | — | 583 | — | — | Warburg 1271) |
| " | ox. | — | — | ø | ø | — | — | 647 | Negelein and Gerischer 880) |

TABLE 8. *(Continued)*

| Substance (Source) | State of Oxidation | Reagent or Temperature | Geen, Yellow, and Red Bands, m$\mu$ | | | | | | Literature |
|---|---|---|---|---|---|---|---|---|---|
| | | | $\beta$ and $\beta^1$ | Cyto-chrome c | b | $a_1$ | a | $a_2$ | |
| | | | | | | | ($\alpha$-Band) | | |
| Respiratory Ferment (Azotobacter) ...... | red. | CO | — | — | — | — | — | 637 | Negelein and Gerischer 880) |
| | red. | HCN | — | — | — | — | — | 632 | |
| | ox. | HCN | — | — | — | — | — | ø | |
| „ (Baker's yeast) ....... | ox. | HCN | — | — | — | 583 | — | — | Warburg and Haas 1260) |
| | red. | — | — | — | — | 583 | — | — | |
| „ (B. pasteurianum) ...... | red. | CO | 524, 540 | — | — | 590/600 | — | — | Kubowitz and Haas 616) |
| „ (Torula utilis) ........ | red. | CO | ø, 540 | — | — | 590 | — | — | |
| „ (B. pasteurianum) ...... | red. | CO (+ 10°) | 512, 540 | — | — | 590 | — | — | Warburg and Negelein (1277) |
| | red. | CO (+ 0,2°) | — | — | — | 590 | — | — | |
| | red. | CO | — | — | — | 593 | — | 640 | |
| | ox. | HCN | — | — | — | ø | — | ø | |
| Spirographis hemin ...... | red. | Pyridine | 509/520 | 550/560 | 580/582 | 584 | — | — | Warburg and Negelein 1275) |
| „ porphyrin ........ | — | — | — | 550/560 | 563/575 | 591/594 | 590—615 | 639/647 | |
| „ hemoglobin ...... | red. | CO | — | 545/557 | 563/575 | 585/603 (594) | 603/617 | 603/617 | |
| Chlorocruorin ........ | red. | — | — | — | 569/577 | — | 600/618 | 600/618 | |
| Pheohemin-b ........ | red. | CO | — | 550/560 | — | 583 | 590—615 | — | Warburg and Negelein 1275) |
| | red. | Pyridine | — | 558 | — | — | — | — | |
| Pheohemoglobin-b ...... | red. | — | — | 557 | — | — | 585—615 | — | |
| Blood hemin ......... | red. | CO | — | ~540 | — | — | 590—605 | — | |
| | red. | Pyridine | — | — | — | — | — | — | |
| | red. | CO | — | — | — | — | — | — | |
| Protoferroheme IX ...... | red. | CO | — | ~530 | 570 | 589 | — | — | Warburg 1270, 1268) / Reid 975) |
| Pyridine hemochromogen ... | red. | CO | 525,4 | 557,8 | — | — | — | — | Warburg 1271) |
| Hemoglobin ......... | — | CO | 527 | — | 557 | 570 | — | — | Roche 992) |
| Peroxidase ......... | red. | (Na₂S₂O₄) | 498 | 548 | — | 583 | — | 645 | Kubowitz 616) / Kuhn 649) |
| | ox. | ph = 10 | ø | 549 | 558 | 583 | — | ø | |
| | ox. | Na₂S₂O₄ | ø | 530,5 | 545,5; 561 | ø | — | ø | |
| | red. | H₂O₂ | ø | — | — | 583 | 594,5 | ø | |
| Catalase (liver, cucumber seeds) ...... | — | — | 500 | 540 | — | — | — | 629*) | Keilin and Mann 551) |
| | — | KCN | ø | ø | 557 | 589 | — | ø | Zeile 1877) / Zeile 1371) |
| | — | — | 506,5 | 544 | — | 570 | — | 629,5*) | Keilin c.s 548) |
| | — | C₂H₅OOH | 505 | 540 | — | — | — | 622*) | Stern 1089, 1088) / Stern 1087) |
| Mesoporphyrin (blood) ...... | — | Ether Sol. | 494,9 | 534,5 | 567,6 | 578,3 | 596,6 | 623,3 | Fischer 344) |
| „ (catalase) ........ | — | ,, | 498,2 | 528,6 / 526,1 | 568,7 | — | — | 624,1 | Stern 1084) |

*) The discrepancies in these data may be due to the fact that different optical methods were employed in these measurements (1086).

Let us first consider the spectrum of the cytochrome system as it has been described in different cells (e.g. bee's wing muscle, bakers yeast, frog muscle, etc.) by D. KEILIN 535) who was following up the early work by the naturalist MacMunn. In anerobic bee muscle he observed a spectrum of reduced cytochrome with four long-wave bands, at about 604, 566, 550, and 521 m $\mu$. He designated these bands with a, b, c and d in the order given. This terminology has created a certain confusion since KEILIN himself could show that the a, b, and c band represent the $\alpha$-band or long-wave band of the three cytochrome components a, b, and c respectively while the d-band probably consists of three fine bands in close proximity or even superposition which represent the $\beta$-bands of the three cytochromes. These relationships are illustrated in Fig. 4. The $\gamma$-or main bands of these catalysts are the three bands found by WARBURG and NEGELEIN in the violet region which have already been mentioned.

EULER and his associates (287)) confirmed the position of the cytochrome bands in yeast cells. They improved the technique of direct examination of these spectra and developed it to a quantitative method for the intracellular determination of cytochrome. The position of the cytochrome bands changes little from one object to the other. The relative intensity however will vary considerably, and in some instances one or the other of these bands seems to be completely absent (KEILIN 535), TAMIYA and YAMAGUTCHI 1135)).

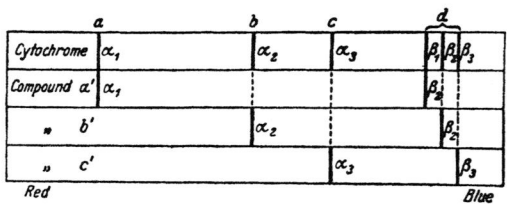

Fig. 4.  Position and Correlation of Cytochrome Absorption Bands in Bakers' Yeast according to KEILIN 540).

From a spectroscopic point of view the component a is at the same time the most complex and the most interesting of the three. In many cells the typical a-band at 605 m $\mu$ is missing; instead, bands around 590 or 630 m $\mu$ are found. These atypical cytochrome-a bands have been designated as $a_1$ and $a_2$-band respectively and have been charted for a number of pathogenic bacteria by FUJITA and KODAMA 381a). The intriguing feature of the $a_1$-band near 590 m$\mu$ is that it is located precisely where WARBURG and his collaborators found the $\alpha$-band of the respiratory ferment by the photochemical method. The $a_1$-band both in baker's yeast (WARBURG and HAAS 1260) and in acetic acid bacteria (WARBURG et al. 1280, 1277) behaves both with regard to its position at 590 to 593 m$\mu$ and to its reaction with CO and HCN as would be expected from the long-wave band of the ferrous form of the respiratory ferment (see Table 8). The $a_2$-band in azotobacter, though differing in position from that of WARBURG's respiratory ferment, being situated in the red instead in the yellow, reacts exactly like the latter enzyme with cyanide and carbon monoxide (880, 1280)) (see Table 8). As a rule the $\alpha$-bands of the ferrous forms of red heme derivatives are found at 570 m $\mu$ or below, those of mixed-colored hemins around 580 to 595 m $\mu$, and those of green hemins beyond 600 m $\mu$. On the basis of this classification the band in azotobacter, ascribed by WARBURG to the respiratory ferment, would be caused by a green hemin derivative. Unfortunately, the photochemical absorption spectrum of the respiratory ferment of azotobacter has as yet not been determined. This experiment should provide decisive evidence for or against WARBURG's claim that it is possible to see the spectrum of the ferment in living cells with the spectroscope. Such evidence is particularly required in view of the criticisms and objections raised by KEILIN 542, 541) against WARBURG's interpretation of his experimentally undisputed observations. KEILIN insists that the various bands observed directly by WARBURG are not those of the respiratory enzyme but those of more or less modified cytochrome-a derivatives. Very recently KEILIN and HARTREE 550) have reported further observations . The new evidence suggests that the band of cytochrome-a at 605 m $\mu$ belongs to two slightly different hemochromogen compounds having the same heme nucleus. One of these compounds, called a, is not autoxidizable and does not combine with KCN or CO, while the other, called $a_3$, is autoxidizable and will combine with both reagents. In this respect it resembles the components $a_1$ and $a_2$ which occur only when component a is absent. The CO-compound of $a_3$ has its absorption band at 593 m $\mu$ and, according to KEILIN, may be responsible for the absorption band found by WARBURG in the photochemical absorption spectrum of the respiratory ferment. However, the cytochrome component $a_3$ cannot be identified with the respiratory ferment (cytochrome oxidase). Three possibilities with regard to the chemical nature of the latter are discussed by KEILIN and HARTREE: (1) The enzyme may be a hematin-protein compound similar to cytochrome-a but spectroscopically invisible due to its low concentration in the living cell; (2) the enzyme may be an iron-protein complex devoid of porphyrin; (3) the enzyme may be a copper-protein. The last assumption, while not supported by direct experimental evidence, appears to be favored by several considerations: Copper salts have been found by KEILIN to be the only simple metal salts capable of rapid oxidation of all three cytochrome components, the addition of copper salts to nutrient media and diets will increase the cytochrome-a and the indophenol oxidase concentration of plant and animal cells (ELVEHJEM 176a)), the intensity of the indophenol reaction given by tissues is roughly paralleled by their copper content, and finally the respiratory enzyme resembles somewhat in its behaviour the polyphenol oxidase of plant cells which has recently been demonstrated to be a copper-protein (see p. 181).

It is interesting to note that ROCHE and BÉNÉVENT 992) have expressed views concerning the cytochrome-a complex quite similar to those put forward by KEILIN and HARTREE. They assume that cytochrome-a consists of three components, viz. a, $a_1$ and $a_2$ and that $a_1$ and a possess a hemin grouping analogous to that of the respiratory enzyme. This would explain why $a_1$ and the ferment show the same typical bands, $\alpha = $ 585 to 590 m $\mu$ and $\beta = \infty$ 525 m $\mu$.

According to WARBURG 1234) it is also possible that the band in the yellow as observed in acetic acid bacteria and in baker's yeast which is slightly shifted by CO to 593 m $\mu$ is due to an intermediary oxygenation product of the ferrous form of the respiratory ferment. Such a complex would be analogue to oxyhemoglobin. This view is supported by the observation made by WARBURG and NEGELEIN 1275) that the addition compounds of oxygen with certain artificial "hemoglobins", e.g. spirographis hemoglobin, do not exhibit two absorption bands in the green like oxyhemoglobin but only one absorption band in the long-wave region. WARBURG assumes that this loose addition compound changes spontaneously into the ferric form of the enzyme (analogous to methemoglobin), rapidly in the case of acetic acid bacteria and slowly in that of baker's yeast. The band in the yellow can hardly be caused by the ferric form of the enzyme itself since all known ferric compounds absorb specifically in the red region. Only in the case of azotobacter do we find the ferrous form showing absorption in the red.

Upon oxidation all of the sharp cytochrome bands fade. There remain faint shadows at 520 to 540 m $\mu$ and at 550 to 570 m$\mu$ (535)) with a maximum at $\alpha = $ 566.5 m $\mu$ and

$\beta = 528.7$ m $\mu$ **(723))**. This is equivalent to saying that upon oxidation only the $\alpha$-band of cytochrome-b and probably its $\beta$-band are retained. It be may added that the b-component, in general, appears to be the most stable and the a-component the most labile against oxidizing agents. URBAN **1130)** reports that in aerated, washed, respiring cells of baker's yeast at 6°, 26° and 34° only $\alpha$-band of cythchrome-b at 565 m $\mu$ and a weak band at 535 to 545 m $\mu$ are visible. Upon adding glucose no change is detected at 6°. If the temperature is raised and the reduction thereby accelerated, at 26° the $\alpha$-band of cytochrome-c and at 34° also that of the a-component will appear. In addition, also the weak $\beta$-bands at 537 and 523 m $\mu$ become detectable while the band of the oxidized cytochrome at 535 to 545 m $\mu$, ascribed to ferricytochrome-c by URBAN, fades away. It would appear, however, that the faint absorption observed in the oxidized state at 535 to 545 m $\mu$ belongs to the b-component rather than to c, inasmuch as KEILIN et al. **214)** have measured the spectrum of isolated ferricytochrome-c and have found bands at 530 and 563 m $\mu$. In the reduced state they found, in addition to the simple $\alpha$-band at 550 m $\mu$, a complex $\beta$-band with a maximum at 520 m $\mu$ (see Table 8). In strongly alkaline solution the spectrum of the ferroform remains unaltered while the ferriform has two bands, at 525 to 545 m $\mu$ and at 570 to 580 m $\mu$ (THEORELL **1156, 1157)**). The bands observed in moulds by TAMIYA **1132)** and ascribed by this author to the free intracellular hematin of KEILIN, may in fact be caused by ferricytochrome-b. The same holds probably for the bands seen in adrenal tissue by ROSENBOHM **1004)**.

It is this difference in stability against oxidation and other agents **293)** and the variation in the relative concentration of these three components which has undoubtedly materially contributed to the reigning confusion with regard to the assignment of certain absorption bands to the various cytochromes or to the respiratory ferment. In general it may be said that the $\alpha$-bands of the reduced forms can today be attributed to individual pigments with a little more assurance than the $\beta$-bands. The bands at 540 m $\mu$ and at 524 m $\mu$ in B. pasteurianum, for instance, have been attributed to the CO-compound of the respiratory ferment by KUBOWITZ and HAAS **616)** though the final proof for this is still lacking.

The spectroscopic study of living cells and of cell-free extracts has lately become even more complicated by the work of R. LEMBERG on bile pigment hemochromogens. He has shown that, in the course of the transformation of blood pigments to bile pigments, the first step consists in the oxidative scission of one methin (—CH—) bridge with the formation of a CO and a COH-group in the two adjacent pyrrol rings **(711a))**:

Hemochromogen    Verdohemochromogen

An open tetrapyrrolic chain, typical for the bile pigments, is thus formed. However, the central iron atom, by virtue of valency forces acting upon the pyrrolic nitrogen atoms, maintains the ring structure of the molecule. These bile pigment hemochromogens behave much like ordinary hemochromogens and have, in fact, been mistaken for hemochromogens. WARBURG and NEGELEIN 1272a), for instance, who obtained a green product from blood hemin by a treatment involving oxido-reduction, considered it to be a green hemin, i.e. a hemin comparable to those obtained from chlorophyll by substituting iron for magnesium. LEMBERG 711a) could show that their compound is actually a verdohemochromogen consisting of the iron complex salt of biliverdin with pyridine as the nitrogeneous base. Corresponding compounds with globin as the nitrogeneous constituent have recently been described by LEMBERG and WYNDHAM 711b) and by EDLBACHER and SEGESSER 228) *).

The opening of the ring structure renders the iron labile in these bile pigment hemochromogens. Once the metal is removed, the pyrrolic structure will "uncoil" into a straight chain, and the properties characteristic for bile pigment derivatives will become manifest.

The occurrence of bile pigment hemochromogens (verdohemochromogens) in various biological materials, e.g. in purified catalase preparations from horse liver, in purified preparations of cytochrome-c from baker's yeast **), and in blood extracts has been demonstrated by LEMBERG and WYNDHAM 711c). The important point here is that most of these pigments have absorption bands in the red region of the spectrum (see Table I. in LEMBERG and WYNDHAM's publication) and that these workers suggest a relationship between this class of compounds and the absorption bands near 630 m $\mu$ which are found in the "cytochrome" spectrum of several bacteria. This is the "cytochrome-$a_2$"-band of Azotobacter, Acetobacter, B.coli and of various pathogenic intestinal bacteria (NEGELEIN and GERISCHER 80), KEILIN 541, 542), FUJITA and KODAMA 381)). LEMBERG and WYNDHAM 711c) point out that the spectrochemical properties of the cytochrome-$a_2$ band bear a striking resemblance to those of biliviolinhemochromogens. B i l i v i o l i n s are a group of isomeric compounds, probably differing only in the situation of the double bonds in the molecule, which arise through dehydrogenation of biliverdins. The correlation of the various groups of bile pigments, then, is as follows:

$$\text{Bilirubin} \xrightarrow{\text{Oxidation}} \text{Biliverdin} \xrightarrow{\text{Dehydrogenation}} \text{Biliviolin.}$$

Like biliverdin, biliviolin will form complex salts with iron and with other metals. The zinc complex salt of biliviolin is characterized by a strong red fluorescence. These metal complexes, in turn, will combine with nitrogeneous compounds like pyridine to form biliviolinhemochromogens. Like the band of cytochrome-$a_2$, the absorption band in the red of the biliviolinhemochromogens is shifted towards the infra-red by oxidation and is immediately destroyed by ammonium sulphide.

It may be, of course that the occurrence of bile pigment hemochromogens in biological sources has no other significance beyond the fact that they represent intermediates in the biological break-down of heme derivatives. On the other hand, the behaviour of the "cytochrome-$a_2$"-band in azotobacter toward CO and HCN (NEGELEIN and GERISCHER 880)) which suggests a close relationship to the respiratory ferment of WARBURG and the interesting catalytic properties of bile pigment hemochromogens in vitro (LEMBERG et al. 711b)) make further studies of the biological function of bile pigment hemochromogens highly desirable.

---

*) See also BARKAN and SCHALES 70a).
**) Such absorption bands in the red had previously been noted by BIGWOOD et al. 114) in solutions of cytochrome-c obtained from bakers yeast by the method of HILL and KEILIN 486).

The position of the absorption bands of c a t a l a s e and of p e r o x i d a s e will be found listed in Tables 7 and 8. Although some of these bands are located in the same region as some of the cytochrome bands, this fact has not yet caused any serious confusion (1) because the concentration of these enzymes in living cells is too low to permit a direct visibility of their spectrum without previous purification and concentration procedures *) and (2) because these enzymes may readily be separated from cytochrome in preparations and may also be identified in mixtures by their typical spectrocopic and catalytic properties.

The spectrum of purified c a t a l a s e preparation from horse liver or cucumber seeds, according to the pioneer work of ZEILE and HELLSTRÖM **1377, 1371)**, shows three absorption bands in the visible, viz.:

I. $\underbrace{620\text{---}649}_{629}$ m $\mu$;   II. $\underbrace{530\text{---}550}_{540}$ m $\mu$:   III. $\underbrace{490\text{---}510}_{500}$ m $\mu$

The position of these bands was determined by visual spectroscopy and by photography. KEILIN and HARTREE **548)** obtained, with the aid of the reversion spectroscope of HARTRIDGE, the values 506.5 m $\mu$, 544 m $\mu$, and 629.5 m $\mu$ for the peaks of the bands, while STERN **1088)**, on the basis of records obtained by HARDY's photoelectric automatic spectrophotometer, assigns to them the positions 505 m $\mu$, 540 m $\mu$, and 622 m $\mu$. The most readily observed band of catalase is the one in the red. It is asymmetrical, the peak being situated toward the short-wave end of the band. The center of the band is stated to shift from 627.5 to 630 m $\mu$ upon increase in concentration (**548)**).

Like all hemin and particularly hemin-protein compounds, catalase has a strong absorption band in the violet region, corresponding to the SORET band of hemoglobin. This band was found in liver catalase solutions by photographic (ITOH **501)**) and photoelectric (STERN **1088)**) spectrophotometry. Its position is given by STERN at 409 m $\mu$, i.e. almost at the same place as that of the hemoglobin band (410 m $\mu$ (cf. ADAMS **4)**). The relative intensity of the various absorption maxima of catalase and their extinction coefficients as referred to the porphyrin-bound iron content of the enzyme solutions have been determined (STERN **1088)**). The average extinction coefficients, as calculated by means of the equation (DRABKIN and AUSTIN)

$$\varepsilon_c = 1 \text{ mM Fe per liter} = \frac{D}{d \times c},$$

where D represents the observed optical density, d the depth of the absorption cell, and c the concentration of the enzyme in terms of mM of the porphyrin-bound iron per liter, were found to be $\varepsilon_{622} \text{ m}\mu = 10.8$ and $\varepsilon_{409} \text{ m}\mu = 145$. The use of the above equation had the advantage that no assumptions as to the molecular weight of the enzyme or to the number of hemin groupings are required.

The u l t r a v i o l e t spectrum of homogeneous catalase has recently been measured (STERN and LAVIN **1096)**). It shows a typical protein band at 275 m $\mu$. While ordinarily the extinction at the peak of the SORET band is much higher than at the maximum of the short-wave ultraviolet protein band, in the case of catalase this ratio is inverted.

A number of qualitative observations concerning the spectrochemical behaviour of catalase preparations will be found in the papers by KEILIN and HARTREE **548)** and by STERN **1089)**.

The catalase spectrum is affected neither by the usual oxidizing or by reducing

---

*) See, however, the observations made on rat liver tissue, freed from blood by perfusion (STERN **1089)**).

agents. Sodium hyposulfite ($Na_2S_2O_4$) for instance which will readily reduce methe-moglobin to hemoglobin or cathemoglobin to globin hemochromogen has no effect on the catalase spectrum. The irreversible slight shift to 624.5 m $\mu$ upon shaking catalase solutions containing $Na_2S_2O_4$ in air is attributed by KEILIN and HARTREE **548)** to the ph-change caused by the formation of sulfurous acid from the hyposulfite by oxidation. The formation of spectroscopically well-defined compounds between catalase and certain inhibitors and substrates will be discussed below (pp. 155, 176).

The spectrum of purified p e r o x i d a s e preparations from horse radish has recently been studied by KEILIN and MANN **551)**. The position of the maxima are given as 498, 548, 583, and 645 m $\mu$ (See also Table 8). A similar band in the red in strong peroxidase solutions from fig sap had previously been described by SUMNER and HOWELL **1110)**. KUHN, HAND and FLORKIN **649)** had failed to notice any long-wave absorption bands in their horse-radish preparations, probably due to an insufficient concentration of the enzyme in their solutions. The most active peroxidase preparations yet recorded in the literature (WILLSTÄTTER and POLLINGER **1336a)**) have apparently not been examined with the spectroscope.

In strongly alkaline solution or after reduction by $Na_2S_2O_4$ the $\alpha$-band in the red at 645 m $\mu$ and the $\beta$-band at 498 m $\mu$ are replaced by bands at about 557 m $\mu$ and at 594.5 m $\mu$ (KEILIN and MANN **551)**; KUHN et al. **649)**). A further band at 527 m $\mu$ observed by KUHN et al. **649)** in reduced peroxidase preparations could not be observed by KEILIN and MANN **551)**. The spectrum changes observed upon adding inhibitors and substrates to peroxidase solutions are discussed on p. 180.

### Reactions of Hemin Systems with Inhibitors:

**Respiratory Ferment:** The ability of carbon monoxide to form a complex with the ferroform of the oxygen transferring enzyme and its reversible photodissociation have made possible the measurement of the absorption spectrum and the classification of this biocatalyst. This has been discussed in detail above (p. 136). The affinity ratios between CO and $O_2$, the relative light-sensitivity, and the effect of temperature have also been dealt with (p. 141). CO has no affinity for the ferriform of the enzyme.

HCN on the other hand, reacts with the tervalent iron of the oxidized form of the enzyme. A HCN-concentration of about $10^{-4}$ M. will completely inhibit the ferment and hence the respiration. Catalyses promoted by nicotine and pyridine hemochromogens are inhibited to an extent of 50 per cent by $2 \cdot 8 \cdot 10^{-5}$ M. HCN. Inasmuch as KREBS **596)** finds spectroscopically that the $Fe^{\cdots}$-CN-complexes are about half dissociated at this HCN concentration and since HCN, at this concentration, will not yet combine with ferrous iron, we have to assume that HCN in this case, just as in that of the respiratory enzyme, blocks the catalysis by reacting with the ferric iron formed in the course of the reaction from the original ferrohemochromogen (see also WARBURG **1228)**). In contrast to the inhibition by CO, the inhibition of respiration by cyanides or free HCN is independent of the oxygen tension **(1223))**.

The essential difference in the action of CO and HCN on the respiratory ferment then, lies in the fact that CO inhibits its oxidation while, HCN inferes with its reduction **(1233))**. That the complex between ferric iron and cyanide is not readily reduced to ferrous iron has been demonstrated in vitro by BARRON and HASTINGS **85)**: cyanide ferri-hemochromogens are only reduced with difficulty by nascent hydrogen.

It can be demonstrated, however, that $CN'$ has also a certain affinity for bivalent iron and that it may partially displace CO and $O_2$ from ferrous complexes. WARBURG,

NEGELEIN and CHRISTIAN 1278), working with ferroheme in borate buffer, found that of 61 cmm. of CO absorbed, 27 cmm. are given off after adding KCN to make the solution 0.01 N. The same workers studied quantitatively the equilibrium between hemoglobin, CO, $O_2$ and methyl carbylamine. The equilibrium constant of the reaction

$$> FeCO + CH_3NC \leftrightarrows > FeCH_3NC + CO$$

at 10° is $K = 5.3 \cdot 10^{-5}$, while that of the reaction

$$> FeO_2 + CH_3NC \leftrightarrows > FeCH_3NC + O_2$$

is $K = 2.5 \cdot 10^{-2}$. It follows that the affinity ratio for hemoglobin between CO, $O_2$ and carbylamine is $19\,000 : 40 : 1$. According to the potential measurements of CONANT and TONGBERG 180), however, the affinity of the CN′ group to ferroheme is still appreciably greater than that of pyridine for the heme.

These reactions appear to have no significance for the special case of the respiratory ferment, since according to WARBURG CN′, within the concentration range tried, will react exclusively with the ferric form of the enzyme. Carbylamines, on the other hand, will not affect respiration or anaerobic fermentation but electively influence the PASTEUR-MEYERHOF reaction, as is evidenced by an increase in aerobic glycolysis (WARBURG 1220)).

Cytochrome: The reports concerning the effect of various inhibitors on the cytochrome system are conflicting. More experimental work appears to be necessary before a clear-cut picture can be given.

Both KEILIN 540) and WARBURG 1224, 1225) agree that, in general, cytochrome will not react directly with HCN or CO under physiological conditions and that the effects observed after adding these reagents to living cells are mainly due to a primary inhibition of the respiratory ferment. This is consistent with the view of these authors that the enzyme is a specific "cytochrome oxidase". If it is accepted that the oxidation of cytochrome in vivo is brought about exclusively by the respiratory ferment while its reduction is effected by dehydrogenase-substrate systems, then any inhibition of the respiratory ferment can only interfere with the reoxidation of ferrocytochrome but not with the reduction of ferricytochrome. This is borne out by the well-known observation that the addition of CO as well as of CN′ to baker's yeast, for instance, causes the appearance of the well-defined absorption spectrum of reduced cytochrome in full strength both under aerobic and under anaerobic conditions. In unpoisoned cells the characteristic ferrocytochrome spectrum is replaced by the indistinct ferricytochrome spectrum upon shaking with air.

For special cases or in isolated sytems, however, the observations and their interpretation are sometimes at variance with the general rule stated above. Pure cytochrome-c, though indifferent against both CO and HCN within the physiological ph-range, will combine with CO in alkaline solutions above ph 11 (KEILIN 540)). While it will not react with molecular oxygen at ph 4 to 10, it becomes autoxidizable below ph 4. This change in behaviour is probably due to more or less irreversible changes in the protein bearer. In certain instances a reaction between CO or HCN and cytochrome components has also been claimed in vivo. According to KEILIN 541, 542) cytochrome-a or its derivatives are sometimes capable of directly reacting with CO. KEILIN attributes this "atypical behaviour" bearer to the lability of the protein of the a-component in certain cells, e.g. in B. pasteurianum, B. coli, B. proteus, and Acetobacter. On this set of observations hinges the controversy between WARBURG and KEILIN concerning the direct visibility of the respiratory ferment. WARBURG wishes to define as respiratory ferment a hemin system reacting with CO and HCN and exhibiting absorption bands at wavelengths identical with those found in the photochemical absorption spectrum. He objects to KEILIN's practice of including both CO- and CN'-insensitive and sensitive hemin systems in his cytochrome scheme.

The observations of Japanese workers may be reconciled still less satisfactorily with the general rule outlined above. According to YAMAGUTCHI 1366) HCN will react with cytochrome-c as well as with b. SHIBATA 1050) still maintains that cytochrome is able to form a loose addition compound with oxygen of the type of oxyhemoglobin. The appearance of the spectrum of reduced cytochrome under aerobic conditions after adding HCN is explained by a displacement of the $O_2$ from the cytochrome molecule by HCN 1053) rather than by a blockade of the respiratory ferment (see also TAMIYA 1134)). While Shibata formerly believed that cytochrome as a whole reacts with $O_2$ and that this reaction is inhibited by CO, he has recently qualified this statement to the effect that only certain components of the cytochrome system, particularly cytochrome-b, will react with $O_2$ and CO respectively (SHIBATA and TAMIYA 1053); YAMAGUTCHI 1366)). FUJITA and KODAMA 382) distinguish between various cell types endowed with qualitatively and quantitatively different cytochrome systems. Like KEILIN, SHIBATA 1053, 1050) wishes to identify certain cytochrome components with the respiratory ferment of WAR-BURG.

Recently, ALTSCHUL and HOGNESS 20a) observed a change in the spectrum of reduced cytochrome c when its solution is saturated with CO, indicating that it forms a complex with CO. This complex formation is said to take place throughout the entire ph range from ph 3.8 to 13.

**Peroxidase:** This hemin enzyme appears to exist in a somewhat stabilized ferric form. Accordingly, WIELAND and SUTTER 1328) found that the enzyme has no appreciable affinity for CO as evidenced by its insensitivity towards this poison. Statements by KUHN, HAND, and FLORKIN 649) to the contrary were refuted by ELLIOTT and SUTTER 248) who showed that the "inhibition" observed by KUHN et al. was actually due to a mechanical damage of the enzyme caused by the gas stream. HCN, on the other hand, will readily inactivate the enzyme. A concentration of $5.10^{-6}$ M. HCN causes an inhibition of 50 per cent which is stated to be irreversible in contrast to the inhibition of the respiratory ferment or of catalase by CN' (400)).

**Catalase:** Active catalase, like peroxidase, contains ferric iron in a somewhat stabilized form (1377, 1089)). Ordinary reducing agents like sodium hyposulfite ($Na_2S_2O_4$) or catalytically activated hydrogen fail to reduce it to the ferrous form *). As would be expected, the enzyme is inhibited by HCN; the complex formed between 1 molecule of HCN and 1 molecule of porphyrin bound iron dissociates upon diluting the system. On the basis of activity determinations in the presence of varying amounts of cyanide and also of crude spectroscopic measurements ZEILE and HELLSTRÖM 1377) estimate the dissociation constant to be $8.10^{-7}$. The inhibition by cyanide is reversible. Similar relationships obtain between catalase and $H_2S$ (1377)).

The effect of CO on catalase has been studied by a number of workers, with varying results. SENTER 1044a) as well as WIELAND and HAUSSMANN 1317a) attributed the decrease of the activity of the enzyme by treatment with CO to a mechanical damage suffered by the catalyst. If catalase-containing extracts of leucocytes are exposed to CO, $CO_2$, $H_2$ or $O_2$ (STERN 1071)), $CO_2$, CO and $O_2$ will somewhat diminish the activity of the enzyme. ph-measurements showed that $CO_2$ as well as $O_2$ cause a certain shift in

---

*) KEILIN and HARTREE 549) find that $H_2O_2$ is a specific reductant of catalase.

ph which may be responsible for the effect observed. While it was at first assumed that the CO-inhibition represents a specific reaction between inhibitor and enzyme, later (STERN **1073**, p. 188)) it was conceded that the phenomenon might perhaps be due to a mechanical damage of the enzyme by the gas stream. Soon afterwards CALIFANO **159)** reported in a short note that purified horse liver catalase is inhibited by CO, provided that oxygen is carefully excluded. A manometric technique was employed. CALIFANO states that the CO-inhibition is partially reversed *) by illumination. This observation, if correct, would strongly suggest the formation of an iron carbonyl complex. The problem was reinvestigated (STERN **1089**)) under experimental conditions, designed to be favorable to an inhibition by CO (low temperature, exclusion of direct light, removal of the oxygen formed in the reaction by a continuous CO-stream). Nevertheless, no consistent and significant inhibitions by CO were observed. Furthermore, attempts to detect the formation of an enzyme-CO complex spectroscopically met with no success. This was also true when CO and $H_2O_2$ were permitted to react simultaneously with the enzyme.

Lately, KEILIN and HARTREE have made interesting observations pertaining to this problem. They found (**548**)) that when $H_2O_2$ is added to catalase which is inhibited by sodium azide ($NaN_3$) or hydroxylamine, the color of the solutions turns from greenish-brown to red and the three-banded methemoglobin-like absorption spectrum is replaced by two strong absorption bands at 590 and 554 m $\mu$. This compound combines reversibly with CO, giving a derivative with bands at 580 and 545 m $\mu$, and reverts to the original azide-catalase only in presence of molecular oxygen, which proves that the iron here is in the ferrous form. Recently, the same workers (KEILIN and HARTREE **549**)) have found that an inhibition of catalase by CO may be due either to the absence of $O_2$ which is necessary for the reoxidation of ferrocatalase to the active ferriform, or to its combination with the divalent iron of reduced catalase. The first type of inhibition would be expected to be light-insensitive while the latter should be light-sensitive. Indeed, some catalase preparations, but not all, were found to suffer a more or less pronounced light-sensitive inhibition by CO. Highly purified catalase preparations are stated to become sensitive to CO after adding traces of sodium azide, cysteine or glutathione, substances which apparently inhibit the reoxidation of reduced catalase. The CO-inhibition thus produced was found to be completely reversible in strong light (visible region of a mercury-vapor lamp).

### b) Constitution and Chemical Properties of Hemin Catalysts.

Many efforts have been made to clarify the relationship of the prosthetic groups of the hemin-containing enzymes to blood hemin (protoferriheme IX). We are faced with the interesting and genetically important question whether all hemin enzymes are derived ultimately from the same mother compound (etioporphyrin) as is the case with hemoglobin and chlorophyll (WILLSTÄTTER) or otherwise which basic configurations correspond to the various catalysts. The solution of this problem is rapidly forthcoming.

Before entering into a discussion of the individual hemin catalysts it may be well to enumerate the more important hemin and porphyrin types and their formulas. Details will be found in the review articles by FISCHER, TREIBS and ZEILE **344)** and by E. BERGMANN **97)**.

---

\*) In a private communication to one of the authors (K.G.St.) Professor CALIFANO states that the extent of the inhibition caused by CO is about 30 per cent and the extent of reactivation by light of the effective wave-lengths is 15—20 per cent.

Hemins are defined as the iron complex salts of porphyrins. The nucleus is formed by porphin, a tetrapyrrolic structure held together by four methin (CH)bridges. Porphyrins are derived from porphin by the introduction of various side chains into the pyrrol rings. It is obvious that the possibility of varying the position of the side chains entails the possibility of the existence of a number of position isomers. An etioporphyrin is a tetramethyl, tetraethyl-porphin. All of the hemin catalysts under consideration may be traced back to etioporphyrin III, $C_{32}H_{38}N_4$, representing 1, 3, 5, 8-teramethyl, 2, 4, 6, 7-tetraethylporphin (Formula A).

By the exchange of two ethyl groups with propionic acid residues 6 isomeric mesoporphyrins, $C_{34}H_{38}H_4O_4$, may be formed. The 1, 3, 5, 8-tetramethyl- 2, 4-diethyl- 6, 7-dipropionic acid-porphin, or mesoporphyrin IX, belongs to the series of blood hemin. If the two remaining ethyl residues are replaced by vinyl groups, protoporphyrin IX, $C_{34}H_{34}N_4O_4$, (Formula B) results. The latter may be prepared synthetically from hematoporphyrin by splitting off two molecules of water. The iron complex of protoporphyrin IX is the blood hemin or protohemin, $C_{34}H_{32}O_4N_4FeCl$ (Formula C). Protohemin (protoferriheme IX) is the prosthetic group or the "agon" of hemoglobin, cytochrome, catalase, and very probably also of peroxidase. These conjugated proteins are probably only different with regard to their bearer proteins. In the case of the cytochromes it may be that a nitrogen base in combination with the heme forms the agon which in turn is tied up with a protein of relatively small molecular dimensions. In contrast to the forementioned biocatalysts the respiratory ferment contains an agon which, though ultimately also derived from etioporphyrin III, differs from protheme IX by one side chain and the distribution of double bonds. The mother substance is probably a rhodoporphyrin. For purposes of orientation in the porphyrin series the following table may be of service:

TABLE 9.

| Nature of side chains in positions 2 and 4 (See Formula B) | Porphyrin Type |
|---|---|
| $CH = CH_2$ | Protoporphyrin |
| $CHOH—CH_3$ | Hematoporphyrin |
| $C_2H_5$ | Mesoporphyrin |
| H | Deuteroporphyrin |
| $CH_2CH_2COOH$ | Koproporphyrin |

### Chemistry of the Cytochromes:

Altogether there exist probably only three well-defined cytochromes: cytochrome-a, b, and c (KEILIN). The intracellular "free hematin" of KEILIN 538, 540), supposedly the mother substance of the cytochromes, has as yet not been clearly defined or identified. The best known and most extensively studied component is cytochrome-c.

Cytochrome-c is the only cytochrome component which has been obtained in solution and in pure state. The others can only be studied spectroscopically within the living cell. Purified cytochrome-c was first isolated by KEILIN 539) form a Delft strain of baker's yeast. His method involved plasmolysis of the yeast by NaCl, rapid heating of the plasmolysate to 90°, extraction of the cytochrome by a sulfite-hyposulfite solution, treatment with $CaCl_2$ and $SO_2$-gas, fractional precipitation with $SO_2$, and washing with little water. Red scales are obtained which dissolve in water at ph 6.5 to form a red solution exhibiting the characteristic absorption bands of cytochrome-c. Modifications of this method have been described by BIGWOOD 114) and by GREEN 415).

ZEILE and REUTER 1379), working with other yeast strains but also with Delft yeast, failed to obtain satisfactory results by KEILIN's method. By removing impurities with kaolin, precipitation in the form of ferricytochrome and drying with acetone 1 gm. crude product with a hemin content of 1.3 per cent was obtained from 12 kg. yeast. A further purification could be achieved by fractional precipitation and adsorption on kaolin and $BaSO_4$; the hemin content rose to 3.2 per cent hemin and the degree of purity was estimated to be 92 per cent. It is worth noting that only oxidized but not reduced cytochrome-c is adsorbed by kaolin.

Pure cytochrome-c was isolated for the first time by H. THEORELL 1156, 1157). His starting material was beef heart muscle which he found to be richer in cytochrome than yeast. The extraction from the defatted material is accomplished by dilute sulfuric acid. After neutralizing, the cytochrome-c is precipitated by saturation with ammonium sulfate at 60°. The resulting product (7.5 gm. from 1 kg. of heart muscle) is 50 per cent pure (0.17 % Fe); it is further purified by adsorption on $BaSO_4$, elution with HCl, and acetone precipitation. The most ingenious step in this method consists in the adsorption of the cytochrome on finely divided cellophane which had previously been found to adsorb hemin pigments in dialysis experiments. Elution is brought about by dilute ammonia. The end-product as obtained by concentration and drying in vacuo over sulfuric acid is described as amorphous. The yield is about 1 gm. from 100 kg heart muscle. The elementary analysis shows that there were present 49.18 % C, 7.33 % H, 14.4 % N, 1.18 % S, 27.5 % O, and 0.34 % Fe; In contrast to earlier findings of Coolidge 184) no inorganic iron is contained in the cytochrome complex (Green 415); Hill and Keilin 487)). The purity of the preparation was concluded from the fact that the percentage of porphyrin-bound iron is the same as that in hemoglobin and that attempts to raise the figure by different purification procedures failed.

The procedure described by Theorell has lately been considerably simplified by Keilin and Hartree 584a). Finely minced ox or, still better, horse heart muscle is extracted with dilute trichloroacetic acid at ph 4. After neutralizing, impurities, particularly hemoglobin, are removed by ammonium sulfate precipitation. From the filtrate the cytochrome-c is precipitated in the oxidized form by shifting the ph from 4.9 to 3.7 by the addition of trichloroacetic acid. The red deposit is washed with saturated ammonium sulfate solution, suspended in distilled water and dialyzed through cellophane at low temperature against 1 % NaCl-solution. Thus 6 kg. ox heart will yield approximately 1 gm. and the same amount of horse heart about 1.6 gm. of pure cytochrome-c with an iron content of 0.34 %.

Inasmuch as attempts to further fractionate cytochrome of this purity by adsorption on calcium phosphate or by fractional precipitation either with trichloroacetic acid in presence of ammonium sulfate or with lead acetate failed to raise the iron content above 0.34 % this value may be accepted as representing the iron content of pure cytochrome-c. Keilin and Hartree also report that hot aqueous extracts of horse heart muscle yield only cytochrome preparations of a low degree of purity (18.3 gm. of 14 % pure pigment per 100 kg. of muscle).

The ultimate preparation of pure cytochrome-c makes this important catalyst available for studies of an analytical chemical character and it has already permitted several controversial points to be settled pertaining to its properties. To mention only a few of these points, Keilin and Hartree show that the two additional absorption bands in the red region of the spectrum described by Bigwood et al. 114) are absent in solutions of pure cytochrome c and hence must be due to impurities, probably to verdohemochromogens (Lemberg and Wyndham 711b)). The transition from reduced to oxidized cytochrome is a true oxidation reaction involving a change in valency of the iron from the divalent to the trivalent state. The view of Shibata (1050)) that cytochrome is able to combine loosely with molecular oxygen in a manner analogous to hemoglobin is erroneous; oxidized cytochrome may be boiled in evacuated tubes in the complete absence of oxygen without suffering reduction. Within the physiological ph-range cytochrome-c is practically non-autoxidizable. Besides a number of inorganic reagents (ferricyanide, copper salts, $H_2O_2$) indophenol oxidase but not catechol oxidase will oxidize reduced cytochrome. When a solution of ferricytochrome-c is heated to the boiling point, it turns yellow brown and the parahematin-like spectrum disappears. On cooling, the original red color and absorption spectrum are restored. This reversible change is a general property of ferrihemochromogens and is attributed to a temporary thermal dissociation of the nitrogeneous bearer from the heme nucleus. Reduced cytochrome does not exhibit this remarkable property.

Cytochrome-c is readily reduced, e.g. by sodium hyposulfite ($Na_2S_2O_4$) or by hydrogen activated by palladium or platinum (Keilin **540**), Green **415**)). One atom of porphyrin-bound ferric iron is reduced by one atom of hydrogen (Theorell **1157**)). The same holds for the reduction by alkaline ferrous tartrate: One cytochrome equivalent is reduced by one equivalent of the ferrous salt. Glutathione, in neutral solution, is also a suitable reducer (Bigwood and Thomas **118**)). Metabolites like succinate will not reduce cytochrome c (Keilin **540**)); the corresponding dehydrogenases have to be present in addition to the substrates (e.g., lactic dehydrogenase plus lactate (Green **415**)).

In vivo, the reoxidation of cytochrome is accomplished by the respiratory ferment of Warburg (pheohemin enzyme). This does not mean that the reaction must be between the respiratory ferment and cytochrome-c. Warburg has proposed a scheme according to which the three cytochrome components are connected in series and not in parallel to each other (**1234**)). Experiments by Haas **430a**) who compared the over-all rate of respiration of yeast with the rate of reduction of cytochrome by the reducing systems of the cell and who found both of the same order of magnitude, speak strongly in favor of this view. His experiments, however, did not indicate the exact position of cytochrome-c in this chain. This problem has recently been taken up by Tamiya and Ogura **1134**) and by Ball in the Laboratory of Warburg **58b**). The Japanese workers conclude from titration experiments with $O_2$, benzoquinone, and sodium hyposulfite that the normal potentials of the three cytochrome components at ph 7.0 decrease from cytochrome-c, over a to b. This would mean that cytochrome-c is closest to the respiratory ferment in the chain, while cytochrome-b would react with the activated metabolite systems. Ball **58b**), on the other hand, observing qualitatively the effect of oxidation-reduction indicators on the intracellular cytochrome spectrum, proposes the sequence cytochrome-a, c and b. An attractive feature of this hypothesis is that the soluble and relatively low-molecular c-component would function as the c a r r i e r  o r  t r a n s p o r t e r  o f h y d r o g e n  equivalents between the other two components which are apparently intimately linked up with the structure of the cell and which otherwise may not be able to react with each other due to spatial separation.

These considerations show the importance of the knowledge of the o x i d a t i o n-r e d u c t i o n  p o t e n t i a l  of the cytochrome system. This, however, is a rather controversial subject. Stone and Coulter **1105**) had estimated the normal potential, $E_o'$, of cytochrome c in certain bacteria to be $+ 0.280$ V. at ph 7.6. Coolidge **184**), working in the laboratory of Conant, found similar values for yeast cytochrome c ($+ 0.260$ V. at ph 7). These determinations were, however, severely criticized by Green **415**) on the basis of electrometric titrations of purified cytochrome c prepared from bakers yeast by a modification of Keilin's first method. Green found: $E_o' = + 0.125$ V. at ph 7.14, $+ 0.082$ at ph 7.43, $+ 0.082$ V. at ph 8.88 and $+ 0.060$ V. at ph 9.2. More recently, De Toeuf **1183**) concluded that the normal potential of the ferro-ferricytochrome c system must be between $+ 0.120$ and $0.180$ V. at ph 7.

In contrast to the findings of Green and of De Toeuf, Ball **58b**) on the basis of his rather crude experiments cited above, estimates the potential of cytochrome c to about $+ 0.270$ V. at ph 7.4. His result is confirmed by Laki **694**) in Szent-Györgyi's institute: If quinhydrone is added to an oxidase preparation from pigeon breast muscle showing the absorption bands of the three cytochrome components and if the PH is varied by suitable buffers, the c-component is only reduced by the quinhydrone at alkaline reaction whereas it is oxidized at acid reaction. At ph 7.4 the cytochrome is present approximately in a half-reduced and hald-oxidized state. At this ph, therefore, its normal potential should be about the same as that of the quinone-hydroquinone system, i.e. about + 0.280 V. When the experiment was repeated with pure cytochrome prepared by H. THEORELL 1157) no reduction could be observed with quinhydrone. This result, to LAKI, seems to suggest that the potential of isolated cytochrome c is more negative than that of cytochrome c linked to the structure of the cell. At this point it must be mentioned that KEILIN and HARTREE 550a) could observe a reduction in vitro of cytochrome c preparations of comparable purity both by hydroquinone and by p-phenylenediamine.

The question has recently been reinvestigated by WURMSER and FILITTI 1358, 1359). These workers prepared cytochrome c from heart muscle according to KEILIN and HARTREE. They then determined the equilibrium potential formed in mixtures of ferro- and ferricytochrome of a ratio determined by spectrophotometry. In another series of experiments, ferricytochrome was mixed with known amounts of reductinic acid the

potential of which had pre- viously been determined by MAYER 763) and which will partially reduce ferricyto- chrome with great rapidity. After attainment of equilibrium the degree of reduction of the cytochrome was ascer- tained by spectrophotometry. WURMSER and FILITTI find that the normal potential of cytochrome c is + 0.253 V. between ph 5 and 8. Their value is therefore in agreement with the early findings of COULTER and of COOLIDGE as far as ph 7 is concerned where as they cannot confirm the statement of COOLIDGE that the potential changes with the hydrogen ion concentration in this ph-range.

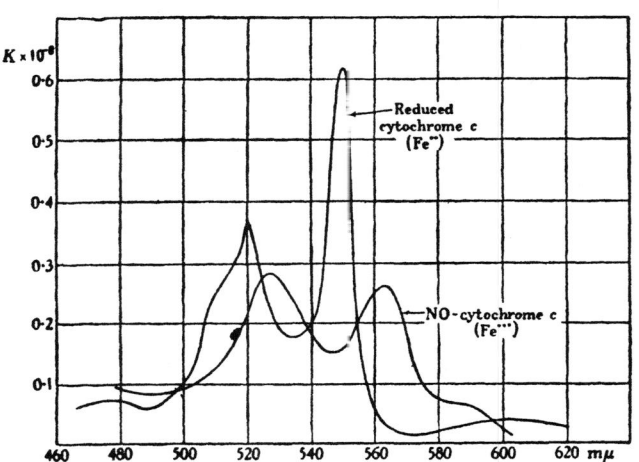

Fig. 5. Absorption Spectrum cf NO-cytochrome complex as compared with that of reduced cytochrome. Ordinate, absorption coefficient per gram atom of cytochrome iron; abscissa, wave-length in millimicrons. (KEILIN and HARTREE 548a)).

About the redox potentials of cytochrome a and b nothing definite is known. Accor- ding to BALL, their behaviour towards indicators suggests a normal potential of cyto- chrome a of about + 0.290 V. and of cytochrome b of about — 0.04 V.

Cytochrome-c does not form compounds with substances such as cyanide, sulfide, fluoride, azide, hydroxylamine or peroxides which are known to combine with other heme derivatives. The only substance with which cytochrome-c reacts in a reversible manner, within the physiological ph-range, is nitric oxide (NC) (KEILIN and HARTREE 548a)). The absorption spectrum of the NO-complex together with that of reduced cytochrome is shown in Fig. 5.

The NO-complex of cytochrome, the spectrum of which is shown in Fig. 5, contains ferricytochrome. Two distinct bands are noted at 565 and 530 m $\mu$. Ferrocytochrome will not combine with NO, but the ferricytochrome-NO compound is reduced by hyposulfite to a complex which may possibly represent the ferrocytochrome-NO combination.

The concentration of cytochrome-c in solution may be determined spectrophotometrically; a convenient wave-length being 549.7 m $\mu$. For this wave-length, the relation holds:

$$\ln \frac{I_o}{I} = c \; d \; \times \; 0.62 \; \times \; 10^8,$$

where c is the concentration of reduced cytochrome-c in terms of gm. atoms of iron per cc., and d the layer of thickness of solution (DIXON, HILL and KEILIN **214**); KEILIN and HARTREE **548a)**).

It is highly significant that a number of independent methods have yielded very similar values for the m o l e c u l a r  w e i g h t of cytochrome-c. ZEILE and REUTER **1379, 1376)**, from the ratio of total cytochrome as determined by the difference of adsorbed material in the reduced and oxidized state (see above) and from the hemin content of their purest preparations (3.5 %) estimate the molecular weight to be 18,000. With his pure preparation from heart muscle THEORELL **1157)** obtained a diffusion constant $D_{20°} = 11.1 \cdot 10^{-7}$ cm²/sec., corresponding to a molecular weight of 16,500 on the basis of the EINSTEIN-STOKES equation. The iron content of 0.34 % yields a minimum molecular weight of 16,500 on the assumption that one iron atom is contained in one cytochrome molecule. PEDERSEN, in unpublished experiments quoted by THEORELL **1157)**, finds a sedimentation constant $S_{20°} = 2 \cdot 10^{-13}$ dynes · cm⁻¹ · sec⁻¹ and a partial specific volume of 0.707 from which a molecular weight of 16,500 is calculated. The much higher value found by ZEILE and REUTER **1379)** in diffusion experiments with ANSON and NORTHROP's porous disc method is probably due to an association phenomenon; free heme also yields values by this procedure which are too high.

This point of view is further strengthened by the results of preliminary determinations of the osmotic pressure of cytochrome-c at ph 6.8, carried out by ADAIR (quoted by KEILIN and HARTREE **548a)**); the provisional results for the molecular weight agree within 10 % with the equivalent weight determined by the iron content, i.e. 16,500.

Contrary to the case of hemoglobin, where 4 "HÜFNER units" of 16,500 each appear united to form one macromolecule of 66.000 molecular weight carrying f o u r heme groupings it would appear then that cytochrome-c in solution has the size of just one "HÜFNER unit" with o n e heme group. It is well to keep in mind, though, that the extraction of the pigment from the cells is carried out with rather drastic reagents, e.g. dilute sulfuric acid, which may cause the breakdown of a larger complex into smaller units.

The i s o e l e c t r i c  p o i n t of cytochrome-c was found by ZEILE **1376)** at ph 8.2 and by THEORELL **1157)** evenmore alkaline, namely at ph 9.82 in borate buffer and at 9.86 in glycine buffer. The electrophoretic mobility has been determined by THEORELL **1157)**. In contrast to ordinary proteins the ionic mobility was found to be independent of the ph between 7.0 and 9.3 as u = $5 \cdot 10^{-5}$ cm²/sec/V.

The last named constants are largely if not solely determined by the protein component of the cytochrome. The relatively easy method of preparation of pure cytochrome-c has enabled KEILIN and HARTREE **548a)** to determine several a m i n o  a c i d s in c y t o c h r o m e  h y d r o l y s a t e s. Table 10 gives the results of the analysis of 1.44 gm. cytochrome-c after hydrolysis with strong HCl.

Tryptophane was also estimated in a tryptic hydrolysate of cytochrome-c. The value obtained, viz. 0.94 %, agrees well with that found in the acid hydrolysate.

Table 10 shows that with respect to the amino acids so far determined in cytochrome-c the differences as compared with hemoglobin are not very large, with the exception of tryptophane.

### TABLE 10.
### (From KEILIN and HARTREE 548a)).

| Amino Acids | In Cytochrome-c per cent | In Hemoglobin (Horse) per cent |
|---|---|---|
| Arginine . . . . . . . . . . . . . | 5.6 | 3.57 |
| Cystine . . . . . . . . . . . . | 1.1 | 0.74 |
| Histidine . . . . . . . . . . . | 7.8 | 8.13 |
| Lysine . . . . . . . . . . . . . | 9.1 | 8.31 |
| Tryptophane . . . . . . . . . . | 0.9 | 2.38 |

The manner in which the protein is linked up with the heme residue in cytochrome-c is still under investigation. The situation appears to be more complex than in the case of hemoglobin. There the consensus of opinion is that the protein is linked to the iron atom of the heme group by a covalent bond. For cytochrome-c three possibilities have been discussed on the basis of the experimental facts: (1) The heme contains a nitrogeneous base in a side chain of the porphyrin structure in the form of a quaternary ammonium salt in addition to the carrier protein which is linked to the cytochrome iron in the usual manner, (2) The nitrogeneous base is a peptide and is linked to the heme in the form of a peptide bond between a carboxylic group of the side chains of the porphyrin and an amino group of the peptide. In addition, there would be the bearer protein in the usual type of linkage. And (3), the entire protein pheron is combined with the heme agon through a peptide linkage so that the Fe atom is not involved at all.

In their early experiments on the nature of the porphyrin contained in cytochrome HILL and KEILIN 486) obtained a so-called "unmodified" c-porphyrin by splitting off the iron with HCl and SO$_2$ in the absence of oxygen. This c-porphyrin differs from all known porphyrins by its solubility in water. ZEILE and PIUTTI 1378) attribute this atypical behaviour to the fact that the mild HCl-SO$_2$-treatment has not split off the nitrogeneous base in the porphyrin side chain. The absorption bands of this "unmodified" porphyrin are located between those of proto- and hematoporphyrin. ZEILE and his collaborators tried to elucidate the structure of this porphyrin by synthesis. The introduction of N-bases into the porphyrin molecule by way of a carboxylic group in the manner of a peptide bond (ZEILE and PIUTTI 1378)) failed to yield derivatives similar in their absorption spectrum or in other properties to the cytochrome porphyrin. On the other hand, the attachment of tertiary N-bases to the unsaturated side chains of protoporphyrin was more successful.

ZEILE and PIUTTI 1378) prepared adducts of glycine, pyridine, and other N-bases to the unsaturated side chains of protoporphyrin with the aid of HBr. After introducing iron into the products obtained with pyridine, collidine and quinoline they showed the characteristic cytochrome absorption band at 550 m$\mu$. The constitution of the pyridine adduct which was prepared in crystalline form is indicated by the formula on p. 164.

It is undecided as yet whether the pyridine nitrogen is attached to the porphyrin side chains 2 or 4. If N-bases containing primary, secondary or acyclic tertiary nitrogen are used, hydrophilic porphyrins with the spectral properties of the c-porphyrin are obtained. However, in this case the N-base cannot be split off with HBr-acetic acid.

Therefore ZEILE assumes that cytochrome-c contains a tertiary cyclic N-base which is combined with the porphyrin in 2- or 4-position through the formation of a quaternary ammonium salt. These synthetic experiments cannot be regarded as furnishing decisive proof for the constitution of cytochrome-c. ROCHE and BÉNÉVENT 991) point out that the synthetic products, in spite of the similar absorption spectrum, exhibit a somewhat different chemical behaviour when compared with the natural unmodified c-porphyrin: The synthetic hematins obtained with pyridine or glycine methylester will combine with globin to form compounds of the type of methemoglobin, whereas cytochrome-c is unable to do so. Cytochrome-c when combined with globin forms a ferrihemochrogen (parahematin) which may be reduced to a ferrohemochromogen (ROCHE and BÉNÉVENT 991)). This indicates that the bridging link between the cytochrome hemin and the protein component is more complex than in other heme proteins. Recently THEORELL has attacked this question by breaking down the natural cytochrome-c in a gradual manner (1160)). He finds that the iron may be removed more simply than by the $HCl-CO_2$ method of HILL and KEILIN, viz, by saturating an aqueous cytochrome solution at 0° with HCl-gas. The interesting point is that the "unmodified c-porphyrin" thus obtained is soluble in aqueous solvents and insoluble in organic solvents; that it exhibits the electrophoretic behaviour of a protein with an isolectric point near ph 7; that it is unable to pass parchment membranes; that its molecular weight as estimated from the ratio heme/dry matter is about 4—5000, and that it contains 11 % nitrogen. Upon hydrolyzing with hydrochloric acid at 100° amino acids are gradually split off from this large molecule. When the molecule has reached a size of 1600—1800 the porphyrin-polypeptide complex begins to be precipitable isoelectrically at ph 4.2. The next lower product had a molecular weight of 1160—1320. All preparations were found to contain sulfur (1 to 1.8 atoms S per molecule). The titration curve traced with the glass electrode revealed 4 basic groups in addition to those present in hematoporphyrin. Three of these groups are $NH_2$-groups (van SLYKE method). There are also present 6 acid groups. By prolonged hydrolysis it was possible to split off still more amino acids until a relatively pure preparation of a molecular weight 1020 was obtained. It contained 6 N-atoms and only traces of amino nitrogen. The titration curve indicates the existence of 5—6 acid groups. The hydrolysis did not affect the characteristic properties of the c-porphyrin. The spectrum, the hydrophilic behaviour and the ability to combine with Fe to form a hemochromogen with bands at 550 and 520 m $\mu$ were retained throughout.

The structure of the porphyrin itself as contained in cytochrome-c has been elucidated by the experiments of ZEILE and REUTER 1379). It had been known previous to their work that cytochrome-c contains iron capable of reversible oxidation-reduction and that the removal of the iron leads to a porphyrin of the hematoporphyrin type (HILL and KEILIN 486)). ZEILE and REUTER subjected cytochrome-c to HBr-acetic acid decomposition. Crystalline hematoporphyrin was isolated as a reaction product just as in the case of blood hemin. After reduction with HJ-acetic acid to the mesoporphyrin stage and esterification with HCl-methanol the mesoporphyrin dimethylester thus obtained was compared spectroscopically and with regard to melting point and mixed melting point with the corresponding ester of synthetic mesoporphyrin IX and was found to be identical with it in every respect. This proves that the heme in cytochrome-c is protoheme IX (blood heme), modified by the introduction of an N-base into the porphin structure and tied up with a protein definitely different from globin.

Recently, H. Theorell 1163) has made another important contribution to the problem of the constitution of cytochrome-c. He could show that the porphyrin-c as obtained by hydrolysis of cytochrome-c with HCl contains a side chain which is split off by HBr-glacial acetic acid and which contains 2 atoms of sulfur, two carboxylic and two amino groups. In both instances the amino groups are in α-position to the COOH-groups. Both S-atoms appear to be present in the form of thio-ether (C—S—C) linkages forming a bridge from the porphyrin nucleus to the carrier protein. On one side, the S-atoms are tied to those C-atoms which after HBr-acetic acid cleavage carry the OH-groups of the hematoporphyrin thereby formed while on the other side the S-atoms are

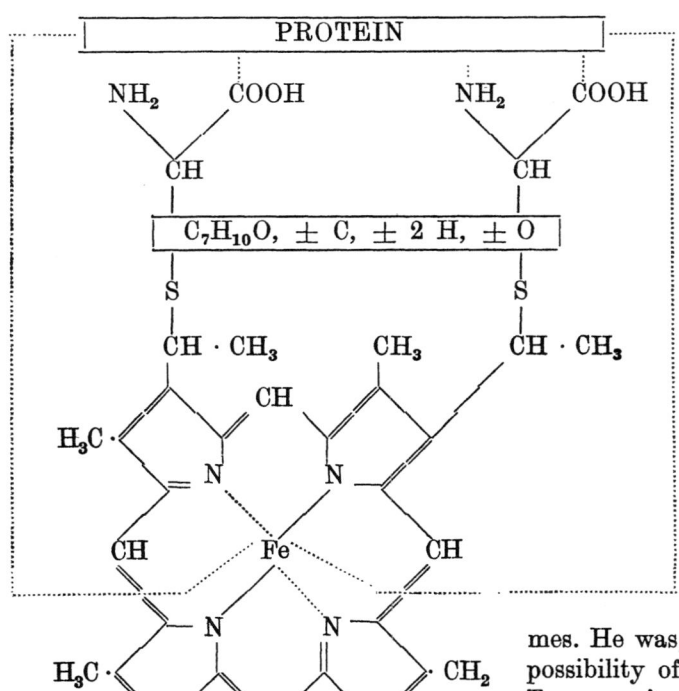

attached to C-atoms in the diamino-dicarboxylic acid side chain of the porphyrin-c.

One or several of these amino and carboxyl groups are linked by peptide bonds to the portion of the protein component which is split off by acid hydrolysis. The stability of the thio-ether bridge accounts for the failure to split cytochrome-c in a reversible manner. It will be remembered that Bersin 101a) some time ago advanced the hypothesis that sulfur atoms might be involved in the linkage between hemin group and bearer protein of certain enzymes. He was, at that time, considering the possibility of a dithio- (—S—S—) linkage. Theorell's new experiments, for the first time, lend experimental support to the theory of the role of sulfur*) as a bridging link, although in a form differing from that visualized by Bersin. The adjacent formula, according to Theorell, represents best our present knowledge of cytochrome-c:

*)  l-Cystine has since been obtained from porphyrin-c (H. Theorell, Enzymologia, 6, 88 (1939)).

We know very little about the chemistry of the c y t o c h r o m e   c o m p o n e n t s   a n d   b*). Neither has as yet been isolated in native state from cells. It is possible that the prosthetic group of cytochrome-a is a mixed-colored or pheohemin just as is the case with the respiratory ferment (ROCHE **990**)). It is possible that the pheohemin isolated by ROCHE **989, 990**) and by NEGELEIN **879**) from heart muscle represents the prosthetic group of cytochrome-a. A material rich in cytochrome-b is the muscle of *Actinia*. ROCHE **795**) obtained from it a pyridine hemochromogen which exhibited the properties of blood hemin. ROCHE assumes that, at least in this particular case, cytochrome-c and b have the same prosthetic group of the protoporphyrin type and perhaps even the same protein bearer; the only difference being the type of linkage between heme and protein. It will be remembered that KEILIN **536**), on the basis of spectroscopic comparison of the cytochrome spectrum with that of other hemochromogens, concluded that they are ultimately derived from the same hemin and that the differences in their absorption spectrum may be due to their different degree of "dispersion" in the cell. It would appear that the a-component has a heme different from the other two components. It is possible that cytochrome-c and b have the same agon but a different pheron (bearer). This will require further experimentation. YAKUSHIJI and MORI (Acta Phytochim. **10**, 125 (1937)) claim to have obtained cytochrome-b in purified form from yeast.

### Respiratory Ferment:

The hemin character of the respiratory ferment was deduced by WARBURG **1233**) from its behaviour towards CO and light, HCN, and from the pattern of the absorption spectrum (see Fig. 3, p. 145) as obtained by the photochemical method. Particularly typical for a heme derivative is the steep $\gamma$-band in the violet region, at 436 m$\mu$. On the other hand, the shift of the enzyme bands towards the red as compared with the bands of blood hemin derivatives suggested that the heme contained in the enzyme is not a r e d hemin. The possibility that the peculiar state of dispersion or distribution of the enzyme within the living cell or that the nature of the bearer protein might be responsible for this red-shift, was ruled out by WARBURG. In his model experiments neither a change in state of aggregation, adsorption, or coupling with bases or proteins led to a shift of the blood hemin bands into the vicinity of the enzyme bands (**1276**). It must be mentioned however that KEILIN (l.c. 540) reports that it is possible to shift the position of the main absorption band of protohemochromogens by suitable choice of the N-bases from 417 to 452 m$\mu$. The "SORET"-band of piperidine hemochromogen lies at 438 m$\mu$ (KEILIN **540**)); so does the band of pyrrolidine hemochromogen (ZEILE **1372a**)).

The absorption spectrum of a hemin is to a certain extent dependent upon its constitution, i.e. of the nature of the radicals present in the side chains of the pyrrolrings or in the methin bridges. While it is possible to detect spectroscopically certain substituents as unsaturated side chains, carboxylic groups, substituents in the methine bridge, etc., a difference between a methyl, ethyl, or even propionic acid grouping, or with respect to position of substituents in isomer porphyrins finds hardly an adequate expression in the absorption spectrum. With this reservation one may state that only such hemins may be considered, when casting about for close relatives of the respiratory ferment, the absorption of which is shifted towards the red relative to blood hemin. It has already been mentioned that WARBURG distinguished between r e d, g r e e n, and m i x e d - c o l o r e d hemins. The r e d hemins are derivatives of the blood pigment hemo-

---

*) See OPPENHEIMER'S „SUPPLEMENT" p. 1656.

globin (protoheme) and of the far-reaching breakdown of chlorophyll. They are spectroscopically characterized by the fact that the main ($\gamma$-) band of their CO-compounds is situated around 420 m $\mu$ and below and that the long-wave ($\alpha$-) band is at 570 m $\mu$ and below. The green hemins, or pheophorbid-hemins, are derived from chlorophyll by hydrolyzing off the phytol group and substituting iron for the magnesium. In contrast to the respiratory ferment they show selective light absorption in the red region of the spectrum. The mixed-colored hemins (p h e o h e m i n s) are intermediate between the red and the green hemins with regard to their spectrum. Besides the respiratory ferment this group includes such pigments as spirographis hemin, pheohemin b, kryptohemin. Inasmuch as no direct evidence concerning the chemical structure of the respiratory ferment is available, we shall have to limit ourselves to a short discussion of the chemistry of the known representatives of the pheohemin group. This is done in the assumption that the data obtained with such pheohemins are more or less representative for the heme group of the enzyme.

## Spirographis Hemin.

This hemin is the prosthetic group of chlorocruorin, the blood pigment of the marine worm Spirographis. Besides the heme there is present a protein which is different from globin (ROCHE **993, 984**)). WARBURG et al. **814**) succeeded in the preparation of pure crystalline spirographis hemin of the empirical formula $C_{32}H_{32}N_4O_5FeCl$ the uncertainty being $\pm$ 1 C and $\pm$ 1 H. It contains 2 C-atoms less and one O-atom more than blood hemin. The molecular weight would be close to 640; the number of acid groups, according to manometric determinations, is 2; the dissociation constants are larger than $10^{-7}$. Four of the oxygen atoms are present as carboxyl groups. The fifth O-atom was considered to be in the form of a keto group (WARBURG, NEGELEIN and HAAS **1276, 1279**)).

WARBURG and NEGELEIN **1275**) were able to split off the iron from spirographis hemin by means of ferroacetate-HCl in an inert atmosphere. The porphyrin was purified by the usual HCl-ether fractionation and finally obtained in pure crystalline form. The empirical formula is $C_{32}H_{31}N_4O_5 \pm$ 1 H. The spectrum in neutral ether shows bands at
I. 509—520; II. 550—560; III. 580—582; IV. 591—594; V. 639—647 m $\mu$, band II. being the strongest. If the porphyrin is treated with hydroxylamine the absorption bands are shifted towards the blue end of the spectrum and a crystalline oxime is obtained. It was this reaction which led WARBURG to believe that the fifth O-atom is contained in a carbonyl group.

The ferrous form of spirographis hemin as obtained by reduction of the ferric form with cysteine in weakly alkaline solution (0.01 N. NaOH) suffers a molecular rearrangement at 75° which may be demonstrated by preparing the pyridine hemochromogen and measuring the position of the long-wave absorption band before and after the rearrangement:

Band before Rearrangement . . 584 m $\mu$
Band after Rearrangement . . 553 m $\mu$.

The interesting point here is that the long-wave absorption band of the hemochromogen shifts from the position typical for a green-red hemin (pheohemin) to a spectral region commonly occupied by red hemins. By linking spirographis hemin to globin the rearrangement is prevented.

The chemical study of spirographis hemin was continued and brought to a successful conclusion by H. FISCHER and SEEMANN **343**). The constitution of the pigment was elucidated with the aid of 139 mg. of this precious material.

The careful comparison of spirographis porphyrin with the spectroscopically and analytically similar oxorhodoporphyrin (1,3,5,8-tetramethyl-2-acetyl-4-ethyl-6-carboxylic acid porphin-7-propionic acid) had the following result. A depression of the mixed melting point of the two dimethylesters precludes an identity of the two compounds. This is further substantiated by the fact that the iron complex salts of the two porphyrins behave differently upon catalytic hydrogenation. On the other hand, both porphyrins will form a monoxime while the spectroscopically similar diacetyldeuteroporphyrin yields a dioxime with hydroxylamine.

Inasmuch as the earlier experiments of FISCHER and BREITNER had demonstrated the presence of a formyl group in position 3 in chlorophyll-b, FISCHER and SEEMANN investigated the possibility that the fifth O-atom of spirographis porphyrin may be present in this form rather than as a carbonyl group. The fact that this group could be oxidized to a carboxyl group proved this working hypothesis. Experiments designed to prove or disprove the structure of spirographisporphyrin as that of a formylrhodoporphyrin, with the formyl group in analogy to chlorophyll-b tentatively in position 3, had no decisive results. At this stage of the investigation, FISCHER and SEEMANN took again recourse to the catalytic reduction of spirographishemin and -porphyrin. It was found that upon hydrogenation the formyl group is completely reduced to a methyl group. This, together with the observation that diazoester reacts with spirographisporphyrin in such a manner that the formyl group yields a ketocarboxylic acid ester residue, suggested that the porphyrin contains 2 propionic acid, 1 formyl, 1 vinyl, and 4 methyl groups in the side chains. By the method of mixed melting point determination the identity of hydrogenated spirographisporphyrin with 1,2,3,5,8-pentamethyl-4-ethylporphin-6,7-dipropionic acid was established. Therefore the vinyl group in spirographisporphyrin must be in 4-position while the formyl residue must be in 2-position. This formulation is further supported by the identity of the product of the resorcinol fusion of spirographis hemin with deuteroporphyrin IX.

The foregoing evidence and other facts permit to assign to spirographis porphyrin the structure of a 1,3,5,8-tetramethyl-2-formyl-4-vinylporphin-6,7-dipropionic acid as shown in the following formula

Spirographis Porphyrin (acc. to FISCHER and SEEMANN **343**)).

This formulation explains the shift of the absorption bands of the spirographis hemochromogen observed by WARBURG and NEGELEIN 1275). These authors had used hydrazine for the reduction of the ferri- to the ferrohemochromomogen. It is probable that under such conditions hydrazone formation occurs at the formyl group which causes a blue-shift of the absorption bands. Spirographishemin is thus characterized as a close relative of protohemin IX (blood hemin) which in turn is derived from etioporphyrin III. It is suggested by FISCHER and SEEMANN that spirographishemin is formed by the oxidation of a vinyl group of protohemin in position 2. This possibility of its biogenesis is supported by the fact that protoheme IX is contained in cytochrome-c (see p. 165) and thus genetically older than spirographis hemin.

It appears very likely that the porphyrin present in WARBURG's respiratory ferment also contains a formyl group in the side chain.

Besides spirographishemin and its combination with the natural protein (chlorocruorin) or with globin (spirographishemoglobin) (see below), WARBURG has compared the spectra and properties of other hemins with that of the respiratory ferment.

When the enzyme spectrum was obtained in 1928—29 by the photochemical method, no hemins were known which, according to their spectrum, could be considered closely related to the enzyme hemin. In the following year H. FISCHER and ZEILE 345) prepared diacetyldeuterohemin and H. FISCHER 342) obtained pheohemin-a by reducing chlorophyll-a with HI and introducing iron into the molecule. Although the absorption bands of these two hemins are somewhat shifted towards the red, i.e. towards the enzyme bands, as compared with blood hemin, their position is still too far towards the short-wave region to make them an effective model for the enzyme. The fact that derivatives of chlorophyll-b have, in general, bands situated more towards the red than those of chlorophyll-a, induced WARBURG and CHRISTIAN 1236) to perform experiments in the chlorophyll-b series. They reduced chlorophyll-b to pheoporphyrin-b by HI. The introduction of iron into the molecule in propionic acid solution yielded pheohemin-b, $(CH_3CH_2CO)C_{32}H_{29}N_4O_6FeCl$. It is a dicarboxylic acid and contains a keto group and a formyl group *). Pheohemin-b is a mixed-colored hemin, it is green in concentrated and red in dilute solution. The fact that the combination between this hemin and globin exhibits an absorption spectrum closely resembling that of the respiratory ferment has led WARBURG to call the group comprising spirographis hemin, the respiratory enzyme and pheohemin the class of pheohemins.

If chlorophyll is treated with strong acid, the phytol ester linkage is hydrolyzed and the magnesium is split out. The resulting compound is called pheophorbide. WARBURG 1232) prepared pheophorbide-b from chlorophyll-b and introduced iron instead of the magnesium into the complex. Pheophorbide-b hemin is a green hemin. Its combination with globin, pheophorbide-b hemoglobin, has absorption bands located further towards the red than the respiratory ferment. This excludes the possibility that the hemin of the ferment is a green hemin **). The structure of pheophorbide-a and -b, suggested by H. FISCHER, is given in the following formula:

---

*) The structure of pheoporphyrins and related compounds is discussed in the recent reviews on the structure of chlorophyll by STEELE 1066) and H. FISCHER 341).

**) This refers to the respiratory ferment of yeast and acetobacter; the enzyme of azotobacter may be a derivative of a green hemin (880)) or of biliviolin (711c)).

Pheophorbide-a and -b (acc. to H. FISCHER **341**))
Pheophorbide-a: X = CH$_3$
Pheophorbide-b: X = CHO

The position of the long-wave absorption bands of the pyridine hemochromogens of some of these hemins is shown in Table 11.

TABLE 11.

(From WARBURG and NEGELEIN **1275**))

Long-wave Absorption Bands of Several Hemins in the Form of Pyridine Ferro Hemochromogens.

| Hemin | Wave-length |
|---|---|
| Protoheme IX (Blood hemin) . . . . . . . . . . . . . . . | 557 m$\mu$ |
| Pheohemin-b . . . . . . . . . . . . . . . . . . . . . . | 584 m$\mu$ |
| Spirographishemin . . . . . . . . . . . . . . . . . . . | 584 m$\mu$ |
| Pheophorbid-b hemin. . . . . . . . . . . . . . . . . . | 600—622 m$\mu$ *) |

If the hemins are permitted to react with native globin in neutral aqueous solution, compounds analogous to methemoglobin are formed. Assuming an equivalent weight for globin of 16,000, WARBURG and NEGELEIN **1275**) find that slightly over one globin equivalent is required for the saturation of one molecule of hemin. This observation suggests a quite tight, i.e. little dissociated, linkage between hemin and protein. The measurement of the combination between the two components is based on the fact that at the wave-length of the absorption maximum of the "methemoglobins" in the blue-violet region the combination process causes an increase in light absorption over that of the free hemin. Figure 6. shows that in the case of blood hemin a constant value of extinction at 405 m$\mu$ is attained, if 28 mg. globin or more are added to 1 mg. hemin

---

*) Band of the ferricompound in pyridine-chloroform. The hemochromogen reaction is indistinct.

SPIROGRAPHIS HEMIN

171

at ph 7.0, corresponding to a ratio of 1.1 equivalents of globin to 1 mol of hemin.

When the "methemoglobins" (ferrihemoglobins) are reduced with cysteine and exposed to oxygen, it is found that bloodhemoglobin, rhodohemoglobin, diacetyldeuterohemoglobin and pheohemoglobin, but not pheophorbid-b-hemoglobin are able to bind oxygen reversibly to form "oxyhemoglobins". The absorption bands of the various complexes are listed in Table 12.

It is remarkable that when oxygen attaches itself to the first three hemoglobins in Table 12, the broad band of the ferrohemoglobin is split up into two sharp bands, while in the instance of pheohemoglobin the band of the ferrohemoglobin is not split up but only shifted. The position of the absorption bands of the CO-complexes will be found in Table 8 (p. 146).

Following the discovery of various hemins resembling more or less that of the respiratory ferment, various workers have searched for such pheohemins (or mixed-colored hemins) in mammalian tissues. NEGELEIN 877), working up pigeon breast muscle, obtained a crystalline porphyrin of the empirical formula $C_{33}H_{32}N_4O_5$. This corresponds to an additional $CH_2$-group in comparison with spirographis porphyrin. After the introduction of iron a compound showing the properties of a mixed-colored hemin was obtained. The same compound, termed kryptohemin by NEGELEIN, could also be obtained from crude blood hemin in a yield of about 0.2 per cent. The biological significance of kryptohemin has become very questionable since NEGELEIN 878) found that it is formed as a by-product in the course of the HCl-fractionation of the blood hemin.

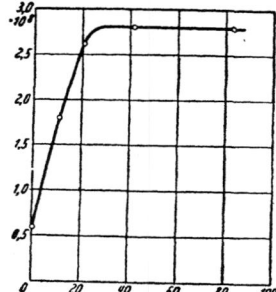

Fig. 6. Curve representing the changes in light absorption occurring upon adding varying amounts of globin to 1 mg. blood hemin dissolved in 200 cc. 0.1 M. neutral phosphate buffer. Abscissa: mg. globin in 200 cc. solution; Ordinate: Lightabsorption coefficient, $\beta$, at 405 m $\mu$. (From WARBURG and NEGELEIN 1275)).

TABLE 12.

Absorption Bands of Synthetic Ferri-, Ferro-, and Oxyhemoglobins (WARBURG and NEGELEIN 1275)).

| | Ferri-hemoglobin | Ferro-hemoglobin | Oxyhemoglobin |
|---|---|---|---|
| Bloodhemoglobin . . | 634 m $\mu$ | 545—570 m $\mu$ | 533—548, 573—583 m $\mu$ |
| Rhodohemoglobin . . | 625 m $\mu$ | 540—560 m $\mu$ | 533—542, 573—582 m $\mu$ |
| Diacetyldeutero-hemoglobin . . . . | No band in the red | 550—595 m $\mu$ | 548—558, 585—595 m $\mu$ |
| Pheohemoglobin-b . . | 642 m $\mu$ (weak) | 585—595 m $\mu$ | 595—605 m $\mu$ |

Shortly afterwards, however, NEGELEIN 879) succeeded in isolating a pheohemin, free from protohemin, from horse heart muscle. It shows a strong absorption band at 587 m $\mu$, i.e. approximately there where the $\alpha$-band of the respiratory ferment is situated; and it is certainly no artefact. More recently, ROCHE and BÉNÉVENT 990, 989) obtained a hemin from heart muscle by a slightly different method. The pyridine hemochromogen shows absorption bands at 425, 530, and 580 m $\mu$. With alkali this hemin is rearranged to the hemin of NEGELEIN. ROCHE and BÉNÉVENT are of the opinion that they are dealing here with the prosthetic group of cytochrome-a rather than that of the respiratory enzyme. It is also possible that these two catalysts have the same agon.

Obviously it will be very difficult to prove that any hemin isolated from tissues actually represents the prosthetic group of the respiratory ferment and not that of other intracellular hemin derivatives, e.g. of cytochrome components. The problem will remain open until such a time when the respiratory ferment may be obtained in solution and apart from other cell constituents (see p. 264 footnote).

### Catalase:

For some time there had been indications that iron and particularly porphyrin-bound iron is a constituent of the enzyme catalase. HENNICHS 475a), working in EULER's laboratory, searched for a correlation between the iron content and the enzymatic activity of his preparations. Although the inhibition by HCN pointed to the presence of heavy metal in the enzyme, he was unable to find a quantitative relationship because he considered the total iron which amounted to as much as 4 per cent in some of his preparations. EULER and JOSEPHSON 296) investigated the degree of sensitivity to HCN in relation to the iron content of more highly purified catalase preparations without arriving at clear-cut results. KUHN and his collaborators, especially BRANN 142), established the catalatic activity of blood hemin. Although one hemin molecule is able to decompose only $10^{-2}$ molecules of $H_2O_2$ per second as compared with catalase, one molecule (or rather one iron atom) of which splits $10^5$ $H_2O_2$ molecules per second, hemin is very active when compared with inorganic ferric iron ($10^{-5}$ molecules of $H_2O_2$ per second per Fe atom). Work along these lines was continued by EULER and his associates, among them by ZEILE 1370).

In 1927, EULER and JOSEPHSON 296) had obtained a catalase preparation from horse liver of the extremely high activity and purity of Kat. f. = 43,000 *) which they preserved in dry form. In 1930, EULER, ZEILE and HELLSTRÖM 323) found that this preparation contained 0.6 per cent hemin, as determined spectrophotometrically. Inasmuch as the method of purification employed in this instance did not exclude with certainty a contamination by hemoglobin, ZEILE and HELLSTRÖM 1377) developed a method which would permit one to obtain highly active liver catalase free from even traces of hemoglobin. They adapted the chloroform emulsion procedure used by TSUCHIHASHI 1187) for blood catalase for their purpose.

Horse liver is finely minced and extracted with water. The extract is treated with alcohol and the heavy protein precipitate formed is removed. The remaining red fluid is shaken with an alcohol-chloroform mixture. At this stage not only colorless liver proteins but also all of the hemoglobin is coagulated. The enzyme remains almost entirely in the supernatant olive green-brown solution. With a pocket spectroscope the following spectrum is observed in such preparations:

I. 650 . . . 646—620 . . . . 610    II. 550—530 . . . . 520 . . . 510—490
629 m $\mu$              540 m $\mu$              500 m $\mu$

This spectrum resembled that of methemoglobin, while the spectrum of alkaline hematin has bands shifted approximately 10 m$\mu$ towards the blue region when compared with the spectrum of the liver pigment. A further increase in purity is accomplished by adsorbing

*) Kat.f. = $\dfrac{k \text{ monomolecular}}{\text{gm. Enzyme in 50 cc. Reaction Mixture}}$

the pigment on calcium phosphate or on aluminium hydroxide gel and by subsequent elutriation by 1% $Na_2HPO_4$. The iron in this hemin compound is in a stabilized trivalent state: strong reducing agents like $Na_2S_2O_4$ are without an effect. By adding alkali as well as pyridine and hyposulfite a typical hemochromogen spectrum is formed which is indistinguishable from that of protohemochromogen. Treatment with glacial acetic acid and hydrazine yields the spectrum of protoporphyrin. The proof that this pigment, which obviously contains a heme either identical with or closely similar to blood hemin (protoheme IX), is the enzyme catalase itself and not an impurity consists in the fact that the ratio of the enzymatic activity, k, to the content in porphyrin-bound iron, $Fe_P$, remains constant for a given liver extract upon fractional adsorption on various adsorbents and aging. For one horse liver preparation the $k/Fe_P$-ratio was found equal to 2500 (ZEILE and HELLSTRÖM 1377)) and for another it was 3200 (ZEILE 1371)). Still other values of $k/Fe_P$ were found for catalase preparations obtained from plant sources, e.g. cucumber seedlings. These discrepancies still remain to be explained. They may be due to slight differences in the protein bearer of the enzyme.

Additional evidence for the hemin nature of the active group of catalase was furnished by ZEILE and HELLSTRÖM 1377) through a study of the reaction of the enzyme with HCN. Upon adding an equimolecular amount of HCN to a concentrated catalase solution a compound with two absorption bands, at 584 and 557 m $\mu$, is formed. The fact that dilution or aeration will partly or even completely restore the spectrum of the free enzyme is proof for the reversibility of the phenomenon. The values obtained for the dissociation constant of the HCN-complex by approximate spectroscopic measurements and by accurate activity determinations agree well with each other (K = $8.6 \cdot 10^{-7}$).

The enzyme will form a similar reversible complex with $H_2S$ which is known to be a strong inhibitor of catalase. The bands of the $H_2S$-catalase complex are at 640 and 580 m $\mu$ with a shadow reaching to 540 m $\mu$.

Finally, the observation of ZEILE 1371) that catalatically active solutions prepared from germinated cucumber seeds exhibit an absorption spectrum identical with that seen in liver extracts and that this pigment forms a HCN-complex with properties analogous to those of the liver pigment, justifies the conclusion drawn by ZEILE that catalase is a hemin containing enzyme. Although the evidence produced by this author pointed to a close relationship or even identity of the catalase hemin with ordinary blood hemin, the experience that isomeric hemins possess an identical absorption spectrum called for direct chemical proof concerning the structure of the prosthetic group of catalase. Such a proof has been furnished (STERN 1081a, 1084)).

Enzyme preparations obtained by the method of ZEILE and HELLSTRÖM are usually too dilute to serve directly for the isolation of the catalase hemin. The enzyme may be greatly concentrated without an appreciable gain in purity by the use of acetone and $CO_2$ at low temperature. This procedure represents one step in the preparation of the yellow enzyme from brewer's yeast (WARBURG and CHRISTIAN 1241)). Concentrates were thus obtained which had an activity, k, equal to 26,000. They will show the enzyme spectrum in layers of 1 mm. thickness. The catalase concentrates were subjected to cleavage by acid-acetone. Upon concentrating the solution after removing the denatured colorless protein, crude hemin crystals were obtained. 50 pounds of horse liver yielded approximately 40 mg. crude hemin crystals. Recrystallization from propionic acid-HCl gave 9 mg. pure hemin. In order to determine the configuration of the side chains in the catalase porphyrin, the hemin was deprived of its iron by treatment with HI-glacial acetic acid. The porphyrin thus formed was extracted with ether and fractionated with

HCl. The mesoporphyrin fraction was esterified with HCl-methanol and the resulting dimethyl ester was isolated and purified by recrystallization from pyridine-methanol (yield 1 mg.). The spectrum of the isolated mesoporphyrin was found to be identical with that of an authentic sample of pure mesoporphyrin IX. The same was true for the spectrum of the dimethyl ester when it was compared with that of pure synthetic mesoporphyrin IX dimethyl ester prepared by H. FISCHER. Ultimate proof of identity of the mesoporphyrin dimethyl ester obtained from catalase hemin with the synthetic FISCHER ester was afforded by the fact that the melting point and the mixed melting point of these samples were the same. The mesoporphyrin esters are especially suited for this purpose, because they have well-defined melting points which differ in the case of isomers and which will show appreciable depressions when mixtures of isomers are tested. This is proof that the catalase hemin is a derivative of etioporphyrin III and that the side chains of the mesoporphyrin are arranged in the pattern IX just as in blood hemin. There still remained to establish the identity of the catalase hemin with protohemin, since hemato-hemin and mesohemin will yield the same mesoporphyrin as protohemin. This was done by combining the enzyme hemin with native globin and by comparing the methemo-globin and the derived hemoglobin and oxyhemoglobin with the corresponding pigments prepared from crystalline horse blood hemin and the same native globin. The position of the bands of these pigments was identical. It should be mentioned, however, that the reduced hemoglobin obtained from the enzyme hemin showed a somewhat better defined absorption band than the blood hemin control. Taken all together, the conclusion appears justified that the prosthetic group of catalase is identical with that of the respiratory protein hemoglobin. Inasmuch as the iron of catalase exists in the ferric state, the enzyme should more appropriately be compared with methemoglobin rather than with reduced or oxyhemoglobin.

Less is known about the protein moiety of catalase. That it cannot be identical with globin is apparent from the very low catalytic activity of methemoglobin. The protein is much more stable than globin, at least in the combined state: chloroform treatment which will rapidly denature hemoglobin leaves the catalase protein intact. Several physical-chemical constants of catalase which are largely due to the protein component have been determined. The isoelectric point was found at ph 5.58 (STERN 1072)) in good agreement with earlier measurements by MICHAELIS and PECHSTEIN 817). The molecular weight has recently been determined with the aid of the analytical ultracentrifuge. The sedimentation constant $(S_{20}°)$ of horse liver catalase was found to be $11 \cdot 10^{-13}$ cm. · sec.$^{-1}$·dynes$^{-1}$ (STERN and WYCKOFF 1098,1099)). The sedimentation constant of purified beef liver catalase was determined as $1 \cdot 21^{-13}$. SUMNER and GRALÉN 1108, 1109), in SVEDBERG's laboratory, found a sedimentation constant of $11.3 \times 10^{-13}$ between ph 6.3 and 9.6. The determination of the diffusion constant $(D = 4.1 \cdot 10^{-7})$ and of the partial specific volume $(v_p = 0.73)$ permitted the calculation of the molecular weight. The value obtained was 248,000. From the iron content of pure catalase it follows that one molecule of the enzyme, just as the four times smaller molecule of hemoglobin, must contain four heme groupings *).

The isolation of the prosthetic group of catalase in crystalline state (724)) has lately been followed by the preparation of the enzyme itself in this form. SUMNER and DOUNCE 1107b), by a method involving fractional precipitation with dioxane and salting-

---

*) Earlier experiments with ANSON and NORTHROP's diffusion method had indicated a molecular weight of catalase of the same order of magnitude as that of hemoglobin (STERN 1076)). This conclusion must be revised in the light of the results obtained with the ultracentrifuge.

out with ammonium sulfate, obtained crystals from beef liver showing high catalatic activity. Eight recrystallizations from phosphate buffer with the aid of ammonium sulfate gave a product of Kat.f.=25,000 to 26,000. The appearance of the crystals is either that of fine needles or of thin plates. Prisms were also obtained. It should be noted that the specific activity of the crystals is somewhat lower than of some highly purified, amorphous, enzyme preparations previously obtained (e.g. by EULER and JOSEPHSON 296)). The iron content of the crystals varied from 0.15 to 0.06 per cent, depending on the way in which they were prepared and the number of recrystallizations. The isoelectric point of the crystalline material as determined by the NORTHROP-KUNITZ apparatus after adsorption on quartz particles as well as the point of minimum solubility were found at ph 5.7. Solutions of the crystals showed absorption bands at 627 and 536 m$\mu$.

The air-driven quantity ultracentrifuge as developed by BEAMS, BAUER and PICKELS and by WYCKOFF, has been found useful in the purification of catalase (STERN and WYCKOFF 1099)). Using chemically purified horse liver catalase of an activity, k, varying from 2100 to 6,500, and of a purity, Kat. f., varying from 4,000 to 8,980, as the starting material, bottom fractions of an activity as high as k = 161,500 and of a purity as high as Kat.f. = 33,400 were obtained. The hemin content of the best preparation was 900 mg. per liter or approximately 1 per cent of total dry weight. The molecular sedimentation not only permitted the concentration of the enzyme to this extent, but it also brought about separation from a much heavier red pigment (Sedimentation constant, $s_{20}° = 65 \cdot 10^{-13}$) with a molecular weight probably of the order of 3 to 4 millions. The best fraction obtained with the ultracentrifuge was not only homogeneous when examined with the analytical ultracentrifuge, but also when subjected to electrophoresis in the apparatus of TISELIUS where the boundaries of the colloids are made visible by TOEPLER's „Schlieren"-method (K. G. STERN, unpublished). Catalase purified in this manner may therefore be designated as homogeneous as judged by two independent criteria, viz. sedimentation in a gravitational field and migration in an electric field.

By a method involving ammonium sulfate fractionation AGNER 14a) has recently obtained catalase preparations of a Kat.f. of 55,000. This is far in excess of the activity of SUMNER's crystalline beef liver catalase fractions, another illustration of the fact that crystalline form, even after repeated "recrystallisations", is not a guarantee for the purity of a protein.

At about 0.4 saturation with ammonium sulfate, according to AGNER, an iron-containing compound is precipitated. From the properties mentioned by AGNER it appears that this substance is identical with the macromolecular red pigment separated from catalase previously by STERN and WYCKOFF 1099) by differential ultracentrifugation. In addition, AGNER finds a low-molecular copper protein in catalase preparations which may be removed by differential precipitation with picric acid. He advances the tentative hypothesis that this copper protein might be indispensable for the activity of the enzyme inasmuch as it may catalyze the reoxidation of the ferrous form of catalase by molecular oxygen as postulated by the scheme of KEILIN and HARTREE 549). The observation of the latter workers, however, that the CO-inhibition of catalase may sometimes be relieved by illumination (see also CALIFANO 159)) appears to militate against AGNER's hypothesis since it is generally assumed that the CO complexes of Cu compounds are not dissociated by light. If it were true that both iron and copper participate in catalase action, than the CO-inhibition could only be relieved by light if both types of CO-complexes were subject to photodissociation.

One of the important points concerning the constitution of catalase is the question whether protoferriheme IX and the protein moiety are the only components of the enzyme whether it contains a second prosthetic group, namely, v e r d o h e m o c h r o m o g e n (see also p. 150). During the acid-acetone cleavage of concentrated catalase preparations a blue-green pigment had been obtained in addition to the catalase hemin (STERN 1084)) which at that time was identified by R. LEMBERG as biliverdin. Later on, LEMBERG and WYNDHAM 711b) showed that the biliverdin did not exist as such in the purified liver extracts but was formed from a precursor by the treatment with HCl-acetone. The precursor was identified as verdohemochromogen which was possibly present in combination with a protein. SUMNER and DOUNCE 1107b) state that even their eight times recrystallized beef liver catalase preparations contain or give rise to the formation of the blue pigment. It may be added that also highly concentrated and purified horse liver catalase solutions prepared by the quantity ultracentrifuge have been found to contain the precursor of the biliverdin (K. G. STERN, Unpublished observations). Obviously, if the precursor of the biliverdin is not a constituent of the enzyme itself, it must share certain physical-chemical properties with the enzyme, e.g. solubility, adsorbability, molecular size, which enable it to follow the enzyme even into extensively purified fractions. Under these circumstances it would appear that decisive proof for or against the presence of the bile pigment in the enzyme molecule could best be furnished by synthesis rather than by analysis of catalase. If catalase were built in a manner strictly analogous to methemoglobin it should be possible to resynthesize the enzyme from the carefully purified protein carrier and the hemin residue obtained by reversible dissociation and, still more conclusive, from the natural protein bearer and natural or synthetic protoferriheme IX (blood hemin). Now it has actually been claimed by K. AGNER 14) that it is possible to dissociate catalase by dialysis against HCl in a manner analogous to the procedure employed in the reversible dissociation of the yellow enzyme by THEORELL 1154) and that the non-dialyzable, colorless, colloidal component (protein) and the dialyzable colored group (hemin) may be recombined with a restoration of the catalatic activity. If this were so, it would only be necessary to substitute pure crystalline protoheme for the original colored group in order to solve the problem outlined above. Unfortunately, AGNER has not yet published a more detailed account of his experiments. Attempts to reproduce his results in other laboratories have failed (TAUBER and KLEINER 1145); K. G. STERN, Unpublished). The main difficulty appears to be the pronounced lability of the bearer protein once the link to the prosthetic group has been ruptured. AGNER does not mention the original activity of his enzyme preparation but only the activity of the "resynthesized" product. A comparison of the activity of the latter with that of liver preparations obtained by methods identical with that used by AGNER indicates that the extent of the reactivation which he achieved was probably less than 1 per cent of the original activity.

M e c h a n i s m  o f  C a t a l a s e  A c t i o n: If monoethyl hydrogen peroxide is added to a highly active solution of liver catalase, the absorption band of the free enzyme at 622 m$\mu$ disappears. Instead, a new absorption band at 570 m $\mu$ appears. At a rate corresponding to that of the decomposition of the substrate by the enzyme, the new band fades and the original band of the free enzyme reappears. The cycle may again be released by adding fresh substrate (STERN 1082)). The unstable compound responsible for the new absorption band has the properties postulated by the classical theory of HENRI and MICHAELIS for an intermediary enzyme-substrate compound. The spectroscopic cycle has been recorded by a spectrograph. The microphotometry of the plates thus obtained shows that a fraction of the enzyme is not regenerated after completion of the reaction, but is destroyed in a side reaction (STERN 1087)). A large excess of substrate, of the order of $10^5$ molecules per enzyme molecule, is required for the complete transformation of the spectrum of the uncombined enzyme into that of the enzyme-substrate complex. This ratio does not imply that 1 molecule of the intermediate consists of 1 catalase and $10^5$ peroxide molecules. It simply means that an excess of substrate of this order is required to shift the equilibrium in the reaction

Enzyme + Substrate $\rightleftarrows$ Enzyme-Substrate Compound

entirely to the right (STERN 1085)). No such intermediate is observed if hydrogen peroxide is used instead of ethyl hydrogen peroxide. Attempts have been made to measure the rate of combination between the enzyme and the alkyl hydrogen peroxide by spectroscopic, spectrographic and photoelectric methods (STERN 1085); STERN and DUBOIS 1091, 1090)). Inasmuch as it is felt that the values obtained so far may be subject to some revision upon further experimentation, they are not included in this discussion. In any event, it has been demonstrated by these experiments that the rate of formation of the enzyme-substrate intermediate is rapid compared with the rate of the over-all reaction. It may be that the rate of decomposition of the intermediate is the rate-determining step in the chain of individual reactions composing the total process. The observation that the data obtained by kinetic studies of the over-all reaction fit the equation for a monomolecular reaction seems to support this tentative conclusion. The cause of the instability of the enzyme-substrate compound is not known. The iron of the enzyme persists in the trivalent state when combined with the substrate (1085)). The constitution of the complex is very probably analogous to that proposed by HAUROWITZ 467) for the methemoglobin-$H_2O_2$ compound which was discovered 38 years ago by KOBERT 584). A corresponding complex between methemoglobin and ethyl hydrogen peroxide ($C_2H_5OOH$) has only recently been described (KEILIN and HARTREE 545); STERN 1082)). Pyridine ferrihemochromogen is also capable of combination with $H_2O_2$ (HAUROWITZ 468, 469)). KEILIN and HARTREE 548) confirmed the observation of the catalase-$C_2H_5OOH$ complex and state that the enzyme will form a similar complex with $H_2O_2$ in the presence of certain inhibitors, like sodium azide and hydroxylamine. They are of the opinion that in the latter complex the enzyme iron exists in the bivalent state. More recently, these workers 549) have reported that while hydrogen peroxide is a specific reductant of catalase, molecular oxygen is required for the reoxidation to the active ferric form. Rapid removal of oxygen from the system will cause a strong inhibition of the reaction with the substrate. KEILIN and HARTREE propose the following scheme for the reaction between catalase and hydrogen peroxide:

$$4\ Fe^{\cdots} + 2\ H_2O_2 = 4\ Fe^{\cdot\cdot} + 4\ H^{\cdot} + 2\ O_2$$
$$4\ Fe^{\cdot\cdot} + H^{\cdot} + O_2 = 4\ Fe^{\cdots} + 2\ H_2O$$

$$\overline{\qquad\qquad 2\ H_2O_2 = 2\ H_2O + O_2 \qquad\qquad}$$

While this schema has a certain degree of probability, final proof for its validity has not yet been offered. In particular it seems to be difficult to understand why such a strong oxidant like $H_2O_2$ should be unable to react with the ferrous form of the enzyme. On the other hand the protein bearer seems to affect the reactivity of the heme grouping in such a drastic manner, as evidenced by the resistance of the enzyme towards reducing agents, that it is possible that the same is true for its behaviour towards oxidizing reagents.

From almost every point of view the problem of the nature of the protein bearer and of the link between the protein and the prosthetic heme group has attained a pivotal position. It is the protein which determines the catalytic specificity and the quantitative activity of the enzyme complex; and it is also the protein which modifies profoundly the general chemical reactivity of the molecule. Further progress in enzyme chemistry is largely bound up with the progress in our knowledge of protein structure in general and of the enzymatic bearer proteins in particular.

The ultraviolet absorption band observed with chemically purified liver catalase preparations (STERN 1074)) is due to the enzyme itself and not to an impurity: homogeneous enzyme preparations obtained by ultracentrifugation exhibit a strong band at 275 m$\mu$ (STERN and LAVIN 1096)). This band, as in all hemin proteins, is due to the protein moiety. The fine structure of this band indicates the presence of tryptophane, tyrosine and phenylalanine in the molecule.

### Peroxidase:

WILLSTÄTTER, in the course of his work on horseradish peroxidase, observed that the iron content of the preparations would at first increase (1336b)) and later on decrease (1336)) upon progressive purification. This finding led him to conclude that iron was not a constituent of peroxidase. However, he considered the possibility that the enzyme may exert its function in a reaction system containing iron.

This work was continued several years later by KUHN, HAND and FLORKIN 649). Instead of determining the total iron content of their peroxidase preparations, these workers estimated the hematin content by a spectroscopic method after conversion into pyridine hemochromogen. They came to the conclusion that there exists parallelism between the very low hematin concentration and the enzymatic activity, expressed as "purpurogallol number" (P.Z.) of horseradish peroxidase solutions. Upon measurement with a photoelectric spectrophotometer a steep absorption band at about 420 m$\mu$, but no long-wave bands were found in the enzyme solution. In view of the formation of pyridine protohemochromogen from peroxidase preparations, KUHN and his associates concluded that the enzyme contains protoheme in combination with a nitrogeneous pheron. Its constitution would then be analogous to that of catalase.

The results obtained by KUHN were re-examined by ELLIOTT and KEILIN 247). While at first a marked parallelism was found between hematin content and enzymatic activity, this relationship broke down in more highly purified products. Furthermore, the amount of hematin-iron present in these preparations of medium activity was much higher than the hematin or even the total iron concentration found by other workers in highly active preparations. The situation at this stage of research is perhaps best illustrated by the diagram (Fig. 7), taken from ELLIOTT and KEILIN's paper.

ELLIOTT and KEILIN point out that the figures given by KUHN, HAND and FLORKIN for the activity of their peroxidase solutions, expressed in P.Z., appear too high when compared with the hematin content of their solutions. While, for example, a peroxidase preparation of ELLIOTT and KEILIN of P.Z. = 818 contained 1.05 per cent hematin, KUHN et al. found only about 0.105 per cent hematin in a product of P.Z. = 3400. ELLIOTT and SUTTER 248) had already noted that the quantities of enzyme mentioned by KUHN et al. as taken for their activity determination were ten times too great. Upon dissolving a purified peroxidase preparation in water, ELLIOTT and KEILIN obtained a distinctly brown solution exhibiting a strong absorption band at 642 m$\mu$ and two other bands at 550 m$\mu$ and at 500 m$\mu$ which were partly masked by a general absorption, due probably to another, yellow pigment present in the preparation. Upon reduction a hemochromogen spectrum with a strong band at 555 m μ was observed. In the presence of more alkali another band at 526 m μ may be detected. The heme present in this compound may readily be transformed into pyridine hemochromogen. Furthermore, TEICHMANN's hemin crystals could easily be obtained from different enzyme preparations. The authors expressed the belief that the compound responsible for the original absorption spectrum in peroxidase solutions is free acid hematin. Upon reinvestigating the question in 1937, KEILIN and MANN 551), while confirming the break in proportionality between the enzyme activity and the concentration of total hematin iron observed by ELLIOTT and KEILIN in the course of the purification of the enzyme, arrive at the conclusion that,

nevertheless, the active group of peroxidase is a hematin compound.

For the purification of the enzyme, the method of ELLIOTT 243) was employed. Non-cultivated horseradish is finely minced and extracted with water. The pulp, when treated in a hydraulic press, yields a crude brownish extract (P.Z. = 0.65). The extract is saturated with ammonium sulfate and the greyish precipitate is filtered off. It is suspended in water and dialyzed through cellophane first against running tap water and then against distilled water. Upon adding two volumes of 90 per cent alcohol a precipitate of low peroxidase activity (P.Z. = 14 to 26) is formed. This is removed in the centrifuge. The addition of 0.7 vol. absolute alcohol yields a small amount of a highly active precipitate (P.Z. = 112 to 300). This fraction could be further purified in various ways, e.g. by alcohol precipitation, adsorption of impurities

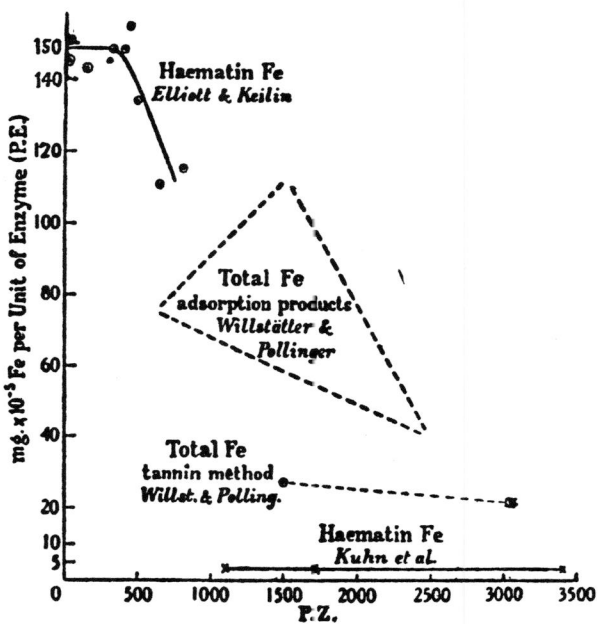

Fig. 7. Diagram showing the amounts of hematin-iron or of total iron per unit of peroxidase activity, in preparations of varying concentration. The total iron figures of WILLSTÄTTER and POLLINGER are scattered within the triangular area indicated in the diagram. (From ELLIOTT and KEILIN 247)).

on tricalcium phosphate gel, or by adsorption of the enzyme on alumina A of WILLSTÄTTER and precipitation by tannic acid. After concentrating the eluate at room temperature a preparation with a P.Z. of 1500 was obtained.

When examined with a HARTRIDGE reversion spectroscope four absorption bands at 645, 583, 548, and 498 m μ are found in all of these preparations. While the fourth band has probably the highest extinction, the first band in the red region which is not masked by other pigments is the most conspicuous absorption band and may readily be detected in preparations of comparatively low activity and even in the horseradish root itself. The band at 583 m μ is very weak and hardly detectable in preparations of low activity. This spectrum which was previously ascribed by ELLIOTT and KEILIN to acid hematin is actually that of the enzyme peroxidase which contains hematin in combination with a protein. Although we are probably dealing here with protoheme IX, i.e.

with blood hemin, chemical evidence of the type procured for the heme of catalase is still lacking. The iron in the active enzyme is in the trivalent state. The spectrum of peroxidase resembles closely that of methemoglobin and of catalase. In contrast to the latter, peroxidase may readily be reduced to the ferrous form by sodium hyposulfite ($Na_2S_2O_4$). Reduced peroxidase shows two bands; a narrow and weak band at 594.5 m $\mu$ and a very strong and broad band at 558 m $\mu$. On oxidation the bright red ferrous form reverts to the brownish red ferric form.

The ferrous but not the ferric form of peroxidase combines reversibly with CO. The two bands of the ferrous form are thereby shifted to 578 and 545.5 m $\mu$. KCN which in 0.001 M. concentration inhibits the activity of the enzyme forms a spectroscopically well-defined complex with the active ferric form. The red solution shows a strong and wide absorption band at 542 m $\mu$ and a weak and narrow band at 581.5 m$\mu$ with a shadow connecting the two bands. In general appearance the spectrum of the KCN-peroxidase complex resembles somewhat that of cyanide methemogoblin. The addition of sodium fluoride to a slightly acid peroxidase solution causes a color change from brownish red to green. This is due to a shift of the four absorption bands of the free enzyme to 615, 561, 529.5 and 496 m $\mu$.

The change in the position and intensity of the first band is clearly visible in slices of horseradish root treated with NaF. The fluoride compound may be used for the spectral-colorimetrical estimation of the enzyme in situ. The inhibition of peroxidase amounts only to 50 per cent by 0.001 M. fluoride and to 25 per cent by 0.003 fluoride. When NO is added to peroxidase in an atmosphere of pure nitrogen, the four-banded spectrum of the free enzyme is replaced by two strong bands at 570.5 and 539.5 m $\mu$. $H_2S$ will combine with the enzyme to form a complex with maxima at 587.5 m $\mu$ (weak) and 549.5 m $\mu$ (strong). Interesting phenomena were observed by KEILIN and MANN 551) when hydrogen peroxide was added to strong peroxidase solutions. The color of the enzyme solution turns red and the enzyme spectrum is replaced by two bands which are situated at 561 and 530.5 m $\mu$. The peroxidase-$H_2O_2$ complex is unstable and breaks down more or less rapidly to yield the free enzyme. Now it is generally assumed that peroxidase is capable of decomposing hydrogen peroxide only in the presence of a suitable donator like leucomalachitegreen or pyrogallol. The authors assume, therefore, that the lability of the enzyme-peroxide complex is due to the presence of a substance in the enzyme preparation which reacts wit the $H_2O_2$ which is activated by the enzyme. They find that one molecule of the peroxide is required to form the enzyme-substrate compound showing the two absorption bands mentioned above. On further addition of $H_2O_2$ to a solution containing this complex the bands at 561 and 530.5 m $\mu$ become diffuse and a new band begins to appear at 583 m $\mu$. When the excess of $H_2O_2$ reaches a concentration of from 15 to 25 molecules per gram atom of iron contained in peroxidase the two bands of the original enzyme-$H_2O_2$ complex are replaced by two very strong bands located at 583 and 545.5 m $\mu$. A maximum of intensity is reached at a ratio of about 100 molecules of substrate to 1 Fe atom of the enzyme. The color of this enzyme-substrate complex of higher order is deeper red than that of the equimolecular compound. The $H_2O_2$ suffers rapid decomposition into water and molecular oxygen. Owing to the rapid break-down of the higher-order complex the authors were unable to ascertain the number of $H_2O_2$ molecules required for its formation. A true peroxidatic reaction may be initiated by adding suitable donators like hydroquinone, pyrogallol, or ascorbic acid. A number of spectroscopic tests warrant the conclusion that in both compounds with $H_2O_2$ the iron of the enzyme remains in the trivalent state.

While it is true that break in proportionality occurs between the hematin content of peroxidase preparations (estimated as pyridine hemochromogen) and their enzymatic activity at certain levels of enzymatic activity and also depending on the method of purification, there exists a strict proportionality between the catalytic activity and the concentration of the intact pigment as measured by the relative intensities of the absorption band at 645 m $\mu$. This holds for preparations covering the wide range between P.Z. = 10 to 1500. The same is true for the intensity of the long-wave band of the fluoride-peroxidase complex. These findings suggest strongly that, while the absorption band at 645 m $\mu$ is solely due to the enzyme, there are other hematin compounds present in the preparations which will yield pyridine hemochromogen. In other words, the peroxidase hematin represents only a portion of the total hematin compounds in the enzyme solutions which are estimated as pyridine hemochromogen. It has not been possible as yet to determine the concentration of active peroxidase hematin in terms of hemin or of iron even in the most active and highly purified enzyme preparations. The reason for this is that it is not known yet whether the peroxidase hematin is the only hematin compound present even in preparations of P.Z. = 1500.

In conclusion it may be added that SUMNER and HOWELL 1110) have found an absorption band at 630 to 640 m $\mu$ in purified peroxidase preparations from fig sap. The hematin content of their preparation of P.Z. = 700 was found to be slightly higher than 1 per cent in agreement with the findings of ELLIOTT and KEILIN in the case of horseradish peroxidase.

While it appears that the prosthetic group of peroxidase is identical with proto-ferriheme IX (blood hemin), nothing is known about the protein component. Work on the protein will have to await the preparation of homogeneous peroxidase.

### Appendix: Copper Proteins (Orthophenol Oxidase):

This enzyme is built in a manner quite different from the hemin-containing enzymes just discussed. The only common feature is that there is also a heavy metal in the molecule. However, the metal is c o p p e r instead of iron. Although the structure of orthophenol oxidase is as yet unknown, the advances which have of late been achieved in this field would seem to warrant a discussion at this time.

Potatoes, mushrooms, and other plant material contain a soluble oxidase. This enzyme which has been called phenol oxidase, polyphenol oxidase, laccase, or catechol oxidase, is able to catalyze the oxidation of phenolic compounds like pyrocatechol by molecular oxygen *). The first reaction product in the case of pyrocatechol is ortho-quinone which may be either reduced back to pyrocatechol in the presence of suitable reducing systems, e.g. ascorbic acid or dihydropyridine nucleotides, or which may be further oxidized by the same enzyme.

KUBOWITZ 613), working in WARBURG's laboratory, has succeeded in purifying orthophenol oxidase from potatoes to a large extent. The procedure employed in the purification consists essentially in alternating acetone and ammonium sulfate fractionations, removal of inert proteins by heat denaturation at ph 7.4 and 9.7 and by precipitation with silver acetate, dialysis, and a final acetone fractionation.

The purest enzyme preparation was weakly yellow. The solution is relatively stable at 0° (ph 7), and still more stable in half saturated ammonium sulfate solution. KUBOWITZ states that the enzyme is a c o p p e r - p r o t e i n compound. The metal cannot be separated

---

*) The earlier literature is reviewed in OPPENHEIMER 902).

from the rest of the molecule by dialysis; but it is removed by the action of acids. The copper content in various preparations is strictly proportional to the activity, as evidenced by the linear relationship illustrated in Fig. 8. The ratio $\dfrac{\text{Mol. Oxygen transferred}}{\text{Mol. Copper} \times \text{Minutes}}$ is found to equal 880 under the conditions of the test devised by KUBOWITZ for the determination of the enzymatic activity.

The copper content of the enzyme preparation showing the greatest specific activity was 0.165 per cent. By comparing the oxidase with hemocyanin the Cu content of which is given in the literature as maximal 0.34 per cent, KUBOWITZ estimated the degree of purity of his best enzyme preparation to be about 50 per cent. The isoelectric point, as determined by electrophoresis, was found near ph = 5.4. The nitrogen content of the purest preparation was about 15 per cent.

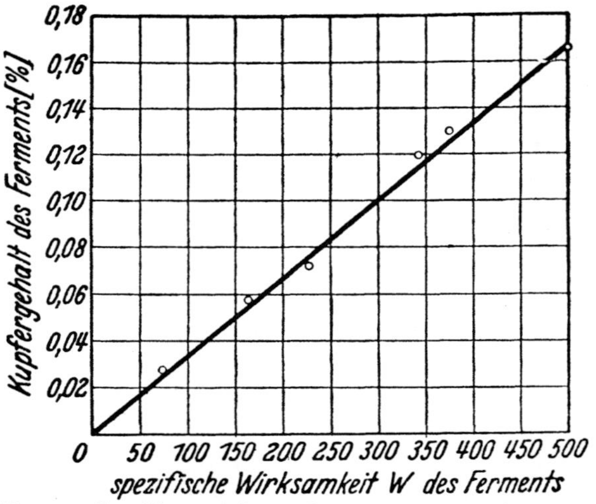

KEILIN and MANN 552) were able to obtain the same enzyme in a perhaps even higher degree of purity from cultivated mushrooms (*Agaricus* or *Psalliota campestris*). The general procedure involved the following steps: fine mincing, pressing out in a hydraulic press after grinding with sand and adding water, precipitation with ammonium sulfate, dialysis, fractional precipitation with lead acetate, adsorption on tricalcium phosphate, followed by fractional precipitation with acetone in presence of lead acetate, another adsorption on tricalcium phosphate and a final fractional precipitation with acetone. The activity of the various fractions was determined manometrically, from the rate of oxygen uptake by catechol, and colorimetrically, from the amount of purpurogallin formed from pyrogallol under specified conditions. The purest enzyme preparation was free from hematin. Confirming KUBOWITZ' discovery it was found that the oxidase is a copper-protein complex. In the range of less pure preparations of low activity there exists no proportionality. This is due to the presence of extraneous copper. A similar observation with respect to the presence of iron in peroxidase preparations had previously been made by WILLSTÄTTER and POLLINGER 1336). Only when the copper content reaches the low and constant value of about 3.2—3.5 $\gamma$ per Enzyme Unit does the Cu content become strictly proportional to the enzyme activity. The purest enzyme preparation obtained by KEILIN and MANN had a Purpurogallin Number of 940 and a copper content of 0.30 per cent. Inasmuch as this is higher than the copper content of crystalline hemocyanin (0.173 to 0.26 per cent (HERNLER and PHILIPPI)), the authors are inclined to consider their best preparation as the pure enzyme. 1 $\gamma$ of Cu contained in the enzyme molecule is able to transfer to catechol, at 20° in 1 min., about 6,000 cmm. $O_2$. 9.6 mg. of this enzyme preparation were obtained from 15 kg mushrooms. While the crude extract of mushrooms would readily attack compounds like catechol, pyrogallol

Fig. 8.   Specific activity and copper content of potato polyphenol oxidase. (From KUBOWITZ 613)). Abscissa, Specific activity;   Ordinate, Copper content in %.

and p-cresol, the enzyme lost the property of oxidizing monophenols upon purification. Both Kubowitz and Keilin find that the enzyme is inhibited by HCN, H₂S and CO. In an atmosphere containing 90 vol. per cent CO and 10 vol. per cent O₂, the rate of oxidation of pyrocatechol is only one half that observed in 90 vol. per cent argon and 10 vol. per cent O₂. Other substances known to form stable complexes with copper, like salicylaldoxime or diethyldithiocarbamate, also inactivate the enzyme (Kubowitz 613)).

Kubowitz 614, 615) reports further interesting results obtained with the potato phenol oxidase. Upon adding HCN to a solution of the enzyme inactivation results and copper diffuses out through a dialyzing membrane. The addition of copper in inorganic form to the non-dialyzable residue restores the activity. From this experiment Kubowitz concludes that Cu in ionized form represents at the same time the prosthetic and the active group of the enzyme. It appears however that his results may be capable of a somewhat different interpretation. This is particularly true in view of the fact that Kubowitz obtained an analogous result with hemocyanin from *Octopus*. Now it is known from the work of Conant et al. 178) that the prosthetic group of hemocyanin consists of copper in complex linkage with a sulfur containing polypeptide. If it is assumed, as a working hypothesis, that the metal is linked up with the rest of molecule through dithiolinkages (—S—S—) *), one would expect the metal to be detached from the large complex by the action of HCN which is known to reduce dithio to sulfhydryl groups. A small molecular and consequently diffusible Cu-SH-complex would thus be formed. An addition of copper salt to the remainder of the enzyme molecule may then serve to reoxidize its SH-groups to S—S-groups, a reaction known to be catalyzed by copper salts, and at the same time provide for a replacement of the copper lost through the dialysis. A difficulty here is that Cu ordinarily will form complexes more readily with SH- than with S-S-compounds. Whatever the true explanation may be, it is felt that the observations of Kubowitz, at the present time, do not seem to prove conclusively his contention that Cu ions are the prosthetic group of phenol oxidase.

Kubowitz was furthermore able to show that the metal undergoes a valency change in the course of the catalysis. When pyrocatechol is added to phenol oxidase under pure CO, one molecule of enzyme will oxidize one molecule of the substrate to o-quinone. The univalent Cu thus formed absorbs the theoretical amount of CO from the atmosphere. It may again be released by adding HCN. The latest preparations of potato polyphenol oxidase obtained by Kubowitz by an improved procedure (615)) contain 0.2 per cent Cu and are considered to represent the pure enzyme. **)

## 2) Vitazymes.

The term "vitazymes" has been proposed by v. Euler 324) for substances which represent vitamines, i.e. essential food factors, and which exert their function in the organism by combining with a colloidal bearer (protein) to form an enzyme. Thus far two cases of this kind have been uncovered, viz. that of the yellow enzyme the agon of which is identical with vitamin B₂ (or vitamin G according to the American nomenclature) and the case of carboxylase (and pyruvic acid oxidase) the prosthetic group or coenzyme of which is identical with vitamin B₁ pyrophosphate. Although a similar

*) See Th. Bersin 101a).
**) According to Keilin and Mann (Nature 143, 23 (1939)) the enzyme Laccase is also a copper-protein. The very strong oxidation of catechol by the blood of some invertebrates is an „pseudo oxidatic" action of the hemocyanin itself (Bhagvat and Richter, Biochem. Jl. 82, 1397, (1938)).

function is being suspected with regard to vitamin C, no convincing proof in support of that view has as yet been furnished. Nevertheless, ascorbic acid shall briefly be treated in this section, because there is evidence to show its participation in oxidative biological processes. At a later time it will probably become necessary to include the pyridine nucleotides in this group (see p. 212).

### a.  Flavinphosphoric Acid and Yellow Enzymes (Flavoproteins) *).

In 1932, WARBURG and CHRISTIAN **1239, 1240)** reported the discovery of a new, hemin-free, oxygen transferring enzyme in bottom yeast and in tissue extracts. In accordance with its color the authors designated it briefly as "yellow enzyme". Although not very satisfactory, this term has been retained by WARBURG and by many workers up to this day. The test system for the activity of the enzyme consisted of hexose monophosphoric acid (ROBISON ester), of a thermostable coferment obtained from red blood cells and of an "activating enzyme" or "zwischenferment" prepared either from red blood cells or from LEBEDEW juice from bottom yeast. Only in presence of the yellow enzyme did oxidation of the substrate to phosphohexonic acid and uptake of molecular oxygen take place. It could be shown that the enzyme was capable of reversible oxido-reduction; and that the reduced leuco form could be readily autoxidized by molecular oxygen (or methylene blue).

The over-all reaction could be subdivided into two steps. In the first step hydrogen was transferred from the substrate (after "activation" by the coenzyme-zwischenferment) to the yellow enzyme which was reduced to the leuco form. In the second step the active form of the yellow enzyme was regenerated by reaction with molecular oxygen. The oxygen is not completely reduced to water, but only partially, namely, to hydrogen peroxide.

Upon warming the enzyme with methanol-water, a yellow, low-molecular pigment with an intense green fluorescence was split off (WARBURG and CHRISTIAN **1245)**). If this pigment was irradiated in alkaline solution with a tungsten filament lamp, a photo-derivative was obtained which was both water and chloroform soluble and which could readily be crystallized. Elementary analysis yielded the empirical formula $C_{12}H_{13}N_4O_2$. After heating with barium hydroxide solution, urea was isolated as the xanthydrol complex. Each of these findings proved of great significance for the subsequent elucidation of the structure of the pigment.

WARBURG and CHRISTIAN were not the first workers to observe the presence of a yellow, green fluorescing pigment as a constituent of biological materials. As a matter of fact, BLYTH, in 1879, described a yellow pigment, lactochrome, which he obtained from milk and which must have consisted, at least in part, of lactoflavin. The study of lactochrome was again taken up in 1925 by BLEYER and KALLMANN **125)**. They did not succeed in obtaining the pigment in pure form. Early in 1932, SZENT-GYÖRGYI and BANGA **1127)**, in the course of the purification of the coferment of lactic dehydrogenase from heart muscle, obtained a yellow pigment which they called c y t o f l a v. Although they ascertained that it was not directly concerned with the dehydrogenation of lactic acid in muscle, the fact that it could be reversibly reduced and reoxidized led them to suspect that cytoflav plays some rôle in cellular respiration. In the same year it was reported

---

*)  Review articles have been written by KUHN, WAGNER-JAUREGG, THEORELL, WEYGAND, WARBURG **627, 1207, 628, 630, 1209, 631, 1161, 684)**.
See also OPPENHEIMER's "SUPPLEMENT", p. 1577—1585.

that a yellow, green fluorescing pigment occurs in purified horse liver extracts and that it appeared to be associated with the catalatic activity exhibited by these preparations (STERN 1077)).

Early in 1933, KUHN, GYÖRGY and WAGNER-JAUREGG 645) and ELLINGER and KOSCHARA 239) described "a new class of biological pigments", designated by the first group of workers as flavins and by the second groups as lyochromes which are obviously identical with the yellow, green fluorescing pigments previously decribed. The most interesting fact discovered by KUHN, GYÖRGY and WAGNER-JAUREGG 646, 647) was the identity of the flavin, isolated in crystalline form from milk whey (lactoflavin), with a component of the vitamin B-complex. It was found capable of promoting growth in rats stunted by a flavin-free diet. More specifically, these workers identified lactoflavin with vitamin $B_2$ (or vitamin G). Later work in other laboratories has shown that lactoflavin and lactoflavin phosphate are not only essential for the rat, but also for the dog where a lack of this vitamin eventually leads to the death of the animals. The essential nature of lactoflavin for human nutrition has not yet been conclusively demonstrated.

**Riboflavin:** In order to proceed from the simpler to the more complex substances we shall first take up the chemical structure of free lactoflavin. Following this its relationship to the yellow enzyme will be discussed.

KUHN et al. 646, 647) obtained crystalline, pure flavins from milk whey as well as from egg white (lactoflavin and ovoflavin respectively). Hepatoflavin from liver was isolated in crystalline form by several workers (STERN 1077b), KARRER et al. 525a), STARE 1064)). Other sources from which flavins were isolated include egg yolk, malt, various plants *) . The empirical formula of pure lactoflavin is $C_{17}H_{20}N_4O_6$ (KUHN et al. 646, 647)). Irradiation in alkaline solution yields a photoderivative identical in every respect with that obtained previously by WARBURG and CHRISTIAN 1245). This compound, called lumiflavin by KUHN and photoflavin by STERN, of the composition $C_{12}H_{13}N_4O_2$, differs from lactoflavin by an amount of $C_4H_8O_4$. KUHN assumed that the latter group represents a hydroxyl-containing side chain (carbohydrate?) which during photolysis is split off from a cyclic system containing weakly basic, probably tertiary N-atoms (KUHN et al. 659, 666, 660)). Lumiflavin contains an alkali-labile ring the constitution of which was exhaustively studied by KUHN and his coworkers. For some time, they were not able to decide whether lumiflavin is an alloxazine or a quinoxaline derivative. The summary formula found for lumiflavin agreed with that of a trimethylalloxazine prepared many years ago by KÜHLING. Alloxazine is an analogon to phenazine, with the difference that one of the two benzene rings is replaced by a pyrimidine ring. Preliminary experiments performed by KUHN and BAER 635) seemed to demonstrate such fundamental differences between lumiflavin and alloxazine that KUHN at that time was ready to discard the idea of a relationship between the two compounds. This inclination was enhanced by the discovery 635) that 2-tetrahydroxybutyl-quinoxaline, upon irradiation in alkaline solution, yields the chloroform soluble, yellow fluorescing, quinoxaline, thereby providing a perfect model for the photolysis of lactoflavin. On the other hand, KUHN and WAGNER-JAUREGG 675) found that lactoflavin shares with alloxazine the property to yield, upon reduction by zinc, tin or sodium amalgam in dilute HCl, a red

---

*) For references the review by THEORELL 1161) should be consulted.

intermediate on the way to the colorless completely reduced form. The red color was attributed to the formation of a semiquinoid monohydro radical of the type studied by L. MICHAELIS. In view of the considerable difficulties encountered by KUHN and his colleagues in their endeavours to determine the structure of lactoflavin and in particular that of the photoderivative by analytical and degradation procedures, STERN and HOLIDAY attacked the problem from a synthetic angle. A comparison of KÜHLING's alloxazine with photoflavin (lumiflavin) showed that these compounds agree with regard to their remarkable stability towards strong acids, their light absorption pattern, and, as KUHN had already found, with regard to the red intermediate upon reduction in acid medium. They differ with regard to solubility, alloxazine being insoluble in water as well as chloroform, and also to fluorescence: alloxazine does not exhibit a green fluorescence comparable to that of photoflavin. Now it occurred to them that phenazine and oxyphenazine differ from pyocyanine (see p. 228) in a way somewhat comparable to the difference between alloxazine and photoflavin. In both instances the biologically interesting pigment contains more methyl groups as compared with the mother substance. In pyocyanine the methyl group is attached to one of the phenazine-nitrogen atoms; and the formal analogy between phenazine and alloxazine suggested as a working hypothesis that in photoflavin a methyl group might be attached to one of the nitrogen atoms in the central pyrazine ring of alloxazine and thus produce the striking changes in properties between the two substances. It was indeed found that methylation of alloxazine under conditions where substitution at the central N-atoms would be expected produces an orange-yellow, strongly green fluorescent, water and chloroform soluble dye with an absorption spectrum closely similar to that of photoflavin both with respect to the pattern and the position of the bands (STERN and HOLIDAY 1091, 1095). A series of such N-alkylated alloxazines was synthesized, and for the "natural" photoflavin the structure of a xyleno-9-methyl-alloxazine was postulated. The synthesis of three of these trimethyl-alloxazines (6, 7, 9-, 7, 8, 9- and 6, 8, 9-trimethyl-alloxazine) was also described. A decision as to which of the three compounds is identical with photoflavin was not made due to their closely resembling properties.

In any event, these experiments proved that photoflavin and therefore also lactoflavin are derivatives of benzalloxazine and furthermore that one of the pyrazine nitrogen atoms carries a substituent which is an alkyl group in photoflavin and a carbohydrate-like chain in lactoflavin. In the course of their synthetic experiments, STERN and HOLIDAY 1095) prepared a number of alloxazines carrying alkyl groups in the benzene ring, among them 6,7-dimethyl alloxazine. KARRER et al. 526, 527, 529) isolated the same compound as a product of the photolysis of lactoflavin in acid and neutral solution and they gave it the name lumichrome. This finding left no doubt that photoflavin is 6, 7, 9-trimethylalloxazine (or 6, 7, 9-trimethylisoalloxazine in the nomenclature proposed by KUHN).

Benzalloxazine      Lumichrome      Lumiflavin (Photoflavin)

Kuhn et al. 656, 681) proved the constitution of lumiflavin by a synthesis, which is more clear cut than that previously described by Stern and Holiday 1095). The various steps of the synthesis (see also Kuhn and Reinemund 655)) are indicated in the following schema:

Synthesis of Lumilactoflavin according to Kuhn.

The interest was now focussed on the carbohydrate-like s i d e c h a i n of lactoflavin. Karrer et al. 527, 529) were the first to introduce hydroxyl-containing side chains in 9-position in alloxazine. The products thus obtained showed a behaviour upon photolysis which was analogous to that of lactoflavin. Both Kuhn and Weygand 683, 685) and Karrer et al. 518) were able to synthesize 6,7-dimethyl-9-(1-l-arabityl)-isoalloxazine which was the first synthetic product to exhibit a certain vitamin B$_2$-activity. Whereas both groups of workers were inclined at first to believe that the side-chain in lactoflavin is also derived from l-arabinose, in other words that the synthetic product was identical with the vitamin, the finding that the biological activity of the araboflavin was $1/2$ to $1/3$ as small as that of natural lactoflavin and that the melting points of the tetraacetyl derivatives did not agree, militated against this view. The search for the nature of the side chain was therefore continued, and about simultaneously Kuhn 658, 667), and Karrer 518) and their associates reported the synthesis of the fully active d-ribose derivative. The identity of the riboflavin with natural lactoflavin was confirmed by v. Euler, Karrer et al. 298) and György 426). The method of synthesis was considerably improved both in Kuhn's and in Karrer's laboratory **524, 633, 682, 576, 631**). In particular, the use of boric acid as a catalyst in the condensation of alloxan with the dimethyl-amino-ribitylamino-benzene has brought about a great increase in yield **686**). One of the recent syntheses of riboflavin (lactoflavin) is illustrated by the following sequence of reactions (Karrer and Meerwein **524**)):

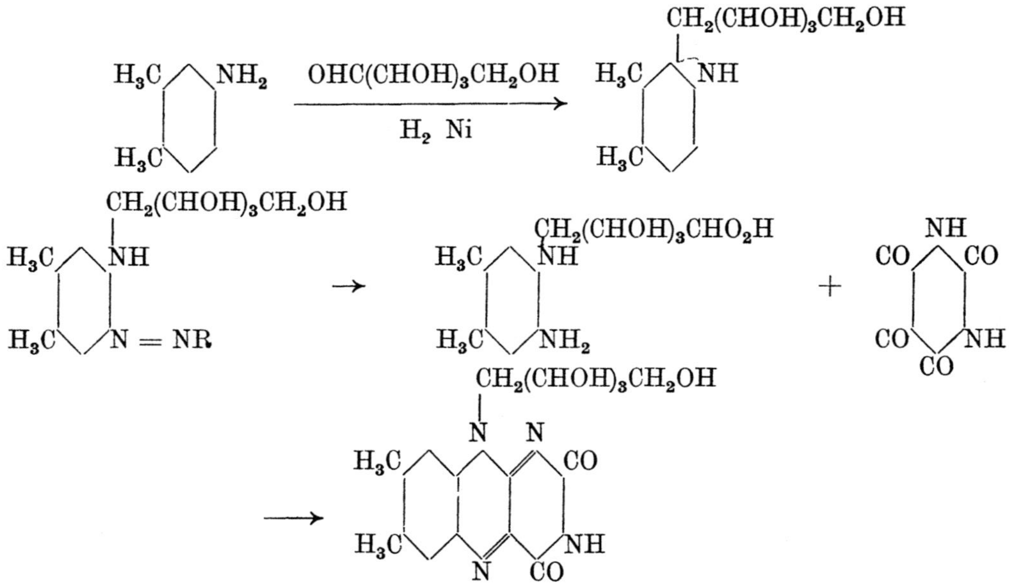

Since then it has become apparent that all the flavins isolated from various plant and animal sources are identical with riboflavin (cf. Kuhn et al. **646, 647**)) with the exception of the uroflavins obtained from Koschara **588**) [from urine which differ distinctly in their behaviour upon chromatographic analysis.

The isolation of riboflavin (lactoflavin) from biological sources is based on the following principles: Riboflavin is readily adsorbed on Fullers earth or Frankonit from mineral acid or acetic acid solution. Elution is accomplished by pyridine-methanol-water mixtures. A precipitate of lead sulfide, formed in the solution, may also be used as adsorbent. Accompanying purine and other bases are removed by picric acid treatment. For the final purification the formation of silver- and thallium salts by riboflavin is utilized with advantage (cf. Kuhn et al. **627, 1207, 628, 630, 1209, 631**)). The most suitable material for the preparation of natural riboflavin in quantity appears to be milk whey. Liver is also relatively rich with respect to the pigment, but the isolation and the separation from the many concomitant substances of similar physical-chemical properties is quite tedious (cf. Stern **1077b**)). While all of the flavin in milk occurs in free, dialyzable form, the contrary is true for liver. Boiling of aqueous liver extracts will readily liberate the flavin from its linkage to protein.

The pigment crystallizes from aqueous or alcoholic solutions in orange colored needles which tend to form clusters. The melting point or, better, the decomposition point of natural riboflavin has been found anywhere between 271° and 293°, coupled with considerable differences in the solubility of the fractions (Kuhn **667**)). Aqueous solution of riboflavin show a brilliant green-yellow fluorescence, particularly when irradiated with long-wave ultraviolet or short-wave violet light. The fluorescence is quenched in strongly acid and alkaline solution. The determination of the intensity of the fluorescence in dependence on the hydrogen ion concentration has been used for the calculation of the dissociation constants of riboflavin (KUHN and MORUZZI 653a)). The values thus derived are $pK_1' = 1.7$ and $pK_2' = 10.2$. The isoelectric point of riboflavin as calculated from these values, ph 6, should be replaced by the term isoelectric zone which, according to electrophoretic measurements of THEORELL 1151), extends from ph 3 to 9. Other physical-chemical properties of riboflavin, i.e. the absorption spectrum and the oxidation-reduction potential, will be discussed below, jointly with those of the other substances of this group.

### Riboflavin Phosphoric Acid and Flavin Adenine Dinucleotide:

Up to 1934 it was generally believed that free lactoflavin (riboflavin) represents the prosthetic or active group of the yellow enzyme of WARBURG and CHRISTIAN. This belief was disproved, when THEORELL 1151) in WARBURG's laboratory, upon cleavage of the yellow ferment with methanol, obtained the colored group combined with a phosphor-containing acid radical. While free lactoflavin, according to THEORELL, will not migrate in an electric field at ph 7.2, the colored group of the enzyme shows an electrophoretic mobility of $u = 16.10^{-5}$ cm$^2$·sec.$^{-1}$·volt$^{-1}$ at the same ph. THEORELL was able to prepare the colored group of the enzyme in crystalline form, as the calcium salt, and to identify it by elementary analysis as the monophosphoric acid ester of lactoflavin. While THEORELL in contrast to claims made by KUHN 685, 664) stated that free riboflavin does not combine with the bearer protein of the yellow enzyme to form a catalytically active compound, he found that riboflavin monophosphate will do so in a quantitative manner.

Lactoflavin monophosphate is not only obtainable from yeast but it has also been prepared from animal tissues, e.g. liver (THEORELL, KARRER et al. 1164)). A synthetic lactoflavin phosphoric acid ester was obtained by treating lactoflavin with POCl$_3$ in pyridine (KUHN and RUDY 662)). The position of the acid radical in the carbohydrate side-chain remained unknown. The product showed a slight coupling with the protein of the yellow enzyme. It may well have represented a mixture of isomeric phosphoric acid esters. Later, KUHN, RUDY and WEYGAND 668) synthesized lactoflavin-5'-phosphoric acid via the 2',3',4'-triacetyl-5'-trityl derivative. KUHN and RUDY 663) could show that the complex obtained by coupling the synthetic ester with purified bearer protein (50 per cent pure) had the same catalytic activity towards ROBISON and NEUBERG ester as the corresponding complex obtained with the aid of natural flavin phosphoric acid. As far as the latter ester is concerned, KARRER, FREI and MEERWEIN 521) were able to exclude the positions 2' and 3' of the carbohydrate chain as the place of attachment of the acid residue: Upon oxidation with periodic acid no formaldehyde is formed which is known to arise from pentose-3-phosphoric acid but not from pentose-5-phosphoric acid. Therefore, only positions 4' and 5' remain as possibilities, the latter of which appears the most likely one. The same workers modified the synthesis of the ester on the basis of KARRER and MEERWEIN's method 524). In view of very recent developments it is of great interest to note that KARRER, FREI and MEERWEIN 521) state that their preparations of lactoflavin phosphoric acid from liver invariably contained adenylic acid up to 60 per cent of the total weight. Although the authors were inclined to regard the adenine nucleotide as an impurity, they did not consider it impossible that there might exist a chemical linkage between the two constituents of their preparations. WARBURG and CHRISTIAN report 1256) that the coenzyme of the d-alanine oxidase, as isolated from

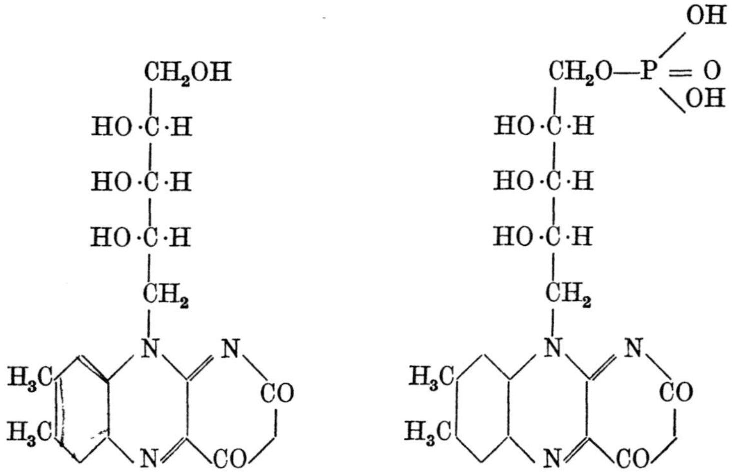

Riboflavin (Lactoflavin)
(6,7-dimethyl-9-d-ribityl-alloxazine)

Riboflavin Phosphoric Acid
(6,7-dimethyl-9-d-ribityl-alloxazine-
5'-phosphoric acid)

horse kidney, has the chemical composition of a flavin-dinucleotide. They intimate that the yellow enzyme which they isolated from bottom yeast in 1932 represents only one member of a whole group of enzymes with prosthetic groups containing flavin. Their finding explains why KARRER and also other workers encountered such difficulties in attempting to prepare pure riboflavin and riboflavin phosphoric acid from animal tissues, particularly from liver. In this connection one might also refer to the observation of GYÖRGY, KUHN and WAGNER-JAUREGG 863) that optimal vitamin $B_2$-activity of lacto-flavin or of the phosphoric acid ester is produced only by simultaneous administration of vitamin $B_4$ concentrates which contain large amounts of adenine as an impurity.

In a recent communication, KARRER et al. 522) report that they have now tested their old riboflavin-phosphoric-adenine-nucleotide preparations obtained from liver for activity in the d-alanine dehydrogenase system. The result is that the impure flavin adenine nucleotide is active as the coenzyme of d-alanine dehydrogenase while further purified nucleotide preparations, showing a ratio of 1 : 1 between flavin ester and adenine, were no longer active. *)

STRAUB 1107) from KEILIN's laboratory reports that highly active d-alanine oxidase coenzyme preparations from kidney contain appreciable amounts of flavin as shown by their optical behaviour (fluorescence).

### The Yellow Oxidation Enzyme:
("Old" Yellow Ferment = Riboflavinphosphate Proteid).

Crude yellow ferment is obtained from brewers yeast according to WARBURG and CHRISTIAN 1239) in the following manner:

LEBEDEW juice from dried bottom yeast is purified by treatment with lead sub-acetate. After removal of the excess of lead with phosphate the enzyme is precipitated at low temperature with $CO_2$ and acetone in the form of a viscous oil. The oil is dissolved in water and the acetone-$CO_2$ procedure is twice repeated. This is followed by repeated reprecipitation with methanol at 0°. The crude enzyme may be stored in dry form.

*) See also E. NEGELEIN and BRÖMEL, Bioch. Zs. 300, 225 (1939).

One of the purification procedures recommended by WARBURG and CHRISTIAN for the crude enzyme preparation consists in the removal of impurities by shaking a solution of the enzyme in NaCl solution with chloroform and octanol for 24 hours at 38°. Even then there remain appreciable quantities of foreign proteins and particularly of polysaccharides. H. THEORELL 1148) was able to remove the tenaciously associated yeast gum from the yellow enzyme by ingenious electrophoretic procedures. Inert proteins are then removed by fractionation with ammonium sulfate. Upon dialysis against ammonium sulfate-containing acetate buffer of ph 5.2 the yellow enzyme precipitates in the form of particles of the size of red blood cells and with straight edges. Some of these crystals showed double-refraction. The content in flavin phosphate was 0.61—0.64 per cent. The minimum m o l e c u l a r w e i g h t, calculated on the basis of one flavin residue per molecule, is 73,000 $\pm$ 4000. Actually, KEKWICK and PEDERSEN 553) obtained the following values upon studying THEORELL's preparation in the ultracentrifuge: 77,000 from the sedimentation equilibrium diagram and 82,800 as calculated from the sedimentation constant (5.76 $\pm$ 0.09 $\times$ 10$^{-13}$ cm/sec), from the diffusion constant (6.07 $\times$ 10$^{-7}$ cm²/sec) and from the partial specific volume (0.731).

Attempts to obtain pure yellow enzyme have also been made in KUHN's laboratory. Purification by the chloroform method of WARBURG and CHRISTIAN, electrophoresis according to THEORELL, and subsequent adsorption on aluminium hydroxide gel C $\beta$ and elution with Na$_2$HPO$_4$ yielded preparation of 50 per cent purity (KUHN and RUDY 663)). More recently, WEYGAND and STOCKER 1296) were able to prepare yellow enzyme of 100 per cent purity by the following steps: adsorption of the enzyme from a dialysed bottom yeast juice by aluminium hydroxide gel C $\gamma$ (at ph 7, in the presence of phosphate), elution with an ammonium hydroxide-diammonium phosphate mixture, repetition of the adsorption and elution procedure, precipitation of the enzyme at the isoelectric point (ph 5.2) by two volumes of saturated ammonium sulfate solution, repetition of the salting-out procedure, adsorption on aluminium hydroxide, elution and dialysis. From 30 kg. dried yeast 4.18 g. pure yellow enzyme were thus obtained.

In pure form the yellow enzyme is much less stable than the preparations of WARBURG and CHRISTIAN which contained protective colloids of polysaccharide nature. It may be stored, however, for some time at 0° under saturated ammonium sulfate solution. The enzyme shows the solubility properties of an albumin: It is soluble in distilled water and does not precipitate at the isoelectric point. Upon addition of ammonium sulfate it begins to precipitate at 55 per cent saturation. The precipitation is complete at 67 per cent saturation. The ph-stability zone, as determined in the ultracentrifuge, has a lower limit at ph 4.5 to 4 (at about 30°). Upon lowering the temperature the stability increases in this range.

The enzyme, upon elementary analysis, yields figures typical for a protein (THEORELL 1162)):

C = 51.5 %; H = 7.37 %; N = 15.9 %; P = 0.043; S = 1.0 %.

The specific rotation, [$\alpha_D$], is —30°.

The analytical study of the yellow enzyme has been extended to its c o n t e n t in a m i n o a c i d s by KUHN and DESNUELLE 640). These authors employed preparations of WEYGAND and STOCKER 1296) which had been further purified by adding ammonium sulfate to 58 per cent saturation. This treatment removed 10—15 per cent of inert protein but did not shift the content in flavin, probably because the inert proteins adsorb some of the pigment upon precipitation. The remaining chromoprotein was sharply and completely precipitated within the range from 58 to 66.6 per cent saturation.

In this range there exists a linear relationship between the logarithm of ammonium sulfate concentration and the amount of protein remaining in solution, a relationship which represents a good criterium for homogeneity of a protein (E. J. Cohn). The figures obtained upon elementary analysis for this chromoprotein were:

C = 51.34, 51.49 %; H = 7.04, 7.21 %; S = 0.48 ; N = 16.27 %.

The only significant difference between these values and those of Theorell is the sulfur value (0.48 % as compared with 1.0 %). The optical rotation was found independent of the ph from 3.8 to 10.4 and somewhat dependent on the wavelength of light used for the determination. The rotation, $\alpha_D$, at ph 7.4 was —0.27° (d = 10 cm, c = 0.70). Yellow enzyme preparations of this purity were dissociated into the colored group (riboflavin phosphate) and the bearer protein either reversibly by dialysis against 0.005 N. HCl at 0° or irreversibly by treatment with methanol. The bearer protein was hydrolyzed by boiling with dilute $H_2SO_4$ and in the hydrolysate a number of amino acids were determined colorimetrically. The values obtained were: Arginine 8.25 %,Histidine 2.75 %, Lysine 13.7 %, Tyrosine 7.75 %, Phenylalanine 5.75 %, Tryptophane 4.86 %, Cystine 0.48 %, Glutamic acid 7.1 %. In a following paper (Kuhn and Desnuelle 642)), the content in aspartic acid was estimated to be approximately 2 %. Furthermore, the three hexone bases, arginine, histidine and lysine, were isolated in form of crystalline derivatives and glutamic acid was obtained as the hydrochloride. It is interesting to note that only 20 per cent of the total sulfur could be accounted for in the form of cystine even if the improved procedure of hydrolysis with formic acid-HCl, developed for insulin by Du Vigneaud and Miller, was followed. There is therefore a strong indication that other sulfur-containing amino acids are contained in the protein. Altogether about 66 per cent of the total nitrogen content of the bearer protein have been accounted for by the amino acid analyses. According to the present status, one molecule of the bearer protein of the yellow enzyme contains, as major constituents, approximately 33 molecules of arginine, 13 molecules of histidine, 66 molecules of lysine, 40 molecules of proline, 30 molecules of tyrosine, 24 molecules of phenylalanine, 17 molecules of tryptophane, 34 molecules of glutamic acid, 12 molecules of aspartic acid and only 1 to 2 molecules of cystine. These figures are based on a molecular weight of the enzyme of 70,000.

### Reversible Dissociation of Yellow Enzyme:

In 1934, H. Theorell 1148) performed an experiment of fundamental importance and of classical simplicity: When a solution of the yellow enzyme is dialyzed against 0.02 N.HCl at 0°, the inside fluid is slowly decolorized. It can be shown that the colored group has passed the membrane and that a colorless protein has remained behind. It exhibits the properties of a denatured protein, e.g. precipitation upon shifting the ph to 7. This "metaprotein" is unable to couple with lactoflavin monophosphate. If, however, the acid is replaced by water and dialysis continued until all traces of HCl are removed, 50 to 70 per cent of the metaprotein is renatured. It is now capable of combination with the flavin phosphate; and the resynthesized complex has the same catalytic activity as the original enzyme. The combination between the prosthetic group and the protein takes place in a stoichiometric manner, as shown in the diagram, Fig. 9.

Saturation of the bearer with the prosthetic group is reached when the molecular ratio between the two components is 1 : 1 (1154)). The test system was that described by Warburg and Christian, viz. Robison ester, coferment, zwischenferment, in an oxygen atmosphere at 38°. The free, native bearer protein is considerably less stable than the whole enzyme complex.

The procedure of THEORELL for the reversible dissociation of the yellow enzyme by dialysis against HCl and subsequently against distilled water which requires several days for its completion has recently been replaced by a still simpler method yielding both components in a good yield within about one hour. WARBURG and CHRISTIAN 1257) add the same volume of saturated ammonium sulfate solution to a solution of yellow enzyme purified by electrophoresis, cool to 0° and shift the ph to about 2.8 by adding 0.1 N.HCl. The colorless precipitate thus formed contains 78 % of the protein bearer in native form while the prosthetic flavin group remains in the supernatant solution. Resynthesis is accomplished by mere remixing of the components. It may be demonstrated not only by the usual activity test with ROBISON ester as the substrate but also visually by the disappearance of the green fluorescence of the flavin component when it

is added to the protein solution under observation with filtered ultraviolet light. The resulting flavin proteid does not fluoresce.

The finding of THEORELL that the bearer protein, after acid dialysis of the enzyme, i.e. in the metaprotein stage, gives a positive test for SH-groups with nitroprusside which is abolished by the renaturation could not be confirmed by KUHN and DES-NUELLE 641). These authors conclude on the basis of titration experiments with the blue porphyrindin radical as the oxidant that neither the reversible combination of the prosthetic group with the protein nor the reversible

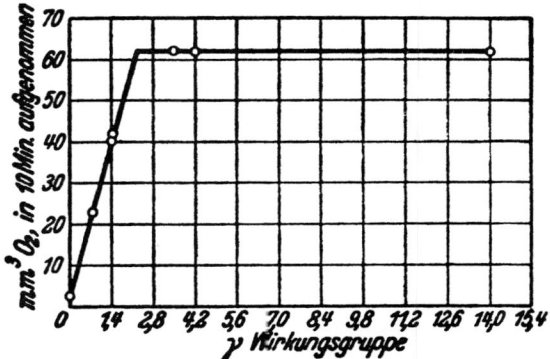

Fig. 9. Resynthesis of the yellow enzyme. Ordinate, catalytic activity in terms of $O_2$-uptake of the test system; Abscissa, amount of flavin phosphate added to constant amount of carrier protein (THEORELL 1154))

transition into the metaprotein or the irreversible heat denaturation of the protein involve the formation or disappearance of sulfhydryl groups.

Of the synthetic complexes of the type of the yellow enzyme, that prepared with the aid of synthetic riboflavin monophosphate has already been mentioned (p. 189). An analogous, catalytically active, complex has been prepared by coupling synthetic l-arabo-flavin with the natural protein component (KUHN et. al. 669)). Just as araboflavin is less active as vitamin $B_2$ than is riboflavin, the artificial enzyme complex also shows an enzymatic activity somewhat inferior to that of the natural yellow enzyme. KUHN and RUDY 664) claim that free riboflavin, if employed in excess, is also able to form an active complex with the protein which is subject to dissociation even in neutral solution.

### Link between Prosthetic Group and Bearer Protein:

In contrast to free riboflavin or riboflavin monophosphate the yellow enzyme does not show a fluoresence (THEORELL (l.c. 116)). The fact that salt formation at the NH-group of the alloxazine nucleus in 3-position will quench the green fluorescence of free flavins induces KUHN and BOULANGER 636) to assume that this imino group is one of the places of attachment of the prosthetic group to the protein in the enzyme. This is supported by the observation that 3-methyl-lactoflavin is devoid of vitamin $B_2$-activity. This is to be interpreted as due to the inability of the substituted flavin to combine with the specific protein in the animal organism. The third argument adduced by KUHN and BOULANGER in favor of this hypothesis is the change in the oxidation-reduction potential of the free flavin phosphate upon combination with the protein (see below). The much higher affinity of the phosphorylated flavin for the protein as compared with the free flavin points to the acid radical as a second bridge between the colored group and the carrier. KUHN suggests the accompanying formula for the yellow enzyme.

The affinity of non-phosphorylated flavin, although small, for the protein and the occurrence of this free form in biological sources, e.g. in the retina (see EULER et al. **267)**, GYÖRGY and KUHN **429)**, KUHN et al. **664, 676, 427, 428))**, suggest the possibility that there may exist also in nature flavoproteins with the imino group in 3-position in the alloxazine nucleus as the only place of attachment. They are probably of secondary importance.

Experiments designed to characterize the link between colored group and protein by the ability of the yellow enzyme to de-ionize silver ions which it shares with saccharase (EULER and SVANBERG) have thus far been unsuccessful (KUHN and DESNUELLE **643))**.

Yellow Enzyme
(acc. to KUHN and BOULANGER).

**Absorption Spectrum:** The absorption spectrum of the yellow enzyme in the visible region was first determined by WARBURG and CHRISTIAN **1246)**. They found an absorption band at 380 m $\mu$ ($\beta$-band) and one at 465 m $\mu$ ($\alpha$-band). After obtaining the enzyme in pure form, THEORELL **1161)** extended the measurement into the short-wave ultraviolet. Here the enzyme shows the typical protein absorption at 270 m $\mu$. Although the absorption bands in the visible and in the long-wave ultraviolet are due to the flavin group of the enzyme, the corresponding bands of the free riboflavin monophosphate (and also riboflavin itself and the photoderivative, lumiflavin) are shifted about 20 m $\mu$ towards the blue end of the spectrum. The absorption maxima of lactoflavin (riboflavin), according to Kuhn et al. **646, 647)**, are at 220, 270, 365, and 445 m $\mu$. THEORELL places the steep band in the short-wave ultraviolet at 290 m $\mu$. In any event, this maximum is almost completely masked in the yellow enzyme by the superimposed protein band. In alkaline solution the peaks of the $\alpha$- and $\beta$- band of lactoflavin are shifted towards the red (4 m$\mu$), the same being true for lactoflavin monophosphate (8 m $\mu$); at the same time, the green fluorescence exhibited by both compounds at neutral solution disappears (KUHN; RUDY **1010)**.

The fluorescence emission spectrum of lactoflavin has recently been measured by EYMERS and VAN SCHOUWENBURG **329)**. The absorption spectra of allo-

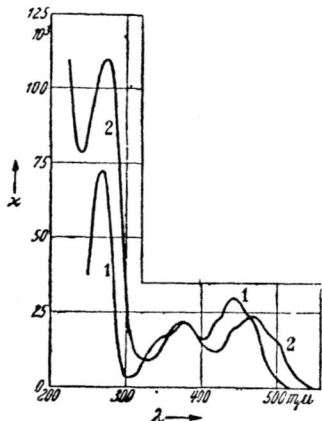

Fig. 10. Absorption Spectra of the Yellow Enzyme (2) and of Riboflavin (1) acc. to RUDY **1010)**.

xazine and of related compounds have been studied by HOLIDAY and STERN 490).

**Oxidation-Reduction Potentials:** In their first communications on the yellow enzyme, WARBURG and CHRISTIAN 1240, 1241) stated that it is capable of reversible reduction and oxidation. The same had been observed by SZENT-GYÖRGYI 1127) with respect to cytoflav which is identical with flavin phosphoric acid. Subsequently it has been shown that all compounds of this class, i.e. not only the enzyme and the flavin group, but also the photoderivative, lumiflavin (photoflavin) and even the simple alloxazines may be reduced by suitable reductants to colorless leuco compounds and reoxidized to the original form. KUHN and WAGNER-JAUREGG 675) made the observation that when lactoflavin (and also alloxazine) is reduced in strongly acid solution, a red intermediate with an absorption band at 490 m$\mu$ is formed in the course of the transformation of the yellow, oxidized, to the colorless, leuco- or dihydroform. They suggested that this red intermediate is a semiquinoid radical of the type extensively studied by MICHAELIS and his associates (see p. 100).

Potentiometric tritrations, performed with purified but still somewhat impure flavin preparations from liver, urine, and malt, confirmed the reversibility of the oxidation-reduction process and indicated a normal potential ($E'_o$) at ph 7 of about —0.200 V. (STERN 1079)). Subsequently, crystalline photoflavin obtained from liver and from yeast (the latter furnished by WARBURG and CHRISTIAN) was subjected to a potentiometric study (STERN 1080)). The normal potential, $E'_o$, at ph 7.0 was found at —0.227 V. While in neutral and alkaline solutions the titration curves are similar to those of systems with an electron number of 2, the slope of the curves becomes increasingly steeper with decreasing ph. At very low ph-values a break in the titration curves, at the point where the red intermediate of KUHN and WAGNER-JAUREGG exists, indicates that here the reduction and oxidation are taking place in two distinct steps each involving the exchange of one electron. Above ph 2, considerable overlapping of the two individual steps occurs; but nowhere is the index potential exactly that of a two-electron system, indicating that throughout the entire ph-range there ought to exist a certain, if only small, concentration of the semiquinoid radical.

The normal potential of lactoflavin at ph 7 was given by KUHN and MORUZZI 654) as —0.21 V. (extrapolated value).

When both the natural and the synthetic members of the flavin group became available in pure form and in sufficiently large amounts, comprehensive potentiometric studies were conducted both in KUHN's and in MICHAELIS' laboratory. The position of the normal potentials with varying ph was determined with greater accuracy and the effect of changes in the molecule, such as substitution in the benzene, pyrimidine and pyrazine ring, was studied. The reader is referred to the papers by KUHN and BOULANGER 636) and by MICHAELIS et al. 822) as to the details of their results. The normal potential, $E'_o$, at ph 7.0, of lactoflavin (riboflavin), natural and synthetic riboflavin phosphoric acid was found between —0.185 and —0.191 V. (KUHN and BOULANGER 636)). It was confirmed by both groups of workers that the red intermediate formed upon reduction in acid solution has the properties of a semiquinoid radical. Furthermore, MICHAELIS et al. 822) could show that the color of the radical in the neutral ph range is olive green as contrasted to the red color of the cationic radical. The most important finding in these studies was that of KUHN and BOULANGER 636) that the potential of the yellow enzyme is much more positive than that of the free riboflavin or of the phosphoric acid ester: $E'_o$ at ph 7.0 was found between —0.059 and 0.066 V., i.e. about 120 millivolts more positive than that of riboflavin alone. The combination of the negative riboflavin system with the protein bearer thus creates a redox system of a potential in the methylene blue range. Furthermore, the protein renders the reactivity of the flavin radical more specific.

The value given above for the normal potential of the yellow enzyme was determined at 38° because the establishment of equilibrium potentials was too sluggish at 20°. KUHN and BOULANGER find in the case of 6, 7, 9-trimethylalloxazine that the potentials between 0° and 38° are proportional to the absolute temperature ($E'_o$ at ph 6.9 was —0.181 V. at 0°, —0.201 V. at 20° and —0.216 V. at 38°) *).

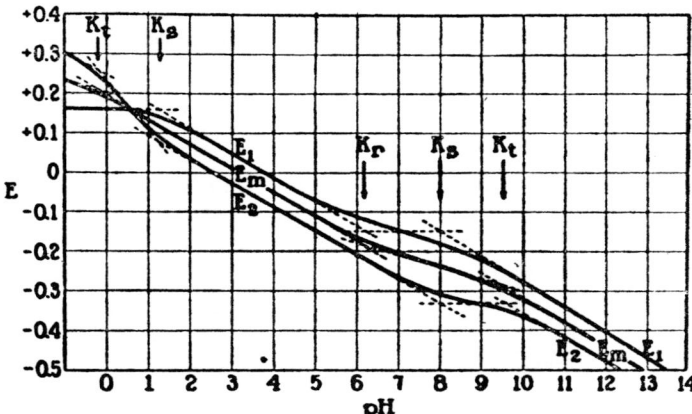

Fig. 11. Plot of the normal potentials of riboflavin against ph. $E_1$, normal potential of mixture of the fully reduced and of the semiquinoid form; $E_2$, normal potential of mixture of the semiquinoid and the fully oxidized form; $E_m$, normal potential of mixture of fully reduced and of fully oxidized form; $K_t$, dissociation constants of oxidized form, $K_s$, dissociation constants of reduced form (MICHAELIS, SCHUBERT and SMYTHE 822)).

The dependence of the normal potentials of riboflavin on ph is illustrated in Fig. 11 (for theoretical treatment see p. 100).

KUHN and STRÖBELE 671) have prepared quinhydrone-like derivatives of riboflavin (rhodoflavin, verdoflavin, chloroflavin) which differ with regard to their level of oxidation, i.e. where the ratios between quinoid and hydroquinoid form are different from 1 : 1. MICHAELIS and SCHWARZENBACH824) have recently reinvestigated the potentials of riboflavin in solution and they find that only the free semiquinoid

radical and its bimolecular dimerisation product but no other species of intermediary oxidation level are formed.

### Other Alloxazine Proteids.

That the original "yellow enzyme" and alanine oxidase are not the only alloxazine proteins is borne out by a preliminary report by BALL 59) from WARBURG's laboratory stating that xanthine oxidase, too, belongs to this group. Purified solutions of xanthine oxidase from whole milk have a strong golden brown color and an absorption band between 400 and 500 m $\mu$. This band disappears if hypoxanthine or xanthine is added under anaerobic conditions. The original color is restored by air. The difference spectrum between the reduced and the oxidized form has two bands at 370 and 465 m $\mu$, i.e. there where the yellow enzyme has specific light absorption. The prosthetic group of the enzyme may be split off by acid or alcohol. The colored solution thus obtained has two bands at 450 and 375 m$\mu$, similar to the bands of free riboflavin, and it exhibits also other properties of flavin preparations. Thus, it may be converted to lumiflavin (or photoflavin) by irradiation in alkaline solution. While it has not yet been possible to effect a quantitative cleavage between prosthetic group and native protein pheron,

---

    *) With regard to the interaction of the yellow enzyme with pyridine coenzymes the notes by HAAS 430b) and by KENNER 561a) should be consulted.

a partial separation has been accomplished. The protein part so obtained shows a 3—4 fold increase in activity upon addition of a solution of the colored group. The latter cannot be replaced by riboflavin mono phosphate or by the flavin-adenine-dinucleotide obtained from alanine oxidase. Its constitution is as yet unknown excepting the demonstration that it contains a flavin component. It is not surprising, therefore, that several years ago GREEN and DIXON 418a) were unable to find an activating effect of riboflavin on relatively impure xanthine oxidase preparations when molecular oxygen was used as the acceptor. They concluded that no flavin or yellow enzyme is required for the reaction of this dehydrogenase with molecular oxygen. The findings of BALL stress the need for exercizing caution in drawing conclusions from experiments with systems the chemical composition of which is unknown.

Recent publications from WARBURG's laboratory make it clear that there exists in Nature a whole class of "yellow enzymes" or, better, alloxazine proteids which differ with respect to the constitution of their prosthetic group as well as that of the bearer protein and which show a well-defined specificity. A "crossing-over" of the prosthetic groups or of the pherons in the test tube will produce synthetic enzymes of equally well-defined properties.

For purposes of classification, WARBURG, in a recent review (1235)), has suggested to index the various enzymes with the aid of the donators and acceptors with which they react under the conditions of the in vitro tests. WARBURG and CHRISTIAN 1257) now propose to furnish the protein component with the same index as that given to the active conjugated protein from which it is obtained by splitting off the prosthetic alloxazine group. They suggest, accordingly, the following names for the yellow enzymes and their proteins thus far isolated in WARBURG's institute. (Table 13.):

### TABLE 13.

| Protein | Enzyme | Other Names |
|---|---|---|
| 1.— Protein $O_2$, Dihydropyridine | Alloxazine Proteid $O_2$, Dihydropyridine | Original Yellow Enzyme |
| 2.— Protein $O_2$, Dihydropyridine | Alloxazine-Adenine-Proteid $O_2$, Dihydropyridine | — |
| 3.— Protein (Methylene blue), Dihydropyridine | Alloxazine-Adenine-Proteid (Methylene blue), Dihydropyridine | — |
| 4.— Protein $O_2$, Amino Acids | Alloxazine-Adenine-Proteid $O_2$, Amino Acids | d-Amino Acid Oxidase |
| 5.— Protein $O_2$, Xanthine | Alloxazine-Adenine-Proteid $O_2$, Xanthine | Xanthine Oxidase (SCHARDINGER Enzyme) |

Enzyme No. 1. is the original yellow enzyme, now called the "old" yellow enzyme by WARBURG. Enzyme No. 2. differs from enzyme No. 1. only with regard to the prosthetic group which in the first case is a simple alloxazine nucleotide and in the other case is an alloxazine adenine dinucleotide. The latter is actually the prosthetic group of d-alanine oxidase (Enzyme No. 4) (WARBURG and CHRISTIAN 1256)). Enzyme No. 2 has not yet been found in nature but is a synthetic product obtained by WARBURG and CHRISTIAN

**1257)** by coupling the native bearer protein of the original yellow enzyme with the prosthetic group of the alanine oxidase. In both instances the alloxazine nucleus may be made to react with dihydropyridine. The reduced (dihydroalloxazine) form of both enzymes reacts with about the same rate with molecular oxygen. The positions of the long-wave absorption bands are:

|                  | Prosthetic Group | Whole Enzyme |
|------------------|------------------|--------------|
| Enzyme No. 1. . . . . . . | 375, 450 m $\mu$ | 380, 465 m $\mu$ |
| Enzyme No. 2. . . . . . . | 375, 450 m $\mu$ | 380, 465 m $\mu$ |

The positions of the absorption bands of the two enzymes are not only identical but the linking of the protein to the prosthetic group will cause the same shift in both cases.

Enzymes No. 2. and 3. have the same prosthetic group, viz. alloxazine adenine dinucleotide, but different protein entities. Enzyme No. 3. has actually been isolated from brewers yeast (Strain R of SCHULTHEISS-PATZENHOFER BREWERY, Berlin) by HAAS **432)**. It differs from the synthetic enzyme No. 2 by the fact that it reacts only very slowly with molecular oxygen but rapidly with methylene blue. In other words, Enzyme No. 3 is only slightly autoxidizable. Like Enzymes No. 1 and 2, Enzyme No. 3 is able to react specifically with dihydropyridine (reduced pyridine phosphonucleotides).

Enzymes No. 3 and 4 too, differ only with respect to the bearer protein. While Enzyme No. 3 reacts only with dihydropyridine, Enzyme No. 4 reacts only with amino acids. The constitution of Enzyme No. 5. does not yet appear to be established with complete certainty (see BALL **59)**). According to STRAUB et al. (p. 227) the diaphorase, too, is an alloxazine proteid (See also STRAUB, Nature **143**, 76 (1939)).

### b) Co-Carboxylase and Carboxylase.

The relationship between cocarboxylase and carboxylase is comparable to that between lactoflavin phosphoric acid and the yellow enzyme: The combination of the coenzyme with a specific bearer protein represents the complete carboxylase system. And, just as in the case of the yellow enzyme, the coenzyme or prosthetic group is the phosphoric acid ester of an essential food factor, in this instance of vitamin B₁. Moreover, it appears that when this vitamin phosphate is combined with other specific proteins enzyme systems of somewhat different character are formed.

#### Chemistry of Vitamin B₁: *)

Isolation: After extensive preliminary work by FUNK, SEIDELL and WILLIAMS, JANSEN and DONATH **502)**, in 1926, were able to isolate from rice polishings a crystalline vitamin B₁ preparation of high biological activity. The compound analyzed as $C_6H_{10}N_2O$. Subsequently, WINDAUS et al. **1341)** isolated a crystalline vitamin preparation from autolyzed press yeast. They obtained at best 250 mg. vitamin from 100 kg. yeast, representing a yield of 10 per cent. Although their material appeared to be essentially identical with that of JANSEN and DONATH, they found it to contain sulfur; and the for-

---

*) Recent reviews: R. GREWE **424)**; C. R. ADDINALL **5)**; R. R. WILLIAMS **1332, 1332a)**; OPPENHEIMER's "SUPPLEMENT", p. 1419—1424.

mula best fitting their analyses was $C_{12}H_{17}N_2OS$. In 1934, WILLIAMS and his associates (1332)) were able to raise the yield of vitamin from yeast to 20 per cent of the total (500 mg. per 100 kg. yeast). Rice polishings yielded 25 per cent of their vitamin content (5 mg. per ton). The highest yield so far obtained is that secured by KINNERSLEY, O'BRIEN and PETERS 563a) by a modification of their original method: 100 kg. yeast gave 600 mg. of the vitamin.

The principal steps in the method worked out by WILLIAMS and put in large scale operation by MERCK consist in adsorption of the vitamin from the crude extracts by Fullers earth, elution with quinine, removal of adventitious matter by benzoylation in soda-alkaline solution and chloroform extraction, precipitation of the vitamin by silver nitrate, barium hydroxide, and phosphotungstic acid. Finally the vitamin chloride hydrochloride is obtained in pure form by recrystallization from HCl-alcohol. In this manner beautiful, colorless monocline needles with oblique ends and mostly arranged in rosettes are obtained, having a melting point of 255° (corrected) with decomposition. The empirical formula of the hydrochloride is $C_{12}H_{18}ON_4SCl_2$. Both chlorine atoms are bound in ionogen form. Various names have been proposed for the vitamin, such as "aneurin" by JANSEN, "catatorulin" by PETERS, and "thiamin" by WILLIAMS. None of these terms has as yet achieved general recognition.

Properties: In agreement with its salt character, vitamin B$_1$ is readily soluble in water, appreciably soluble in methanol, little soluble in ethanol or acetone, insoluble in ether or benzene. The isoelectric point is at about ph 9.2. It is optically inactive. The aqueous solution is most stable at ph 6.5. While the vitamin is rather heat-stable in acid solution, it is readily destroyed in alkaline solution (ph 9) when heated to 100°. The free vitamin base may be obtained in amorphous form by treatment with silver oxide. By concentrated HCl it is converted into a compound of the formula $(C_{12}H_{16}N_3SOCl)Cl_2$ which indicates the presence of an aliphatic hydroxyl in the vitamin. It interacts slowly with nitrous acid with the formation of nitrogen. Alkali at higher temperatures liberates $H_2S$ without changing the nitrogen content. The reaction which has been most important for the elucidation of the structure of the substance is the scission by sulfite at ph 5 and at room temperature according to the equation

$$C_{12}H_{18}ON_4SCl_2 + H_2SO_3 \rightarrow C_6H_9ONS + C_6H_9O_3N_3S + 2\ HCl$$

which was established by WILLIAMS (l.c. 1332), following his observation that vitamin extracts from rice polishings are readily inactivated by sulfurous acid. The vitamin is not attacked by benzoyl chloride in soda alkaline solution nor by ozone or tetranitromethane. The inertia towards the last two reagents indicates the absence of ordinary double bonds.

Chemical Constitution: Upon oxidation of vitamin B$_1$ with nitric acid two crystalline decomposition products are obtained (WINDAUS et al. 1341)), an ethyl ester, $C_7H_{11}O_5N_3$, and an acid, $C_5H_5O_2NS$. Like the intact vitamin the latter substance liberates 1 mol $H_2S$ when heated in alkaline solution. This finding seemed to indicate the presence of a sulfhydryl group in the vitamin molecule. However, WILLIAMS 1333) was able to show that not only thiols but also compounds containing sulfur in heterocyclic linkage may give rise to $H_2S$ under these conditions. He identified the acid $C_5H_5O_2NS$ as 4-methylthiazole-5-carboxylic acid:

$$\begin{array}{ccc} N & \!\!\!\!-\!\!\!\!- & C\!-\!CH_3 \\ \| & & \| \\ HC & & C\!-\!COOH \\ & \diagdown S \diagup & \end{array} \qquad \text{I.}$$

The same acid is obtained by oxidation with nitric acid of the basic split product, $C_6H_9ONS$, arising in the sulfite scission of the vitamin (see equation above). This base must therefore represent a thiazole derivative carrying a methyl group at carbon atom 4 and an hydroxyethyl residue, $C_2H_5O$, at carbon atom 5. Since the vitamin is optically inactive, the hydroxyl group is probably in β-position. The formula of the base,

$$
\begin{array}{c}
\text{N}\!\!-\!\!-\!\!-\!\!\text{C}\!\!-\!\!\text{CH}_3 \\
\parallel \qquad \parallel \\
\text{HC} \qquad \text{C}\!-\!\text{CH}_2\!-\!\text{CH}_2\text{OH} \quad \text{II.} \\
\diagdown\,\text{S}\,\diagup
\end{array}
$$

was subsequently confirmed by synthesis (CLARKE and GURIN 172)). The mild conditions under which this 4-methyl-5-hydroxyethyl-thiazole is formed from the vitamin leave little room for doubt that the vitamin contains a thiazole ring system. It could be shown that the free OH-group of this base does not arise in the course of the sulfite scission but that it exists already in the intact vitamin.

The next step was to ascertain the manner in which this thiazole system is linked up with the remainder of the vitamin molecule. Inasmuch as the OH-group has to be ruled out as this point of attachment, the heterocyclic N-atom of the thiazole base would appear to be the most likely participant in this link. It is obvious that the attachment of a residue R to the tertiary N-atom of the thiazole base II will yield a quaternary ammonium salt (III) in accordance with the scheme:

$$
\text{R—X} + \text{N} \!\!<\!\!
\begin{array}{c}
\text{CH}_3 \\
| \\
\text{C}\!=\!\!=\!\text{C—CH}_2\!\cdot\!\text{CH}_2\text{OH} \\
| \\
\text{CH}\!-\!\!-\!\!\text{S}
\end{array}
\;\leftrightarrows\;
\begin{array}{c}
\text{X} \quad \text{CH}_3 \\
\backslash \quad | \\
\text{R—N}\!\!<\! \text{C}\!=\!\!=\!\text{C—CH}_2\!\cdot\!\text{CH}_2\text{OH,} \\
| \\
\text{CH}\!-\!\!-\!\!\text{S}
\end{array}
$$

II                                III

where X is an electronegative residue, e.g. halogen.

Among the evidence favoring a configuration of the vitamin of the kind shown in formula III the following facts may be mentioned: In quaternary ammonium salts of the type III the components would be expected to form a reversible equilibrium system. Both the instability of the vitamin in aqueous solution as demonstrated by the changes in the absorption spectrum (see below) and the ease of the sulfite cleavage of WILLIAMS suggest such a thiazolonium salt structure for the vitamin. The iodomethylate of compound II (X = I, R = CH$_3$), prepared by CLARKE and GURIN 172), proved to be a good model for the vitamin with regard to its behaviour towards bromine and also NaOH, thereby further strengthening the formulation of the vitamin as a thiazolonium compound. The stability of the model towards sulfite is no hindrance because in contrast to the vitamin the equilibrium between the components of the quarternary ammonium salt, indicated in the last equation, leads to soluble cleavage products. The sulfite cleavage of the vitamin goes nearly to completion only because one of the products is constantly removed from the reaction system owing to its insolubility (compound $C_6H_9O_3N_3S$ in equation p. 199). The thiazolonium hypothesis received further support from the form of the electrometric acid-base titration curve of the vitamin (WILLIAMS and RUEHLE 1334a)): 2 molecules of alkali are used by one vitamin molecule besides the 1 mole base required for neutralization of the HCl present in the hydrochloride. After the titration (end-point, ph 9), the vitamin may be regenerated by the addition of acid. The titration process may be satisfactorily interpreted as the primary formation of a carbinol base followed by the reversible opening of the thiazole ring with the formation of the sodium salt of a formyl compound *).

If it be accepted that the vitamin is indeed a thiazolonium salt, there remains the question as to the nature of the second part of the molecule combined with the thiazole half through the N-atom. It has been mentioned above that the nitric acid breakdown of the vitamin yielded an ethylester $C_7H_{11}O_5N_3$ besides the methylthiazole carboxylic acid; and that the sulfite cleavage yielded an insoluble acid, $C_6H_9O_3N_3S$, aside from the substituted thiazole base. The spectrographic comparison of the former with imidazole and pyrimidine derivatives clearly indicated the presence of a pyrimidine ring in the ethyl ester. The same is true for the acid of WILLIAMS, which is a sulfonic acid of the empirical formula, $C_6H_8N_3$—$SO_3H$. The stability of this substance defeated attempts to elucidate its precise structure by further breakdown. New hopes were roused when WINDAUS, et al. 1341) were able, by careful permanganate oxidation, to isolate a base, $C_6H_{10}N_4$, as a reaction product the character of which pointed to its origin from the still unknown pyrimidine portion of the vitamin molecule. While originally the constitution of a diamino-dimethyl-pyrimidine (IV) was attributed to this base, later work and particularly the discovery of thiochrome (see below) have led to a revision of this formula in the manner indicated in formula V:

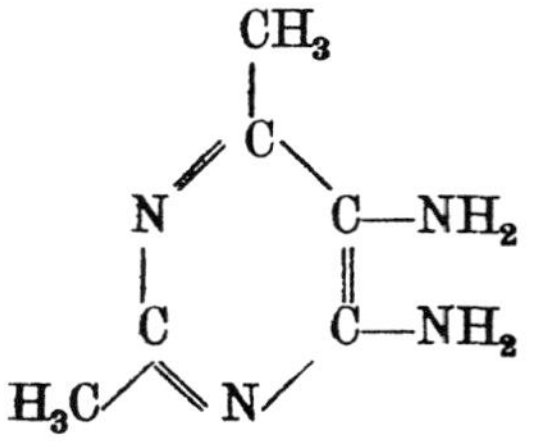

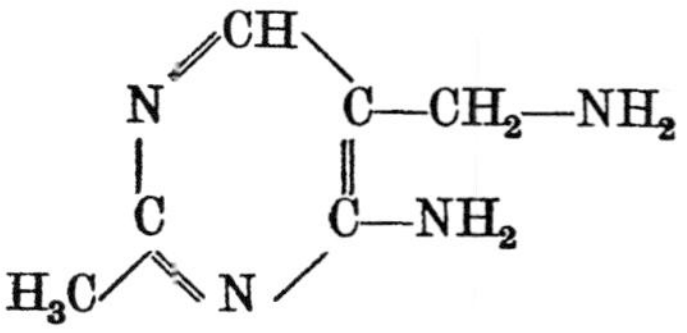

The compound V. has been synthesized by several workers (ANDERSAG and WESTPHAL 23); GREWE 422); TODD and BERGEL 1179)). It proved to be the key for the final elucidation of the structure of the vitamin. The NH$_2$-group in the side chain of the diamine is derived from the thiazole ring, while the CH$_2$-group is the bridging link between the two parts of the molecule. This leads to the structural formula VI for the vitamin:

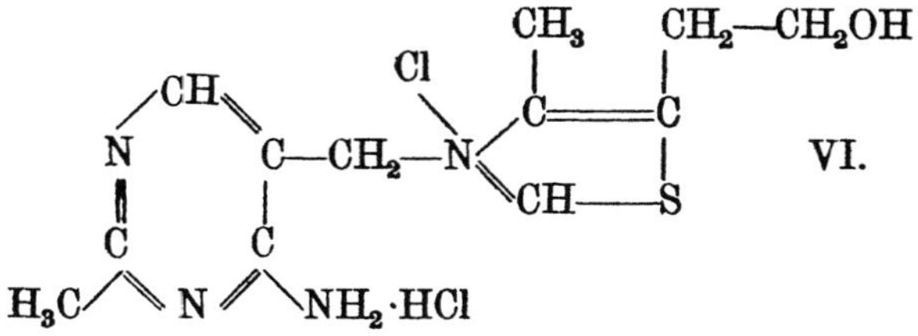

The first to postulate a methylene bridge between the thiazole and the pyrimidine portion of the molecule on theoretical grounds were MAKINO and IMAI 748, 500). Shortly afterwards WILLIAMS 1331), too, arrived at the correct formulation of the vitamin.

Synthesis: The synthesis of vitamin B$_1$ has afforded ultimate proof for the correct-

---

*) For a fuller discussion of the titration curve the paper by WILLIAMS and RUEHLE 1334a) and the review by GREWE 424) should be consulted.

ness of the structural formula. It was accomplished independently by ANDERSAG and WESTPHAL 23), WILLIAMS 1334) and TODD and BERGEL 1179). The synthesis of WILLIAMS (cf. CLINE, WILLIAMS and FINKELSTEIN 173)) consists in the condensation of 2-methyl-5-bromomethyl-6-aminopyrimidine hydrobromide with 4-methyl-5-β-hydroxyethyl-thiazole by heating in butanol and subsequent exchange of the bromine for chlorine with AgCl. The individual steps are shown in the following flow chart.

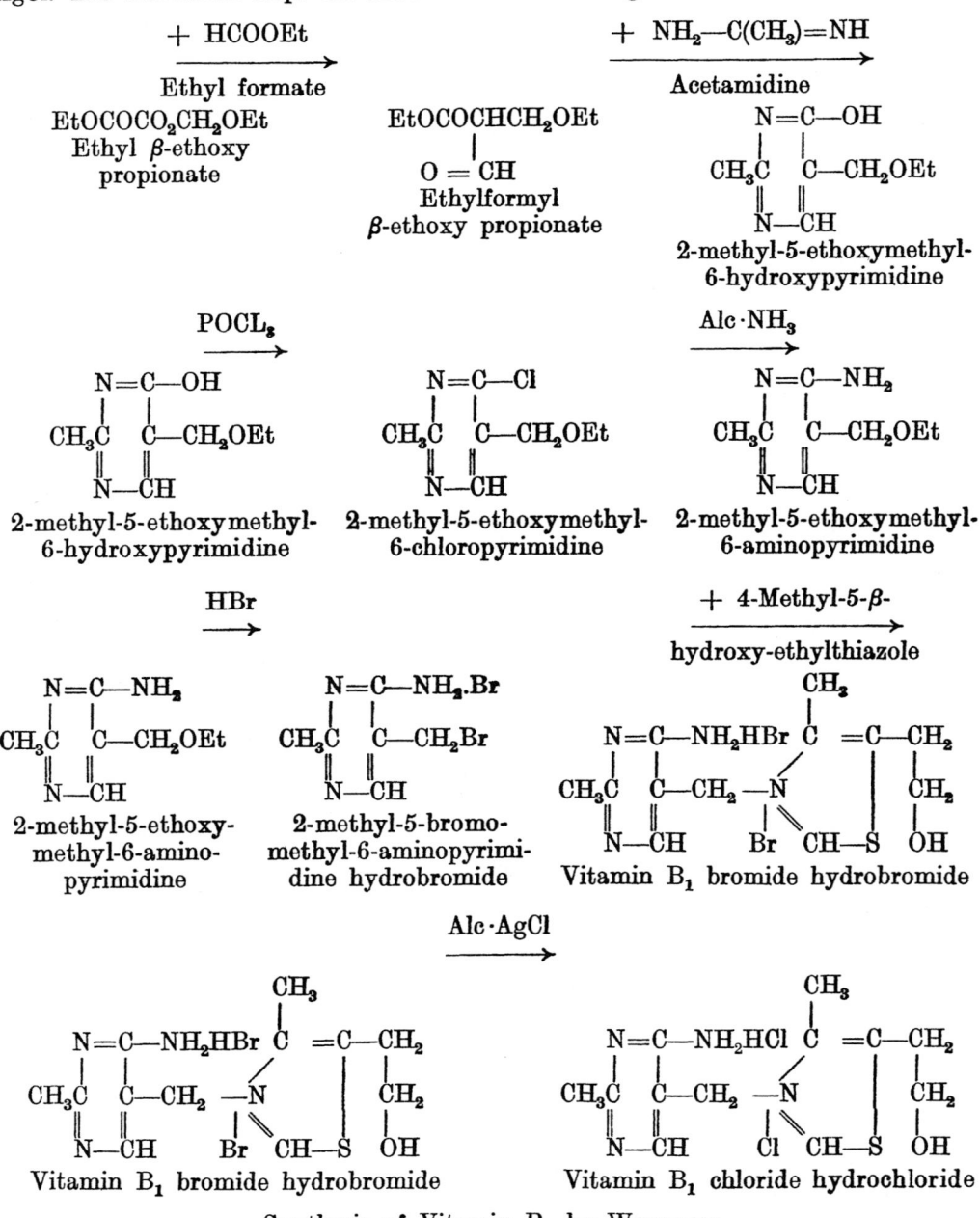

Synthesis of Vitamin B₁ by WILLIAMS.

The procedure developed by ANDERSAG and WESTPHAL 23) is based on the condensation of 2-methyl-4-amino-5-bromomethyl-pyrimidine dihydrobromide with 4-methyl-5-hydroxyethyl-thiazole by heating for 30 mins. at 120—130° and converting the resulting vitamin B$_1$ hydrobromide to the corresponding hydrochloride through the picrate by treatment with HCl and recrystallization from dilute alcohol. The third method which was found independently by ANDERSAG and WESTPHAL and by TODD and BERGEL 1179) depends on the condensation of 4-amino-5-thioformamido-methyl-2-methylpyrimidine with α-chloro-γ-hydroxypropyl ketone or one of its derivatives. The synthesis of vitamin B$_1$ has made possible its industrial preparation in large amounts.

Oxidation of Vitamin B$_1$: Thiochrome: In 1935, KUHN, WAGNER-JAUREGG, VAN KLAVEREN and VETTER 677) isolated a yellow pigment from yeast which was distinguished by a strong blue fluorescence in neutral or alkaline solution. The empirical formula, $C_{12}H_{14}ON_4S$, suggested a close relationship to vitamin B$_1$. They found thiochrome to be reversibly reducible and reoxidizable. Shortly afterwards, BARGER, BERGEL and TODD 69) were able to prepare thiochrome by oxidizing vitamin B$_1$ in alkaline solution with potassium ferricyanide. A number of other oxidizing agents, e.g. $H_2O_2$, selenium dioxide, permanganate, porphyrindin, and porphyrexide will transform the vitamin to a greater or smaller extent into thiochrome (KUHN and VETTER 673)); and even molecular oxygen will do so provided that alcoholic or alkaline aqueous solutions of the vitamin are exposed to the atmosphere for a sufficient length of time (KINNERSLEY, O'BRIEN and PETERS 564)). The structure of thiochrome, shown in formula IX, was first suggested on theoretical grounds by MAKINO and IMAI 748) and later proved by synthesis by TODD, BERGEL, FRAENKEL-CONRAT and JACOB 1180). These workers brought the dichloropyrimidine VII into reaction with the aminothiazole derivative VIII which in turn had been prepared from the corresponding chloroketone and thiourea:

VII.

VIII.    →    IX. Thiochrome.

KUHN and his colleagues (677, 673)) consider it not impossible that the thiochrome which they obtained from yeast might have been formed from a precursor, presumably vitamin B$_1$, in the course of the preparation. While it is thus possible that thiochrome is only an artefact, it has acquired considerable importance as a specific derivative of vitamin B$_1$ into which the latter is converted for purposes of assay. This method which has been developed mainly by JANSEN and his associates WESTENBRINK and GOUDSMIT, while important for the science of nutrition, is outside the scope of this monograph.

Reduction of Vitamin B$_1$: LIPMANN 721, 726a) reports that vitamin B$_1$, dissolved in phosphate buffer of ph 7.5, will absorb 0.94 mol $H_2$ per mol vitamin in the presence of platinum black. Sodium bisulfite, $Na_2S_2O_4$, too, reduces the vitamin in bicarbonate solution, at ph 7.5 and 25°. The reduction process, resulting in the formation of dihydrovitamin (X), is tentatively formuled as follows:

VI.
Vitamin $B_1$

X.
Dihydrovitamin $B_1$

Furthermore, a "very distinct green yellowish colour is stated to appear when vitamin $B_1$ (synthetic, I. G. FARBEN IND.), in a 0.5—1 per cent solution, is reduced with hyposulfite" (LIPMANN 723)). The same transient green color was observed by LIPMANN when the vitamin is reduced with zinc dust in normal hydrochloric acid. LIPMANN suggests that the colored intermediate may be a half-reduced thiazole (cf. ERLENMEYER et al. 255)) and thus a semiquinoid radical of the kind studied by MICHAELIS. The reduction could not be shown to be reversible (726a) *).

Absorption Spectrum: The absorption spectrum of the vitamin, when measured in dilute HCl (0.005 N), exhibits a characteristic maximum at 247 m$\mu$ (See Fig. 12). PETERS and PHILPOT 925)). If, on the other hand, the spectrum of the vitamin hydrochloride is measured in water as the solvent two maxima are found which may be separated as much as 50 m$\mu$, the definition and separation being somewhat dependent upon the preceding treatment of the vitamin sample and the technique employed in the

---

*) Mr. J. W. HOFER, in this laboratory, failed to obtain evidence for the reversibility of the reduction of vitamin $B_1$ by hyposulfite by reductive potentiometric titration. Mr. J. L. MELNICK was able to confirm LIPMANN's observation concerning the reduction of vitamin $B_1$ by activated hydrogen as carried out in WARBURG-BARCROFT manometers. The reduction by hyposulfite fell somewhat short of the extent reported by LIPMANN (2.76 mols. of acid are stated to be produced in bicarbonate solution, 2 mols. of which are ascribed to the oxidation of the hyposulfite by the vitamin while the additional acid is supposed to be derived from the reduction process itself). Color tests for vitamin $B_1$ and the behaviour towards ferricyanide in alkaline solution (thiochrome reaction) after the reduction indicated the complete absence of unchanged vitamin $B_1$. If, therefore, the dihydrovitamin should be capable of reversible reoxidation molecular oxygen seems to be an unsuitable oxidant. Experiments on $B_1$-avitaminotic pigeons indicate that the reduction product obtained from vitamin $B_1$ by hyposulfite has no antineuritic potency in amounts as high as 200$\gamma$ (effective dose of vitamin $B_1$, 10 $\gamma$). The green-yellow "intermediate" described by LIPMANN was not detectable when synthetic vitamin $B_1$ from various sources was reduced with hyposulfite in n e u t r a l solution. Only when the bicarbonate concentration was insufficient to prevent acidification by the HCl contained in the vitamin and the acid formed during the reduction, such a color could be observed (near ph 3). However, hyposulfite alone will give rise to a green-yellow tint at such hydrogen ion concentrations.

In contrast to the free vitamin, synthetic cocarboxylase (vitamin $B_1$ pyrophosphate) was found to be highly active after reduction, both in polyneuritic pigeons and in the yeast test system for cocarboxylase activity. Like the dihydrovitamin, dihydrococarboxylase is not autoxidizable. Alkaline ferricyanide converts it into thiochrome pyrophosphate.                    K. G. STERN.

measurement (Smakula 1058)). Wintersteiner, Williams and Ruehle 1342) found two bands, at 235 m $\mu$ and at 267 m $\mu$, both in alcoholic and aqueous solution *). According to Heyroth and Loofbourow 483) this change of the absorption curve with ph suggests that in aqueous solution the vitamin dissociates in smaller units and that it also suffers partial deamination. The fact that simple aminopyrimidines behave in a similar manner seems to militate against the above interpretation.

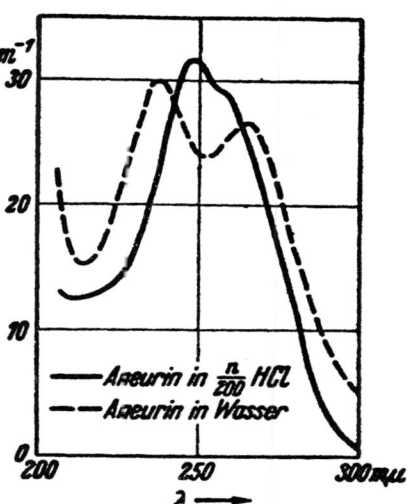

Fig. 12. Absorption Spectrum of Vitamin E₁ (from Grewe 424)).

### Co-Carboxylase:

The zymase or "holozymase" complex of yeast contains thermolabile, colloidal components as well as thermostable, low-molecular constituents. The first thermostable activator or coenzyme of this system to be discovered was the cozymase of Harden and Young (in 1904). There followed the recognition of magnesium in inorganic form as an indispensable constituent (Lohmann, 1931). One of the component enzymes of the zymase complex is carboxylase which was discovered in 1911 by Neuberg 884) and which is able to convert pyruvic acid into acetaldehyde and $\alpha$-keto carboxylic acids in general into the corresponding lower aldehydes by decarboxylation. This reaction is not affected by the presence or absence of oxygen. Whereas in the earlier stages of the work on carboxylase it was held that carboxylase did not require a coenzyme for its action, Auhagen, in 1932, demonstrated (44, 45, 47, 46)) that a coenzyme, called cocarboxylase, is necessary for the function of the enzyme. He showed that washing of dried yeast with weakly alkaline phosphate will remove a thermostable activator of carboxylase. His coenzyme preparation consisted undoubtedly mainly of adventitious material and of only a few per cent of cocarboxylase.

Auhagen's work was continued and brought to a successful conclusion by Lohmann and Schuster 737, 738) in Meyerhof's laboratory. The isolation of the cocarboxylase was accomplished as follows:

Boiled yeast juice, obtained from brewer's yeast (Löwenbräu, Munich, strain RS) is purified by treatment with alcohol. The crude cocarboxylase is precipitated from the alcohol-water solution by Ba-acetate in alkaline solution. The coenzyme is eluted from the precipitate by $HNO_3$—$H_2SO_4$ and the eluate is brought to weakly acid reaction by NaOH. Large amounts of organic impurities and inorganic phosphates are removed by precipitation with ethanol and reprecipitation with methanol. The purified coenzyme is absorbed on frankonite KL or clarit and eluted with hot dilute pyridine. After acidifying with $HNO_3$ a further fractionation is performed by precipitation with methanol-ether. Impurities are now removed in the form of their picrolonates. Following this, adenine

---

*) According to measurements by Mr. I. L. Melnick the position of the two maxima at ph 7.4 is at 235 and 265 m $\mu$. The reduced form of the vitamin, as obtained by treatment with platinum-activated hydrogen, shows two maxima, at 237 and 280 m $\mu$ at the same ph. The spectrum of cocarboxylase changes in a similar manner upon reduction.

nucleotides are precipitated by Ba (OH)$_2$, and an intensively greenish white fluorescing substance is removed as the Ag-salt. Subsequently, the cocarboxylase is precipitated as the Ag-salt after adding NH$_3$ to ph 7. The neutral silver salt is decomposed by H$_2$S and the cocarboxylase is precipitated by phosphotungstic acid from dilute HCl. Upon treatment with acetone the phosphotungstate is split into acetone-soluble phosphotungstic acid and the free coenzyme which is partly precipitated in crystalline form. The precipitate is dissolved in dilute HCl and further purified by precipitation with acetone. The crude crystals are again dissolved in HCl, precipitated with acetone, and twice recrystallized from dilute HCl with alcohol.

Crystalline cocarboxylase forms needles with a melting point of 242—244°. It is readily soluble in cold water. At neutral and acid reaction the aqueous solution is color-less while at alkaline reaction it is yellow. Upon acidifying the color disappears and the biological activity remains unaffected. From 100 kg. yeast 700 to 800 mg. crystalline cocarboxylase hydrochloride are obtainable, i.e. about 15 per cent of the amount present in the original boiled yeast juice.

Elementary analysis yields the following values:

C, 30.85 %; H, 4.53 %; N, 11.72 %; P, 13.00 %; S, 6.78 %; Cl, 7.41 %. This analysis fits best the empirical formula C$_{12}$H$_{21}$O$_4$N$_4$P$_2$SCl. Further chemical study has shown that the cocarboxylase represents the pyrophosphoric acid ester of vitamin B$_1$. The formula, as developed by LOHMANN and SCHUSTER, is:

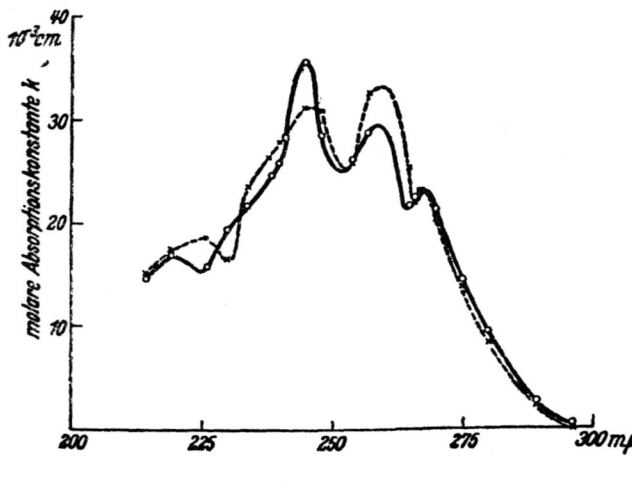

Upon sulfite cleavage a diphosphorylated thiazole and a phosphorous-free pyrimidine are obtained. Titration curves and the result of alkaline hydrolysis support the formulation of the coenzyme as an asymmetrical pyrophosphoric acid ester. Upon suitable acid hydrolysis the corresponding monophosphoric acid ester is obtained. In aqueous solution, one OH-group of the phosphoric acid radical forms a salt linkage with the primary amino group of the pyrimidine. The absorption spectrum of the pyrophosphoric acid ester (cocarboxylase) and of the monophosphoric acid ester is shown in Fig. 13.

X‒ ‒ ‒X Cocarboxylase,
O———O Monophosphoverbindung.

Fig. 13. Absorption Spectra of Cocarboxylase and Vitamin B$_1$ monophosphate (LOHMANN and SCHUSTER (738)).

Synthesis of Cocarboxylase: Cocarboxylase has been synthesized in vitro from synthetic vitamin $B_1$ and $POCl_3$ (STERN and HOFER 1093)). The amount of conversion in these experiments was rather small but quite distinct as shown by activity determinations in a test system containing washed dry yeast and pyruvic acid (see below). Higher yields may be procured by the use of pyrophosphoryl chloride (LOHMANN *)) or of sodium pyrophosphate and orthophosphoric acid (TAUBER 1143)). A procedure has been described whereby crystalline and allegedly pure coenzyme preparations may be obtained by the last mentioned method (WEIJLARD and TAUBER 1286a)). In several laboratories it has also been possible to convert vitamin $B_1$ enzymatically, at least in part, into cocarboxylase (EULER and VESTIN 320); TAUBER 1142, 1141); LOHMANN and SCHUSTER 738); KINNERSLEY and PETERS 565)). In computing the extent of conversion the recent report of OCHOA 892) from PETERS' laboratory should be taken into consideration: while vitamin $B_1$ alone is inactive as cocarboxylase, it augments the activity of the coenzyme in mixtures of the two substances.

Biological Activity of Cocarboxylase: Both the natural coenzyme and synthetic preparations will enable yeast, inactivated by washing with alkaline phosphate buffer, to decarboxylate pyruvic acid. The washed yeast contains a specific bearer protein which is supplemented by the coenzyme to form the active "holo-enzyme" or symplex. Furthermore, alkaline washed yeast regains the power to ferment sugar after adding the following substances: hexosediphosphate, magnesium, adenylpyrophosphate, cozymase and cocarboxylase (AUHAGEN 45); LOHMANN and SCHUSTER 738)). The latter authors, furthermore, find that cocarboxylase as well as the monophosphate has high vitamin $B_1$-activity when tested in pigeons; and that cocarboxylase may replace vitamin $B_1$ in PETERS' catatorulin test (pyruvate oxidation by brain tissue of avitaminotic pigeons). The possibility that vitamin $B_1$ pyrophosphate may be the coenzyme of other enzymes besides carboxylase is strongly indicated by the finding of LIPMANN 722, 726) that the respiration of alkaline washed B. Delbrückii with pyruvate as the substrate is increased by cocarboxylase and still more by cocarboxylase plus riboflavin phosphate. The reaction products are acetic acid and $CO_2$, vitamin $B_1$ has no effect. Inasmuch as no anaerobic decarboxylation takes place in this system, LIPMANN concludes that the vitamin $B_1$ pyrophosphate here plays the rôle of the coenzyme of p y r u v i c a c i d d e h y d r o-g e n a s e rather than carboxylase. This is important, because the power of mammalian tissues to decarboxylate pyruvic acid anaerobically appears to be much smaller than their capacity to oxidize pyruvic acid aerobically. More specifically, the removal of pyruvic acid by brain tissue of avitaminotic pigeons after addition of vitamin $B_1$ (catatorulin test of PETERS) has been shown to be a reaction occurring under uptake of oxygen which would not be the case with a purely carboxylatic mechanism **). While in yeast the vitamin $B_1$ pyrophosphate undoubtedly fulfills its main function as the coenzyme of carboxylase, in animal cells it may rather function as the coenzyme of pyruvic dehydrogenase. Further work in this field will be required, particularly with reference to the specific bearer proteins with which the coenzyme combines to form active enzyme systems. However, there is little room for doubt that vitamin $B_1$ is intimately connected with carbohydrate metabolism in general and with the breakdown or transformation of pyruvic acid in particular.

---

*) Private communication.
**) See the recent review by R. A. PETERS 322) and the paper by F. LIPMANN 725).

**Carboxylase: \*)**

This enzyme, consisting of a specific protein and cocarboxylase, has as yet not been obtained in pure or nearly pure state. For the purpose of testing the activity of the coenzyme it is sufficient to remove the natural coenzyme from dried yeast by treatment with alkaline phosphate; the remaining bearer protein is capable of recombining with the coenzyme within a ph-range of about 6 to 7. Although it is possible to obtain the entire enzyme complex in homogeneous solution by macerating dried yeast with water or phosphate buffer of ph 6 at 30° for 1 to 3 hours (LEBEDEW juice), most workers have reported a rapid inactivation of the dissolved enzyme. This fact above all seems to have hindered the progress of our knowledge of the protein component.

LANGENBECK and his associates have attempted a purification of the enzyme. Bottom yeast (Germania Brewery, Münster, and Hackerbräu, Munich) was worked up at low temperature. The solutions were prevented from freezing by the addition of 0.5 vol. methanol. This permitted working at —20°. Adventitious proteins were precipitated by tannin and the excess tannin was in turn removed by adsorption on hide powder. Following this, the enzyme was adsorbed on alumina gel B according to WILLSTÄTTER and KRAUT and eluted with primary phosphate. LANGENBECK, JÜTTEMANN, SCHAEFER and WREDE 703) state that a 1000 fold purification of carboxylase could be achieved in this manner, and they believe their purest preparations to be protein-free. The work was continued by LANGENBECK, WREDE and SCHLOCKERMANN 704) who perfected the low-temperature purification process with the use of methanol-water mixtures further. The earlier observation that tannic acid will remove impurities without precipitating the enzyme could not be reproduced with later samples of dried yeast. It was possible to purify the enzyme somewhat by dialyzing out impurities through cellophane at low temperature (—17 to —20°) in water-methanol mixtures \*\*).

SCHOENEBECK 1032a) has also tried to purify the enzyme. He was able to stabilize it by the addition of glycerol. AXMACHER 48a) effected a purification of carboxylase by fractional acetone precipitation.


## c) Ascorbic Acid.

Ascorbic acid (Vitamin C) was isolated independently by SZENT-GYÖRGYI 1117) from the adrenal cortex, from orange juice and cabbage and by KING from orange juice. The substance was first designated as hexuronic acid (1112)) and later renamed $l$-ascorbic acid (1121)). Its empirical formula is $C_6H_8O_6$. From a simple hexose this com-

---

\*)  For historical references on carboxylase see C. OPPENHEIMER 902, p. 1556.

\*\*)  Experiments by Mr. J. L. MELNICK, in progress in this laboratory, show that carboxylase preparations obtained from another strain of bottom yeast (FROHBERG strain, HULL BREWERY, New Haven) are not as labile as those studied by LANGENBECK. It is quite possible to handle the preparations at room temperature (better at 0°). While it has been possible to adsorb the enzyme on various types of aluminium hydroxide gels (B, C$\gamma$) the elution, even after adding glycerol, is effected only with appreciable losses. In a maceration juice, clarified by alumina gel (EIMER & AMEND), spinning in an air-driven quantity ultracentrifuge for 60 mins. at 512 r.p.s. (65,000 g at bottom of tubes, angle of inclination 25°) produced an appreciable concentration gradient with respect to enzymatic activity; the bottom layer being almost twice as active as the top layer, while the middle third layer had an activity approaching that of the original solution. By adding CO-hemoglobin to yeast juice and by comparing the relative distribution of this pigment and of the enzyme activity after ultracentrifugation the molecular weight of carboxylase was found to be about 140,000.

K. G. STERN.

pound is distinguished by a system of conjugated double bonds. Its constitution, illustrated by the following formula, has been elucidated by the efforts of HAWORTH and HIRST et al. (470)), MICHEEL et al. 829, 830, 831), OHLE 897), KARRER et al. 517) and REICHSTEIN 969). The final proof for the structure of the vitamin was furnished by its synthesis by REICHSTEIN 969) and by HIRST.

Strictly speaking, ascorbic acid is not an acid but a lactone with strongly acidic properties. The first dissociation constant, $p_{Ka1}$, is 4.17, the second, $p_{Ka2}$, equals 11.57. A third dissociation constant is larger than 14 and corresponds to the dissociation of the alcoholic OH-groups (121)). RANGANATHAN et al. 956) find that ascorbic acid, upon electrophoresis, migrates to the anode in the entire ph-range from 1.0 to 13.0, and they conclude that this points to the presence of a free carboxyl group in the vitamin. The melting point is given as 190—191° (896)) and 188° (475)). Ascorbic acid is very soluble in water. The aqueous solution shows an optical rotation of $[\alpha]_{5780\,\text{Å}}^{18°}. = +24°$ (896)). *l*-Ascorbic acid crystallizes in well-formed monocline sphenoidal crystals (Cox 186)). The absorption curve shows only one maximum in the short-wave ultraviolet region, at 265 m$\mu$, which is not affected by irradiation (BOWDEN and SNOW 140)). The fact that the peak of the absorption band is shifted towards the long-wave region as compared with that of simple hexoses, is correlated by RUDY 1010) with the existence of conjugated double bonds in the vitamin. Upon oxidation the intensity of the bands decreases rapidly, a phenomenon more pronounced in neutral solutions than in acid solutions where the vitamin is more stable. By the addition of KCN the stability is considerably increased together with the molecular extinction (RUDY 1010)). For further details concerning the properties, the determination, occurrence and elucidation of the structure of *l*-ascorbic acid and its homologues the reader is referred to the review articles by OHLE 896), MICHEEL 828) and WILLSTAEDT 1335).

d-Ascorbic acid and its homologues are of little physiological interest (cf. OHLE 896)).

The reversibility of ascorbic acid as an oxidation-reduction system has already been discussed on p. 94. The first oxidation product of ascorbic acid is dehydroascorbic acid; and it is generally agreed that it may reversibly be reduced to the fully active vitamin. At higher ph dehydroascorbic acid, in turn, shows reducing properties of even stronger character than ascorbic acid. Oxidation in this instance leads to 2,3-diketo-l-gulonic acid. The latter forms a reversible system with dehydroascorbic as the reduced form only at ph 5.5 to 6.5. At ph 5.5 where both reduced forms are stable, the redox potential, according to BORSOOK et al. 136), of the ascorbic acid ⇌ dehydroascorbic acid system is $E_0' + 0.112$ V. and of the dehydroascorbic acid ⇌ diketogulonic acid system it is $+0.015$ V. (at 35°). In accordance with the high sensitivity of ascorbic acid against heavy metals it is to be expected that the potential, especially at high ph-values, will be somewhat affected by the electrode material (FRUTON 379)). For this reason FRUTON chose the indicator method. He as well as WURMSER and MAYER 1364) and BALL 58a) affirmed anew the reversibility of the system ascorbic acid⇄dehydroascorbic acid. According to WURMSER and LOUREIRO 1360), there exist, however, two potential steps as would be postulated from the results of BORSOOK. BALL 58a) disagrees with BORSOOK on this point. In acid solution, particularly in the absence of traces of heavy metal, ascorbic acid is quite stable. At ph > 5.5 ascorbic acid suffers irreversible breakdown leading to a loss of its antiscorbutic and catalytic properties. This destruction process is not an oxidation (BORSOOK 136, 137)), since its rate is unaffected by the presence or absence of air, and since it cannot be prevented by the presence of $H_2S$ or glutathione. Alkaline solutions (ph 8.5 to 9) of ascorbic acid, however, absorb large amounts of oxygen (in the presence of heavy metal ions) (EULER, 265)). The following agents bring about a reversible oxidation of ascorbic acid: oxidation by $O_2$ in acid solution in presence of ions of heavy metals or on charcoal (350)), also within the physiological ph-range, up to ph 7.6 (KELLIE 554)), by iodine and bromine if a large excess is avoided (cf. ROE 994a)) and finally the oxidation by peroxide activated by peroxidase (TAUBER 1140)). The latter process occurs in the presence of quinones. The existence of a specific a s c o r b i c  o x i d a s e in plant extracts has been claimed, but there seems still to be the possibility that this "enzyme" is in fact copper in loose combination with unspecific plant proteins.*) Conversely, an enzyme has been found in blood which reduces dehydroascorbic acid (994a)). The reaction between methylene blue and ascorbic acid has a ph-optimum at 5—6 (EULER 265)). In all cases SH-groups either protect ascorbic acid against oxidation or they reduce the dehydro-ascorbic acid formed back to ascorbic acid (cf. TAUBER 1140)). Ascorbic acid, obtained in this manner by reduction with $H_2S$, is stable for several days under $CO_2$ at low tempera-ture (350)). In general, l-ascorbic acid exhibits rather strongly reducing properties, e.g. towards cupric ions in acid solution or dyestuffs (303)). Nitrites but not nitrates are re-duced (519)). The kinetics of the oxidative catalysis by heavy metal ions has been care-fully studied by BARRON et al. 79). The temperature coefficient of the oxidation of as-corbic acid by $O_2$ as catalyzed by $Cu^{++}$, $Q_{10}$, within the range 27 to 37°, is 1.65. Other details concerning the reaction will be found on page 67.

According to PANTSCHENKO-JUREWICZ and KRAUT 916) and MOSTERS 835), ascorbic acid is probably the agon of esterase, a claim that still remains to be proven.

*Appendix*:
**Reductones:** The reductones were discovered by WURMSER 37, 1356, 993, 1361, 1360) who described them under the name "glucide X" and later "redoxin". They were subsequently studied by EULER and his associates and renamed reductones (302)). These substances are related to ascorbic acid both in structure and properties.

WURMSER et al. observed that buffered sugar solutions will develop very negative potentials when stored under anaerobic conditions (p. 59). Furthermore, EULER and KLUSSMANN 301) found that the reaction of alkali with sugars with a free carbonyl group (aldoses and ketoses) at elevated temperatures yields strongly reducing substan-ces for which they coined the collective term reductones. EULER and MARTIUS 305) were able to isolate the reductone from glucose, g l u c o-r e d u c t o n e, in crystalline form. It has the empirical formula $C_3H_4O_3$ and a molecular weight of 88. It shows the same elementary composition as pyruvic acid or ascorbic acid. The compound decomposes be-

---

*) Details will be found in OPPENHEIMER's "SUPPLEMENT", p. 1587.

tween 200 and 220° and is distinguished by its high reducing power, particularly in acid solution. Solutions of $Ag^+$- and $Cu^{++}$-salts, permanganate or $HI_2$-I are reduced almost instantaneously, while methylene blue in phosphate buffer of ph 7.2 is rapidly reduced to the leuco base. The absorption spectrum is similar to that of ascorbic acid: the peak of the absorption band is found at 268 m$\mu$ at ph 2—4 and at 287 m$\mu$ at ph 5.7 and 10. Upon oxidation the maximum shifts slightly to 280 m$\mu$ (WURMSER 1365)). the solutions react strongly acid and lend themselves to acidimetric titration (305)). Inasmuch as the reductones do not contain free carboxyl groups, EULER 306) attributes their strongly acidic character to their particular enolic configuration. When studied manometrically in alkaline solution, the reductones will show an oxygen uptake at least equal to that of ascorbic acid (265)).

The constitution of reductone is that of an enol-tartronic dialdehyde, HO—CH =, C(OH)—CHO (305, 306, 970)). It forms a reversible oxidation-reduction system (1365)).

While reductones are most readily formed from dihydroxyacetone and from glyceric aldehyde, they may also be obtained in yields from 5 to 8 per cent from glucose, maltose, fructose, arabinose, mannose, galactose (EULER et al. 303)) and also from xylose (301, 305)) by heating for 2 mins. to 90° in alkaline solution. Although the reductones show a chemical behaviour quite similar to ascorbic acid, they have no antiscorbutic activity (303)). Their occurrence in biological material suggests a physiological function of the reductones, perhaps in oxido-reductive processes in muscle (WURMSER) or in carbohydrate breakdown in yeast where reductones have been found by EULER and KLUSSMANN 300). In model systems with methylene blue as acceptor the reductones react as powerful donators. The reduction of methylene blue by reductones is accelerated by KCN and phosphate (264)), while their oxidation by $O_2$ in presence of heavy metals is inhibited by these agents, as would be expected (764)).

Analogous substances with a similar absorption spectrum (maxima between 265 and 290 m$\mu$) are produced by irradiating aqueous solutions of glucose, fructose, arabinose, glycerol, sorbit and glucosamine with ultraviolet light at temperatures below 20° (HOLTZ 493)). They are likewise reversible redox systems and show the same behaviour towards dyestuffs and physiological donators as the reductones (493, 491, 113)). According to EULER 264), carboxyl-free dicarbonyl sugars, formed in the course of biological carbohydrate breakdown, may also act as reductones. A similar substance, "reductinic acid", was obtained by REICHSTEIN et al. (970)) upon energetic acid treatment of pectin and galacturonic acid in beautiful crystals which had the empirical formula $C_5H_6O_3$, melting point 213°, and also strongly reducing properties. The compound probably has the structure.

$$
\begin{array}{cc}
\text{OH} & \text{OH} \\
| & | \\
\text{C} =\!=\!= & \text{C} \\
\end{array}
$$

$H_2C$    $C$=H

$CH_2$

which is analogous to that of ascorbic acid. Another, similar substance has been isolated by OTT et al. 911) from adrenal cortex (beef). It simulates ascorbic acid in its reducing and general chemical properties and is stated to show very faint antiscorbutic activity; there is an appreciable difference, however, with regard to chemical composition: C, 25,86 %; H, 5.97 %; N, 5,2 %; P, 0.55 %, Rest O.

### 3)  Nucleotide Coenzymes and Enzymes.

Introductory remarks: Just as in the case of the hemins, it is proposed to discuss in this section several chemically related substances in spite of the fact that, when combined with various specific bearer proteins, the enzymes thus formed will exhibit sometimes a very different type of specificity.

The chemical relationships between the agons of certain dehydrogenases and phosphorylases have only recently been uncovered by WARBURG and by v. EULER and their colleagues. The common feature of the enzymes in this group is that they consist of an adenine nucleotide as coenzyme in combination with a specific protein. The coenzyme may represent an adenine mononucleotide or an adenine dinucleotide, containing nicotinic acid amide or flavin as additional important components:

TABLE 14.

| Mononucleotide . | Adenine- (furano) ribosido-5-tri-phosphoric acid | Cophosphorylase (MEYERHOF-LOHMANN's Coferment) |
|---|---|---|
| Dinucleotides . . | Adenine- (furano) ribosido-5-di-phosphoric acid-(furano) ribosido-nicotinic acid amide | Codehydrogenase I. (Cozymase) |
| | Adenine- (furano) ribosido-5-tri-phosphoric acid-(furano) ribosido-nicotinic acid amide | Codehydrogenase II. (WARBURGS Coferment) |
| | Adenine-flavin-dinucleotide | Coenzyme of d-Alanine oxidase |

Even in the most recent literature, there exists a considerable confusion with regard to nomenclature, almost each school or laboratory using their own terms to the point of exclusion of others. In the present text the new nomenclature of EULER 282, 283) will be preferred in general, although other terms will also be given.

According to EULER, the total enzyme complex or holodehydrogenase consists of the apodehydrogenase (pheron=bearer protein=zwischenferment of WARBURG = dehydrogenase of the English authors in earlier publications) and of a codehydrogenase (agon = coenzyme or coferment, in some instances identical with cozymase).

In the future, the pyridine enzymes may have to be classified as vitazymes in view of the recent discovery that nicotinic acid or its amide or a closely related substance represents the anti-pellagra principle of the vitamin B complex (ELVEHJEM, SMITH). According to EULER and coworkers, cozymase is a water-soluble vitamin factor promoting the growth of rats. A discussion of these interesting relationships is outside the scope of this monograph. For the present, the grouping of the pyridine enzymes together with the simple adenine nucleotide enzymes has been retained on account of their close chemical relationship.

## TABLE 15.

### Survey:

| Coenzyme | Protein Bearer | Enzyme Complex |
| --- | --- | --- |
| Diphosphopyridine nucleotide (cozymase, coferment I) | + Protein A = | Alcohol dehydrogenase |
| | + Protein B = | Lactic acid dehydrogenase |
| | + Protein C = | Glyceric aldehyde phosphoric acid dehydrogenase |
| Triphosphopyridine nucleotide (Coferment II) | + Protein D = | Hexosemonophosphate (ROBISON ester) dehydrogenase |
| | + Protein E = | Glucosedehydrogenase of HARRISON (?) |
| Adenine-flavin-*) dinucleotide „ | + Protein F = + Protein C = | d-Alanine oxidase Dihydropyridine dehydrogenase |
| Flavin nucleotide*) | + Protein H = | Xanthine oxidase |
| Adenosine triphosphorice acid (adenyl pyrophosphoric acid, cophosphorylase) | + Protein I = | Phosphorylase |

### a) Pyridine Coenzymes.

#### Diphosphopyridine Nucleotide **) (Cozymase, Codehydrogenase I):

As shown in the above schema, this nucleotide is the agon of several hydrogen-transferring enzymes. While it is now mainly designated as cohydrogenase I or diphosphopyridine nucleotide, the best-known term for this substance is c o z y m a s e, referring to its essential role in alcoholic fermentation as a supplement to the zymase complex. It was discovered in 1904 by HARDEN and YOUNG. Other synonymous names are coferment of fermentation (WARBURG) and coreducase (EULER) (see 314). The hydrogen transfer is brought about by the reversible reduction and reoxidation of the pyridine moiety of the coenzyme.

The compound has as yet not been synthesized. Natural preparations of a high degree of purity have been obtained, e.g. from yeast (EULER) and heart muscle (OCHOA 891a)).

---

*) These flavin-containing coenzymes have already been discussed on p. 189 and 197.

**) Reviews: HARDEN 450); MYRBÄCK 849, 856); EULER 260). For a discussion of the function of cozymase see also OPPENHEIMER's "SUPPLEMENT", p. 1404—1410.

While WARBURG 1253) prefers red blood cells to yeast as the starting material, EULER and SCHLENK 317) consider yeast as the best source. The coenzyme isolated from animal organs is identical with that present in yeast 853)). Cozymase is a chemical individual, independent of the source (cf. 289)) It is a colorless and readily water-soluble substance (317)).

**Purification** (cf. EULER 260)): Both bottom and top yeast may be used as starting material, provided that the cozymase content is not too low: The ACo of the fresh yeast should not be lower than 150 (see below for explanation of term ACo).

The washed and pressed yeast is extracted by stirring it into warm water and maintaining a temperature of 80° for about 5 mins. This treatment thermoinactivates enzymes capable of splitting cozymase but it does not harm the coenzyme. After cooling, the suspension is filtered or centrifuged. The first step consists in a precipitation of inert protein material by lead acetate (HARDEN). The supernatant solution is freed from lead by $H_2S$ and the latter is driven off by a neutral gas. The coenzyme is precipitated by mercuric nitrate and the Hg-complex is decomposed by $H_2S$. The next step consists in precipitation of cozymase by phosphotungstic acid. After decomposing the precipitate with $H_2SO_4$-containing amylalcohol-ether, the liberated coenzyme is precipitated by $AgNO_3 + NH_3$. The silver salt is decomposed by $H_2S$ and the $Cu^+$-salt is formed by adding CuCl in HCl. The copper salt is decomposed by $H_2S$ and the coenzyme is precipitated in highly purified form by alcohol. At this stage the ACo-value is about 400,000 and the loss of coenzyme from the beginning amounts to 75 %. Still further purification may be accomplished by fractional precipitation with alcohol, adsorption on activated $Al_2O_3$ and elution, precipitation from HCl-methanol solution by ethyl acetate or treatment with barium hydroxide plus lead acetate. The last steps serve for the removal of small amounts of codehydrogenase II and cophosphorylase. The adsorption method increases the activity maximally to ACo = 600,000 which represents the most active cozymase preparation yet obtained.

The purity of cozymase is still preferably determined with the aid of the classical method of activation of fermentation (846)) in the presence of an excess of normal apozymase, the $CO_2$ liberation at 30° being proportional to the coenzyme concentration. In view of the fact that the result is somewhat affected by the properties of the particular apozymase used (310)), MYRBÄCK 867) prefers to ascertain the degree of purity of cozymase by chemical methods, eg. by the hypoiodite consumption. The cozymase unit (Co), as defined by EULER and MYRBÄCK, is that amount of coenzyme which under specified conditions will liberate 1 cc. of $CO_2$ per hour. The degree of purity is expressed by

$$A\ Co = \frac{cc \cdot CO_2/\text{hour}}{g \cdot \text{dry weight}}.$$

The purest cozymase preparations of EULER and his associates had an A Co value of 400,000 (282, 283, 259)) and more recently even of $\infty$ 650,000. The latter preparation was completely free from adenosine phosphoric acids and codehydrogenase II (triphospho-pyridine nucleotide).

**Chemical Constitution:** The elucidation of the structure of cozymase has progressed through the isolation of the various constituents in the form of smaller or larger breakdown products of the coenzyme. Adenine was demonstrated in the molecule by EULER and MYRBÄCK 309) who found at first 28 % and later 19 % (cf. 259). MYRBÄCK and ÖRTENBLAD 865) discovered the presence of two further N-atoms besides those of adenine. WARBURG and CHRISTIAN 1253, 1250) and about simultaneously ALBERS,

SCHLENK and EULER 19, 282, 283, 259) identified cozymase as a dinucleotide with one purine (adenine) and one pyridine (nicotinic acid amide) nucleus. Cozymase contains 8 % P, the ratio between P and mole carbohydrate being 1 : 1 and the number of P-atoms in cozymase being 2 (1250, 309, 282, 283, 259)). Contrary to an earlier suggestion (EULER 269)) the molecule is free from sulfur. The carbohydrate is a pentose (308)) and, more specifically, d-ribose since deamination yields inosinic acid (847)). Upon hydrolysis 2 moles of ribose phosphoric acid are liberated per molecule of cozymase (SCHLENK 1016)). According to EULER, KARRER et al. 299) the phosphoric acid is linked to the ribose in 5-position in both instances.

The ultimate proof for the detailed structure of cozymase is still lacking. EULER and SCHLENK 316, 1022) give the empirical formula as $C_{21}H_{27}N_7P_2O_{14}$ and suggest the following structural formula (1022)) as the most probable one:

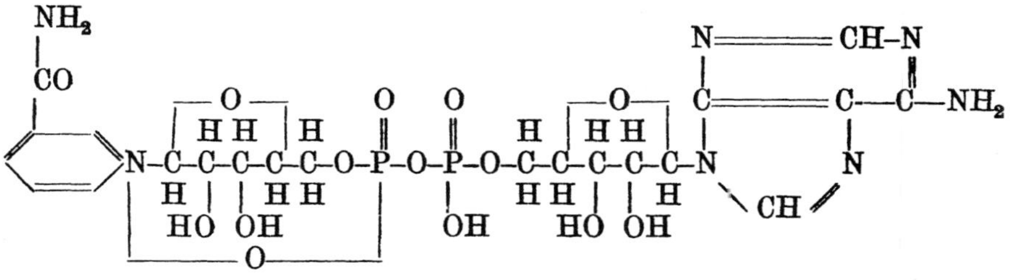

```
 NH₂
 |
 CO
              ┌──O──┐   O   O   ┌──O──┐        N═══════CH─N
              | H H | H   ‖   ‖  H | H H |      |               |
   N─C─C─C─C─C─O─P─O─P─O─C─C─C─C─N   C═══════C─C─NH₂
      H | | H H |       |   |     H H | | H        N
      HO  OH    |      OH  OH        HO  OH      ╲ CH ╱
              └─────O─────┘
```

The essential feature of this particular formula is that it is assumed that one phosphoric acid residue forms an inner salt with the basic N-atom of the pyridine ring with the aid of an O-atom. This would explain why cozymase appears to be monobasic. Furthermore, this formulation would be in agreement with the assumption of MYRBÄCK 855) that cozymase contains a lactone ring.

MYRBÄCK 856) agrees with the formula of EULER and SCHLENK in principle, but he prefers to write it slightly differently, with an open lactone ring, O⁻ and N⁺:

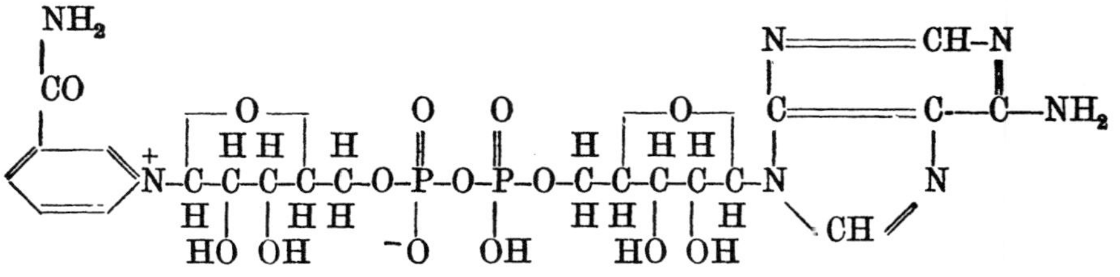

```
 NH₂
 |
 CO
              ┌──O──┐   O   O   ┌──O──┐        N═══════CH─N
              | H H | H   ‖   ‖  H | H H |      |               |
  ⁺N─C─C─C─C─C─O─P─O─P─O─C─C─C─C─N   C═══════C─C─NH₂
      H | | H H |       |   |     H H | | H        N
      HO  OH    ⁻O      OH        HO  OH       ╲ CH ╱
```

The same way of writing the formula has now also been adopted by EULER and SCHLENK 317). On the basis of the observation that acid is liberated upon reversible hydrogenation of cozymase, WARBURG and CHRISTIAN 1255) suggested the following structure for the pyridine nucleotide moiety of the coenzyme:

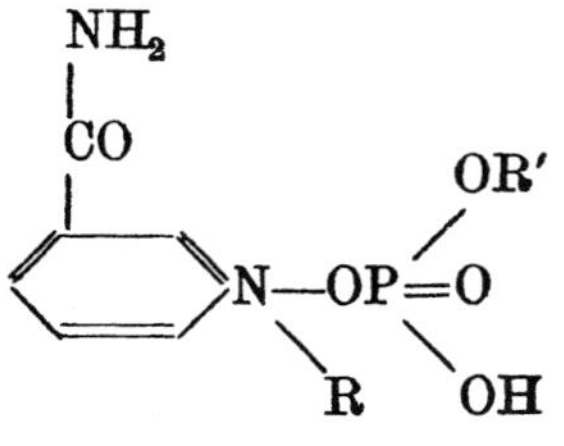

```
       NH₂
       |
       CO
                          OR'
                         ╱
     ╲N─OP═O
       |       ╲
       R       OH
```

where R represents the pentose residue and R' another sugar residue. This formulation has at the present only historical interest, particularly since KARRER et al 525), with the aid of model experiments, arrived at the following formulation of the reversible reduction:

This formulation takes into account that the reduced form is capable of existence only in alkaline solution.

The existence of a pyrophosphate link in cozymase as shown in the formula of EULER et al. has recently been experimentally proved by the isolation of adenosine diphosphoric acid from cozymase after alkali treatment (EULER, SCHLENK and VESTIN 318)).

Ultimate proof for the structure of the coenzyme will consist in a synthesis of the compound by an unambiguous method.

**Reversible Hydrogenation:** The phenomenon of reversible hydrogenation of cozymase is of fundamental importance for the catalytic action of the pyridine enzymes. WARBURG discovered (1251, 1255)) the reversible oxidation-reduction of the pyridine nucleus contained in the coenzyme. Upon hydrogenation the quarternary nitrogen atom of the pyridinium base is reversibly changed into trivalent nitrogen, the $N=C$ double bond being saturated (255)). After preliminary model experiments by KARRER and WARBURG 532), KARRER and his students 525, 528, 520, 530, 531) have carried out extensive studies of this reaction in a number of model systems. The most important model substance is the nicotinic acid amide iodomethylate, which behaves like cozymase and also like codehydrogenase II with regard to light absorption, fluorescence, color, etc. (530)). Both the oxidized and the orange colored dihydro form were prepared in pure state by KARRER to formulate the structure and the reduction of the pyridine part of cozymase in the manner indicated above. It is noteworthy that the free nicotinic acid amide with tertiary nitrogen is not reduced by $Na_2S_2O_4$, in contrast to cozymase, the iodomethylate, or trigonellin which is the methylbetain of nicotinic acid:

The dihydroform of trigonellin, just like that of cozymase, is reoxidized by the yellow ferment (WARBURG 1255)).

In the course of the reversible reduction of cozymase by $Na_2S_2O_4$ in weakly alkaline solution an orange colored substance is formed (530, 1255)) which upon dehydrogenation may be changed back into the original cozymase. This orange colored form was observed by WARBURG and KARRER during the reversible reduction of model compounds and subsequently by EULER et al. 11) upon reduction of cozymase with $Na_2S_2O_4$ in bicarbonate buffer (ph = 7—8) and of holozymase (or cozymase + apohydrogenase) by suitable donators. According to EULER et al. 11), it is important whether the reduction is carried out in weakly or in strongly alkaline solution: in the first case the orange form is only an intermediate step in the reduction process, while in the latter it is a stable end-product. Perhaps the orange form represents a semiquinoid radical which in acid or neutral solution yields unchanged cozymase. HELLSTRÖM 474), in agreement with this view, considers the yellow stage to be monohydro-cozymase.

The formation of the transient yellow stage could not be observed in the case of the model compounds of KARRER 520). His orange and yellow dihydropyridine compounds are irreversibly changed upon acidification (525)) as evidenced by the changes in the absorption spectrum (see below). Dihydrocozymase, too, will yield only 20 % unchanged cozymase upon acidifying the solution after reduction, while 80% is irreversibly destroyed. A discussion of the relationship of the various forms, as suggested by EULER and his associates, will be found in MYRBÄCK's article 856).

Dihydrocozymase may be reoxidized to cozymase by acetaldehyde in the presence of its bearer protein (WARBURG), by the yellow enzyme + $O_2$, and by the "coenzyme factor" of DEWAN and GREEN (see below). It is stable towards methylene blue (268)). According to ADLER, HELLSTRÖM and EULER 11) dihydrocozymase formed by the action of $Na_2S_2O_4$, but not that produced by enzymatic reduction, is stated to be slightly autoxidizable with a ph-optimum at about 6. It may be, however, that the reoxidation in this case is brought about by peroxide-like substances formed by the oxidation of $Na_2S_2O_4$.

Cozymase is irreversibly reduced with catalytically activated hydrogen (MYRBÄCK 854)). Under these conditions 6 hydrogen atoms are absorbed per molecule of coenzyme (KARRER and WARBURG 532)).

GREEN and DEWAN 417) have studied the oxidation and reduction of cozymase on different bearer proteins and by various reagents by means of a spectrographic method. They find that dihydrocozymase is not autoxidizable and that it is dehydrogenated by flavin and the yellow enzyme (flavoprotein). They estimate the position of the normal potential $E'_0$ at ph 7.7 as close to —0.270 V.

KARRER et al. 530) report for the reversible reduction product of nicotinic acid amide iodomethylate the extremely negative potential $E = -0.450$ V. at ph 8.53 and 20°. At lower ph-values decomposition takes place. The attachment of a sugar residue diminishes the reducing power of nicotinic acid amide considerably.

CLARK 171) has recently used the data of EULER et al. 272) on the establishment of an equilibrium between cozymase and the lactate system for a calculation of the approximate placement of the potential curve of the coenzyme on the redox scale. Although he admits that the selection of the lactate system as a reference system is not very favorable for this purpose, an indication of the validity of the treatment is found in the fact that the slope of the $E'_0$/ph-curve is roughly that to be expected of the process $Ox^+ + 2\varepsilon + 2H^+ \rightleftharpoons$ Reductant. The calculation yields a value of —0.325 V. for the normal potential, $E'_0$, of cozymase at ph 7.4 and of —0.361 V. at ph 8.7. The change in potential per ph-unit averages to 0.028 V. as compared with 0.029 V. for a two-electron system. In analogy to other chromoproteids, the attachement of cozymase to a protein bearer will probably render the potential still less negative. The sugar residues themselves in cozymase show no reducing action: they are firmly bound and are split off only upon energetic acid hydrolysis (MYRBÄCK 852)). They can, therefore, affect the potential only indirectly, by changing the energy of the whole molecule. The behaviour of various cozymase models towards oxidizing agents is described in KARRER's 525, 530, 11) papers. According to HELLSTRÖM 474), monohydrocozymase has a potential more negative than hydrosulfite. This renders potential measurements in cozymase-hydrosulfite systems very difficult.

**Absorption Spectrum and Fluorescence:**

Cozymase has a well-defined absorption maximum at 260 m$\mu$ (WARBURG and CHRISTIAN 1255); MYRBÄCK and EULER et al. 859, 279)). Hydrogenation causes the appearance of an additional broad band at 320—360 m $\mu$ with a maximum at 340 m $\mu$ which disappears again upon dehydrogenation. It does not matter whether the reduction and reoxidation is brought about by chemical reagents ($Na_2S_2O_4$) when working with pure cozymase or by biological donators (carbohydrates) and acceptors (acetaldehyde) when working with the entire enzyme system (coenzyme plus specific protein) (WARBURG and EULER).

The peak of the "dihydro-band" of the iodomethylate is situated at 360 m $\mu$ and that of trigonellin at 350 m$\mu$. In all instances the new absorption band disappears not only upon reoxidation but also upon acidification. Under these conditions the iodomethylate, trigonellin, and codehydrogenase II will exhibit a new band at 295 m $\mu$ (KARRER 530); HAAS 431)). Arabinosido- anf xylosido-dihydro-nicotinic acid amide, the "dihydrobands" of which are at 340 m $\mu$, show a band at 275 m $\mu$ in acid asolution (525)). All this points to an irreversible change suffered by the dihydroforms upon acidifying their solutions (WARBURG, KARRER, EULER). EULER et al. 11) failed to observe the band at 295 m $\mu$; instead they found a small maximum at about 300 m $\mu$. This may partly be due to the fact that, according to HAAS 431), the band at 295 m $\mu$ is observed only in the presence of sulfite. The "yellow stage" shows a broad band at about 360 m $\mu$ 11). The "dihydro-band" of cozymase diasppears upon dehydrogenation by yellow enzyme and by acetaldehyde (in the latter case only in presence of apodehydrogenase) (EULER et al. 279)).

While the pyridine compounds in the oxidized form show no fluorescence, all dihydropyridine compounds exhibit a strong whitish fluorescence when irradiated with filtered ultraviolet light. This makes it possible to follow the oxidation-reduction of the pyridine compounds by carrying out the reaction before an analytical mercury vapor lamp (WARBURG 1251)). Acidification and the consequent destruction of the dihydroforms abolishes the fluorescence.

**Physical and Chemical Properties of Cozymase:**

The m o l e c u l a r w e i g h t of cozymase is as yet not definitely established. WARBURG 1255) estimates it to be close to 700. The empirical formula given by EULER and SCHLENK (see above) corresponds to a molecular weight of 663. Experimental determinations yield somewhat lower values (850, 861, 858, 313, 312)).

Cozymase passes the usual dialyzing membranes (313, 889, 450, 308)). It is not adsorbed by silicic acid (889)) nor by $Al(OH_3)$ from acid solution, but it is strongly adsorbed by the latter at ph > 7 (889, 311)). $Al_2O_3$ adsorbs cozymase less strongly than codehydrogenase II (276)), thus enabling a separation of the two closely related substances (see below). Cozymase is not precipitated by most heavy metal salts or alkaloid reagents (cf. MYRBÄCK 856)).

Cozymase is weakly optically active: the specific rotation is about — 70° for the green mercury line (862)) and about —20° for the red cadmium line (859)). Thermal inactivation (heating for 1 hour) decreases the optical activity to —10° (859)).

The coenzyme is strongly acidic: it titrates as a monobasic acid (v.EULER and SCHLENK 1022, 317); MEYERHOF et al. 797)) contrary to earlier findings of EULER et al. 259)). The i s o e l e c t r i c p o i n t of cozymase is at ph 3.1. Below ph 3.1 cozymase will migrate to the cathode and behave as a monovalent base with a $pK_B$ of about 10 just like adenine. Cozymase exhibits, therefore, amphoteric properties (797)) and may be treated as an acid only above ph 3.1. Upon heating in alkaline solution a second acidic group is uncovered (864, 851, 865)).

In alkaline solution, particularly at elevated temperatures, the coenzyme is rapidly destroyed (866)) whereas dihydrocozymase remains unchanged when heated for 30 mins. to 100° in 0.1 N. NaOH (11)). In acid solution, cozymase is quite stable (854)) while dihydrocozymase is instantly destroyed by acid (see above). This difference in stability of the two forms towards acid and alkali may serve for the determination of the equilibrium between the oxidized and reduced form in living cells (7)). The optimum of stability at room temperature is at ph 3—4 (289)). At higher temperatures (up to about 70°) cozymase is quite stable, particularly in acid solution; at still higher temperatures and particularly at higher ph the coenzyme is inactivated (cf. 856)). At about 80°, and especially in weakly acid solution, cozymase is markedly more stable than codehydrogenase II (290)). When kept at higher temperatures and for a longer period (1 hour at 100° at ph 3.5 or for a shorter period in alkaline solution) cozymase decomposes (857)), perhaps with the formation of adenosine phosphoric acid since EULER et al 290, 322) observed an increase in the effect on glycolysis and "umphosphorylierung" with a simultaneous decrease in cozymase activity (see also 270) and 291)).

It appears that under the influence of alkali cozymase breaks down into nicotinic acid riboside and adenosine diphosphoric acid (1201, 316, 318)). Thus the "cophosphorylase" formed by alkaline inactivation of cozymase (1024, cf. p. 223) is very probably adenosine diphosphoric acid. Whether the reaction products formed by acid hydrolysis of cozymase may also act as cophosphorylase (1021)) and whether the active component is different from adenosine phosphoric acid (272)) is not quite clear (1021)). (See also the hydrolysis experiments by EULER and GÜNTHER 291)). In any event, pure cozymase itself is probably inactive as cophosphorylase, contrary to earlier findings by VESTIN 1199). The same holds for the substance formed by acid inactivation of dihydrocozymase (272, 277)) *).

Ultraviolet light inactivates cozymase rapidly in dilute solution (1014)) with preservation of the cophosphorylase activity. The coenzyme is relatively stable against oxidizing agents, even $H_2O_2$, in acid solution. $H_2O_2$ in presence of traces of iron, hypobromite (856)) and also hypoiodite inactivate cozymase rapidly, 6 atoms of I being absorbed by 1 molecule of coenzyme (852)). Iodine itself, in acid solution, is unable to attack cozymase (855)). Nitrite causes slow deamination (848, 858, 857)). When the $NH_2$-group on the adenine part of cozymase is removed by treatment with $HNO_2$, the resulting desaminocozymase shows 1/3 of the activity of cozymase (1025a)).

While cozymase is stable towards most enzymes (863)), it is readily attacked by phosphatases (848, 867, 847)). SCHLENK, GÜNTHER and EULER 1025) have recently subjected cozymase to the action of emulsin preparations containing nucleotidases with a ph-optimum in weakly acid solution and free from nucleosidases (prepared according to BREDERECK and HELFERICH). The reaction yields two nucleosides, namely adenosine and a nicotinic acid amide nucleoside (probably nicotinic acid amide riboside). This nucleoside is unable to replace cozymase in alcoholic fermentation, and attempts to effect an enzymatic resynthesis of cozymase from the nucleoside plus adenosine, adenylic acid, adenosinedi- and triphosphoric acid in presence of apozymase have thus far failed. Enzyme preparations from pancreas and castor oil seeds but not the lipase contained in the latter will also split cozymase (262)). After the death of the cell cozymase is rapidly

---

*) See, however, OPPENHEIMER's "SUPPLEMENT" p. 1402.

destroyed by autolytic enzymes. This enzymatic breakdown proceeds in a manner different from that in alkaline solution and goes probably further than the liberation of adenylic acid which itself is also decomposed in tissues (292)). Further literature on this subject will be found in MYRBÄCK's compilation (856)).

### Triphosphopyridine Nucleotide:

### (WARBURG's Coferment, Codehydrogenase II)*)

This coenzyme, like cozymase, owes its ability to transfer hydrogen to the presence of a pyridine nucleus in its molecule. According to WARBURG 1255, 1253), it is a triphosphopyridine nucleotide and thus distinguished from cozymase by the presence of a third phosphoric acid residue (1259)). It was first isolated from the red blood cells of the horse where it is combined with a specific protein ("Zwischenferment") to form a hydrogen transferring enzyme capable of dehydrogenating hexose monophosphate (ROBISON-ester) (1259, 1258)). It is probably identical with HARRISON's glucose phosphate dehydrogenase.

Constitution: The first structural formula proposed for codehydrogenase II by WARBURG 1255) was as follows:

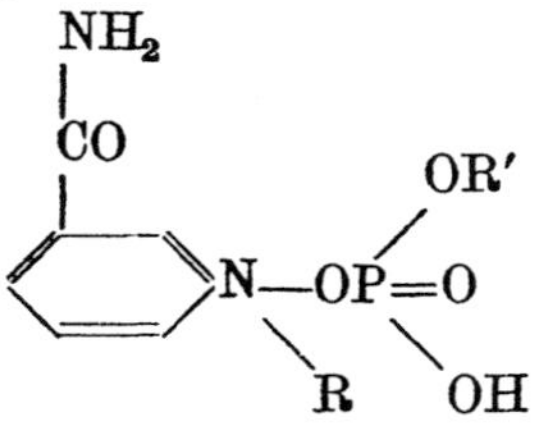

$$
\begin{array}{c}
NH_2 \\
| \\
CO \\
|
\end{array}
\qquad
\begin{array}{c}
OR' \\
\diagdown \\
N{-}OP{=}O \\
\diagup \quad \diagdown \\
R \qquad OH
\end{array}
$$

R and R′ designate pentose residues. EULER and SCHLENK 317), on the other hand, formulate the coenzyme in analogy to cozymase in such a manner that all three phosphoric acid groups are lined up in a triphosphate linkage:

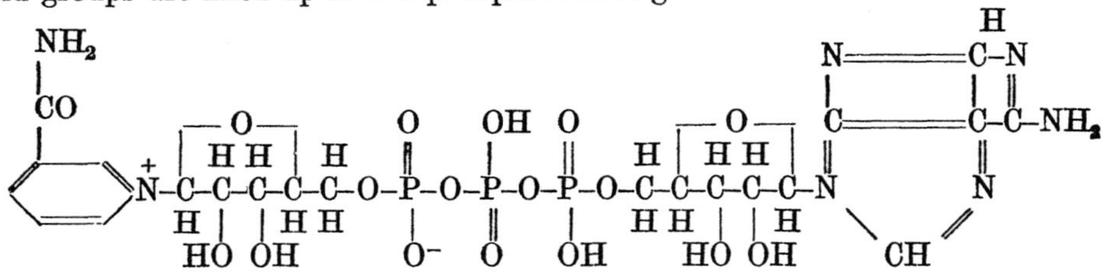

$$
\begin{array}{c}
NH_2 \\
| \\
CO
\end{array}
\quad
\overset{+}{N}{-}\underset{H}{\overset{H}{C}}{-}\underset{HO}{\overset{H}{C}}{-}\underset{OH}{\overset{H}{C}}{-}\underset{H}{\overset{H}{C}}{-}O{-}\underset{O^-}{\overset{O}{P}}{-}O{-}\underset{O}{\overset{OH}{P}}{-}O{-}\underset{OH}{\overset{O}{P}}{-}O{-}\underset{H}{\overset{H}{C}}{-}\underset{HO}{\overset{H}{C}}{-}\underset{OH}{\overset{H}{C}}{-}\underset{H}{\overset{H}{C}}{-}N
$$

Inasmuch as adenosine triphosphoric acid has not yet been isolated as a breakdown product the formula is still hypothetical.

The determinations of the electrophoretic mobility of codehydrogenase II by THEORELL 1149) indicated only two dissociation constants: $p_{K_1} = 1.8$ and $p_{K_2} = 6.1$. According to these findings, the coenzyme is only dibasic and one of the free OH-groups of the phosphoric acid residues is either linked to the pyridinium nitrogen atom, as is probably the case in cozymase, or to the NH₂-group of the adenine nucleus. The latter possibility is supported by the failure of THEORELL 1149) to detect free basic groups in the molecule and by the resistance of the coenzyme towards nitrite (WARBURG and CHRISTIAN 1259)). Inasmuch as an addition of boric acid does not increase the acidity of the molecule, conjugated OH-groups are probably absent (THEORELL 1149)). Other measurements of THEORELL 1150), on the other hand, indicate that each stage of dis-

*) Reviews: WARBURG 1235); W. SCHLENK 1017).

sociation corresponds to 2 H-ions which would suggest that the whole molecule is tetrabasic.

**Absorption Spectrum:** Codehydrogenase II. has the same characteristic band at 260 m$\mu$ as has cozymase. There is also complete analogy in the spectral changes accompanying the reversible reduction by $Na_2S_2O_4$ or by hexosemonophosphate in presence of the bearer protein (Zwischenferment of WARBURG). The "dihydro-band" is situated at 345 m$\mu$ (**1295, 1253**)) and the dihydroform exhibits a white fluorescence when irradiated with filtered ultraviolet light (**1255**)). Historically speaking, all important features of these pyridine coenzymes where first discovered in the case of the codehydrogenase II and most of the subsequent work on cozymase by WARBURG and by EULER was *"eine Analogiearbeit"* (WARBURG **1235**)). Upon irreversible hydrogenation with Pt-$H_2$ in alkaline solution 6 H-atoms are absorbed per coenzyme molecule and no "dihydro-band" appears (**1259**)): The pyridine nucleus is reduced to piperidine which has no selective absorption in this range; this reduces the height of the band at 260 m$\mu$ to the value corresponding to the adenine component only (**1259**)). Upon destruction of the dihydroform by acid the "dihydro band" is abolished and a band at 295 m$\mu$ appears (KARRER **530**); HAAS **431**)). This band, too, is stable only in presence of sulfite (**431**)).

**Preparation:** The coenzyme was first isolated from red blood cells by WARBURG and CHRISTIAN **1259**). The centrifuged cells are cytolyzed with water. An acetone fractionation is followed by precipitation with mercuric-acetate and as the barium salt. The latter step permits the removal of most of the accompanying adenylic acid and cozymase. Further purification is accomplished by precipitation of the codehydrogenase II by ethyl acetate from HCl-methanol solution and by fractionation with lead acetate. The fluorescence originally observed (**1149**)) is due to an impurity (**1247**)) which may be removed by oxidation with $Br_2$ (**1253**)). Attempts to effect a purification by electrophoresis were not successful (THEORELL **1149**)).

Codehydrogenase II may also be obtained from yeast by working up the mother liquor remaining after precipitating the cozymase as the cuprous salt (EULER and ADLER **276**)). A separation of the two coenzymes may also be brought about by chromatographic analysis on an $Al_2O_3$ column: codehydrogenase II is more strongly adsorbed than cozymase and accumulates in the upper layers of the column from which it may be eluted by treatment with m/50 to m/100 $KH_2PO_4$ at elevated temperature (**276**)). Furthermore, codehydrogenase II, in contrast to cozymase, is precipitated by lead acetate (**259**)). Another possibility is the difference in solubility of the Ba-salts in water after adding alcohol (WARBURG). In this manner it is also possible to separate the two pyridine coenzymes from cophosphorylase (**1253**)).

The degree of activity and of purity of the codehydrogenase is determined by comparison with a standard preparation in a test system containing the specific protein bearer, ROBISON ester and yellow ferment, or by determining the amount of hydrogen taken up upon irreversible hydrogenation in m/10 borate buffer with platinum black as catalyst. Another method consists in determining the extinction at the maximum of the "dihydro-band" after reversible hydrogenation (**1259**)).

**Composition and chemical properties:** Elementary analysis yields the empirical formula $C_{21}H_{28}N_7O_{17}P_3$ (WARBURG and CHRISTIAN **1259**)), corresponding to 1 mole adenine, 2 moles pentose, 3 moles phosphoric acid, 1 mole nicotinic acid amide, minus 6 $H_2O$. The content in adenine, as determined directly, is about 17 %, in nicotinic acid amide 15—16 % (**1255**)). The latter was first isolated in the form of the picrolonate and later as the hydrochloride (1248)). Another empirical formula put up for discussion by WARBURG and CHRISTIAN 1253) is $C_{22}H_{32}O_{19}N_7P_3$ where one of the sugar residues would be a pentose and the other a hexose. The molecular weight as calculated from the first formula is 743 and from the second 791. The value actually found by cryoscopy is 875 (WARBURG and CHRISTIAN 1259)).

The coenzyme is readily soluble in water and in anhydrous methanol in presence of strong acids. It has not yet been obtained in crystalline form (1259)). The various precipitation reactions are listed by WARBURG and CHRISTIAN 1259) and SCHLENK 1017). The compound is stable in alkaline solution but it is almost instantly destroyed in acid solution (1259)). Ultraviolet light causes rapid destruction (1249)). In muscle tissue about 85 % of the codehydrogenase is inactivated in 1 hour and 94 % in 3 hours (273)). The split products thus formed have not yet been identified.

Preparation of Codehydrogenase II from Codehydrogenase I (Cozymase):

EULER and his associates have lately described experiments indicating a transformation of cozymase into codehydrogenase II. When cozymase is treated with $POCl_3$ (SCHLENK 1020)), a product is obtained capable of replacing codehydrogenase II in the test system for the dehydrogenation of ROBISON ester. The extent of conversion of coenzyme I into coenzyme II is small under those conditions. An enzymatic transformation with the aid of dried yeast (apozymase) and adenosine triphosphoric acid as phosphoric acid donator is indicated by the experiments of VESTIN 1200), EULER and VESTIN 321) and EULER and BAUER 284). The authors are careful, however, in considering their evidence merely as strongly suggestive but not absolutely final. Ultimate proof will consist in the isolation of pure codehydrogenase II from the reaction mixtures. The issue is important because the relative ease with which an additional phosphoric acid residue appears to be attached to the cozymase molecule throws doubt upon the formulation of codehydrogenase II as containing a chain of three phosphoric acid radicals in series (see above), as SCHLENK 1020) points out. If the conversion of cozymase into codehydrogenase II can be verified, a revision of the formula proposed for the latter by SCHLENK and EULER will probably be necessary: It would appear, then, that one of the phosphoric acid residues is contained in a side chain rather than in the main chain of the bridge between the two nucleotides.

### b) Cophosphorylase.

Cophosphorylase, a coenzyme first described by MEYERHOF and LOHMANN, is the active group or agon of phosphorylase. It is very probably identical with adenosinetriphosphoric acid = adenylpyrophosphoric acid. There is a possibility, according to WARBURG 1253), that cophosphorylase is the diadenosine-pentaphosphoric acid described by OSTERN 907) and BEATTIE et al. 94). LOHMANN and SCHUSTER 736), however, consider this compound to be a decomposition product of adenosinetriphosphoric acid*). The catalytic function of adenosine 5-triphosphoric acid may be represented by the following reversible reaction:

Adenosinediphosphoric acid $+$ $H_3PO_4$ $\leftrightarrows$ Adenosinetriphosphoric acid.

The substance has been isolated in pure form from animal tissues, from blood and also from yeast (cf. 1019, 192)). In working up material for adenosinetriphosphoric acid,

---

*) For data concerning adenosine-5-phosphoric acids see SCHLENK 1019).

e.g. heart muscle, it is important to prevent autolysis by removal immediately after the death of the animal and by working at low temperature. Otherwise diadenosine-pentaphosphoric acid is formed (736)). OSTERN and BARANOWSKI 909) succeeded in preparing adenosinetriphosphoric acid by enzymatic synthesis. The separation from the pyridine nucleotides is accomplished by conversion into the barium salts (WARBURG and CHRISTIAN 1253)). A substance with strong cophosphorylase activity is formed by the decomposition of cozymase by alkali (SCHLENK, EULER et al. 1024)).

The structural formula given by LOHMANN 734) corresponds to an adenylpyrophosphoric acid with a free amino group:

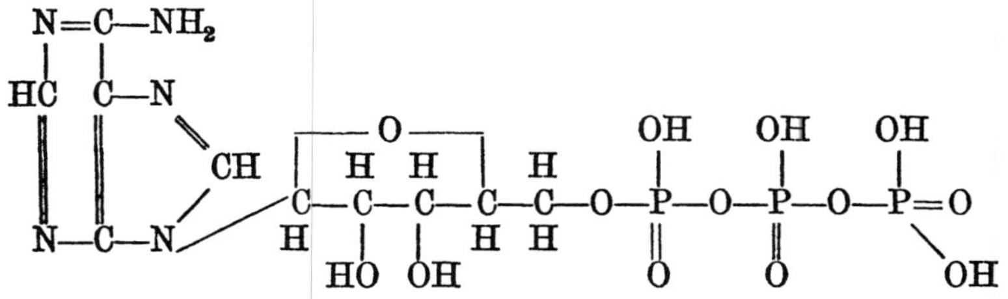

This formula has been confirmed by MAKINO 747). The solutions of adenosintriphosphoric acid are very unstable: even at room temperature, particularly in acid and alkaline solution, phosphoric acid is split off (729)). In the tissues (crustacean muscle) the dephosphorylation takes place in two steps (LOHMANN 733)): first adenosinediphosphoric acid and then adenosinemonophosphoric acid (adenylic acid), is formed. Living as well as toluene-treated B. coli will attack adenosinetriphosphoric acid both under aerobic and anaerobic conditions, causing dephosphorylation and deamination. According to LUTWAK-MANN 741), hypoxanthine, ribose and $NH_3$ are the end-products of this breakdown.

The absorption spectrum is similar to that of cozymase: in both compounds the adenine nucleus is responsible for an absorption band at 260 m$\mu$ which is somewhat broader in the case of cophosphorylase (859, 860, 201)).

The ease of transformation of the various adenosine-5-phosphoric acids into each other and the fact that according to the equation given above adenosindiphosphoric acid gives rise to the triphosphoric acid renders the task of assigning the cophosphorylase function to any one of these compounds very problematic. In fact, in the literature adenylic acid as well as the di- and triphosphoric acid are indiscriminately called cophosphorylase. The term may perhaps be better restricted to the two immediate reaction partners, viz. to the di- and triphosphoric acid, bearing in mind that in the presence of suitable enzymes any other adenosinephosphoric acid may be utilized as precursor of the coenzyme.

In this connection diadenosine-5′5′-tetraphosphoric acid should also been mentioned. It was recently isolated from yeast by KIESSLING and MEYERHOF 563) and at first erroneously identified as a cozymase pyrophosphate (MEYERHOF and KIESSLING 790)). In its activity as cophosphorylase it behaves similarly to adenylpyrophosphoric acid. Probably the largest fraction of the adenosine-5′-phosphoric acid in yeast is present in the form of diadenosine-tetraphosphoric acid.

## c) Protein Bearers of Nucleotide Coenzymes.

Only a few of the more important bearer proteins or mixtures of bearer proteins required to supplement the nucleotide coenzymes to form active holoenzymes (or symplexes) shall be mentioned in this section. This is justified because these proteins have only in isolated instances been obtained in pure form. For a fuller discussion the reader is referred to the review articles by EULER 260), THUNBERG 1174), MEYERHOF 788) and to the "Supplement" of OPPENHEIMER (p. 1373—1388).

### Apozymase:

Apozymase, according to EULER's definition, is holozymase minus coenzymes, in other words the sum of the bearer proteins of the fermentation enzyme complex. Washed dried yeast or dialyzed yeast maceration juice may thus be used as apozymase preparations (LOHMANN 730)). It is very difficult, however, to obtain reproducible results with such insufficiently defined preparations. The same cozymase preparation when added to various crude apozymase preparations will produce fermentation of appreciably differing intensity. If care is taken to employ always the same strain of yeast and to standardize all operations, quantitatively reproducible results may be secured (OHLMEYER 898)). OHLMEYER uses rapidly dried yeast. It is washed 8 times with water to remove most of the cozymase. MYRBÄCK 849) employs exclusively bottom yeast, because he finds that cozymase cannot be completely removed from top yeast. The washed brewers yeast is pressed, driven through sieves and dried at the air. The dry yeast is treated with water 5—10 times and each time centrifuged off. The resulting apozymase is spread in thin layers on glass plates and dried in air.

Such preparations are stated to be stable for about 4 weeks and may be standardized with cozymase preparations of known purity. The separation of an "oxido-reducase", decolorizing methylene blue when mixed with boiled yeast juice containing cozymase, from yeast maceration juice has been effected by ammonium sulfate precipitation (LEBEDEW 706)).

The criteria for the absence of cozymase from apozymase are as follows (MYRBÄCK 849)): Addition of sugar, or phosphate and sugar, or acetaldehyde and hexosephosphate must not give rise to $CO_2$-formation. After treatment with water at 80° the extract thus obtained must not promote fermentation after adding apozymase, sugar, phosphate and hexosephosphate. An activity in the latter case would indicate the presence of small amounts of cozymase contained in the apozymase preparation. On the other hand, addition of cozymase together with sugar and phosphate should give rise to active fermentation.

## Fractionation of Apozymase.

Fermentation Test Proteins of WARBURG:

WARBURG and CHRISTIAN 1253) and NEGELEIN 879a) isolated two protein fractions called A- and B-protein, from yeast maceration juice. In the presence of both protein fractions plus cozymase, adenylpyrophosphate, Mg, Mn and inorganic phosphate, hexosemonophosphate will react rapidly with acetaldehyde in accordance with the over-all equation:

$$2 \text{ Hexosemonophosphate} + 1 \text{ Phosphoric acid} + 2 \text{ Acetaldehyde} =$$
$$1 \text{ Hexosediphosphate} + 1 \text{ Pyruvic acid} + 1 \text{ Phosphoglyceric acid} + 2 \text{ Alcohol.}$$

In the absence of one of the two protein fractions or one of the two coenzymes no reaction takes place. The accumulation of pyruvic acid is due to absence of carboxylase in the system. The detailed mechanism of the above reaction is discussed in the papers by WARBURG and CHRISTIAN and in MEYERHOF's review 788). According to MEYERHOF, KIESSLING and W. SCHULZ 791) the A-enzymes fraction contains the phosphorylating which, in cooperation with the adenylic acid system, are responsible for the "*Umphosphorylierung*" of phosphopyruvic acid and for the synthesis of hexosediphosphate from hexosemonophosphate plus phosphopyruvic acid or plus phosphoric acid. The B-protein fraction contains all the other enzymes required for the fermentation process except carboxylase.

The main steps in the preparation of the A-protein (NEGELEIN 879a)) are the following: Fresh LEBEDEW juice from bottom brewers yeast is brought to ph 7.8 to 8.0. The heavy protein precipitate formed is discarded. The supernatant solution is adjusted to ph 6.5 and the A-protein is precipitated by adding the same volume of saturated ammonium sulfate. After repeating the ammonium sulfate fractionation three times, other inert proteins are removed by heating the solution to 40° for 30 mins. The A-protein is precipitated from 0.35 saturated ammonium sulfate solution by acetic acid (ph 4.35). After dissolving the A-protein at ph 6.6 at 0° alcohol is added to a total concentration of 20 % for stabilization purposes and a further purification is brought about by precipitation of the active fraction at ph 5.2 and at 0° and by redissolving at ph 6.38. The A-protein may be stored for 3 months at 0° under half-satured ammonium sulfate without appreciable loss in activity. In aqueous solution the protein is rapidly denatured.

The B-protein fraction is larger than the A-fraction. It is obtained by removing the A-fraction by acid, neutralizing and precipitating with 75 % acetone (WARBURG and CHRISTIAN 1253)).

### Protein of Acetaldehyde Reducase (Acetaldehydrese).

It has already been mentioned that the pyridine coenzymes, when acting together with specific protein bearers, form a number of holodehydrogenases (p. 212). One of these dehydrogenases is the enzymatic component of the zymase complex which is responsible for the reduction of acetaldehyde to the end-product of yeast fermentation, viz. ethyl alcohol. Like all dehydrogenases, the same enzyme is capable of catalyzing the reverse reaction, i.e. the oxidation of alcohol to acetaldehyde. Both the forward and the back reaction may be used to test the activity of the enzyme. The coenzyme in this case is cozymase, i.e. diphosphopyridine nucleotide. The protein has been obtained in pure, crystalline form by NEGELEIN and WULFF 883).

Preparation: Starting material is LEBEDEW juice from bottom (brewers) yeast. The degree of purity of the protein in the whole juice is 0.0065. Inert proteins are partially removed by denaturation at 55°. Fractionation with acetone increases the degree of purity to 0.022. The active protein is precipitated by ammonium sulfate at 0.62 saturation and subsequently at 0.50 saturation. After dissolving in water the degree of purity is 0.28. By three alcohol fractionations the latter is raised to 0.69. From 28 kg. dried yeast 5.8 g. of protein of this purity were obtained. The protein may now be crystallized from 0.4 saturated ammoniumsulfate solution at 38° (yield 81 %). The fact that the activity per weight remains unaltered after three crystallizations indicates that the crystals represent the pure protein of the dehydrogenase. The salt may be removed by dialysis and the protein may be obtained in dry and stable form by drying the solution at 0°.

Properties: The protein crystallizes in very thin hexagons. Microanalysis yields the following values: C 52.8 %, H 6.96 %, N 16.54 %, S 1.21 %. The absorption spectrum shows the typical protein band at 280 m $\mu$. The protein is relatively stable at 0° and neutral reaction. It is rapidly inactivated at ph < 5.0 and > 8.5. It is very sensitive towards heavy metal ions, particularly copper. The protective action of yeast gum or glycocoll is ascribed to complex formation with heavy metals. The activity of the protein is determined in a system containing pyrophosphate buffer, yellow enzyme, cozymase, alcohol and semicarbazide-HCl which is employed to remove the acetaldehyde formed during the reaction. As the acceptor is pure oxygen, the rate of oxygen uptake is a measure of the activity of the protein under these conditions. Another method consists in following the rate of the reduction of the coenzyme by alcohol by means of a photoelectric method.

Of the remarkable observations concerning this crystalline protein recorded in the paper by NEGELEIN and WULFF 883) only the most important may be mentioned. It was found that the affinity of the coenzyme to the protein depends on its state of oxidation-reduction: The reduced pyridine coenzyme has an affinity to the protein three times higher than that of the non-reduced coenzyme. The nucleotide concentration at which half the holoenzyme is dissociated into nucleotide and protein is $3 \times 10^{-5}$ M. in the case of the hydrogenated coenzyme and $9 \times 10^{-5}$ M. for the nonhydrogenated coenzyme. It should be mentioned that in the instance of a triphosphopyridine nucleotide -protein (hexosemonophosphate dehydrogenase) the two forms of the coenzyme have the same affinity for the protein component (NEGELEIN and HAAS 882)). The rate of reduction of the acetaldehyde is greater than that of the oxidation of alcohol. Assuming a molecular weight of the protein of 70,000, one protein molecule is able to catalyze the reaction of 28,500 molecules dihydropyridine nucleotide with acetaldehyde per minute (at high aldehyde concentration) and of 17,000 molecules pyridine nucleotide with alcohol (at high alcohol concentration). In equilibrium half of the pyridine nucleotide not bound to protein is hydrogenated if the alcohol concentration amounts to 1350 times that of the aldehyde concentration.

The capacity of the protein to catalyze the reaction of many coenzyme molecules with substrate molecules per unit time and the dissociation of the holoenzyme into coenzyme and apoenzyme (pheron) have undoubtedly a biological significance. The concentration of the coenzymes in the cell is so great that they must be far in excess of the concentration of the corresponding protein bearers. In contrast to chromoproteids of the type of hemoglobin or yellow enzyme where prosthetic group and protein are linked with considerable affinity and where each prosthetic group requires the presence of one protein molecule, the dissociable dehydrogenase systems operate with a few protein molecules which bring a large number of coenzyme molecules to reaction with the substrate.

According to MEYERHOF 788) the amount of adenosintriphosphoric acid and of cozymase present in rabbit muscle, for example, would correspond to more than the total protein content of the muscle if a molecular ratio of 1 : 1 for protein/coenzyme were maintained. This fact illustrates that actually a large excess of coenzymes over the corresponding proteins exists in the tissue.

### Diaphorase (Coenzyme Dehydrogenase):

Until very recently it was generally held that the reoxidation of the hydrogenated pyridine nucleotides is usually accomplished by the yellow enzyme which is thereby reduced to the leuco-flavoprotein. The latter reacts with molecular oxygen or with hemin systems in the ferric form to regenerate the oxidized form of the yellow enzyme.

The recent findings by ADLER 10) and by DEWAN and GREEN 199) indicate that, at least in animal tissues, not the yellow enzyme but a different enzyme, called diaphorase*) by ADLER and coenzyme factor by GREEN, represents the specific cozymase

---

*) From the Greek $\delta\iota\alpha\phi\acute{e}\rho\epsilon\iota\nu$ = transfer.

dehydrogenase. The diaphorase, in turn, forms the coupling link between the pyridine enzymes and the hemin systems of the cell (cytochrome-cytochrome oxidase).

Preparation: So far it has not been possible to obtain diaphorase free from all dehydrogenases. Whether rabbit skeletal muscle or hog heart is chosen as starting material depends on which dehydrogenases one wishes to avoid in the final product (DEWAN and GREEN 199)).

Fresh muscle tissue is finely minced, washed 6 times with tap water, ground with sand and m/50 phosphate buffer, ph 7.2, and pressed through cheese cloth. The filtrate is freed from coarse particles by centrifuging and then brought to ph 4.6 with acetic acid. The precipitate is suspended in m/50 phosphate, centrifuged, washed three times with the buffer, and finally suspended in m/10 phosphate, ph 7.2. In the course of the purification the enzyme seems to undergo aggregation. In any event, the final preparations form fine suspensions with a strong TYNDALL effect. Diaphorase shares this property with cytochrome oxidase (see p. 264).

EULER and HELLSTRÖM 294) use the following procedure for the preparation of purified diaphorase from hog heart and skeletal muscle:

Heart muscle brei is alternately washed with water and 2 per cent saline until the mass has become almost colorless. It is then ground with phosphate buffer and centrifuged. The solution is cooled to 0° and the enzyme is precipitated by cold acetone + $CO_2$. The precipitate is dissolved in 0.1 N. $NH_3$. Addition of 0.1 to 0.2 saturated ammonium sulfate solution yields a precipitate containing 2/3 of the original activity. It is redissolved in 0.1 N·$NH_3$ and treated with $CO_2$ at low temperature in such a manner that the ph does not fall below 7. The solution of the purified enzyme thus obtained contains 70 per cent of the total activity. The enzyme may be precipitated by further treatment with $CO_2$. Dry preparations, however, are not very stable.

Properties: The crude enzyme preparations are quite stable. The purified solutions, too, will retain their activity for some time when stored at 0°. The enzyme is remarkably stable in 0.1 N. $NH_3$. The same holds for ph 10, while 95 per cent of the activity will disappear within 5 mins at ph 4.13. (EULER and HELLSTRÖM 294)). Upon warming the solutions to 45° at ph 7.5, 66 per cent of the activity is destroyed in 30 mins. At ph 9.0, 10 mins. heating to 53° cause 70 per cent inactivation.

The absorption curve shows the usual protein band at 260—270 m $\mu$. There is no specific absorption at wavelengths longer than 300 m $\mu$. In comparison with other enzymes the electrophoretic mobility of diaphorase is relatively high and increases rather steeply at ph 9. According to EULER and HELLSTRÖM this behaviour is indicative of a dissociation constant situated near ph 9.5—10.

The ph-optimum is located in the alkaline region: At ph 6—7 the activity is still relatively small. It increases steeply with increasing ph, showing a sharp maximum at ph 10 and declining rapidly at still higher ph-values. Borate, in contrast to phosphate, inhibits the enzyme activity. It is remarkable that the alkaline ph-optimum corresponds to the alkaline optimum of ph-stability of the diaphorase as well as of free dihydrocozymase (p. 219). Pyrophosphate activates the enzyme while Cu-ions inhibit it markedly. Neither iodoacetate nor malonate have an effect *).

Function: The function of diaphorase, as has already been stated, consists in the dehydrogenation of dihydrocozymase. According to GREEN 199) the dihydroform of

---

*) According to a recent note by STRAUB, CORRAN and GREEN (Nature 148, 119 (1939)) diaphorase belongs to the group of yellow enzymes, the prosthetic group being a flavin adenine nucleotide.

codehydrogenase II. is also dehydrogenated by the enzyme, whereas EULER claims absolute specificity for cozymase as the substrate (294)). The nature of the apodehydrogenase has no effect on the reaction of the dihydrocoenzyme with diaphorase (DEWAN and GREEN 199); ADLER and HUGHES 12)). In vitro, methylene blue and pyocyanine may function as acceptors (ADLER et al. 10)). The acceptor, in vivo, is the cytochrome system (a and b). No direct reaction between diaphorase and molecular oxygen has been observed.

Distribution: The occurrence of diaphorase has not yet been systematically studied. Besides its presence in various animal tissues, such as heart and skeletal muscle, brain, kidney and liver of mammals, DEWAN and GREEN 418) have found it in yeast.

### Phosphorylase:

For a discussion of the complex situation with regard to phosphorylase systems the reader is referred to OPPENHEIMER's Supplement to his book "Die Fermente", p. 1391—1403.

It is probable that there is no separate phosphorylase as a synthesizing enzyme, as distinct from phosphatase. Phosphorylation is probably the back reaction catalyzed by phosphatase.

Pure phosphorylase (phosphatase) has not yet been obtained. EULER and ADLER 275) have purified monophosphorylase starting with crude *"Zwischenferment"* preparations from yeast. Their method consists in a series of adsorptions on Al(OH)$_3$ (Cγ) and elutions. If the elution is carried out with acetic acid at ph about 5,5, monophosphorylase may be obtained in partially purified form. The enzyme is destroyed by alkaline eluents.

WINBERG and BRANDT 1340) obtained a stable "heterophosphatese" preparation from yeast by a procedure similar to that used by SCHÄFFNER and BAUER 1015) for the purification of the ordinary yeast phosphatase. The enzyme was purified and stabilized by dialysis against glycerol. Their preparation showed a ph-optimum of about 7 for the phosphorylation of glucose and maltose. The quite flat optimum is similar to that of the adenylpyrophosphatese of muscle.

### 4)    Quinoid Intermediary Catalysts *).

### a) Pyocyanine.

Occurrence: Pyocyanine has thus far been found in nature only in B. pyocyaneus (also called Pseudomonas pyocyanea). The bacterium synthesizes the blue pigment when grown in suitable media such as ragitbouillon MERCK at ph 7.8—8.0 (WREDE and STRACK 1347)) or, better, in peptone-gelatin medium (ELEMA and SANDERS 238)). The pigment is secreted from the microorganisms into the surrounding medium from which it may readily be obtained in crystalline form. Under optimal conditions of culture and with a suitable strain of bacteria, pyocyanine concentrations of 260 mg. per liter medium have been recorded (ELEMA and SANDERS).

Isolation: The cultures are extracted with chloroform and the blue chloroform solutions are treated with dilute HCl which takes the pigment up with red color. The HCl solution is made alkaline by NaOH and the pigment is re-extracted with chloroform. Now either the free base may be obtained in crystalline form by evaporating the chloro-

---

*) The biological effects of these substances, particularly on tissue metabolism (acceptor respiration) have been discussed in a previous section (p. 108).

form, dissolving the amorphous residue in hot water and cooling to 0°, or else the hydrochloride may be obtained by taking up the residue remaining after evaporating the chloroform with HCl and slow concentration in vacuo. Inasmuch as the free blue base suffers decomposition to yellow and colorless products upon storage, it is preferable to prepare and store the red pyocyanine hydrochloride.

Constitution and Synthesis: The empirical formula of the pigment is $(C_{13}H_{10}ON_2)_x$. At first, WREDE and STRACK were inclined to double this simplest formula. Alkali in presence of air yields hemipyocyanin, $C_{12}H_8ON_2$, which could be identified with synthetic α-hydroxyphenazine (I).

I. OH        II. O

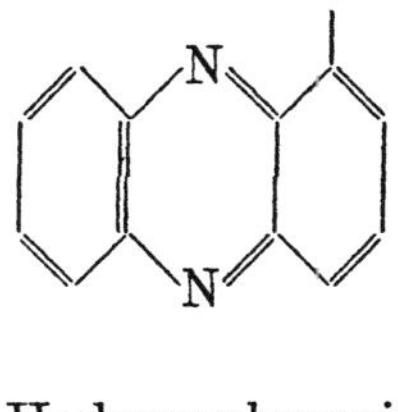

α-Hydroxyphenazine      Pyocyanine (N-Methyl-α-oxyphenazine)

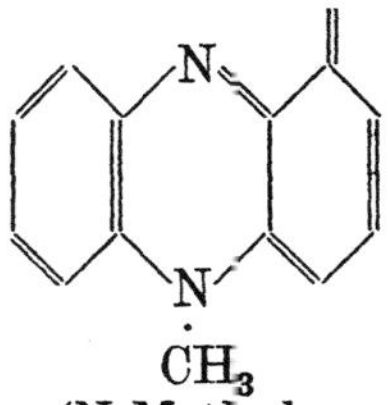

The remaining C-atom is present as a methyl group which is attached to one of the phenazine nitrogens inspite of the ease of its detachment by HI and by oxygen in alkaline solution. This has been proven by the synthesis of pyocyanine by WREDE and STRACK 1348): when α-hydroxyphenazine is heated for a few minutes with dimethylsulfate to 100°, an adduct is formed which upon treatment with alkali yields pyocyanine. The pigment is therefore identical with N-methyl-α-oxyphenazine (II). The synthesis is formally analogous to the preparation of N-methyl-phenazoniumiodide by HANTZSCH from phenazine and methyliodide.

The synthesis of WREDE and STRACK starts with pyrogallol-l-methyl ether. It is oxidized with lead dioxide and coupled with o-phenylenediamine to form α-methoxyphenazine. Treatment with fuming HBr yields α-hydroxyphenazine. The latter is converted into pyocyanine by dimethylsulfate. MICHAELIS, HILL and SCHUBERT 816) have improved the process in several details. They confirm that the synthetic product is identical in all its properties with the natural one. Synthetic pyocyanine hydrochloride is now available commercially (e.g. from HOFFMANN LA ROCHE).

The work of WREDE and STRACK did not permit one to decide whether the methyl group is attached to the phenazine nitrogen nearest to the oxygen atom or to the one distant from it. This point has recently been decided by HILLEMANN 488) in favor of the formula shown above (II): The reaction of leuco-pyocyanine with oxalylchloride yields a labile oxalylderivative with the participation of the oxygen and one nitrogen atom of pyocyanine. This can only be the case, if the N-atom close to the OH-group in leucopyocyanine is not methylated. As HILLEMANN points out, it is as yet not possible to decide between the two possibilities indicated in the formulations IIa and IIb:

IIa.                IIb.

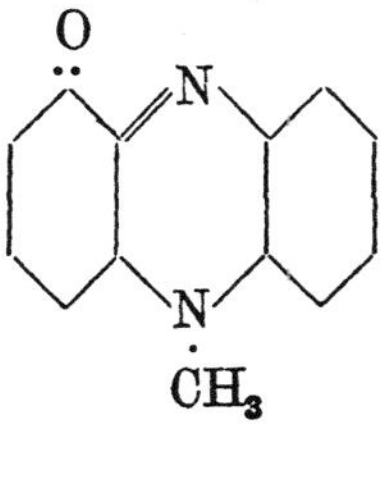

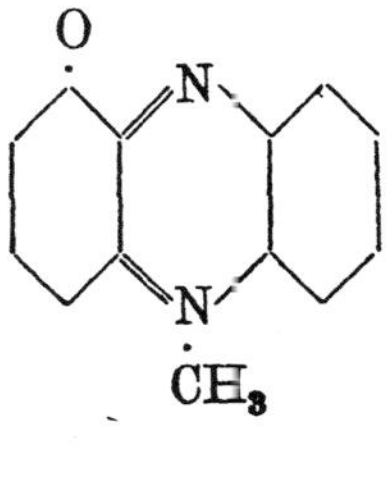

O                O

CH₃             CH₃

The chemical properties of pyocyanine and of the hydroxy-phenazine methosulfate appear to favor formula IIb, i.e. the "phenol betain" structure.

Molecular weight: Molecular weight determinations in alcohol or chloroform yield no reproducible values. From freezing point determinations in glacial acetic acid WREDE and STRACK 1347) concluded that the oxidized form of the dye has double the molecular weight of the reduced form. They assign, therefore, to pyocyanine the empirical formula $C_{26}H_{20}O_4N_2$. Both FRIEDHEIM and MICHAELIS 376) and ELEMA 233) conclude from their potentiometric studies that the dye has half the molecular weight given by WREDE and STRACK and consequently the formula II which had been doubled by the latter workers. KUHN and VALKÓ 672) confirmed the findings of MICHAELIS and of ELEMA by diffusion measurements in water and in benzene.

Chemical Properties: Pyocyanine crystallizes from chloroform-petrolether in thin dark blue needles. It is sparingly soluble in cold water, benzene, ether and petrol ether; and readily soluble in warm water, chloroform, warm alcohol, nitrobenzene, pyridine and phenol. The solutions show no well-defined absorption bands. The pigment shows a sharp melting point at 133° without decomposition. Dry distillation yields a sublimate of yellow crystals of α-hydroxyphenazine. 2 N.NaOH destroys it.

A distinct fluorescence of pyocyanine in solution has been described by DHÉRÉ. It has already been mentioned that the free pyocyanine base which exists in alkaline solutions is blue and that the cation (or hydrochloride) is red.

Oxidation-reduction behaviour: Reducing agents transform pyocyanine into the colorless dihydro-pyocyanine. The reduction is readily reversible. Dihydro-pyocyanine is autoxidizable (376, 233)). In alkaline solution the reduction takes place in one step involving the uptake of two electrons. Consequently, the color change is here from blue directly to colorless (cf. 362)). In acid solution, on the other hand, i.e. at ph below 6, the reduction proceeds in two steps which are indicated by a color change from red over green to colorless (FRIEDHEIM and MICHAELIS 376), ELEMA 233)). The result of the careful potentiometric studies of pyocyanine by MICHAELIS and by ELEMA is summarized in Fig. 14 showing the dependence of the normal potentials on the hydrogen ion concentration. The ordinate gives the potentials as they were directly measured, i.e. referred to the calomel electrode. The normal potential of pyocyanine, as referred to the standard hydrogen electrode, $E'_o$, is —0.039 V. at ph 7.0. This is slightly more negative than the potential of methylene blue.

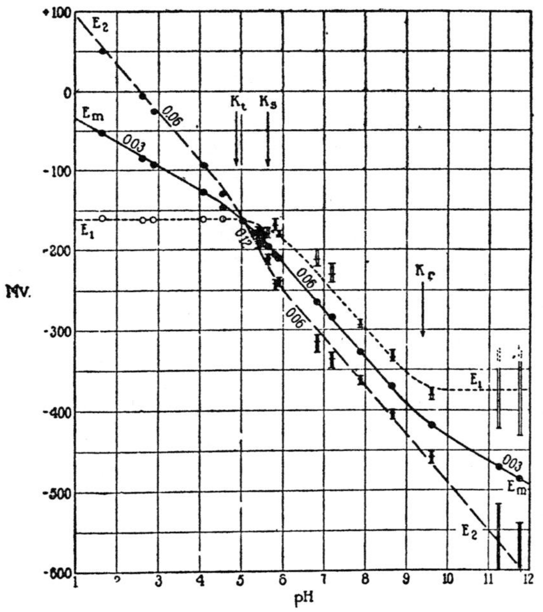

Fig. 14. E/ph-Diagram of pyocyanine acc. to MICHAELIS 808).

The diagram, Fig. 14, shows the three normal potentials $E_1$, $E_2$, and $E_m$ of the various steps of the reduction-oxidation process as calculated from the index potentials (see p. 101). At ph above 6 the two steps show marked overlapping. MICHAELIS has calculated that at ph 7 there may exist as much as 10 per cent of the semiquinoid radical form. Dissociation constants are indicated at ph 4.85 (oxidized form), ph 5,7 (semiquinone) and ph 9.26 (reduced form).

The semiquinoid radical could be isolated by KUHN and SCHÖN 670) in the form of the perchlorate as obtained by treating leuco-pyocyanine with $ClO_4''$ in ether. The titration of the compound with sodium stannite shows that the dye is contained in it in the monohydroform. A further proof of the radical nature of the intermediate form is the paramagnetism of the perchlorate (KATZ 533)). The formula of the radical as proposed by MICHAELIS 805, 801, 803) is as follows:

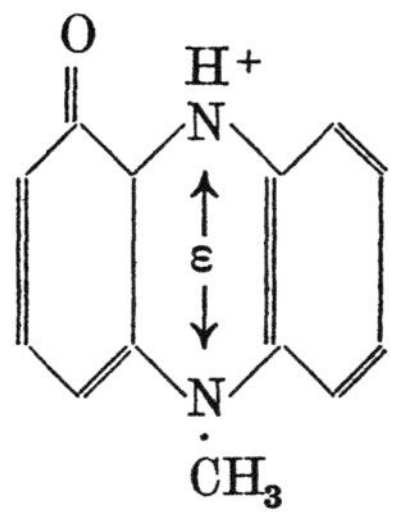

The electron is thought of as oscillating between the two azine nitrogen atoms. Conversely, the positive charge will also oscillate in accordance with the electrostatic requirements of the resonance phenomenon.

In dilute alkali, under pure nitrogen, the radical will suffer disproportionation into the fully oxidized and the fully reduced form:

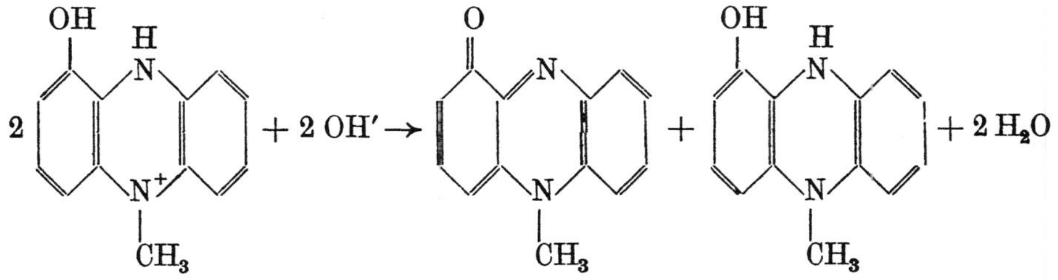

Upon admission of air the leuco form is reoxidized and consequently the color strength and stannite amount required for reduction are doubled (670)).

Precursors, Isomers, and Homologues of Pyocyanine:

The mother substance of pyocyanine, α-hydroxyphenazine, is also a reversible redox system (MICHAELIS, HILL and SCHUBERT 816)). The oxidized form is lemon-yellow between about ph 1 to 11; it turns red at ph < 1, and cherry-red at ph > 11. The semiquinoid form is green at every ph where it is capable of existence. The normal potential at ph 7 is about 100 millivolts more positive than that of pyocyanine. The E/ph-curve of this dye will be found in the original paper as well as in the review article by MICHAELIS 805). As in the case of pyocyanine, the ordinate is plotted in terms of the potentials measured against the calomel electrode and not as referred to the hydrogen electrode. In order to obtain the corresponding $E_o'$-values, about 240 millivolts must be added to these values.

WREDE and STRACK 1348) have prepared the ethyl and the propyl homologues of pyocyanine. Their potentials have as yet not been measured. However, their increased solubility in water and their greatly increased stability towards alkali and oxygen as compared with pyocyanine may render these dyes useful in biological experiments. Qualitatively, the two homologues show an oxidation-reduction behaviour similar to pyocyanine.

The potentials of $\beta$-hydroxyphenazine and of N-methyl-$\beta$-oxyphenazine (methyl aposafrone) have recently been studied by PREISLER and HEMPELMANN 937). Both systems show clearly the phenomenon of two-step reduction in solutions more acid than ph 3 and an overlapping of the two steps at higher ph-values. The $E_o'$-values at ph 7.0 are about —0.200 V. for $\beta$-hydroxyphenazine and about —0.165 V. for N-methyl-$\beta$-oxyphenazine. These potentials are, then, considerably more negative than that of pyocyanine. The semiquinones are yellow green in acid solution. Only the N-methyl compound seems to show promise as a redox indicator and ph indicator as well.

It should be possible to cover the physiologically important range of the redox scale by compounds related to pyocyanine.

### b)  Chlororaphine*).

Chlororaphine, the pigment of the Bacillus chlororaphis G. and S., was first studied from various points of view by LASSEUR in 1911. The constitution of chlororaphine was elucidated by the recent work of KÖGL et al. 586a). They showed it to be identical with monohydrophenazine-$\alpha$-carbonamide.

Preparation: Bacillus chlororaphis is grown according to LASSEUR on a liquid "synthetic" medium containing asparagine, glycerol, and inorganic salts. The pigment is secreted from the bacteria in crystalline form. The green crystals may be separated from the cells by fractional centrifugation. Besides the green chlororaphine, the bacteria produce a yellow pigment, called oxychlororaphine. The third form described by LASSEUR, called xanthoraphin, was found to be identical with oxychlororaphine. Chlororaphin is freed from adventitious oxychlororaphin by washing with chloroform. It is then recrystallized in an atmosphere of nitrogen from acetone-water. One liter of culture medium yields about 37 mg. chlororaphin.

Constitution: KÖGL received 240 mg. of the original preparation of LASSEUR in the form of oxychlororaphine. Its empirical formula is $C_{13}H_9ON_3$. Upon heating with KOH one mole $NH_3$ is split off and a carboxylic acid of the empirical formula $C_{13}H_8O_2N_2$ is formed. The latter is decarboxylated to phenazine by distillation with soda lime. Oxychlororaphine was found to be identical with synthetic phenazine-$\alpha$-carbonamide (I). The latter, upon mild reduction, e.g. with zinc dust in boiling water, yields a derivative identical with chlororaphine (II). Chlororaphine is therefore a quinhydrone-like derivative

I.
Phenazine-$\alpha$-carbonamide
(Oxychlororaphine
= Xanthoraphine)

II.
Semiquinone
(Chlororaphine)

III.
Dihydro-form.

*)  Reviewed by KÖGL 586) and MICHAELIS 805).

of oxychlororaphine. The original view of KÖGL that chlororaphine is a molecular compound of the fully oxidized and the fully reduced phenazine-carbonamide has been modified in accordance with the views of MICHAELIS to the effect that it is now considered to be the semiquinoid radical of oxychlororaphine both in the crystal and in mineral acid solutions. At other ph-values the radical disproportionates into the quinoid and benzenoid forms.

Properties: Chlororaphine crystals melt in $N_2$ at 228—230ᶜ (uncorrected). They are insoluble in water, chloroform, petrolether, benzine, toluene and in alkali. They are sparingly soluble in alcohols and readily soluble in acetone, phenol, aniline, glacial acetic acid and mineral acids. When exposed to air the crystals are slowly transformed into oxychlororaphin (phenazine-α-carbonamide). In acetone-HCl, the pigment shows absorption bands at 705, 638, 587 and 538 m μ.

Oxychlororaphine (phenazine-α-carbonamide) crystallizes from methanol in pale-yellow needles (M.P. 241°). The crystals are insoluble in petrolether, little soluble in water, ether and alcohols, somewhat better soluble in chloroform, acetone, and glacial acetic acid.

Oxidation-Reduction Potential: The semiquinoid nature of chlororaphine was fully confirmed by the potentiometric study of ELEMA 234). He found that at very low ph-values all three oxidation-reduction stages (see above) are sufficiently soluble to permit potentiometric titrations in a homogeneous system. Upon reduction the color of oxychlororaphine turns from light yellow to an intense emerald green and then to orange yellow, the latter representing the color of the dihydrocompound. Fig. 15. shows the three normal potentials of the chlororaphine system plotted against ph, according to ELEMA.

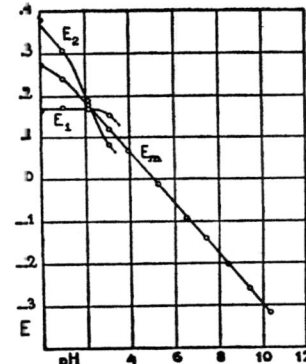

Fig. 15. $E_o$/ph-curve of Chlororaphine (ELEMA 234)).

The crossing-over of the three potentials is located at ph 2.1 (semiquinone constant at this point = 1). At the right side of the crossing the formation constant appears to remain constant, although small, over a certain ph-range. The radical might thus be formed to a small extent even at higher ph-values just as in the case of pyocyanine and riboflavin. The extremely low solubility of the semiquinoid form (chlororaphine) in the physiological ph-range accounts for its formation in crystalline form in the bacterial cultures. $E_o{}'$ at ph 7 is about —0.130 V., i.e. 100 millivolts negative to pyocyanine.

### c)  Toxoflavin. *)

Toxoflavin is the name conferred by VAN VEEN and MERTENS 1193, 1194, 1195) upon the prosthetic group of a yellow pigment which arises in cultures of Bacterium bongkrek under aerobic conditions. The native pigment complex together with another colorless substance, is held responsible by these workers for a number of cases of food-poisoning among the natives of Java. Whereas the native pigment complex is highly toxic when administered perorally or intraperitoneally to monkeys, pigeons and rats, the crystalline prosthetic group, toxoflavin, retains its high toxicity only for rats when injected intraperitoneally, 5 to 25 γ representing the lethal dose. In the monkey, 1 to 2 mg. per os will induce vertigo and sleep. The decrease in toxicity in the course of the

*)  See also p. 116.

purification of toxoflavin is attributed by the authors to the detachment of the yellow dyestuff from a natural colloidal bearer, presumably of protein nature, but also apparently containing lipids (phosphatides).

Preparation and properties of native pigment complex: Young glycerol-water cultures of B. bongkrek which are intensely yellow are treated with basic lead acetate. This reagent precipitates impurities. The yellow filtrate is freed from excess lead by adding phosphate (ph 7). The yellow pigment may be further purified by dialysis. The dry residue obtained by evaporation in vacuo at low temperature is strongly yellow colored and contains lipids as well as proteins besides toxoflavin. It is soluble in water but insoluble in absolute alcohol at 0° indicating the linking of the dyestuff to a protein. The whole complex may be precipitated by phosphotungstate at weakly acid reaction. On the other hand treatment of the aqueous solution of the pigment with alcohol at 30° causes a partial breakdown into an alcohol-soluble yellow pigment and an insoluble protein. However, the alcohol-soluble part does not yet represent free toxoflavin, being still precipitable with phosphotungstate. If now the aqueous solution is treated with $Na_2SO_4$ and if the salted-out pigment is extracted with chloroform, free toxoflavin passes into the solvent. Accordingly no precipitate is formed with phosphotungstate. Further proof for complex formation under biological conditions is furnished by the inability of the native complex to pass dialyzing membranes while the free toxoflavin is readily diffusible.

Isolation and properties of crystalline toxoflavin: A liquid culture medium containing glycerol, peptone and inorganic salts dissolved in water (ph 7) is spread in thin layers to ensure saturation with oxygen and is inoculated heavily with B. bongkrek. After 45 hours at 28° the culture has assumed a deep yellow color and contains a maximum of the pigment. It is saturated with anhydrous $Na_2SO_4$ at 0° and extracted with cold chloroform. After adding an equal volume of petrolether to the yellow-green, weakly fluorescing chloroform solution the toxoflavin, which is present in free form, is taken up by water. The aqueous solution is saturated with $Na_2SO_4$ and the pigment is extracted with ethyl acetate. After adding petrolether ice-water will again extract the toxoflavin. The concentrated solution is brought to dryness in a high-vacuum over $P_2O_5$ and adventitious fatty material is partly removed by pretolether and ether extraction. The residue is dissolved in a little water and once more evaporated to dryness in a high-vacuum. The almost pure, crystalline toxoflavin is now dissolved in chloroform, concentrated in vacuo and induced to crystallize by adding dry ether. One liter of culture medium will yield about 15 to 20 mg. toxoflavin in the form of small yellow crystals (partly flat needles and partly platelets). The crystals melt sharply at 171—172° with slight decomposition.

Toxoflavin is readily soluble in water and alcohol, less soluble in ethyl acetate and chloroform, and almost insoluble in ether, benzene and petrolether. The green fluorescence is more pronounced in amyl alcohol than in water. Light does not affect the pigment under varying conditions. Toxoflavin dialyzes rapidly through parchment membranes. It is adsorbed on norite and fullers earth and may be eluted from the adsorbates by dilute alcohol and pyridine (or ammonia). While the yellow color is quite stable in water at ph 3 to 7, it is slowly abolished at ph 1 and 9. Only at the latter ph the decolorization is reversible for a short time.

Light-absorption: According to measurements by E. R. HOLIDAY (cf. 1083) there exists a general resemblance between the absorption patterns of toxoflavin and of the flavin pigments. In aqueous solution (ph 6.5) a steep absorption band at 260 m$\mu$

is accompanied by an inflection at 310 m $\mu$ and by a lower and rather broad band centred at 405 m $\mu$. The molecular extinction coefficients calculated for a molecular weight of 166 (see below) amount to about one-half of the corresponding values for flavins. After standing for a short time at ph 3 or ph 11 the fading of the yellow color is accompanied by an almost complete disappearance of specific light absorption. At ph 9 the heigth of the long wave band decreases and a distinct band is formed instead at 320 m $\mu$. This band also disappears on standing.

Structure: Inspite of many chemical details brought to light through the work of van Veen and Mertens 1195) the problem of the exact chemical constitution of toxoflavin is still unsolved. The only conclusion that may be drawn at the present time with assurance is that the original assumption of van Veen and Mertens that the pigment may belong to the class of flavin pigments studied by Warburg, Ellinger, Kuhn, Karrer, Stern and others is untenable. The name toxoflavin chosen on the basis of this hypothesis is therefore a little misleading.

Microanalysis yields 43.5—43.26 % C; 3.57—3.98 % H; 34.04—33.9 % N, corresponding best to the empirical formula $C_6H_6N_4O_2$. There is one N-methyl group present in the molecule which is removed only with great difficulty by HI. According to Rast's campher method and to kryoscopic determinations in water the empirical formula given above is not to be doubled. Accordingly, toxoflavin would be an isomer of methylxanthine which is difficult to reconcile with the general properties of the pigment.

Toxoflavin is very stable against oxidizing agents like $H_2O_2$, $HNO_3$, $Br_2$. With bisulfite a brick-red addition compound is formed which is decomposed and decolorized by dilute acids. Heating with $Ba(OH)_2$ to 100° does not yield urea in contrast to photoflavin. Toxoflavin is devoid of acidic groups. Upon heating with $KClO_3$ plus HCl a strong murexide test is obtained when $NH_3$ is added. Alloxan (or methylalloxan) was identified as a product of oxidation under these conditions, besides methylamine and $NH_3$. Toxoflavin does not react with $HNO_2$ or bromine. Careful treatment with N.HCl yields a colorless compound the crystals of which melt at about 250° with decomposition. Microanalysis indicates that one $H_2O$ has entered the toxoflavin molecule. This would suggest that the ring present in toxoflavin has been opened to form an isomer of a methylated pseudo-dihydroxypurine. However, concentrated HCl even at 100° fails to cause ring formation. The derivative no longer gives the murexide test with $KClO_3 + HCl$. The solution in alkali is intensely yellow.

In a recent communication, van Veen and Baars 1192) assign the following structure to toxoflavin:

$$CH_3-N-CO-C=N-CH_2$$
$$OC-NH-C\overline{\quad\quad}N$$

The close relationship of this formula to that of methylxanthine is obvious. The place of attachment of the $CH_3$ group has been chosen arbitrarily. Ultimate proof of this constitution may probably be afforded only by synthesis.

Oxidation-Reduction Behaviour: Van Veen and Mertens 1195) made the following observations: In weakly acid or alkaline solution toxoflavin is reversibly reduced by sulfurous acid to a colorless compound. Upon shaking with air or addition of $H_2O_2$ or bromine water the yellow color is restored. This may be repeated at will. Upon hydrogenation in ethyl acetate with platinum black as the catalyst about 1 molecule $H_2$ is taken up and a brown product results. In acetic acid with platinum oxide as the catalyst, on the other hand, about 3 molecules $H_2$ are absorbed and a colorless solution results. In both instances shaking with air will restore the original yellow color. Young, actively growing bouillon cultures of B. bongkrek as well as liver brei in phosphate buffer at ph 7 will also slowly decolorize a dilute solution of toxoflavin. The process may be accelerated by adding glucose. Again, shaking with air restores the yellow color.

A gift of crystalline toxoflavin by Dr. A. G. van Veen made possible a potentiometric study of the pigment (Stern 1083)). It was established that toxoflavin represents the oxidant of an oxidation-reduction system which is fully reversible and electro-active between ph 4 and 8. The normal potential, $E_o'$, as referred to the normal hydrogen electrode, is —0.049 V. at ph 7.0. The slope of the titration curves is atypical throughout the ph-range investigated and shows an increasing steepness towards the alkaline range. Unfortunately the lability of toxoflavin in alkaline solutions did not permit one to advance into a ph-range alkaline enough to show, perhaps, a distinct break in the titration curves and visible semiquinone formation. However, it was felt that the consistent increase in steepness of the titration curves towards the alkaline range was sufficient basis for a tentative treatment of toxoflavin as a two-step redox system in the light of the theory of Michaelis and of Elema. The diagram shown in Fig. 16 contains individual titration curves on the left hand side and the resulting $E_o'$/ph-curve on the right hand side. The latter curve indicates dissociation constants of the reduced form at ph 5.8 and 7.2.

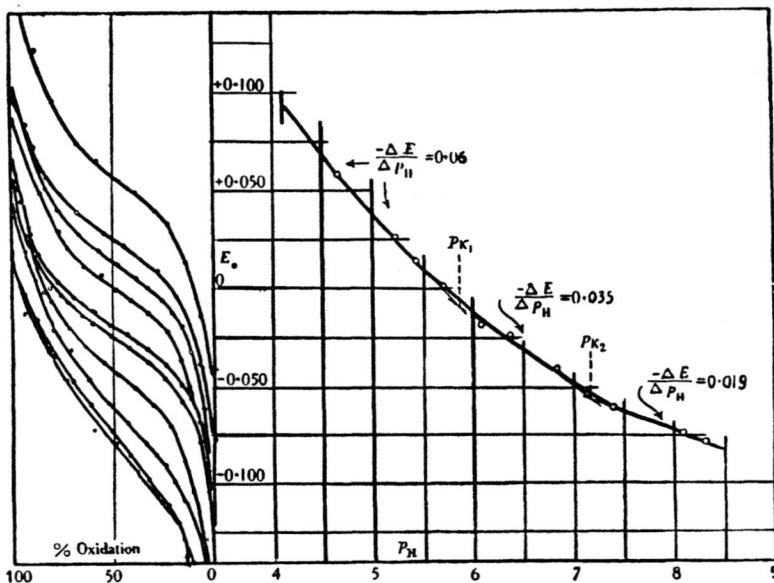

Fig. 16. Potentials of Toxoflavin (Acc. to Stern 1083)).

A scrutiny of the data concerning the flavins, (and more particularly photoflavin which shares several properties with toxoflavin) and toxoflavin results in the conclusion that the similarities noted by van Veen and Mertens are of a rather superficial nature and that they are outweighed by far by fundamental differences between the two types of pigments (Stern 1083)). The only known natural pigments to which toxoflavin may perhaps have a structural relationship are the pterins, the wing pigments of certain butterflies. Xanthopterin, the pigment of *Gonepteryx rhami*, is somewhat similar to toxoflavin with respect to elementary composition, absorption spectrum, reversible oxidation-reduction, murexide reaction indicating a purine nucleus, and fluorescence. *) Unfortunately, inspite of great efforts, the constitution or molecular

*) A discussion of the chemistry of these interesting pigments on the basis of the work of Hopkins, Wieland, Schöpf and other workers will be found in E. Bergmann's review article (97)).

weight of the pterins are likewise still obscure, all definite formulas which have been proposed having been refuted or withdrawn.

### d) Phthiocol and other Naphthoquinones.

Isolation of Phthiocol: When the human type of tubercle bacillus is grown on the LONG synthetic medium, the organism grows on the surface of the liquid in the form of a faintly cream-colored pellicle. ANDERSON and NEWMAN 24) discovered in alcohol-ether extracts of such cultures a pigment, phthiocol, which is yellow in acid and red in alkaline solution and which appears to be specific for tubercle bacilli. The pigment is largely associated with the acetone-soluble fat fraction of the bacteria. After saponification with KOH the pigment was separated from the crude fatty acid fraction and from anisic acid. The most important step in this separation process is the precipitation of the fatty acids in the form of their lead salts. Further purification procedures include extraction with amyl alcohol from the alkaline solution, steam distillation and removal of the admixed anisic acid by extraction of the mixed potassium salts with acetone. 260 gm. tubercle fat yielded 0.17 gm. of pure recrystallized phthiocol. The yellow prisms melt at 173—174°.

Properties: Phthiocol is only slightly soluble in water but it is easily soluble in organic solvents, except petroleum ether, giving yellow colored solutions. In dilute alkalies the pigment dissolves to yield deep red solutions. Acidification changes the color to yellow.

Microanalysis yields the empirical formula $C_{11}H_8O_3$ and molecular weight determination by RAST's method gives values 202—206) which are slightly higher than the minimum molecular weight (188)).

Constitution: Direct acetylation yields a monoacetate which, on reductive acetylation is converted into a triacetate (ANDERSON and NEWMAN 25)). This indicates a quinone linkage in the original substance and thus 3 O-atoms are accounted for, 1 as an OH-group, and the other 2 as a quinone. The yellow color and the volatility of the substance indicate an α-quinone rather than a β-quinone structure. On oxidation with hydrogen peroxide in alkaline solution phthalic acid is obtained. ANDERSON and NEWMAN conclude that phthiocol is 2-methyl-3-hydroxy-1,4-naphthoquinone:

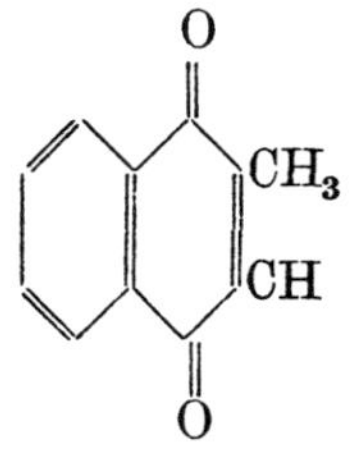

Synthesis: Several methods for the synthesis of phthiocol have been worked out. The first one by ANDERSON and NEWMAN 26) starts with 2-methylnaphthalene. The yield, however, was low, the maximum being 12 per cent. MADINAVEITIA 744) obtained the pigment from 2-methyl-1,4-naphthoquinone by the action of calcium hypochlorite followed by treatment of the reaction product with $H_2SO_4$. The yield was over 60 per cent. Finally, NEWMAN, CROWDER and ANDERSON 886) have described a third synthesis consisting in the decarboxylation of 3-hydroxy-1,4-naphthoquinone-2-acetic acid. The reaction is carried out by heating this compound, dissolved in diphenyl ether, in the presence of the copperbarium chromite catalyst of CONNOR, FOLKERS and ADKINS. The yield of phthiocol or 2-methyl-3-hydroxy-1,4-naphthoquinone was about 41 per cent of the theoretical.

Oxidation-Reduction Potentials: A careful potentiometric study of the potentials of phthiocol has been carried out by BALL 57), in the laboratory of MANSFIELD CLARK. He found both the oxidized and the reduced forms (colorless) perfectly stable throughout the usual ph range. At all ph values a change involving two electrons was found to occur in a single step in contrast to the two-step oxidation encountered in the case of other bacterial pigments (pyocyanine, chlororaphine). The pigment is among the most negative redox systems of biological origin. The normal potential, $E'_o$, at ph 7.0 is about —0.2 V. The following dissociation constants are assigned: for the oxidant $K_o = 8.32 \times 10^{-6}$, for the reductant $K_{r1} = 1.26 \times 10^{-9}$ and $K_{r2} = 2.88 \times 10^{-12}$.

Cultures of tubercle bacilli, according to AKSIANZEW, register potentials ranging from + 0.300 to + 0.005 V. If such potentials may be taken as an indication of the potential extant within the cell the phthiocol system would be maintained completely in the oxidized state. This is supported by the fact that colonies of the organism are described as possessing a yellowish tint.

## Other Naphthoquinones.

A number of other naphthoquinones have been found in nature. All, except phthiocol, occur in plants. Some of them, like juglon, the pigment of walnut shells, and lawson, the pigment of *Lawsonia*, are simple hydroxy-naphthoquinones, while others, e.g. plumbagin, a pigment from *Plumbago*, lapachol, occurring in the Lapacho tree and in other plants, have in addition alkyl side chains. All these substances represent reversible oxidation-reduction systems. The structural formulas and the normal potentials, $E_o$, at ph 0, as referred to the normal hydrogen electrode, will be found in Table 16. which was compiled from papers by BALL 57, 58).

BALL failed to observe any two-step oxidation processes in these pigments. However, inasmuch as quinones of the type of hydroquinone and phenanthrenequinone will form semiquinones in alkaline solution (anionic semiquinones) (MICHAELIS 805)), radical formation might reasonably be expected also in the case of the naphthoquinones provided that the potentials are studied at sufficiently alkaline ph values.

### e) Hallachrome.

This pigment was isolated by MAZZA and STOLFI 770, 771) from a marine polychete worm, *Halla parthenopea*. FRIEDHEIM 371) utilized for the isolation the property of hallachrome that the red, oxidized form at ph < 8.53 is soluble in amyl- or isobutyl-alcohol while it is insoluble in these solvents at ph > 8.53 and now is soluble in water. At the latter ph the oxidized form is green. In still stronger alkaline solution it is irreversibly transformed into a stable yellow compound.

Whereas MAZZA and STOLFI 771) and RAPER 958) consider hallachrome to be identical with the "red body", formed as an intermediate in the tyrosine-tyrosinase reaction, FRIEDHEIM 374) declares this view to be erroneous. The constitution of hallachrome is therefore not yet definitely established.

The leuco-form arising upon reduction in mineral acid solution is colorless. With increasing ph the system assumes a dark olive brown tint. When air is admitted the red,

## TABLE 16.

Structure and Normal Potential of Naphthoquinones.

| Compound | Source | E° of System Temperature | | Reference |
|---|---|---|---|---|
| | | volt. | °C. | |
| Phthiocol | As oxidant in human tubercle bacillus. | 0.2987 | 30 | BALL 57) |
| Lapachol | As oxidant in a variety of woods. | 0.3000 | 30 | BALL 58) |
| Lomatiol | As oxidant in seeds of *Lomatia ilicifolia.* | 0.3000 | 30 | BALL 58) |
| Lawsone | As oxidant in henna plant. | 0.352 | 20 | FRIEDHEIM 373) |
| Juglone | As reductant in walnut shells. | 0.435 | 20 | FRIEDHEIM 373) |

oxidized form is regenerated. FRIEDHEIM finds that the individual titration curves are atypical throughout the physiological ph-range and that they become increasingly steeper towards the alkaline region. Although the color changes are not clear cut, the author interpretes this phenomenon as due to two-step reduction in the alkaline range. Hallachrome, in this respect, behaves similar to the later studied toxoflavin (see p. 236). The over-all normal potential, $E_m$, at ph 7 is $+ 0.022$ V., i.e. close to that of methylene blue. Accordingly, the pigment exerts biological effects similar to those of methylene blue (see p. 116). In the worm, the pigment probably acts as an oxygen store for emergency purposes: Worms which are buried deep under the bed of the sea or which are asphyxiated in air-free water reduce the pigment and are thus able to contract a considerable oxygen debt.

# E. General Biological significance of Desmolysis.

## 1) General Picture of Desmolysis.

A comprehensive treatment of the historical and fundamental aspects of desmolysis will be found in OPPENHEIMER's book "Die Fermente" (902)). Inasmuch as the fundamentals of desmolysis have not suffered drastic changes since the time at which that book was written, the present discussion aims not at repetition but rather at amplification and some redistribution of emphasis.

The time is past when fermentation and oxidation were considered to be two quite distinct types of biological processes and when the enzymes active in fermentation, the zymases, and those active in vital oxidation, were treated as entirely unrelated. Today we speak of one great system of enzymes catalyzing the over-all phenomenon of energy production in the cell, termed desmolysis. Fermentation no longer designates an isolated type of metabolism of micro-organisms but it now covers the total anoxybiontic metabolism of all organisms. And we know that this metabolism is fundamentally the same and that it is promoted by the same type of enzymes no matter whether it remains the only type of metabolism in the cell or whether it merely represents the preparation for the terminal oxidation in oxybiosis. If the anoxybiontic metabolism does not pass over into oxybiosis, certain reactions take place leading to stabilization of the anaerobically formed compounds, and lactic acid or ethyl alcohol appear as the end-products. The correlation between respiration and fermentation or rather between oxygen tension and fermentation is maintained by the so-called PASTEUR-MEYERHOF reaction, the mechanism of which is still largely obscure.

If we disregard these special processes, there is one uniform mechanism operative in both phases (fermentation and respiration) of desmolysis, namely, the transfer of metabolic hydrogen. In anaerobiosis it terminates in lactic acid or alcohol and in aerobiosis it terminates in water. The hydrogen transfer is effected by two kinds of reversible oxidation-reduction systems, viz. by the catalysts in the narrow sense of the word (enzymes and mesocatalysts) and by certain metabolites performing catalytic functions at certain stages of the breakdown process (e.g. fumaric acid, oxaloacetic acid). Recent work has tended to obliterate the line of demarcation between these two groups to a certain degree. It only remains well defined there where the metabolites undergo irreversible changes, e.g. oxidation to $CO_2$ and water. In that case the metabolites cannot act as redox systems and are therefore no longer catalytically active. If we consider the system

Succinic acid $\rightleftarrows$ Fumaric acid $\rightleftarrows$ Malic acid $\rightleftarrows$ Oxaloacetic acid,

we have to assume that only a fraction of this reaction occurs in a reversible manner, whereas the bulk of the components is further broken down into pyruvic acid and ultimately completely oxidized. The reversible fractions of desmolysis, on the other hand, are the only ones capable of making available the energy required for the maintenance of life, e.g. for chemical synthesis, and physical chemical work such as osmotic work at semipermeable membranes.

The characterization of desmolysis as a hydrogen shift towards oxygen implies that the second end product of oxybiosis, $CO_2$, does not arise through a direct oxidation of carbon atoms but by catalytic decarboxylation or similar reactions just as in anoxybiosis. THUNBERG 1770) has lent particular emphasis to this point of view. In contrast to other metabolites, the catalytic $CO_2$ production from hexoses, and thus for the major part of energy-yielding desmolysis, may find its simplest expression in the schema

$$C_6H_{12}O_6 + 6 \ H_2O \rightarrow 6 \ CO_2 + 12 \ H_2.$$

There is a question whether this reaction is possible from an energetic point of view. WURMSER 1354, p. 23) has tried to estimate the change in free energy involved in this over-all process. The result, based on not-too-well established data, is that this idealized process involves almost no change in free energy. He found a small loss in free energy, $\triangle F = -6277$ cal. On the other hand, the balance with regard to heat of reaction is strongly negative, the reaction is endothermal: $\triangle H = 145222$ cal. It might be mentioned that the non-validity of BERTHELOT's principle is encountered in all anaerobic sugar transformations. The transition of glycogen into lactic acid, for instance, yields almost 50 % more free energy than heat (BURK 156)). In any event, the energy change is almost zero and it does not matter whether $\triangle F$ is actually a small positive or negative value, particularly because the reaction is completely schematized. In reality the well studied individual reaction, $CH_3 \cdot CO \cdot COOH \rightarrow CH_3 \cdot CHO + CO_2$, yields an appreciable amount of free energy. The main purpose of the above schema is to show that THUNBERG's view is not untenable on the basis of thermodynamical considerations. Energetically it is to be interpreted in such a manner that under anaerobic conditions the latent energy of the reaction $6 \ C + 6 \ O_2 \rightarrow$ is just sufficient to liberate again the hydrogen from C-H-compounds which during their formation from $H + C$ have released energy. The energy required for this H release is called d e h y d r o g e n a t i o n  e n e r g y. In the strongly schematized process written down above all of these partial reactions have been combined.

By characterizing the m a i n  c h a i n  o f  c e l l  r e s p i r a t i o n as an oxidation of metabolic hydrogen we arrive at a unified and comprehensive conception of its nature.

## 2) The Stages of Breakdown.

### a) General.

It is timely to subject the concept of "stages of breakdown" to scrutiny in the light of newer knowledge. One has to consider to what extent substances, isolated from biological systems and chemically well defined, do actually represent intermediates in desmolysis. The belief of the occurrence of chain reactions in biological processes gains ground steadily. This may mean a number of things. The way is open to interpretation of the reactants as "reactive molecules in statu nascendi" or as "activated molecules" or "free radicals". Up till now the only means of testing the rôle of a compound as an intermediate consisted in its chemical preparation and in determining whether it is actually attacked and changed in a typical manner when added to the biological system in question. Only a positive result carries weight here, because the absence of a reaction under such conditions may be explained by assuming that the substance in question will react only if present not in the common stable form but in "hydrate form" or in the "activated state". This term covers molecules of higher than average energy content, i.e. molecules which still carry the critical energy absorbed during their formation, as well as free radicals. The role of methyl glyoxal in fermentation, as postulated by NEUBERG, has been discussed from this point of view. The importance of these questions has repeatedly been stressed by WIELAND.

To mention a few specific examples, WIELAND and BERTHO 1305) in their studies of the oxidation of alcohol via acetaldehyde to acetic acid by bacteria encountered such unusual kinetic relationships between the first and the second stage of the process that they were led to postulate a much more rapid reaction of acetaldehyde in statu nascendi as compared with the usual, stable form of acetaldehyde. It is possible that the acetaldehyde formed in the first step remains linked to the enzyme, that it is particularly reactive in this state and that it is thus enabled to react rapidly to form acetic acid. QUASTEL and WOOLDRIDGE 952), in this connection, speak of the "polarizing field" of the catalysts on the surface of resting bacteria.

Analogous observations on the course of the aerobic transformation of acetic acid into succinic acid have been instrumental in arriving at a description of enzymatic stepwise processes. WIELAND and SONDERHOFF 1327) showed that THUNBERG's hypothesis of the transition of acetic acid to succinic acid by means of dehydrogenation reactions is so difficult to confirm experimentally, because the "activated" succinic acid in statu nascendi is very rapidly broken down into $CO_2$ and $H_2O$ as the final products. This is the reason why only a small fraction of the total succinic acid can be detected along with the remainder of the acetic acid, whereas "stable" succinic acid is attacked only one third as rapidly as acetic acid. As far as this consideration is concerned, it does not matter that the succinic acid perhaps does not arise exclusively from acetic acid but according to KREBS from acetic acid plus pyruvic acid (p. 274). Similarly, WEIL-MALHERBE 1287) finds that succinic acid when added to animal tissues behaves differently from that actually formed in these tissues. The same holds for other intermediates, e.g. fumaric acid and malic acid. Accordingly we have to assume that only that fraction of the intermediates can be isolated in stable form which has lost its activation energy. The activated molecules, on the other hand, facilitate synthetic processes. According to VIRTANEN 1204) and WIELAND and SONDERHOFF 1327), the combination of activated oxaloacetic acid with acetic acid gives rise to citric acid. The same product results from the collision of activated acetic acid with oxaloacetic acid (KNOOP and MARTIUS 583)). FRANKE discusses the acetoin formation ("carboligase reaction") under the same point of view. These thoughts are undoubtedly also of value when applied to synthetic reactions of an higher order in the cell such as the formation of fats from carbohydrates.

The differentiation between ordinary, stable molecules, activated molecules and free radicals leads with necessity to the generalization that in colloidal biological systems there exist no chemically distinct individual compounds as intermediates, but that we are dealing with equilibrium mixtures of all possible phases and types of molecules at varying levels of energy. Chemically well-characterized individual substances may be obtained from such systems by preparative methods which imply the shifting of the equilibrium towards the formation of the substance to be isolated. It is precisely this situation which renders the a priori decision between several possible schemes of biological processes unpractical. At best, one may say that a certain schema represents the simplest way of describing a given set of reactions.

### b) Anoxybiontic Breakdown of Carbohydrates *).

We have to distinguish between three phases of sugar desmolysis: the phase of the "first attack", anaerobic cleavage, and aerobic terminal oxidation.

Whereas it was formerly believed that the sphere of phosphorylation processes was limited to the phase of the first attack in the case of free hexoses as well as in that of glycogen, the recent work of EMBDEN, MEYERHOF, LOHMANN, ROBISON, DISCHE and PARNAS has shown that phosphorylation reactions also dominate almost the entire field of the anaerobic cleavage reactions. Only the very last steps take place without the participation of phosphoric acid. They concern the transformation of pyruvic acid into lactic acid or into ethyl alcohol via acetaldehyde.

It should be mentioned, however, that certain animal cells and protozoa are able to metabolize glucose without the help of phosphate. Examples are brain tissue (ASHFORD 33)), trypanosomes (REINER and SMYTHE 976)), tumor tissue (BUMM 155), ELLIOTT 246)) and embryonic tissue of the chick (NEEDHAM et al. 870)).

If we disregard these atypical systems for the present, it may be stated that the most profound change in the recent views on carbohydrate breakdown concerns the rôle of methylglyoxal. Under the leadership of NEUBERG this compound had come to be regarded as one of the intermediates occupying a key position of decisive importance. Methylglyoxal was held to be an intermediate in the formation of alcohol via pyruvic acid and acetaldehyde as well as in that of lactic acid. Today the place of methylglyoxal has been taken by pyruvic acid. Its change to lactic acid involves hydrogenation in contrast to methylglyoxal which in NEUBERG's schema forms lactic acid by hydratation. However, the existence in animal tissues of an enzyme, ketonealdehyde mutase or glyoxalase, catalyzing this reaction is proof in itself that methylglyoxal must have some biological significance; its independence of phosphorylation reactions suggests its importance in those cells which metabolize glucose partly or wholly without the participation of phosphoric acid. Incidentally, the glyoxalase reaction is the only instance in which the necessity of glutathione for carbohydrate breakdown has thus far been demonstrated: According to LOHMANN, glutathione in the SH-form is the activator of glyoxalase. It is therefore not surprising that NEEDHAM et al. 1115a) find glutathione to be indispensable for phosphorus-free carbohydrate desmolysis.

The current belief is that pyruvic acid arises first in the form of phosphopyruvic acid and more especially as the phosphorylated enol form:

$$\begin{array}{c} CH_2 \\ \parallel \\ C\!-\!O\!-\!PO_3H_2 \\ | \\ COOH \end{array}$$

The phosphopyruvic acid exists in equilibrium with 2-phosphoglyceric acid. The enzyme catalyzing this equilibrium is called enolase. Phosphopyruvic acid is the last phosphorylated compound in the chain of breakdown. Its phosphoric acid radical is transferred to glucose or glycogen by the action of the cophosphorylase (p. 222). The free pyruvic acid is then either reduced to lactic acid in animal tissues or it is decarboxylated to acetaldehyde in alcoholic fermentation.

---

*) *Recent reviews*: MEYERHOF 788) and PARNAS 917). See also OPPENHEIMER's "SUPPLEMENT", p. 1352—1365.

A further important development in the field of phosphorylation concerns the nature of the substrate which stands at the beginning of the anoxybiontic carbohydrate breakdown. This substrate now appears to be g l y c o g e n rather than a free hexose. For certain animal tissues and cells such as muscle and erythrocytes this fact has been recognized for some time. WILLSTÄTTER and ROHDEWALD 1337) show that the same is true for yeast and leucocytes which synthesize glycogen from glucose previous to the phase of the "first attack". This development is also of interest from an enzymatic point of view. Glycogen is already "phosphorylated" and PARNAS and WILLSTÄTTER have demonstrated for the case of muscle and yeast respectively that the "first attack" on glycogen does not yield P-free hexoses but "active" sugar phosphates. OSTERN 908) has shown the same to hold for starch as the substrate. The first attack on glycogen is not a hydrolysis, but a p h o s p h o r o l y s i s where glycosidic linkages are dissolved by the attachment of phosphoric acid residues (PARNAS). Where free hexoses are directly attacked we have to assume that the process starts with a intramolecular rearrangement of a stable hexopyranose into a r e a c t i v e   f o r m. The latter which is not capable of existence as such is stabilized by phosphorylation. In all of these reactions there are always several phosphoric acid ester formed simultaneously: The monoesters glucopyranose-6-phosphoric acid, fructopyranose-1-phosphoric acid, fructosefuranose-6-phosphoric acid and glucopyranose-1-phosphoric acid (CORI and CORI 185)) and the furanoid fructose-diphosphoric acid. According to the present views the latter is the decisive constant intermediate of the desmolysis proper demonstrating the necessity of the rearrangement of the pyranoid into the furanoid structure. The only uncertainty concerns the very first product arising from glucose. There is a distinct possibility that enolization of the glucose takes place to form the configuration $C_1(OH) = C_2(OH)$ . . . . . which has been proposed by a number of authors, e.g. HARDEN, NEUBERG, OHLE, and NILSSON 890). If this is the case, the enzyme catalyzing the rearrangement would be the hexokinase of MEYERHOF. If the latter is a phosphorylating enzyme, the first step would probably consist in the phosphorylation of the enol form in 6-position and the second in a rearrangement into one of the furanoid hexosemonophosphates. One could of course also visualize an enzyme the only function of which would be to change glucose into the enol form. The latter, being unstable, may then be temporarily stabilized by spontaneous reaction with phosphoric acid. Glycogen, as has already been mentioned, yields hexosemonophosphates by direct phosphorolysis at the glycoside linkages in 4-position (PARNAS). It was formerly assumed with EULER that the monophosphates are further split into two $C_3$-fragments one of which is changed to methylglyoxal without the aid of phosphorus, while the other which contains P is resynthesized to zymophosphate (hexosediphosphate) by a secondary reaction. In the new schema of fermentation the furanoid z y m o p h o s p h a t e occupies the starting position. It arises by the reaction of hexosemonophosphate with phosphopyruvic acid. The first step in the breakdown is the formation of the equilibrium system zymophosphate $\rightleftarrows$ 2 triosephosphoric acid set up with the help of the enzyme aldolase (MEYERHOF and LOHMANN). The triosephosphoric acids, in turn, form an equilibrium system, Glyceraldehyde phosphoric acid $\rightleftarrows$ Dihydroxyacetone phosphoric acid, which is shifted far over to the right hand side.

According to the original view of EMBDEN and MEYERHOF, this equilibrium system of triosephosphoric acids would give rise to the formation of glycerophosphoric acid and 3-phosphoglyceric acid:

$$CH_2OH \cdot CHOH \cdot CH_2O \cdot PO_3H_2 + HOOC \cdot CHOH \cdot CH_2O \cdot PO_3H_2.$$

Although such a dismutation is quite feasible and certainly does occur under certain conditions, it is no longer believed to be an important intermediary step in fermentation. It appears that glycerophosphoric acid is not regularly formed. The reaction has been somewhat relegated to the background in favor of a direct dehydrogenation of phosphoglyceraldehyde by pyruvic acid which was first postulated by DISCHE 206) in the case of erythrocytes and which has now been accepted by MEYERHOF also for the case of muscle.

In any event, at this stage 3-phosphoglyceric acid is formed as the essential intermediate. According to MEYERHOF, it is transformed into 2-phosphoglyceric acid and then to phosphopyruvic acid by the enzyme enolase with the loss of one molecule of water:
$$H_2PO_3 \cdot O \cdot CH_2 \cdot CHOH \cdot COOH \rightarrow CH_2OH \cdot C(O \cdot PO_3H_2)H \cdot COOH \rightarrow CH_2 = C(O \cdot PO_3H_2)COOH$$
Without liberation of inorganic phosphorus phosphopyruvic acid looses its phosphoric acid to hexosemonophosphate which becomes hexosediphosphate (zymophosphate). This transfer is a direct one in the case of yeast and an indirect one in the case of muscle. The free pyruvic acid thus formed is either hydrogenated by glycerophosphoric acid or triosephosphoric acid to lactic acid with the formation of phosphoglyceric acid or it is decomposed by decarboxylation. Depending on the enzyme system in operation, either acetaldehyde or lactic acid is thereby produced. As has already been indicated, the phosphoric acid residues combine with fresh carbohydrate molecules which furnish the hydrogen required for the reduction of acetaldehyde to ethyl alcohol. This process in turn yields phosphoglyceric acid and later pyruvic acid via triosephosphoric acid, and so on. The coenzyme system active in these phosphorylation reactions is adenosinetriphosphoric acid + magnesium (MEYERHOF-LOHMANN): Adenylic acid accepts the phosphoric acid from the phosphopyruvic acid and the adenosinetriphosphoric acid thus formed in turn yields phosphoric acid to glucose or to hexosemonophosphate (PARNAS). In muscle, the process is somewhat complicated by the fact that the adenosinetriphosphoric acid will transfer phosphoric acid first to creatine to form phosphocreatine (phosphagen) and that it is the latter which phosphorylates the sugar.

How do the pyridinenucleotides (e.g. cozymase) and their specific bearer proteins fit into the picture? The present consensus of opinion (EULER 272), MEYERHOF 795)) appears to be that a pyridine ferment participates in the dehydrogenation of triosephosphoric acid, probably of phosphoglyceraldehyde, by being reduced to the dihydropyridine form. Pyruvic acid functions as the acceptor of the hydrogen of the dihydropyridine. The products of the reaction are then lactic acid and phosphoglyceric acid. In muscle, again, a synthesis of phosphocreatine with the aid of adenylic acid interposes itself at this stage. According to EULER, equilibria are set up: the hydrogenated cozymase is not only dehydrogenated by pyruvic acid but also by glyceraldehydephosphoric acid. The combination of the two reactions is equivalent to EMBDEN's "dismutation" of triosephosphoric acid into phosphoglyceric and glycerophosphoric acid. In the event of the catalytic decarboxylation of pyruvic acid the latter is replaced by acetaldehyde as the acceptor of the pyridine hydrogen. The end product in this case is, of course, ethyl alcohol. The pyridine enzyme responsible for this reduction is the a c e t a l d e h y d r e s e which is identical with the well-known alcohol dehydrogenase. Another component of WARBURG's fermentation test system, the A-protein of NEGELEIN, is necessary for the phosphorylation of hexosemonophosphoric acid to hexosediphosphate by phosphopyruvic acid. It appears that the pyridine enzymes react only with triosephosphoric acid formed from zymophosphate. In this sense, the hexosemonophosphates are not yet "reaction forms" since they are not attacked by the pyridine enzymes of anoxybiosis (MEYERHOF).

The question of the coupling between oxido-reduction processes and phosphorylation reactions in anaerobic carbohydrate breakdown has recently been further investigated by MEYERHOF and his associates. The correlation of analytical figures for phosphorus distribution over the various fractions obtained from fermentation systems (inorganic phosphorus, readily and difficultly hydrolysable phosphorus, etc.) with spectrographic studies of the cozymase-dihydrocozymase system during fermentation led MEYERHOF, OHLMEYER and MÖHLE 798) to propose the following alternative schemes for the esterification of inorganic phosphate in fermentation:

1 Phosphoglyceraldehyde + 1 Cozymase + 1 Adenosinediphosphoric acid + 1 Phosphoric acid $\rightleftarrows$ 1 Phosphoglyceric acid + 1 Dihydrocozymase + 1 Adenosinetriphosphoric acid,

or,

2 Glyceraldehyde phosphoric acid + 2 Cozymase + 1 Adenosinemonophosphoric acid (Adenylic acid) + 2 Phosphoric acid $\rightleftarrows$ 2 Phosphoglyceric acid + 2 Dihydrocozymase + 1 Adenosinetriphosphoric acid.

According to these schemes, the coupling between the hydrogenation of cozymase by triosephosphoric acid and the addition of inorganic phosphate to adenine nucleotide is a reversible process. In accordance with suggestions made by EULER and also OSTERN, the workers consider it possible that the molecule common to the two processes — i.e. phosphorylation and oxidoreduction — may be cozymase itself.

The preparation of the radioactive phosphorus isotope $P_{15}^{32}$ (written $P^a$ for purposes of simplicity), has made it possible to "label" the phosphoric acid radicals and to follow their fate in fermentation processes. Thus, HEVESY, PARNAS and their collaborators 479) synthesized radioactive adenylic acid from $Na_2HP^aO_4$ + Adenosine and added it to fermenting yeast. After incubation an appreciable fraction of the radioactive phosphate was recovered in the sugarphosphoric acid ester fraction.

MEYERHOF et al. 796) have applied the same powerful tool to the study of some intermediary reactions of glycolysis. Due to the reversibility of the coupling between hydrogenation of cozymase by triosephosphoric acid and the combination of inorganic phosphate with adenine nucleotide (798)), $P^a$ will invariably enter the adenosinetriphosphate molecule during oxidoreduction. From there, $P^a$ goes to other phosphate acceptors by rephosphorylation. Under physiological conditions the "Wechselzahl" of the readily hydrolyzable phosphate in adenosinetriphosphoric acid amounts to a few seconds.

Cozymase does not bind inorganic phosphate either in the course of the hydrogen transfer or of phosphate transfer. Similarly, all pure re-esterification reactions, e.g. the PARNAS reaction when taking place without oxidoreduction, occur without the participation of inorganic phosphate. The same holds for the intramolecular shift of phosphoric acid groups, e.g. during the equilibrium reaction 3-Phosphoglyceric acid $\rightleftarrows$ 2-Phosphoglyceric acid.

Of all rephosphorylations involving adenine nucleotide the splitting of phosphopyruvic acid only is irreversible. The authors, therefore, attribute to this reaction a function comparable to that of a safety valve in anaerobic carbohydrate breakdown.

There can no longer be any serious doubt that the new scheme of anaerobic carbohydrate breakdown as sketched above is correct. The main question which remains open is whether it is the only scheme of desmolysis. One of the most important arguments brought forward against this hypothesis is that in certain instances cells are able to metabolize glucose in the presence of fluoride which is known to block EMBDEN-MEYERHOF's scheme at the stage of phosphoglyceric acid. This is true for certain bacteria and

A Working, Fairly Stabilized, Schema of Intermediate Carbohydrate Fermentation, with Energetics (EMBDEN-MEYERHOF-PARNAS Schools. (MEYERHOF. 1936; PARNAS, 1935; KERMACK, 1935; STEWART and STEWART. 1935; LOHMANN, 1934; MEYERHOF, 1935; MEYERHOF. 1936a))

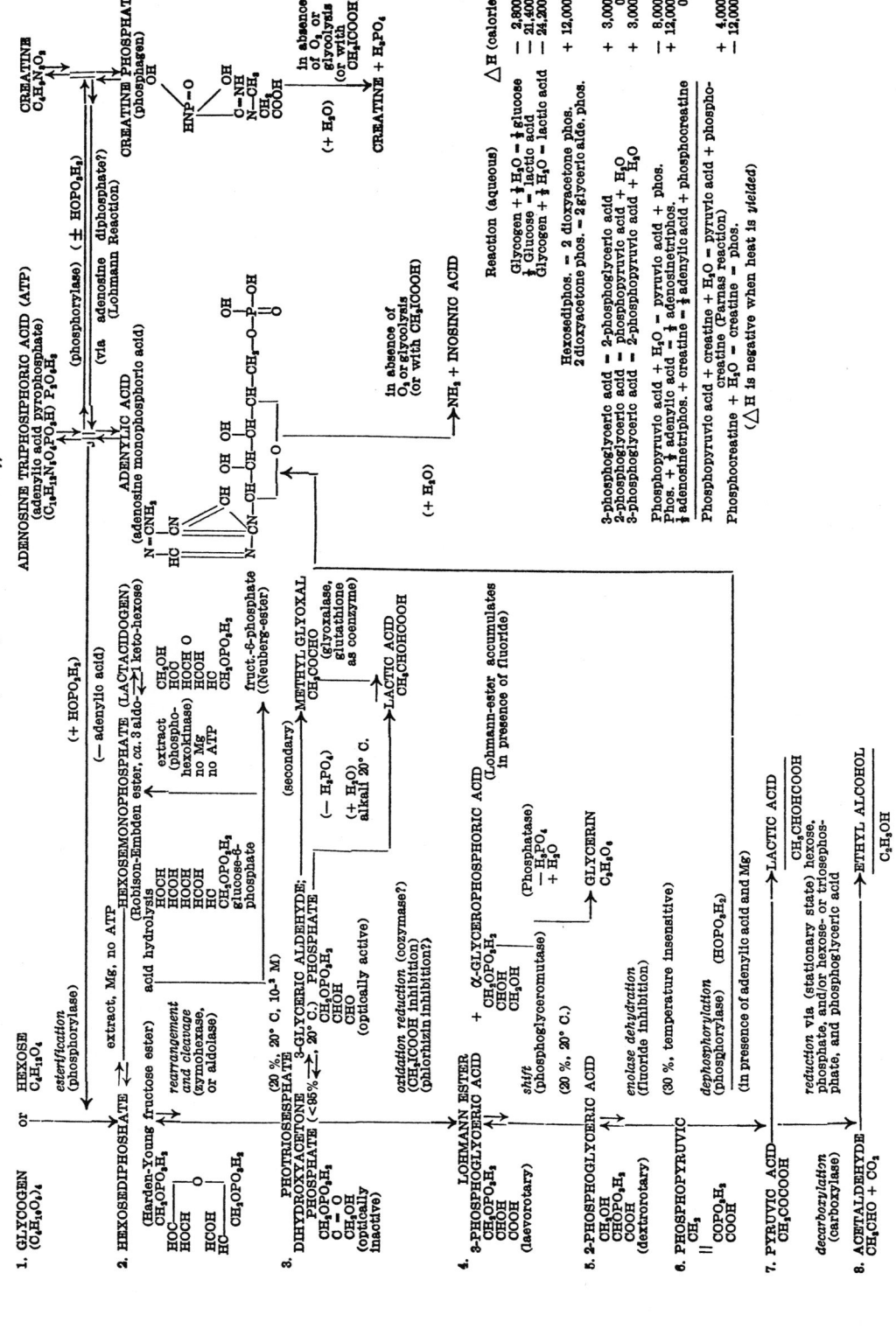

This represents a fairly recent schema of fermentation and glycolysis which has been compiled by D. BURK 157b) on the basis of the results of EMBDEN, MEYERHOF, PARNAS and other workers.

yeast strains (WERKMAN 1294)) and also for embryonic tissues (NEEDHAM 871)) where both mechanisms, viz., the "phosphorylation scheme" and the "phosphorus-free breakdown", seem to be operative. As has already been said, the main intermediate in the latter type of desmolysis is believed to be methylglyoxal.

GADDIE and STEWART 384), on the basis of experiments with glutathione as the activator of the ketonealdehyde mutase (glyoxalase), conclude that methylglyoxal is actually formed in muscle and that it is not to be considered as an artefact. They find that in the absence of glutathione methylglyoxal is formed and that in its presence lactic acid arises. Their experiments would indicate furthermore that glutathione also affects the processes involving pyruvic acid. Acetone-dried preparations of muscle and "THUNBERG"-muscle extracts will form methylglyoxal, but will not attack pyruvic acid even when glycerophosphoric acid is added (AUBEL and SIMON 42)). Glutathione causes an increase of lactic acid formation at the expense of methylglyoxal which is amenable to analysis. There is, of course, also the possibility that there exist avenues of desmolysis leading neither through phosphoglyceric acid nor methylglyoxal. For details concerning the „P-free" breakdown of carbohydrates see OPPENHEIMER's „SUPPLEMENT" p. 1477.

## c) The Oxybiontic Terminal Desmolysis*).

Ordinarily, the oxybiosis of sugars involves as substrates the products formed by the anoxybiontic desmolysis. The key substance here appears to be pyruvic acid, no matter whether it is primarily produced by the main schema of anaerobiosis drawn above or from methylglyoxal or in an indirect manner by dehydrogenation of lactic acid. Sometimes oxidation may begin also with precursors of pyruvic acid, e.g. with phosphoglyceraldehyde or, according to JOST 510) with phosphopyruvic acid. There is, of course, also the approach from ethyl alcohol leading to acetaldehyde and acetic acid. It is not known whether this occurs under physiological conditions except in microorganisms, although added alcohol is oxidized by animal tissues. The principle that the anoxybiosis is the quantitatively most important preparatory stage for oxybiosis is now generally accepted, largely due to the efforts of KLUYVER (cf. 578) and his coworkers (cf. HOOGERHEIDE 496)). Keeping this in mind we may look into other possibilities of oxybiontic carbohydrate breakdown, although their chemistry and their actual import for vital processes are largely obscure. For example, we have knowledge of the beginning of the oxidation of the unshortened carbon chain of hexoses. There is the oxidation of glucosemonophosphate to phosphohexonic acid by an enzyme consisting of codehydrogenase II (triphosphopyridine nucleotide) and of a specific bearer protein (p. 213). In presence of yeast proteins, ROBISON ester is oxidized with the uptake of 3 $O_2$ (WARBURG and CHRISTIAN 1254)). According to LIPMANN 720), the process involves an oxidative decarboxylation: 1.5—2 molecules of $CO_2$ are produced per 1 molecule of oxygen absorbed, no $CO_2$ being liberated under anaerobic conditions. In addition, we know of a process whereby glucose itself is oxidized to gluconic acid. The enzymes catalyzing this reaction are the glucose dehydrogenase of HARRISON 459) which occurs in animal tissues and also in acetic acid bacteria (TANAKA 1137)) and the glucose oxidase of MÜLLER which is found in moulds. According to TAKAHASHI 1129), *Rhizopus* transforms gluconic acid into alcohol, formic acid, acetic acid, succinic acid and fumaric acid with tartaric acid as a hypothetical intermediate.

The possibility of a different type of sugar desmolysis has also been suggested by

_____
*) The literature up to 1933 has been reviewed by LOHMANN 732).

attempts to demonstrate a respiration with sugar as substrate which is independent of fermentation. We refer to the well-known experiments of LUNDSGAARD 740) according to which iodoacetic acid, at least at ph higher than 4.5, suppresses the fermentative metabolism of yeast and cold-blooded animals without affecting appreciably the respirations. There are also the observations of TRAUTWEIN et al. 1186) that yeast, at ph higher than 8, will no longer ferment but continues to respire. The statement of these workers that maltose is oxidized by yeast strains lacking maltase has been criticized by KLUYVER and HOOGERHEIDE 579) on the ground that the maltose used contained fermentable impurities. HOOGERHEIDE 496) has shown that yeast is only able to oxidize those sugars which are also fermented. This problem of an atypical oxidation of sugar involving no lactic acid or other products of fermentation is still at a controversial stage. In the case of plants and particularly the slightly fermenting moulds in particular such an atypical metabolism is quite possible (BOYSEN-JENSEN 141)). The beginning of sugar breakdown may well be made here by the glucose oxidase of MÜLLER and the gluconic acid thus formed may well be desmolyzed in a manner totally different from that of lactic acid. *) There exist also some experiments concerning animal tissues pointing in this direction, e.g. those of LOEBEL 727) who finds that fructose is not glycolyzed by brain slices but that it is completely burned by them, and the analogous experiments of SHERIF 1048) and QUASTEL 947) with galactose. It is not known whether the oxidation of fructose or galactose is accomplished by the main chain of respiration comprising the hemin systems or by the mechanism of the secondary or residual respiration. For a critical evaluation of the effect of iodoacetic acid on fermentation (glycolysis) and respiration reference is made to the papers by MEYERHOF 789), KREBS 599), YAMAMOTO 1368) and to the review article by KREBS 602).

In any event, the key substance of the beginning aerobic breakdown in n o r m a l cells is p y r u v i c  a c i d. In those cells which contain carboxylase the path leads over acetaldehyde to acetic acid which is dehydrogenated to succinic acid. But also in cells which contain no true carboxylase, acetic acid is probably an intermediate. LIPMANN 726) has shown that lactic acid bacteria change pyruvic acid into acetic acid by oxidative decarboxylation, and COOK 182) has demonstrated the presence of acetic acid in fresh liver tissue. The transformation of acetic acid into succinic acid goes probably by way of α-ketoglutaric acid. Succinic acid and ketoglutaric acid may also arise under anaerobic conditions from substances like glutamic acid, aspartic acid, and also sugars (KREBS). Beyond the stage of succinic acid, however, there is no possibility of further anaerobic dehydrogenation. From there on the terminal hemin system and oxygen as acceptor are apparently required.

The decisive key reaction consisting in the dehydrogenation of a c e t i c  a c i d to succinic acid has long been a theoretical postulate. WIELAND and SONDERHOFF 1327) have shown that this was due to the fact that activated succinic acid, i.e. succinic acid in statu nascendi, is rapidly further oxidized. The analytical detection of succinic acid has been possible in the case of *Rhizopus* (TAKAHASHI 1129)), of *Mucor* (BUTKEWITSCH 158)), of "depleted" yeast (WIELAND and SONDERHOFF 1327)) and *Fusarium lini Bolley* (ROTINI et al. 1008)).

Depleted yeast transforms added acetic acid under aerobic conditions mainly into $CO_2$ and $H_2O$; but a few per cent of succinic acid and also citric acid may also be detected. In the event that acetic acid appears as an intermediate in the oxidation by depleted yeast, it can also be demonstrated. This is true for the oxidation of alcohol (WIELAND

---

*) Details will be found in OPPENHEIMER's „SUPPLEMENT" p. 1455.

and WILLE 1329)) and of lactic acid and pyruvic acid (WIELAND et al. 1309)). In both instances, the oxidation proceeds via acetaldehyde, acetic acid and succinic acid. Methylglyoxal is also oxidized via acetaldehyde although only slowly. For animal tissues the change of added acetic acid to succinic acid has as yet not been demonstrated. This may be due to a decreased reactivity of "stable" acetic acid as compared with acetic acid in statu nascendi. Such a reactive form might conceivably be produced in vivo from acetoacetic acid (KÜHNAU 623)).

All the steps following succinic acid have been shown to be enzymatic catalyses (HAHN et al. 440), F. G. FISCHER 337), QUASTEL and WHEATLEY 946) and other authors). These consecutive steps are:

Succinic acid → Fumaric acid → Malic acid → Oxaloacetic acid → Pyruvic acid.

From pyruvic acid the cycle starts anew. We see that the shortening of the carbon chain is accomplished by decarboxylation and not by direct oxidation of the carbon. At the end of a complete cycle, i.e. from pyruvic acid to pyruvic acid, one pyruvic acid equivalent has been completely oxidized while the second has been recovered.

At the stage of oxaloacetic acid a secondary path leads to c i t r i c  a c i d (WIELAND 1309)). Citric acid may represent a normal metabolite. Under anaerobic conditions, yeast ferments citric acid to acetone dicarboxylic acid + formic acid (WIELAND and SONDERHOFF). The former is then hydrolytically decomposed into acetic acid and $CO_2$. The oxybiontic breakdown of citric acid has recently acquired a considerable interest. Upon dehydrogenation by citrico dehydrogenase with methylene blue (MARTIUS and KNOOP 755, 756) and also with oxygen (KREBS 611)) as acceptor, α-ketoglutaric acid is formed and thus the path to the succinic acid system is thrown open. K e t o g l u t a r i c  a c i d is now recognized as one of the most important key substances of oxybiosis. In the breakdown of citric acid, the following compounds play a role besides ketoglutaric acid: isocitric acid, oxalosuccinic acid, and cis-aconitic acid (MARTIUS and KNOOP). All of them are readily oxidized and are also partly changed back into citric acid by surviving tissue. According to KREBS, the formation and the breakdown of citric acid represents a process which in its partial reversibility is somewhat analogous to the fumaric acid cycle of SZENT-GYÖRGYI (see p. 268). Accordingly, KREBS speaks of citric acid as a respiratory catalyst. The "citric acid cycle" is blocked by arsenite at the level of ketoglutaric acid (KREBS).

The difficulty with regard to the first stages following pyruvic acid in animal tissues has been the absence of an active carboxylase in all organs except muscle. The same holds for lactic acid forming bacteria such as B. delbrücki (DAVIS 196)) and B. coli (e.g. MAZZA 769)). On the other hand, the new schema of anaerobic carbohydrate breakdown in such cells postulates the absence or at least only a small activity of a typical carboxylase: If pyruvic acid is to be hydrogenated to lactic acid as the end product of glycolysis, it must not be previously decarboxylated. Kinetic considerations are of course also important. In the reverse process, namely, the dehydrogenation of lactic acid by butyric acid bacteria, WIELAND and SEVAG 1326) observed retardation of decarboxylation as compared with the rate of formation of pyruvic acid. In oxybiosis it is quite possible that the hydrogen originally mobilized for the hydrogenation of pyruvic acid to lactic acid is rapidly used up by a stronger acceptor and that the pyruvic acid in this case might suffer a non-enzymatic breakdown to acetaldehyde of the type observed by WIELAND and WINGLER 1330) at palladium surfaces. According to HAHN 440), oxaloacetic acid suffers β-decarboxylation in this way, even heated tissue being active. The latter statement is contested by SZENT-GYÖRGYI (p. 268). It should be noted that acetaldehyde has not yet been detected as an intermediate in the aerobic breakdown of pyruvic acid in animal tissues (e.g. ELLIOTT et al. 244)). CEDRANGOLO 166), on the other hand, postulates acet-aldehyde as an intermediate in liver metabolism on the basis of inhibitor experiments with dimedone.

Until recently, all schemes proposed for an oxidation of pyruvic acid without a transition to acetaldehyde were purely speculative. They included the condensation of two pyruvic acid molecules to form acetoacetic acid (GORR 410)), parapyruvic acid (ANNAU et al. 29)) or diketoadipinic acid (TOENIESSEN 1182); MAZZA 769); POSSENTI 935)). WEIL-MALHERBE 1287) postulates a dehydrogenative condensation of 2 molecules of pyruvic acid followed by direct decarboxylation to α-ketoglutaric acid (see below). The same author discusses the production of an unstable intermediate in this process (1288)).

It would appear now that all these complex schemes might perhaps be dispensed with. We have to recognize that there exists a direct o x i d a t i v e  d e c a r b o x y l a t i o n of pyruvic acid to acetic acid + $CO_2$. Three different interpretations of the mechanism of this reaction have been proposed which are not altogether mutually exclusive.

There is the explanation by SEVAG 1045) that pyruvic acid in statu nascendi is rapidly oxidized by freshly produced $H_2O_2$. This oxidation of pyruvic acid to acetic acid + $CO_2$ by $H_2O_2$ has been known for a long time and it has recently been investigated from a kinetic point of view by WASSERMANN 1254). SEVAG finds that pneumococci are unable to attack pyruvic acid, whereas lactic acid is transformed into acetic acid + $CO_2$. He assumes that in the dehydrogenation of lactic acid pyruvic acid and $H_2O_2$ are formed which will react with each other instantaneously, even in presence of catalase. According to WIELAND, a catalase resistant peroxide of the type $CH_3$—C—COOH is probably

$$\underset{\text{OH O·OH}}{\overset{\wedge}{\phantom{x}}}$$

formed. It is subsequently decomposed into acetic acid, $CO_2$ and $H_2O$.

It is questionable whether these observations made with anaerobes may legitimately be applied to aerobic organisms, particularly in view of the fact that animal cells and also lactic acid forming bacteria are equipped with an enzyme, p y r u v i c  d e h y d r o g e n a-s e, catalyzing the dehydrogenation of pyruvic acid in a specific manner. The enzyme is closely related to carboxylase, sharing with it the agon, viz. vitamin $B_1$-pyrophosphoric acid (p. 205). One ought to speak of a g r o u p of pyruvic dehydrogenases rather than of t h e pyruvic dehydrogenase inasmuch as there are indications that the bearer proteins to which the prosthetic group or coenzyme is attached in the various cells show minor differences finding expression in a certain difference in properties. BARRON and MILLER 87) have found a pyruvic dehydrogenase in gonococci and they have given it the name α-ketooxidase. The system consists of an anoxytropic dehydrogenase which may utilize as acceptors dyestuffs of suitable potential, e.g. cresyl blue, and of an "oxidizing" iron or hemin system which enables the enzyme to utilize molecular oxygen as acceptor. A similar enzyme which appears to contain a flavoprotein instead of the iron system as the oxygen transferring component has recently been studied by LIPMANN 726) in B. delbrücki. DAVIS 196) had previously shown that this system catalyzes the oxidation of lactate and pyruvate to acetic acid + $CO_2$. LIPMANN was able to show that the dehydrogenase has the same agon as the carboxylase from yeast: after removing the natural coenzyme by suitable washing procedures he could replace it by a crystalline cocarboxylase preparation of LOHMANN and SCHUSTER 738) from bottom yeast. The difference between carboxylase and the B. delbrücki enzyme is that the latter will decarboxylate pyruvic acid only if 2 hydrogen atoms are detached at the same time. According to

LIPMANN **1164)**, pyruvic acid may also be oxidatively decarboxylated by ferric iron and methylene blue in light.

How can we visualize the chemistry of an oxidative dehydrogenation of pyruvic acid to acetic acid and $CO_2$? One would have to assume that pyruvic acid exists in the

form of an ortho-acid, $CH_3 \cdot CO \cdot C \overset{\diagup OH}{\underset{\diagdown OH}{—OH}}$, in the dehydrogenation of which the $CO_2$

becomes labile and is split off. If this is the case the enzyme responsible would be very similar in its action to the aldehydrases which will also attack an ortho-form of the aldehyde, $R \cdot CH(OH)_2$. Another possibility would be that the process takes place in two steps the first of which is a straight decarboxylation of the pyruvic acid and the second of which consists in a rapid oxidation of the acetaldehyde thus formed to acetic acid. The inhibition of the oxidation of pyruvic acid by liver tissue in the presence of dimedone, a reagent for aldehydes appears to favor the second alternative (CEDRANGOLO **166)**).

A related enzyme system is stated to effect a breakdown of pyruvic acid in anima tissues which begins with a "dismutation" reaction (KREBS **605, 606, 609, 610)**; LIP-MANN **724)**). Two molecules of pyruvic acid react to form acetic acid, lactic acid and $CO_2$:

$$CH_3 \cdot CO \cdot COOH + CH_3 \cdot CO \cdot COOH \overset{+ H_2O}{\longrightarrow} CH_3 \cdot COOH + CO_2 + CH_3 \cdot CHOH \cdot COOH \quad (1)$$

Thereby half of the pyruvic acid is transformed into the next lower stage of desmolysis, acetic acid, and the lactic acid is perhaps changed back to pyruvic acid by dehydrogenation. This reaction occurs under anaerobic conditions. Succinic acid is also formed under these circumstances. KREBS assumes that it arises through the oxidative decarboxylation of α-ketoglutaric acid. He finds that the acceptor in this anaerobic transformation is a keto-acid, e.g. pyruvic acid, which in turn is changed into an hydroxy-acid. He furthermore assumes that the same reaction takes place in the course of the desmolysis of pyruvic acid alone. Following upon reaction (1) in which acetic acid is formed, it is assumed that 1 molecule of pyruvic acid + 1 molecule of acetic acid as the joint donators will react with a keto-acid as the acceptor in such a way that ketoglutaric acid plus a hydroxy-acid are formed. In other words, KREBS postulates a dehydrogenation of two methyl groups in the sense of WIELAND:

$$\begin{matrix} CH_3 \cdot COOH \\ \\ CH_3 \cdot CO \cdot COOH \end{matrix} + R \cdot CO \cdot COOH \rightarrow \begin{matrix} CH_2 \cdot COOH \\ | \\ CH_2 \cdot CO \cdot COOH \end{matrix} + R \cdot CHOH \cdot COOH \quad \dots \quad (2)$$

The third stage of the reaction, the oxidative decarboxylation of ketoglutaric acid by "dismutation" corresponds to reaction (1):

$$\begin{matrix} CH_2 \cdot COOH \\ | \\ CH_2 \cdot CO \cdot COOH \end{matrix} + R \cdot CO \cdot COOH \rightarrow \begin{matrix} CH_2 \cdot COOH \\ | \\ CH_2 \cdot COOH \end{matrix} + CO_2 + R \cdot CHOH \cdot COOH \quad \dots \quad (3)$$

The enzyme system catalyzing this reaction would have to be classified as a mutase.

Since any keto-acid may function as acceptor, the hydrogenation of acetoacetic acid (cf. EDSON **229)**) as well as that of oxaloacetic acid might conceivably be accomplished in this manner. This would lend a new aspect to the theory of SZENT-GYÖRGYI (p. 268) which postulates a hydrogenation of oxaloacetic acid to malic acid which in turn exists in an equilibrium with fumaric acid catalyzed by fumarase. In these reactions, too, vitamin $B_1$ or rather its pyrophosphoric acid ester acts as a coenzyme. The scheme outlined above appears to have general validity: any α-keto-acid may either react with itself or with another α-keto-acid to form a simple acid + $CO_2$ plus an α-hydroxy-acid or it may react with a β-keto-acid to yield a simple acid + $CO_2$ plus a β-hydroxy-acid. In the course of such dismutations acetoacetic acid also arises. Consequently, acetoacetic acid may be derived from carbohydrate in the following manner: the sugar is broken down to pyruvic acid; pyruvic acid and acetic acid form an a c e t y l p y r u v i c  a c i d which through an oxidative or hydratizing decarboxylation breaks down into the equilibrium system acetoacetic acid + $H_2 \rightleftarrows$ β-hydroxybutyric acid:

$$CH_3 \cdot CO \cdot CH_2 \cdot CO \cdot COOH \begin{array}{l} \xrightarrow{\;+\,O\;} CH_3 \cdot CO \cdot CH_2 \cdot COOH + CO_2 \\ \xrightarrow{\;+\,H_2O\;} CH_3 \cdot CHOH \cdot CH_2 \cdot COOH + CO_2 \end{array}$$

Liver transforms the acetylpyruvic acid aerobically to acetoacetic acid and anaerobically to β-hydroxybutyric acid.

As far as pyruvic acid itself is concerned, KREBS assumes that this anaerobic desmolysis is not the only mechanism. Whereas in brain and testis the breakdown of pyruvic acid is identical under anaerobic and aerobic conditions, in the case of kidney slices there exists also an additional aerobic mechanism which might perhaps be explained by BARRON's and LIPMANN's scheme.

The concept of the dehydrogenation of pyruvic acid by biological acceptors with a simultaneous liberation of $CO_2$ promises to throw light on some other, still more complex, fermentation reactions. According to WOOD and WERKMAN 1343), for instance, the propionic acid formation by certain bacteria may be due to the functioning of methylglyoxal as an acceptor of the hydrogen of the pyruvic acid. The reduction of the methylglyoxal would yield propionic acid. As a matter of fact, the presence of pyruvic acid and its decomposition into acetic acid + $CO_2$ have been established in these bacteria. On the other hand, propionic acid is probably also formed via succinic acid and by the hydrogenation of pyruvic acid. Again, in Clostridium butyricum, pyruvic acid serves as donator and butyric acid as acceptor with the resulting formation of butanol (BROWN et al. 151)).

### d)  Amino Acids and Fatty Acids.

These two groups of substances will be discussed here under a common heading, since the main reaction of their desmolysis consists in the breakdown of the carbon chains of varying length into products which might be termed "sugar decomposition products" from an enzymatic point of view. Before entering into a discussion of this process, we have to take up the question of the primary deamination of the amino acids which presents a problem of its own.

We have to distinguish between the „independent" or "isolated" deamination of amino acids by certain catalytic mechanisms and the enzymatic desmolysis of amino acids where the deamination is only one of the various phases of breakdown. The former may be brought about by "unspecific" heavy metal catalysis, by specific dehydrogenases, by quinone catalysis and, perhaps, by an anaerobic hydrolytic deamination. The latter is now known to consist in an enzymatic dehydrogenation of the amino acid into the corresponding imino acid and a subsequent hydrolysis to an α-keto-acid with the liberation of $NH_3$. The experimental proof for this schema has been furnished by the experiments of KREBS 600, 601, 608) who was able to stabilize the corresponding keto acids in kidney slices by poisoning their respiration with arsenite (HCN is not suitable in this case). The formation of a peroxide at the nitrogen atom postulated by KREBS has already been characterized as an unnecessary complication (p. 121). The hydrogen peroxide which is formed as a by-product causes secondary oxidations. KREBS was furthermore able to elucidate the fate of the individual amino acids. In the case of glutamic acid, for instance, he showed that the desmolysis leads over α-ketoglutaric acid and succino-semialdehyde to succinic acid:

$$
\begin{array}{c}
\underset{\big|}{\overset{CH(NH_2)-COOH}{\diagup}} \\
CH_2 \\
\underset{CH_2 \cdot COOH}{\diagdown}
\end{array}
\rightarrow
\begin{array}{c}
\underset{\big|}{\overset{CO-COOH}{\diagup}} \\
CH_2 \\
\underset{CH_2-COOH}{\diagdown}
\end{array}
\xrightarrow{-CO_2}
\begin{array}{c}
CH_2 \cdot C \overset{H}{\underset{O}{\diagdown}} + H_2O \\
\big| \\
CH_2 \cdot COOH
\end{array}
\xrightarrow{+ \text{Acceptor}}
\begin{array}{c}
CH_2 \cdot COOH \\
\big| \\
CH_2 \cdot COOH
\end{array}
$$

It is noteworthy that this schema which was proved by KREBS for the case of tissue slices has previously been postulated by F. EHRLICH and verified by NEUBERG for yeast. The last step, the dehydrogenation of the aldehyde to succinic acid, may also proceed anaerobically in the presence of a suitable hydrogen acceptor. Other keto acids, also pyruvic acid, appear to fulfill this acceptor function by their reduction to hydroxy acids, e.g. lactic acid. In this event the aldehyde stage might be dispensed with (KREBS 605, 606, 609)) inasmuch as the keto acid is directly, oxidatively decarboxylated according to KREBS. The above equation would then be simplified as follows:

$$
CH_2 \overset{CO \cdot COOH}{\underset{CH_2 \cdot COOH}{\diagdown}} + H_2O + \text{Acceptor} \rightarrow
\begin{array}{c}
CH_2 \cdot COOH \\
\big| \\
CH_2 \cdot COOH
\end{array}
+ CO_2 + Acc.H_2 \cdot
$$

In this manner succinic acid may be formed anaerobically, e.g. in muscle (MOYLE-NEEDHAM 841)). $NH_3$ is not liberated as such, but is probably utilized in statu nascendi for the synthesis of other amino acids. This assumption of MOYLE-NEEDHAM has been experimentally confirmed by BRAUNSTEIN and KRITZMANN 144). These authors could show that in the dehydrogenation of glutamic acid pyruvic acid functions as $NH_3$-acceptor and is thus converted into alanine. In an analogous manner alanine is directly changed into pyruvic acid, while oxaloacetic acid represents an intermediate in the corresponding conversion of aspartic acid. Succinosemialdehyde arises also from formylglutamine, the first product of the hydrolysis of histidine by histidase, by the action of $H_2O_2$ (EDLBACHER 227)). In this manner the amino acids just mentioned are directly subjected to the schema of desmolysis proposed by THUNBERG. The other amino acids with longer C-chains are converted into fatty acids via the keto acids. If a true carboxylase is present, their decomposition leads through the aldehyde stage; if it is absent, one has to consider an oxidative decarboxylation of the kind discussed for the example of pyruvic acid.

For the purpose of primary deamination, there are special dehydrogenases available. Inasmuch as these enzymes, except that acting on glutamic acid, can probably utilize only oxygen as acceptor they are called amino-oxhydrases (603, 604, 35, 573)). Although the specificity of these enzymes has not been clarified as yet, there appear to exist several amino-oxhydrases, e.g. a special tryptophane dehydrogenase. Furthermore, there seems to be a soluble enzyme attacking exclusively d-amino acids (perhaps both α- and β-acids) and another enzyme which is tied up with the cell structure and which reacts with the naturally occurring *l*-amino acids. Both types occur in micro-organisms and in the kidney and liver of higher animals.

The enzymes concerned with the further breakdown, e.g. opening of the ring of cyclic amino acids, can only be mentioned here: The proline nucleus as well as the proteinogen amines, histamine and tyramine, are attacked oxidatively (cf. KREBS 603)). Histaminase (p. 94) has now been recognized as an oxidizing enzyme, histidase splits the histidine nucleus hydrolytically. The manner of desmolysis of the diamino acids is not known. It is probable, however, that the α-amino group is first split off and that the remainder is broken down by β-oxidation. This at least is the process encountered by KEIL 534) in the instance of the unphysiological δ-aminovalerianic acid which is converted into 4-aminobutanone-(2).

Until recently,. the **desmolysis of fatty acid chains** was no enzymatic problem, since an enzymatic cleavage of longer C-chains was unknown. Today we know of dehydrogenases of fatty acids which are present in animal tissues and which bear the features of anoxytropic dehydrogenases: while methylene blue is readily utilized as acceptor, the terminal respiratory system is required for the interaction with oxygen. Consequently, the oxidation with air is cyanide-sensitive, and it will only proceed with liver slices but not with liver brei (QUASTEL and WHEATLEY 948)).

The mechanism of these enzyme reactions is not yet fully elucidated. We must, therefore, for the present rely on observations made in investigations of intermediary metabolism*). In principle, there are three ways in which the fatty acid chains are attacked. The most important reaction is KNOOP's 581) β-oxidation whereby an acetic acid residue is split off from the fatty acid molecule. This is probably also the manner in which the dehydrogenases break down fatty acids. The action of these enzymes appears to be limited to the dehydrogenation of the $CH_2$-groups close to the carboxyl group. At the double bond thus formed the subsequent β-oxidation, consisting in hydroxylation and further dehydrogenation, takes place. This schema has long been considered as the most likely for butyric acid which is dehydrogenated to crotonic acid and then probably hydrated to β-hydroxybutyric acid by an enzyme analogous to fumarase (see below). However, this mode of desmolysis seems to have general significance:

$$R \cdot CH_2 \cdot CH_2 \cdot COOH \rightarrow R \cdot CH = CH \cdot COOH \rightarrow R \cdot CHOH \cdot CH_2 \cdot COOH \rightarrow R \cdot CO \cdot CH_2 \cdot COOH$$
etc.

A second mechanism which has been widely discussed but not conclusively proven is the dehydrogenation in the middle of the C-chains which might lead to further desmolysis by the attachment of $O_2$ to the double bond and the formation of an intermediate peroxide. It might also be that $H_2O_2$ in conjunction with heavy metals (pseudo-peroxidases) brings about a secundary oxidation of the unsaturated fatty acid. The model experiments by SMEDLEY-MACLEAN 1059) with $Cu + H_2O_2$ and unsaturated fatty acids are of interest in this connection. On the other hand, the relative stability of highly unsaturated fatty acids in intermediary metabolism suggests that they are no intermediates in desmolysis, but that they exert a specific function in the cell.

A third mode of fatty acid breakdown is the ω-oxidation (1185, 1186). Here the terminal methyl group is oxidized. The length of the carbon chains subject to this oxidation is from 7 to 12 C-atoms both in the case of triglycerides and sodium salts of the fatty acids. Undecylic acid will yield undecandicarboxylic acid in this manner. The

---

*) For details the reviews by OPPENHEIMER (l.c. 905, p. 142) and by KÜHNAU 624) should be consulted, as well as OPPENHEIMER's "SUPPLEMENT" p. 1510.

latter is subsequently attacked from both ends until the stage of heptandicarboxylic acid is reached. The quantitative significance of the $\omega$-oxidation in higher animals is probably very small (see also FLASCHENTRAEGER 347) and MAZZA 766)). In this manner, butyric acid could conceivably be converted into succinic acid. The same is perhaps true for acetoacetic acid which according to KÜHNAU 623) also yields succinic acid. However, it is equally possible that the acetoacetic acid is first split hydrolytically into acetic acid and that the latter is transformed into succinic acid in the usual manner (KÜHNAU). According to KREBS 605), the succinic acid arises through interaction of acetoacetic acid and ketoglutaric acid, the $\beta$-hydroxybutyric acid being aerobically reconverted into acetoacetic acid and thus acting as an intermediary catalyst. This $\omega$-oxidation of fatty acids is only a special case of the quite general phenomenon of "methyloxidation" which has been encountered in various classes of substances such as terpenes (e.g. citral, geraniol) and camphor (KUHN 650)). KUHN is inclined to assume a direct oxidation whereby the methyl group is oxidized to $\cdots CH_2 \cdot OH$ and a subsequent dehydrogenation to $\cdots COOH$ with the participation of the usual redox systems such as the yellow ferment.

Even in the case of the biologically well established $\beta$-oxidation, the knowledge of the enzymes involved is quite scarce. It would appear that the fatty acid dehydrogenases found in the pancreas and the intestine (TANGL and BEHREND 1138)), in fatty tissue and liver (QUAGLIARIELLO 941, 942); MAZZA 767)) and in plant seeds (GRANDE 412)) are merely preparing the $\beta$-oxidation proper by creating a double bond between the carbon atoms 1 and 2 which is subsequently oxidized, probably to a keto acid, leading to breakdown. In the course of this $\beta$-oxidation eventually acetoacetic acid is formed. While this substance is normally further oxidized in the presence of carbohydrate decomposition products, it appears as an end-product in diabetes mellitus. The action of the carbohydrate decomposition products may be due either to direct reaction with acetoacetic acid to form ketol, $CH_3 \cdot CO \cdot CHOH \cdot CH_2 \cdot CO \cdot CH_3$, which in turn is readily further oxidized (HENZE 476)) or to their poising effect on the redox potential of the cell as suggested by KÜHNAU 622).

The further desmolysis of acetoacetic acid is pictured by KÜHNAU 623) as follows:

$$
\text{I} \begin{array}{l} C(OH)\!-\!CH_2 \\ \| \qquad\quad |\; {\diagup}OH \\ CH \!-\!\!-\! C\!\!\stackrel{\diagup}{\diagdown} \\ \qquad\qquad OH \end{array}
\rightarrow
\text{II} \begin{array}{l} CH_2 = C(OH)_2 \\ \\ CH_2 = C(OH)_2 \end{array}
\rightarrow
\text{III} \begin{array}{l} CH = C(OH)_2 \\ | \\ CH = C(OH)_2 \end{array}
\rightarrow
\text{IV} \begin{array}{l} CH_2\!-\!COOH \\ | \\ CH_2\!-\!COOH \end{array}
$$

This conversion to succinic acid may also take place anaerobically with methylene blue as acceptor.

It may well be that acetoacetic acid is not directly decomposed in this way, but that it re-enters the oxidation-reduction continuum in the cell by acting as an acceptor in the manner suggested by KREBS (p. 274). EDSON and LELOIR 229) formulate this concept by assuming a dismutation reaction between acetoacetic acid and succinic acid yielding as the primary product $\alpha$-ketoglutaric acid which subsequently is converted into succinic acid. It is difficult at present to follow this hypothesis through to the last consequences.

According to QUASTEL 511), acetoacetic acid is probably largely converted to $\beta$-hydroxybutyric acid. The latter represents about 70 per cent of the total acetone bodies in all types of ketoses which have been observed.

# F. Cell Respiration.

## I. Main Respiration and Accessory Respiration.

### 1) General Considerations.

Cell respiration may be defined as the sum total of all oxidative processes in the cell which lead to the complete desmolysis of the metabolites and which involve the uptake of oxygen as well as the evolution of carbon dioxide. The bulk of the substrates of cell respiration is formed by sugars or their decomposition products. Consequently, the respiratory quotient (R Q) of most tissues approaches unity under normal conditions. The burning of deaminized amino acids and of fatty acids tends to lower the R Q somewhat. Processes with a greatly different R Q do not come under the scope of cell respiration proper; but are either anoxybiontic reactions or experimentally isolated partial stages of oxybiosis such as pure dehydrogenations by oxygen (e.g. of succinic acid) or mere additions of oxygen (e.g. to unsaturated fatty acids) or the oxidative deamination of amino acids. All these reactions form no $CO_2$ and have therefore no R Q. In some instances where a dehydrogenation with a predominant decarboxylation is studied the R Q may approach the value 2. On the other hand, it has already been pointed out (p. 241) that the path of cell respiration is essentially characterized by the transfer of metabolic hydrogen; and that the production of $CO_2$ is automatically coupled with this main process by the fact that there arise in the course of dehydrogenation certain carboxylic acids which are capable of decarboxylation. The latter may either occur as a pure carboxylase reaction or it may be bound up with a further dehydrogenation (p. 252). The basic feature of cell respiration, then, is an interplay between hydrogen donators and hydrogen acceptors, including oxygen, organized and regulated by a series of oxidation-reduction catalysts.

After many detours and excursions we have returned to the concept of BATTELLI and L. STERN (see OPPENHEIMER 902)) who recognized the existence of a main respiration ("Hauptatmung") which is linked up with the structure of the cell and an accessory respiration ("Nebenatmung") which is caused by enzymatic and non-enzymatic systems which may be separated from the cell. This situation has been obscured for some time by the controversy between WIELAND and WARBURG who postulated the unitary nature of cell respiration, the former recognizing only the hydrogenation of oxygen by the dehydrogenases and the latter attributing all purely oxidative reactions to one structure-linked respiratory ferment ("Sauerstoff-übertragendes Ferment der Atmung").

The main respiration is nothing but the integer of the migration of metabolic hydrogen from the first donator to the terminal respiratory system. It is the respiration which is manifested by intact organs and tissues (also tissue slices) and which ceases soon after the death of the cells as well as after mechanical damage (crushing). It is blocked by cyanide in low concentrations ($10^{-3}$ M.). The "first hydrogen donator" is probably a sugar decomposition product, equivalent to lactic acid or pyruvic acid, arising through the typical anoxybiontic, preparatory stage of fermentation (glycolysis). That there are certain sidepaths, such as the direct oxidation of carbohydrate via gluconic acid or of fatty acids, has already been noted. The sensitivity towards cyanide suggests strongly that this main respiration proceeds practically in its entirety through the cytochrome-respiratory ferment system of WARBURG-KEILIN (p. 41). In the instance of bakers yeast, this has been experimentally proved by HAAS in WARBURG's laboratory (cf. WARBURG 1234)): if the respiration is to proceed completely through the hemin systems, the oxygen consumption, A, must be equal to the product of concentration of the hemin system, c, and the rate of the valency change $Fe^{++} \rightleftarrows Fe^{3+}$ per unit time ("Wechselzahl") divided by 4, because each valency change of the iron involves $\frac{1}{4}$ molecule of $O_2$ per 1 Fe (2 FeO $+ 2\dfrac{O_2}{4} = Fe_2O_3$). The relationship $A = k \cdot c \cdot \frac{1}{4}$ could be confirmed by the comparison of manometric and photoelectric data.

The alternative to the main respiration is the **accessory respiration** which has been characterized by BATTELLI and L. STERN by the fact that it is not dependent on the intact structure of the cell and by its relative resistance against organic solvents and proteinases. These authors attributed the phenomenon of accessory respiration to the action of those enzymes which they were able to separate from the tissues, viz., to uricase, aldehydrase, and alcohol dehydrogenase. The progress in our knowledge since that pioneer work requires a new definition of acessory respiration. We may say that it encompasses all those oxygen consuming processes which do not involve the respiratory ferment. Such processes which make use of other autoxidizable systems are:

I. **Dehydrogenating Oxidations:**
   1) by oxytropic dehydrogenases (e.g. SCHARDINGER Enzyme),
   2) via the yellow enzyme,
   3) by oxhydrases and oxidases (oxidative deamination of amino acids, proteinogen amines, tyrosine, uric acid),
   4) quinone catalyses by autoxidizable chromogens,
   5) secondary oxidations by $H_2O_2$
      a) by direct reaction with the substrate,
      b) via peroxidases and thermostable iron systems.

II. **Non-dehydrogenating Oxidations:**
   Oxidation of unsaturated fatty acids,
   $\omega$-oxidation, opening of rings (e.g. of proline).

One might state in general that observations made on crushed tissues or cell extracts are not suitable to permit an evaluation of the quantitative significance of such oxidations for the course of normal cell respiration. Systems which under such artificial conditions will catalyze independently a reaction involving atmospheric oxygen may well cooperate with the catalysts of main respiration in the intact cell. This is largely true for the yellow enzyme which may promote an oxygen uptake under "unphysiological conditions", but which within the cell of higher animals at least will probably react preferably with cytochrome rather than with $O_2$ directly. Uricase appears to represent one of the few enzyme systems which react directly with oxygen in vivo and in vitro by virtue of a separate heavy metal component. However, the amount of oxygen consumed for the oxidation of uric acid is a quantitatively insignificant fraction of the total $O_2$-uptake of the cell. The same holds probably for the primary oxidation of amino acids under normal conditions. If however the turnover of amino acids increases due to lack of carbohydrate, the amount of $O_2$ thus absorbed may attain appreciable proportions since $\frac{1}{2}$ $O_2$ is used up per 1 $NH_2$. Such an accessory respiration with amino acids as substrate becomes manifest when the main respiration of tissues particularly able to deaminize amino acids (kidney, liver, retina) is damaged in any way. This is observed, for instance, upon keeping such tissues in RINGER solution (KISCH 567)) and also when the anaerobic carbohydrate breakdown is interfered with by iodoacetic acid (785)). In accordance with the greater activity of the $d$-aminoaciddehydrogenases, the unnatural $d$-amino acids are faster attacked than the $l$-amino acids (572)). DICKENS and GREVILLE 204) observed a considerable formation of ammonia in tissues (kidney, spleen, JENSEN sarcoma, embryonic tissue) which had been depleted of glucose. With the exception of kidney slices, the $NH_3$-production was inhibited by a supply of glucose and fructose (see however ELLIOTT and BAKER 246)). One will also have to consider the contribution to the accessory respiration made by the oxidation of the aromatic and heterocyclic nuclei of protein decomposition products (proline, tyrosine) and the oxidation of histamine and tyramine. Other mechanisms are indicated in the above schema. The accessory respiration of certain bacteria by means of dyestuffs (pyocyanine, chlororaphine) and the "unphysiological" respiration of facultative anaerobic bacteria via the yellow enzyme are related phenomena.

## 2) The Cyanide-Resistant Residual Respiration.

The complete inhibition of normal cell respiration by HCN and $H_2S$ was one of WARBURG's main arguments in support of his postulate that the entire respiration is catalyzed by iron in the form of his hemin enzyme (1230)). One of the exceptions found by WARBURG himself (1216)) was the case of the alga *Chlorella*: in sugar-free media the respiration is not affected by cyanide. EMERSON 253) has subsequently shown that the addition of glucose increases the rate of respiration up to 20 times and that this "glucose respiration" is readily poisoned by HCN. It is probable that in sugar-free solutions amino acids or fats are burned and this "accessory respiration" is not inhibited by cyanide, because no iron containing catalysts are involved. The cyanide-resistant residual respiration of algae is stimulated by methylene blue (WATANABE 1286)). The insensitivity of the respiration of yeast maceration juice and of facultative anaerobic bacteria towards cyanide is explained by the discovery of the yellow oxidation enzyme and of its carrier function by WARBURG and CHRISTIAN (p. 42). Since then a number of other organisms with a largely cyanide-resistant respiration have been described, e.g. *Paramaecium* (396, 1054)), *Colpidium* (921, 931)), *Planaria* (152)), *Nitella* (1007)), *Ulva* (1286)), *Sarcina lutea* (394)), *Ascaris* (978)), *Melanoplus* eggs during the diapause (131)), and tape worm larvae (375)). All of these organisms appear to possess a rather atypical respiratory system. In the case of cells and tissues with a "normal" respiration involving the burning of sugar, such as liver, kidney, top yeast, *Acetobacter*, etc. hardly any cyanide-resistant residual respiration is observed. It is only when the cells are damaged in some way or when instead of sugar other metabolites are burned that such a residual respiration appears. The appreciable HCN-resistant residual respiration found by DIXON and ELLIOTT 213) was caused by their working with phosphate-RINGER media. ALT 20) and WARBURG 1230) showed that the same tissues (rat liver and kidney) which in the presence of phosphate had a residual respiration of 33 per cent had practically none (1 to 4 per cent) when examined in bicarbonate solution. An exception is retina the respiration of which is entirely unaffected by HCN in bicarbonate-RINGER, only the PASTEUR reaction being inhibited (LASER 705b)). The results of FIELD 333) and of STARE and ELVEHJEM 1065) who found a residual respiration of 13 to 60 per cent in various tissues are probably also due to the use of phosphate. The harmful effect of phosphate on tissues has been confirmed by VAN HEYNINGEN 481) for the range of higher concentrations. According to this author $10^{-2}$ M. phosphate has practically no effect. Several workers (BANGA et al. 66), KISCH 569), TORRES 1184), ENGELHARDT 254)) found a residual respiration varying with the nature of the tissue and of the time lapsed between the death of the animal and the experiment. Their findings might perhaps be explained by assuming that the objects studied had suffered more or less damage before or during the experiment. The importance of the type of substrate oxidized has already been stressed in connection with this phenomenon.

The cyanide-resistant residual respiration is n o t  i d e n t i c a l  w i t h  t h e  a c c e s-s o r y  r e s p i r a t i o n. This is due to the fact that HCN poisons not only the respiratory ferment but also enzymes like uricase, peroxidase, heavy metal-containing oxidases, which play an important rôle in the accessory respiration. On the other hand, in the presence of HCN certain reactions not belonging to the accessory respiration but ordinarily suppressed during the normal main respiration are accentuated, e.g. oxygen transfer via oxytropic dehydrogenases and the yellow enzyme.

## II. The Intermediary Catalysts (Mesocatalysts).

### 1) General.

The problem of prime importance in the field of cellular respiration is the elucidation of the nature and the sequence of the individual steps through which the metabolic hydrogen passes from the first donator on its way to the terminal respiratory system and thus to ultimate oxidation. In dealing with this problem we have to distinguish between the main respiration, comprising essentially the combustion of sugar breakdown products by the hemin system, and the independent accessory respiration. Let us first sketch the main respiration after making certain simplifying assumptions. One of these is that the terminal oxidation is not complicated by the fact that the oxygen is first reduced to hydrogen peroxide and that the further reduction of the latter is possibly accomplished by a special mechanism. Furthermore, it shall be assumed that the preparatory anoxybiosis is already completed, in other words, that pyruvic acid or acetaldehyde are already available as substrates of oxybiosis, instead of being stabilized to the end products of fermentation, lactic acid and ethyl alcohol respectively. Finally, the pyridine coferment involved in fermentation is assumed to exist in the reduced form. The first catalyst of oxybiosis is probably a yellow ferment (flavoprotein) in the oxidized form. It dehydrogenates the pyridine enzyme and hands on the hydrogen to the cytochrome system(p. 41). Let us consider a c e t a l d e h y d e  as  t h e  f i r s t  d o n a t o r  of oxybiosis. The next acceptor is acetaldehydrase which in the oxidized form reacts with the aldehyde to form acetic acid. It is not certain which substance acts as the acceptor for the dehydrogenation of acetic acid. According to REICHEL 966) the acceptor may be a yellow ferment but this is not proven. According to SZENT-GYÖRGYI (p. 268), the fumaric acid system enters the picture at this stage, or more correctly, at the equivalent stage of pyruvic acid. If we now insert pyruvic acid into this schema, this substance may also function directly as a donator (KREBS-LIPMANN): An enzyme system related to aldehydrase catalyzes a dehydrogenation or a dismutation of pyruvic acid resulting in the formation of acetic acid $+ CO_2$ or lactic $+$ acetic acid $+ CO_2$ respectively. The lactic acid reverts to pyruvic acid through the action of lactic dehydrogenase with the aid of an as yet unknown acceptor which might be a hemin system. From here on it is only possible to enumerate the individual stages of breakdown without being able to assign specific acceptors to the various dehydrogenase-substrate systems.

### Reactions with Molecular Oxygen:

The inability to place the subsequent catalysts of oxybiosis in their correct order is intimately connected with the problem concerning the stage of oxybiosis at which the molecular oxygen enters the picture as a chemical reactant. Theoretically speaking this could be the case at any link in the respiratory chain of catalysts and intermediates which represents an autoxidizable redox system. The fact, however, that in cells, respiring in the normal way at the expense of carbohydrate, practically the entire main respiration is blocked by HCN supports the view of WARBURG that the whole of the metabolic hydrogen reaches the respiratory ferment (cytochrome oxidase) as the last acceptor and that the sole function of the oxygen consists in the dehydrogenation of the ferrous form of the hemin enzyme. In other words, it appears that even in oxybiosis all of the intermediary catalysts of main respiration, including the cytochromes, operate "anaerobically" even if they are autoxidizable. Perhaps the most pertinent example is that of the yellow enzyme. In isolated form it is readily autoxidizable. Under the conditions of normal cell respiration, however, it is largely if not entirely dehydrogenated by cytochrome or other acceptors instead of molecular oxygen (p. 266). THEORELL 1158) has shown that this is simply a matter of rates of reaction: the reduced flavin enzyme is much more rapidly reoxidized by ferricytochrome c than by oxygen at the tensions prevailing in the living cell (approx. 40 mm. partial pressure). Even oxytropic dehydrogenase systems such as the SCHARDINGER system may, in vivo, utilize other acceptors than oxygen, e.g. the yellow enzyme. The dependence of the important dehydrogenase systems (e.g. succinic or lactic dehydrogenase) on the presence of "oxidases" is well known.

Oxybiosis and reaction with molecular oxygen are, therefore, no longer synonymous terms. Oxybiosis is meant to designate a biological phenomenon as constrasted to "fermentation" or stabilized anoxybiosis. But it is now evident that the normal oxybiosis of main respiration incorporates by necessity anaerobic individual reactions and only one aerobic process, namely, the reoxidation of the reduced form of the respiratory ferment by molecular oxygen. According to this terminology the logical way to apply the terms "anaerobic" and "aerobic" is the consideration whether oxygen is actually involved as a chemical reactant in the individual process in question. The mere presence of oxygen no longer appears to constitute a sufficient criterion.

What has been said about the main respiration, of course does not apply to the accessory respiration which has actually been defined as the sum of those processes not involving the terminal hemin system. Here molecular oxygen will probably intervene indiscriminately where ever an autoxidizable system exists.

If the terminal respiratory system is either lacking or has been put out of commission by mechanical or chemical damage, the chain of oxybiosis shrinks to a small number of links, because the higher steps leading to the cytochrome system can no longer be dehydrogenated. In that event the yellow enzyme becomes probably the only coupling link between the anaerobic reactions and the molecular oxygen.

## 2)  The Function of the Mesocatalysts.

When considering the contribution of the carrier systems or intermediary catalysts to respiration we have to make a sharp distinction between the phenomena of main and of accessory respiration. If it is agreed that the main respiration proceeds entirely via the terminal hemin system and in view of the fact that the latter, according to WARBURG, is able to deal with any amount of metabolic hydrogen offered to it, the sole function of the carrier systems can consist only in a s p e e d i n g  u p of terminal oxidation, but not in an i n c r e a s e of the total turnover of substrate. In the accessory respiration, on the other hand, the intermediary catalysts may control the volume of the process.

The acceleration of main respiration by the complex system of carriers present in the cell may be due to two different mechanisms one of which is a matter of reaction kinetics, whereas the second is topochemical in nature. It has been pointed out on several occasions in this monograph (especially on p. 90) that the rate of an acceptor reaction depends on the difference in potential in such a manner that too great a difference in the redox potential may slow up either the reduction of the oxidized form of a catalyst or the oxidation of its reduced form by the next higher redox system. Inasmuch as the slowest step determines the rate of the over-all process, maximum velocity of the latter can be expected only if the two interacting redox systems are not too far apart with regard to their normal potentials. Unless a favorable relationship exists between the two oxidized and the two reduced forms concerned, one of the two phases of the reaction, viz. the reduction of the catalyst or the dehydrogenation of the substrate, will be impeded (MICHAELIS 804) p. 146)). The importance of the intermediary catalysts for the bridging of the potential gap between metabolites and oxidation system is also stressed by SZENT-GYÖRGYI 1122).

It may well be that the main task of the yellow enzymes and of the three cytochromes consists in providing a finely graded "p o t e n t i a l  l a d d e r" from the negative region of the hydrogen activating systems to the positive potential of oxygen.

In this connection, a thought of WARBURG 1234) may be mentioned. He sees in the subdivision of the terminal oxidation system into four steps (3 cytochrome components and the respiratory ferment) a measure to guarantee the maximum efficiency of the system. The efficiency of cell respiration is due to its maintenance at an almost constant level: it is independent of the $O_2$-tension as well as of the substrate concentration within wide limits. The independence from $O_2$-concentration is assured by the rapid reoxidation of the respiratory ferment which keeps it almost completely in the ferric form and thus maximally active even at a low $O_2$-tension. If the ferric form of the enzyme were to react directly with the substrate the rate of its reduction would depend on the substrate concentration. In order to avoid a possible delay on this score, the cytochrome system interposes itself between the enzyme and the substrate. Inasmuch as the cytochrome system is maintained essentially in the ferrous form during respiration, the reduction of the hemin enzyme by cytochrome is also effected at the optimum rate. The cytochrome system in turn is kept reduced by the substrate-dehydrogenase systems.

The other mechanism capable of explaining the function of the intermediary catalysts is one which takes the names "c a r r i e r" and hydrogen "transport" literally. When THUNBERG coined the term "hydrogen transport", he visualized a transport through space rather than through a series of reaction steps. This concept has been adopted and refined by WURMSER 1354, p. 349), RAPKINE and WURMSER 963), BORSOOK 138, 135) and other workers. There is no doubt that the individual reactions take place at different locations within the cell. Dehydrogenations, for example, may occur at any place in the microheterogeneous system of cytoplasm with the aid of the freely diffusible carriers. The terminal oxidation of main respiration, on the other hand, is restricted entirely to the "iron centers" which are catalytically active "patches" embedded at certain points of the cell surface and interfaces. The terminal oxidation becomes thus a locally defined t o p o c h e m i c a l  r e a c t i o n of the type of a heterogeneous catalysis. If this picture is correct, the metabolic hydrogen has to be literally transported to these iron centers. This carrier function is executed by the intermediary catalysts which are at the same time freely mobile and able to combine reversibly with the metabolic hydrogen. In this manner, two "incomplete enzyme centers" are supplemented to form a complete and active system (BORSOOK) *).

This concept is amenable to further elaboration: QUASTEL 952) assumes that the catalysts of anoxytropic dehydrogenations, too, are partly fixed at surfaces. If this is true, the carrier function of the intermediary catalysts assumes an even greater importance: These mobile redox systems (dyestuffs, metabolites, etc.) would then transport hydrogen not only from a fixed dehydrogenase to the iron centers but also from dehydrogenase to dehydrogenase. The finding of BORSOOK 135) that a suitable dye will cause a transfer of hydrogen from the system formic acid + formic dehydrogenase to pyruvic acid or from lactic acid to fumaric acid in the presence of toluene-treated resting B. coli may be explained in this manner.

EULER and ADLER 10) and GREEN and DEWAN 199) have recently gone even further: They report that specific proteins are required for the hydrogenation as well as the dehydrogenation of diphosphopyridine nucleotide (cozymase). They suggest that the coenzyme shuttles back and forth between these bearer proteins. GREEN and DEWAN 199) depict the oxidation in animal tissues as being catalyzed by a series of immobile centers such as the respiratory ferment, cytochrome a and b, the coenzyme factor (or d i a p h o - r a s e), etc., and a number of freely mobile carriers such as the pyridine coenzyme and cytochrome c.

The subdivision of active enzymatic oxidation systems into homogeneously dissolved "activators" and into large "granules" which may be separated out by centrifuging and which are indispensable for the catalysis has been reported by a number of workers, e.g. EULER and HELLSTRÖM 294), GREEN and DEWAN 199), BANGA 63), GREVILLE 420).

---

*) The hypothesis of the anchoring of the respiratory ferment to the architecture of the cell has been developed by WARBURG to explain why all attempts to obtain the ferment or rather a normal respiration in homogeneous solution have met with failure. The usual oxidase preparations are either minced and washed tissues or turbid suspensions obtained from them by treatment with alkaline phosphate. The loss of oxidase activity of such extracts upon clarification shows that the active principle or principles are contained in the suspended particles. The assumption has been made explicitly and implicitly that the latter represent fragments of the cell with no defined or uniform structure. Heart muscle oxidase preparations of this kind have now been subjected to fractional ultracentrifugation (STERN, HORWITT and SCHEFF, unpublished). It was found that the enzymatic activity (cytochrome oxidase, cytochrome a and b, succinic dehydrogenase), is contained in that fraction which is sedimented at the lowest speed selected (60 r.p.s.). Measurements in the analytical ultracentrifuge indicate a sedimentation constant of approximately $500 \times 10^{-13}$ and a surprising degree of homodispersity. It does not appear impossible that these particles are large protein molecules or particles containing the various active centers in a well-defined spatial arrangement on their surface.

K. G. STERN.

These findings may readily be explained on the basis of the "center-carrier" concept discussed above.

### 3) Place of Individual Mesocatalysts in Desmolysis.

In order to obtain full insight into cell respiration as a series of chemical reaction steps we would have to find out which intermediary catalysts of enzymatic or non-enzymatic nature are involved in passing the metabolic hydrogen from the substrate on to the first cytochrome component and the order in which these coupling links are arranged. Such an insight is at present little more than a hope. We do know, however, a part of the chemistry of oxybiosis with regard to the individual stages of breakdown. It is the well-known schema: acetic acid — succinic acid — fumaric acid — malic acid — oxaloacetic acid — pyruvic acid and acetaldehyde respectively (But see the opposing views of SZENT-GYÖRGYI (p. 268)). We also know most of the specific dehydrogenases concerned, e.g. acetaldehyrase, pyruvic dehydrogenase, succinic dehydrogenase, malic dehydrogenase, etc. Only the acetic dehydrogenase is not yet defined as an individual enzyme. If we accept the latter as a reality, the path of the metabolic hydrogen would lead from acetaldehyde over the following oxidation-reduction systems: Acetaldehyrase, acetic acid $\rightleftarrows$ acetic dehydrogenase $\rightleftarrows$ succinic acid, succinic acid $\rightleftarrows$ succinic dehydrogenase $\rightleftarrows$ fumaric acid, malic acid $\rightleftarrows$ malic dehydrogenase $\rightleftarrows$ oxaloacetic acid. The further change is carboxylatic in nature. If this simple schema would hold, it would mean that the system malic acid + dehydrogenase would be the redox system closest to cytochrome. Actually we are still ignorant of the nature of the coupling link with the cytochrome system and generally of the exact nature of the acceptors responsible for the dehydrogenation of the reduced form of the dehydrogenases. It is not even known whether the cytochrome component concluding the chain $O_2 \rightarrow$ respiratory ferment $\rightarrow$ cytochrome (a) $\rightarrow$ cytochrome (c) $\rightarrow$ cytochrome (b) has an absolute specificity for one metabolite-dehydrogenase system just as the respiratory ferment (cytochrome oxidase) represents a specific dehydrogenase of cytochrome, or whether it may react with several systems in vivo just as it is reduced in vitro by malic acid + malic dehydrogenase, succinic acid + succinic dehydrogenase and acetaldehyde + acetaldehyrase (BIGWOOD 115)). There is, furthermore, the possibility that the cytochrome system does not interact directly with a redox system of the type formed by intermediary metabolites and their corresponding dehydrogenases, but only through the medium of another intermediary catalyst which is of a different nature. We know three such carriers, e.g. the yellow enzyme, ascorbic acid and glutathione. Of these three the only one about which specific and pertinent information in this respect is available is the yellow enzyme. Its interaction with cytochrome has been studied by THEORELL 1158). At least one of the sources is known from which the yellow enzyme accepts hydrogen. It may act as the dehydrogenase of the pyridine enzymes (p. 212). While this mechanism has undoubtedly great importance for the metabolism of microorganisms like yeast, GREEN and DEWAN 199) claim that in the tissues of higher animals this function of the yellow enzyme is taken over by the "coenzyme factor" (EULER's diaphorase) (p. 226)*). The reaction Dihydropyridine + Flavoprotein $\rightleftarrows$ Pyridine + Leucoflavoprotein appears to be a reversible one under certain circumstances. The flavin enzyme might therefore play some rôle in fermentation in addition to its function in oxybiosis where, at least in yeast, it is instrumental in coupling the anoxybiontic preparatory stages with the system of terminal oxidation. The enzyme is qualified for the function of an important intermediary catalyst of oxybiosis by the fact that it is

---

*) Diaphorase has now been shown to be also an alloxazine protein (see Footnote, p. 227).

autoxidizable and that it is also dehydrogenated by cytochrome c (THEORELL 1159)). Whether the enzyme will react in vivo with molecular oxygen or with the hemin system (provided the latter is available) depends largely on the partial pressure of oxygen.

Models demonstrating catalysis by the yellow enzyme and regeneration of the active form by autoxidation are the WARBURG-CHRISTIAN system (hexosemonophosphate-dehydrogenase (carrier protein + codehydrogenase II) — yellow enzyme — $O_2$) and the analogous EULER-ADLER 6a) system (Alcohol — alcohol dehydrogenase — yellow enzyme — $O_2$). In hemin-free cells, too, e.g. in facultative anaerobic lactic acid bacteria, the "respiration" is catalyzed by the yellow ferment which establishes direct contact with molecular oxygen (WARBURG 1234)).

If normally respiring cells, e.g. bakers yeast or acetic acid bacteria, are poisoned by HCN, it can be shown by photoelectric methods (WARBURG 1234)) that the yellow enzyme dehydrogenates its substrate and is in turn reoxidized by $O_2$. This does not prove, however, that the yellow enzyme, under normal conditions, functions independently of the hemin system. On the contrary, the reoxidation of the flavoprotein is normally accomplished by the terminal oxidation system (THEORELL 1158)). If the latter is blocked, the yellow ferment is forced to react directly with molecular oxygen. Since this reaction is relatively slow at the $O_2$-pressure obtaining within the cell, the cyanide resistant residual respiration amounts to only a small fraction of the full hemin respiration (0.5 per cent in bakers yeast and 0.1 per cent in acetic acid bacteria). The low efficiency of the "yellow enzyme respiration" is best demonstrated in the case of the hemin-free lactic acid bacteria. Here, 1 flavoprotein molecule transfers only 100 $O_2$-molecules per minute, or, in terms of the actual experiment, 1 cc. of cells absorbs only 1 cc. $O_2$ per minute. For this result the pigment is reduced and oxidized 30 times per minute ("Wechselzahl"). Inasmuch as the reduction is a rapid reaction, the limiting step is the reoxidation by $O_2$ which is relatively slow due to the low affinity of the enzyme for oxygen.

Another indication that the yellow enzyme may partly substitute for the hemin system in emergencies may be seen in the results of PETT 925a) who finds that the flavin concentration in yeast increases, if cyanide is added to the culture medium; whereas it decreases if the normal respiration is stimulated.

In normally respiring cells there is a certain competition between the hemin system and the yellow enzyme for the oxygen. Due to its far greater $O_2$-affinity the respiratory ferment wins. THEORELL 1159)) has shown that at a partial oxygen pressure of 38 mm about 80 per cent of the oxygen will go through the cytochrome oxidase-cytochrome channel. Inasmuch as the $O_2$ tension in the cell is probably still lower, it may be accepted that under normal physiological conditions practically the entire reoxidation of the dihydroform of the yellow enzyme is accomplished by the hemin apparatus. According to the measurements of HAAS (cf. THEORELL 1158)) the "Wechselzahl" of the yellow enzyme would have to be 4800 in bakers yeast and 19,000 in B. pasteurianum, instead of 30, if the whole respiration were to proceed via the flavoprotein. Even in pure oxygen at 38°, the "Wechselzahl" of the pure ferment is only 55 (THEORELL 1154)). Inasmuch as the reoxidation of the yellow ferment by cytochrome is also far from being a rapid reaction, THEORELL 1161)) concludes that the fraction of total respiration involving the enzyme as a carrier between dehydrogenase systems and cytochrome is hardly greater than the cyanide resistant residual respiration. This would mean that the yellow enzyme, at best, fulfills a highly specialized rôle in normal cell respiration by coupling certain early oxidative stages, perhaps those catalyzed by pyridine enzymes, directly with the terminal respiratory complex. It is, of course, quite possible that the yellow enzyme not only transfers hydrogen to cytochrome but that it may also react with other acceptors, e.g. fumaric acid.

There is little to say at the present time about the function of glutathione as an intermediary catalyst and about its relation to cytochrome or any other component of the respiratory chain. It is true that ferricytochrome is rapidly reduced, in vitro, by SH-glutathione (BIGWOOD 118)), but this is not enough to prove the actual physiological function of glutathione. It is puzzling that in the living cell the glutathione is present practically entirely in the reduced form (BIERICH et al. 112)). If the tripeptide should be able to transfer metabolic hydrogen, e.g. to cytochrome, it must be so rapidly reduced by more negative systems that the equilibrium is always shifted far to the side of the thiol form. The addition of glutathione to living cells has failed to provide much useful information: the respiration of yeast or of animal tissues is affected neither by SH-glutathione nor by SS-glutathione. Most tissues, e.g. liver, brain, erythrocytes, sarcoma cells, keep glutathione in the reduced state. Kidney will oxidize slowly both glutathione and cysteine (ROSENTHAL and VOEGTLIN 1005)). According to the findings of ELLIOTT 240), it is improbable that glutathione acts as a carrier between the succinic dehydrogenase and the cytochrome system. OGSTON and GREEN 895) find hardly any interaction between the succinic acid + succinic dehydrogenase system and glutathione. Taken all together, we are unable to assign a place to this much discussed peptide in the framework of cell respiration. It may be, that, after all, it exerts no general but only certain special functions, such as regulation of protein metabolism through interaction with the sulfur containing groups in cathepsin (WALDSCHMIDT-LEITZ) or a possible function in the PASTEUR-MEYERHOF reaction (p. 98) due to its action as "coferment" of the ketonealdehyde mutase (glyoxalase). It may also remove traces of heavy metals by forming complexes with them and thus prevent them from oxidizing other important substances; or $H_2O_2$ might be "detoxified" by the dehydrogenation of SH-glutathione (FUJITA 383)). One might also mention the recent observations of TAIT and KING 1128) according to which glutathione catalyzes the oxidation of phosphatides in vitro. In general it may be stated that the difficulty with regard to the rôle of glutathione is no longer the question of its reduction by metabolic systems but rather that of its rapid reoxidation for the purposes of an oxidative catalysis (cf. OGSTON and GREEN 895)).

Still less is definitely known about the function of ascorbic acid. At one time SZENT-GYÖRGYI 1119) postulated a direct oxidation of metabolites by a system consisting of ascorbic acid and a specific oxidase. While such a concept is no longer tenable and has even been abandoned by SZENT-GYÖRGYI for the case of animal tissues, it is quite possible that ascorbic acid and its oxidase constitute an important oxidation system in plants. The existence of ascorbic oxidase as a specific enzyme is more or less generally accepted. Since its original discovery by SZENT-GYÖRGYI, it has been encountered by ZILVA 1382) and several other workers, e.g. TAUBER. JOHNSON and ZILVA 505) have been able to differentiate between the enzymatic oxidation of ascorbic acid and that of a quinone catalysis where the vitamin is oxidized in a secondary reaction by the oxidized form of certain phenols, while the reduced form of the phenols is reoxidized either peroxidatically or by polyphenol oxidase. Ascorbic acid oxidase is a typical, oxytropic dehydrogenase and is stated to be cyanide resistant, although the presence of copper in the molecule is strongly indicated. The steric specificity of the catalyst (only l-ascorbic acid is attacked) supports the idea of a specific enzyme rather than that of an unspecific heavy metal catalyst (ROSENBERG 1003)). For details see OPPENHEIMER's „SUPPLEMENT" p. 1588.

## 4) Fumaric Acid Catalysis *).

Development of SZENT-GYÖRGYI's Theory:

One of the facts which induced SZENT-GYÖRGYI to investigate the general importance of succinic and other $C_4$-dicarboxylic acids for cellular respiration was the observation that of all dehydrogenase systems examined only succinic dehydrogenase + succinic acid will rapidly reduce ferricytochrome c. Whereas lactic dehydrogenase + lactic acid can readily reduce dyestuffs of the type of methylene blue, this system cannot react with ferricytochrome. OGSTON and GREEN 895) find that of eleven donator-dehydrogenase systems studied only the succinic system will reduce cytochrome. Inasmuch as it is generally accepted that in cellular respiration the cytochrome oxidase-cytochrome (or WARBURG-KEILIN) system brings about the terminal oxidation of the various metabolites the hydrogen of which has been "activated" or labilized by specific dehydrogenases, SZENT-GYÖRGYI began his search for the missing link between the dehydrogenases on the one hand and the WARBURG-KEILIN system on the other hand. The possibility suggested itself that succinic acid in conjunction with its dehydrogenase acts as this catalytic hydrogen carrier. The following facts support this hypothesis: Succinic dehydrogenase, the enzyme capable of catalyzing the rapid dehydrogenation of succinic acid to fumaric acid by suitable redox systems, e.g. ferricytochrome, is present in all animal tissues and also in yeast and bacteria. The addition of fumaric acid to tissues has long been known to increase their respiration (THUNBERG, GRÖNWALL, etc.). Gözsy and SZENT-GYÖRGYI 404a) showed that this effect is catalytic in nature, since the addition of small amounts of fumaric acid to minced pigeons breast muscle will increase the respiration without being used up itself. The effect of the fumarate actually is not so much an increase in rate of respiration, but a "stabilization" of the respiratory rate: the addition of fumarate to muscle in vitro prevents the falling off of the oxygen uptake usually observed after about 30 mins. and also the simultaneous decrease of the respiratory quotient from unity to 0.8 (BANGA 1126)). Finally, the respiration is strongly inhibited by malonic acid which had previously been shown to poison succinic dehydrogenase, allegedly, in a specific manner (QUASTEL et al. 943, 953)). The presence of an excess of fumaric acid, however, renders the tissue insensitive towards malonate (BANGA 1126)). This and another fact appeared to militate against the simplest hypothesis, viz. that the succinic acid-dehydrogenase-fumaric acid system alone fills the gap between the WARBURG-KEILIN system and the metabolites of the cell. The other fact is the presence of an active enzyme, called fumarase, in all cells which catalyzes the hydratation of fumaric acid to malic acid in a perfectly reversible manner. The function of this ubiquiteous catalyst would be left unexplained by the simple hypothesis. As a matter of fact, the rapid transformation of fumaric acid into malic acid by this enzyme should interefere with the catalytic role of the succinic-fumaric cycle. Incidentally, DAS 194), working in SZENT-GYÖRGYI's laboratory, has since shown that the relief of the malonate inhibition of respiration by fumaric acid does not rule out the simple succinic acid hypothesis: Malonate may compete successfully with succinic acid but not with fumaric acid for the active center of succinic acid dehydrogenase. Once fumaric acid has been attached to this center it may accept hydrogen atoms from other metabolites and hand them on to cytochrome without interference by malonic acid. Only

---

*) The results and interpretations of SZENT-GYÖRGYI up to August, 1937, are summed up in his excellent review in the Acta Med. Szeged 1125) which has also been published in book form.

those molecules leaving the enzyme surface in the form of succinic acid will be prevented from further interaction with the enzyme by the inhibitor. This slow, secondary process loses much of its importance in the presence of added fumaric acid. SZENT-GYÖRGYI initiated a search for other $C_4$-dicarboxylic acids, representing a higher state of oxidation as compared with succinic and fumaric acid, which might act as hydrogen transmitters to the WARBURG-KEILIN system. BANGA and LAKI (l.c. 1126)) showed that such $C_4$-acids of higher degree of oxidation could be represented by the fumaric-hydroxyfumaric acid system. The enzyme active here would be a fumaric dehydrogenase. The hydroxyfumaric acid is unstable and is assumed to undergo a molecular rearrangement to oxaloacetic acid. Now muscle contains an enzyme system which will change fumaric acid rapidly into oxaloacetic acid. GREEN 416) held the view that the enzyme acts on malate which arises through the hydratation of fumarate by fumarase and not on fumarate directly. LAKI 693) proved the view of GREEN to be correct. Consequently, the system fumaric-hydroxyfumaric acid is to be replaced by the system malic-oxaloacetic acid. The function of the fumarase would consist in the provision of malic acid from fumaric acid, the equilibrium concentration in the presence of the enzyme being 3 : 1 in favor of malic acid. BANGA 1126) demonstrated that malate is oxidized to oxaloacetate by the muscle enzyme at a rate comparable to the rate of total respiration. On the other hand, oxaloacetate was rapidly reduced when added to the tissue. According to SZENT-GYÖRGYI 1125), oxaloacetate represents the physiological H acceptor and a determination of the amount of oxaloacetate reduced to malate by a tissue is a true measure of the amount of metabolic hydrogen activated in the tissue. However, it was evident that a simple substitution of the malic-oxaloacetic acid system for the original succinic-fumaric acid system in the schema of cell respiration was equally insufficient to take into account all the facts known about the importance of $C_4$-dicarboxylic acids for cell respiration. Of the three most important facts, viz., the existence of succinic dehydrogenase, the existence of fumarase, and the rapid reduction of oxaloacetate by tissues, the succinate-fumarate theory can explain only the first and the malate-oxaloacetate hypothesis only the last, while the existence of fumarase is not sufficiently explained by either hypothesis. This state of affairs led SZENT-GYÖRGYI to his "united" theory which attempts to fill the gap in the respiratory chain by assuming a cooperation of both systems:

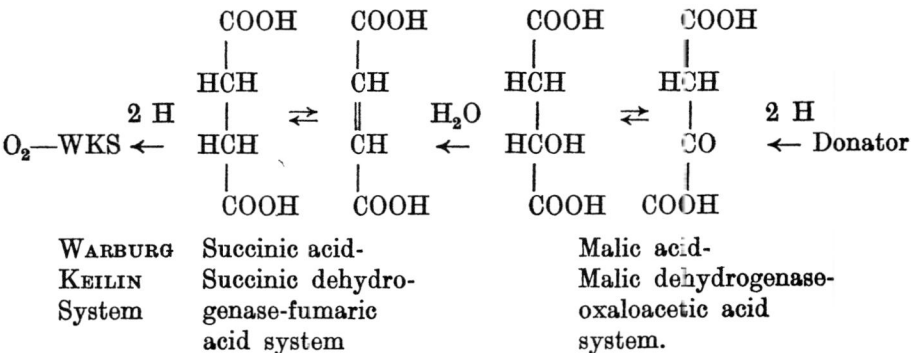

| WARBURG KEILIN System | Succinic acid-Succinic dehydrogenase-fumaric acid system | Malic acid-Malic dehydrogenase-oxaloacetic acid system. |

This schema would furnish an adequate explanation for the existence of fumarase which would have the function to maintain a proper balance between the concentrations of the various $C_4$-acids. The scheme, furthermore, is thermodynamically possible: The normal oxidation-reduction potential of the malate-oxaloacetate system is $E'_o = -0.169$ V.

at ph 7 and 37° (LAKI **692**)) and that of the succinate-fumarate system is $E'_o = -0.018$ V. at ph 7.2 and 37° (LEHMANN **708**)). The malate-oxaloacetate system should, therefore, be able to reduce the succinate-fumarate system. GREEN **416**) was able to show that such an interaction will actually occur provided that an electroactive coupling link or H carrier is present. STRAUB **1106**) finds that malate will reduce cytochrome only via the fumarate-succinate system. The hydrogen coming from other donators may reach cytochrome only by the double system of hydrogen transfer indicated above.

Although BANGA **64**) washed her preparations containing the succinic and malic dehydrogenase extensively, they were still able to transfer hydrogen from donator systems to the cytochrome. This is due to the fact that the enzyme preparation retains tenaciously traces of the $C_4$-acids which suffice for the purpose of the catalysis. SZENT-GYÖRGYI **1125**) concludes from this observation that the enzymes do not utilize the entire amount of $C_4$-acids present in the tissues for the catalytic H transfer but only a

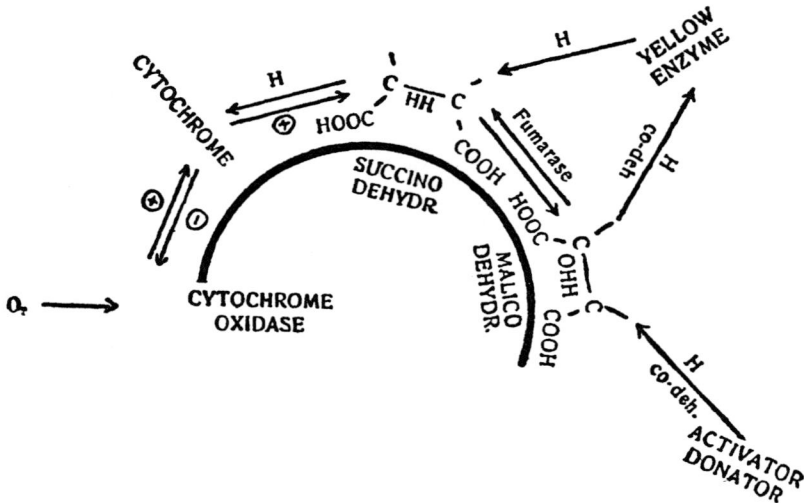

Fig. 17. Schema of Cell Respiration according to SZENT-GYÖRGYI **1125**). The semicircle represents a granule of the muscle cell. The picture is somewhat simplified by not writing the three cytochromes in series.

few molecules which are rather firmly attached to the enzyme surface in an activated condition. He actually goes so far as to designate these $C_4$-acid molecules as prosthetic groups of the enzymes in question. In this terminology, succinic dehydrogenase, when dialyzed free from all traces of succinic and fumaric acid, would represent a specific bearer protein requiring some succinic or fumaric acid for the formation of the active enzyme complex.

Recently, SZENT-GYÖRGYI and his collaborators have extended their analysis of the chain of respiratory catalysts. The aqueous, turbid extracts obtained by treating pigeon breast muscle with water exhibit a respiration of the same order as that of the whole muscle. The turbidity of the fluid is caused by the presence of microscopic granulae which can be separated in the centrifuge and which contain the WARBURG-KEILIN (cytochrome oxidase-cytochrome) system, succinic and malic dehydrogenase. Neither these granules nor the remaining clear fluid show any appreciable oxygen consumption. By remixing, however, the original respiration is restored. It was found that the clear fluid contains, in addition to donators and coenzymes, a thermolabile substance of protein nature which is necessary for the oxygen uptake of the entire system (BANGA 63)). BANGA 63) could show that the donator present in the clear fluid is hexosemono- and diphosphate. Experiments by LAKI, GÖZSY and by MORUZZI indicate that hexosephosphate is not dehydrogenated as such but that it is split into triosephosphate prior to dehydrogenation. The protein present in the clear tissue fluid appears to be the bearer protein of the triosephosphoric acid dehydrogenase. For its functioning cozymase is required (STRAUB, quoted by SZENT-GYÖRGYI 1125)). When an attempt was made to reconstruct the whole system by mixing the microscopic granules, containing the WARBURG-KEILIN system and the dehydrogenases, with coenzymes, fumaric acid, donator and the thermolabile activator, the ensuing oxygen uptake does not usually reach the level obtained in normal respiration. This was ascribed to the lack of sufficient amounts of an electroactive carrier connecting some of the active centers embedded in the surface of the granules. In the instance of the cytochrome oxidase and succinic dehydrogenase centers cytochrome c fulfills this role. It was found that the lack in "contact" was between the succinic dehydrogenase and the malic dehydrogenase centers and that it may be remedied by adding some yellow enzyme. The redox potential of this soluble enzyme, $E'_o = -0.060$ V. at ph 7 and 38° (KUHN and BOULANGER 636)) makes it a suitable carrier between the two dehydrogenases. It can be shown that the yellow enzyme is actually reduced by the malate system and that it is oxidized by fumarate activated by succinic acid dehydrogenase (LAKI 691)). The complete system which shows a rapid oxygen uptake, is represented by the schema in Fig. 17 (From SZENT-GYÖRGYI 1125)).

### Applicability of the Theory to other Tissues and Donators:

The theory of SZENT-GYÖRGYI was worked out with minced breast muscle of the pigeon as the biological material. The respiratory quotient of muscle (just as that of brain) is close to unity and thus indicative of a pure carbohydrate combustion. F. J. STARE 1064b) finds that the $C_4$-cycle operates also in minced kidney and liver tissue. This is borne out by model experiments with dyes (STRAUB 1106)). GREVILLE 419) reports that the pure carbohydrate respiration of brain cortex and retina tissue is inhibited by malonate, but that fumarate will only partially relieve this inhibition. Respiration artificially accelerated by dinitro-o-cresol in brain cortex and tumor tissue is strongly inhibited by malonate, whereas that produced in tumor tissue by a redox indicator was not. He points out that observations made on malonate inhibition cannot confirm any particular theory, since the effect of malonate is not strictly specific. The same author (GREVILLE 420)) was able to confirm many of the original observations of SZENT-GYÖRGYI and his associates with regard to the effect of fumarate and malonate on pigeon breast muscle and also BANGA's findings concerning the distribution of the active system over freely dissolved and granular material.

Certain tissues are deficient in the SZENT-GYÖRGYI (succinic-malic dehydrogenase) system. BOYLAND and BOYLAND 140a) find that cancer tissue shows a very small response to the addition of fumarate and malonate in the same direction as muscle tissue. BANGA (quoted by SZENT-GYÖRGYI 1125)) failed to observe an appreciable reduction of oxaloacetate or pyruvate by rapidly growing malignant tumors. BREUSCH 148) confirms the existence of a partial disturbance in the hydrogen transport mechanism by $C_4$-dicarboxylic acids in mouse tumor tissue (skin carcinoma of the Royal Cancer Institute, London). This tumor is unable either to form or to reduce oxaloacetic acid. The power to dehydrogenate succinic acid is about 20 per cent of that of normal muscle tissue. This is ascribed to a deficiency in the activity of the WARBURG-KEILIN system rather than in that of the succinic dehydrogenase present. The reversible transformation of fumaric acid into malic acid, on the other hand, is undisturbed in mouse tumor tissue. Contrary to normal muscle tissue, tumor tissue is stated to be unable to desmolyse slowly $C_4$-dicarboxylic acids added in excess, even in presence of arsenite which in normal muscle will enhance this process. While the total respiration of tumor tissue is markedly increased by the addition of cozymase, the defect in the hydrogen transport by the $C_4$-dicarboxylic acid system is thereby not remedied. It is very interesting to note that embryonic tissue behaves very similar in this respect to rapidly growing malignant cells (SZENT-GYÖRGYI, I. BANGA and A. BLAZSÓ (cf. 1126); BREUSCH).

Another exception is lung tissue: neither fumaric nor malic acid have an effect on its respiration. Succinate is also only slowly attacked due to the low activity of the succinic dehydrogenase present (BREUSCH 149)). Oxaloacetic acid is just as slowly reduced as by tumor or embryonic tissue.

The $C_4$-system participates also in the oxidation of substrates other than hexosephosphate. For example, the reduction of methylene blue by lactate is inhibited by malonate in the presence of muscle enzyme. Other donators for the $C_4$-system are ethyl alcohol, citric acid, pyruvic acid, lactic acid. In these cases, the whole $C_4$-system is not necessarily required. Certain donators with a normal potential not sufficiently negative to reduce oxaloacetic acid may possibly reduce only fumarate with the aid of the yellow enzyme. This must be so with lactic acid as donator which has the same potential as malic acid and cannot be expected, therefore, to reduce oxaloacetate (LAKI 692)). The absence of malic dehydrogenase in tumor and embryonic tissue suggests that these tissues utilize as main donator a substance reducing fumarate but not oxaloacetate. Experiments by BREUSCH (cf. SZENT-GYÖRGYI 1125)) suggest that lactic acid is the donator here. Actually, both tumor and embryonic cells will produce lactic acid in appreciable amounts.

The $C_4$-dicarboxylic acid system is not involved in the oxidative deamination of amino acids (DAS 193)). Dehydrogenases which are able to react with molecular oxygen directly, e.g. SCHARDINGER's enzyme, xanthine oxidase, do not depend on the $C_4$-system for their activity.

The question of the specificity of inhibitors in connection with the problem of the $C_4$-dicarboxylic acid catalysis deserves further scrutiny. LELOIR and DIXON 711) have recently tested the effect of pyrophosphate on a number of dehydrogenase systems. They find that this substance specifically poisons succinic dehydrogenase. This fact, when considered in the light of SZENT-GYÖRGYI's theory, may well explain the strong inhibition of tissue respiration by pyrophosphate. WEIL-MALHERBE 1287), on the other hand, points out that malonate not only inhibits succinic dehydrogenase but other dehydrogenase systems as well.

Arguments against SZENT-GYÖRGYI's Theory:

One possible argument against SZENT-GYÖRGYI's theory has already been mentioned (p. 268): if succinic dehydrogenase is of vital importance for cellular respiration, then malonic acid which strongly inhibits this enzyme should always inhibit the oxygen uptake of tissues. The explanation given by SZENT-GYÖRGYI for the fact that this is not always the case, is that the $O_2$ uptake is the final result of the function of a long chain of individual steps. The rate of the over-all reaction is naturally limited by the slowest step. Malonic acid may, therefore, be expected to interfere with the normal oxygen consumption, if the hydrogen transfer by the succinic dehydrogenase is or can be made the limiting factor in the chain by the addition of malonate. Addition of fumarate, on the other hand, may only be expected to increase respiration, if the concentration of $C_4$-dicarboxylic acids is the limiting factor. The observation of conditions under which neither fumarate nor malonate will affect the rate of oxygen consumption does, consequently, not constitute proof against the existence of the SZENT-GYÖRGYI cycle.

Another, apparently more serious, objection is that it is well known that $C_4$-dicarboxylic acids are intermediary products in carbohydrate metabolism and that they thus serve as a fuel for the cell. The inference is that substances which are irreversibly broken down by the cell cannot be considered to be true catalysts. Now it is true that fumarate when added to tissues in large amounts is burned (THUNBERG) and that this process has a respiratory quotient above unity (GRÖNWALL) which indicates irreversible combustion of the fumarate. INNES 500a) reports a rapid oxidation of fumaric acid by muscle via malic acid. She finds the extra oxygen uptake accounted for by the amount of fumarate added and therefore does not concede a catalytic effect of this substance. Succinic acid when fed to animals is not excreted (FLASCHENTRAEGER) and is thus probably burned. SZENT-GYÖRGYI, while confirming these experimental observations, differs in their interpretation. He distinguished between the fate of small and of large concentrations of added fumarate. At low concentrations, the effect must necessarily be a catalytic one, because the amount of $O_2$ taken up exceeds several times the amount needed for combustion of the fumarate added and particularly because STARE and BAUMANN 1064b) were able to recover the fumarate added at the end of the experiment. Furthermore, the respiratory quotient does not exceed unity but will be raised only from 0.8 to 1.0. Large amounts of fumarate, on the other hand, are slowly and irreversibly burned by the tissue with the exception of the very small amounts of $C_4$-acids needed for the upkeep of the catalytic cycle and strongly adsorbed at the enzyme surface.

On general principles this argumentation of SZENT-GYÖRGYI is to be supported. It is one of the major recent developments of cellular physiology that not only enzymes, i.e. substances specially constructed by the cell for catalytic purposes, but also certain metabolites may serve the double function of an intermediate in desmolysis and of a catalyst during a certain phase of the total process. All one has to do in order to avoid confusion, is to distinguish between the reversible and the irreversible phases of the over-all process.

Another objection has been raised by KNOOP 1125). He points out that a great number of substances, e.g. imino acids, aldehydes, ketones, compounds with double bonds between C atoms, are reduced by the cell. He would prefer the view that the metabolic hydrogen, after being split off from the first donator, may find its way to the oxygen by a great number of possible routes one of which is that through the $C_4$-acids. KREBS, too, regards the reduction of oxaloacetate by tissues as one of many examples of the reduction of CO groups. SZENT-GYÖRGYI's 1125) answer to this criticism is that experience shows that the main process of respiration follows a strictly defined path. One of the restrictions is that any metabolic hydrogen in order to be eventually oxidized by the WARBURG-KEILIN system must pass through the succinate-succinic dehydrogenase link. Oxidation-reduction potentials alone or even primarily do not prescribe the path of the hydrogen for this is largely determined by conditions of chemical affinity.

A critical discussion of SZENT-GYÖRGYI's theory may be found in OPPENHEIMER's "Supplement", p. 1281.

## 5) Citric Acid Catalysis *).

KREBS' Theory: This theory which has recently been put forward by KREBS constitutes a modification and extension of SZENT-GYÒRGYI's $C_4$-acid theory as developed in the preceding section.

KREBS and JOHNSON 611) observed that the addition of minute quantities of citric acid to muscle tissue will accelerate the combustion of carbohydrates in a manner similar to the effect of succinic acid. The experimental analysis of this phenomenon has led to the following results. In agreement with the findings of KNOOP and MARTIUS 583), citric acid is transformed into α-ketoglutaric acid by living cells. The proof rests on the use of arsenious oxide as inhibitor of the enzyme or enzymes normally attacking the ketoglutaric acid without interfering with the oxidation of the citric acid. Ketoglutaric acid, in turn, upon oxidation yields succinic acid and $CO_2$ (see p. 253). By adding malonic acid to muscle tissue, the desmolysis of citric acid may be stabilized at the succinic acid stage.

According to KREBS, the original succinic acid cycle of SZENT-GYÖRGYI, schematically represented by KREBS as follows

a)  Oxaloacetic acid + "hydrogen of organic molecules" → succinic acid

b)  Succinic acid + $O_2$ → oxaloacetic acid + $H_2O$,

although undoubtedly occurring, cannot account for the whole catalytic effect because of the slowness of step a). Casting about for an alternative mechanism, the experiments described above suggested citric acid as a component in the catalytic cycle. To this end, it would be necessary to demonstrate a regeneration of citric acid from one of its breakdown products. Actually it is found that addition of oxaloacetic acid to muscle tissue causes the synthesis of citric acid in comparatively large amounts. This synthesis which occurs under anaerobic conditions, is tentatively depicted by KREBS as follows:

```
COOH
|
CO              COOH            CO2
|               |               +               COOH
CH3             CO              COOH             |
|               |               |                CH2
+               CH2             CH2              |   OH
       -H2O     |        + O    |        +H2O  C<
OH      --->    C—COOH   --->   C—COOH   --->   |   COOH
|               ||              ||               CH2
C—COOH          CH              CH               |
||              |               |                COOH
CH              COOH            COOH
|
COOH

Pyruvic acid    Oxalo-          „Cis"-          Citric
  +             mesaconic       Aconitic         Acid
Enol-Oxalo-     Acid            Acid
acetic Acid
```

The removal of citric acid synthesized in this or in another manner is an oxidation

---

*)  Reference is made to the review article by KREBS 607)).

by molecular oxygen. The whole schema, called the "citric acid cycle", is represented in the following way:

$$\longrightarrow \text{Oxaloacetic Acid} + \text{Carbohydrate derivative (Pyruvic Acid?)}$$
Citric Acid
α-Ketoglutaric Acid
Succinic Acid
$$\longrightarrow \text{Fumaric Acid}$$

Each phase of this cycle is an oxidative process. KREBS believes that this scheme outlines the principal pathway of the oxidative breakdown of carbohydrate in muscle tissue. Ketone bodies may arise under special conditions through an alternative path branching off from the pyruvic acid (KREBS and JOHNSON 609, 610)):

a)  $CH_3 \cdot CO \cdot COOH \rightarrow CH_3 \cdot COOH + CO_2$

b)

$$
\begin{array}{cccc}
CH_3 & CH_3 & & \\
| & | & +O & \\
COOH & CO & \xrightarrow{-CO_2} & CH_3 \cdot CO \cdot CH_2 \cdot COOH \\
+ & | & & \text{(Acetoacetic Acid)} \\
CH_3 & CH_2 & \xrightarrow{+H_2O} & \\
| & | & -CO_2 & CH_3 \cdot CH(OH) \cdot CH_2 \cdot COOH \\
CO & CO & & \text{($\beta$-Hydroxybutyric Acid)} \\
| & | & & \\
COOH & COOH & & \\
\text{Acetic Acid} & \text{Acetopyruvic} & & \\
+ & \text{Acid.} & & \\
\text{Pyruvic Acid} & & &
\end{array}
$$

This is in agreement with the observation of EMBDEN and OPPENHEIMER in 1912, that the addition of pyruvic acid to liver in excess will give rise to increased ketone body formation. The rate of these reactions is stated to be small compared with the normal rate of the citric acid cycle. They may be of importance, however, for the explanation of diabetic acidosis. KREBS concludes (607)) that "it is now evident that carbohydrates as well as fats may be the source of ketone bodies."

Objections against KREBS' Theory: The theory of KREBS has recently been subjected to a critical reinvestigation in SZENT-GYÖRGYI's laboratory. BREUSCH 147) began by checking the results of KNOOP and MARTIUS. He was able to confirm their finding that citric acid is changed by tissues into α-ketoglutaric acid and also that citric acid, isocitric acid and cis-aconitic acid are reversibly transformed into each other in cells. This is even true for mouse tumor tissue, although the tumor cells are unable to break down citric acid. Citric acid when added to normal tissues in vitro disappears both aerobically and anaerobically, but more rapidly under aerobic conditions.

Theoretically, there exist 3 possibilities for a catalytic function of citric acid as hydrogen transporter: 1) by virtue of a reversible change from tricarballylic acid to cis-aconitic acid, 2) by an equilibrium between ketotricarballylic acid and isocitric acid, and 3) by the citric acid cycle as postulated by KREBS. Process 1) which is analogous to the dehydrogenation of succinic acid does not occur in tissues (liver, muscle) to any appreciable extent. Reaction 2) is unlikely since crude ketotricarballylic acid was not reduced to isocitric acid. BREUSCH maintains that the third possibility, i.e. KREBS' citric acid cycle, is also not a reality under physiological conditions. He believes that citric acid is not formed from oxaloacetic acid in the tissue. If the latter is added to tissues, the main fraction is converted into $l$-malic acid while the amount of citric acid present, instead of increasing in accordance with KREBS' claim, decreases through oxidative breakdown. BREUSCH summarizes his findings by stating that so far no valid proof has been offered to the effect that the $C_6$-tricarboxylic acid system of citric acid and related compounds fulfills a function in oxygen- or hydrogen transport analogous to that performed by the $C_4$-dicarboxylic acid system.

# BIBLIOGRAPHY

1) *E. Abderhalden*, Einfluss von Ascorbinsäure auf Tyrosinase. Fermentforsch. **14**, 367 (1934). — 2) *H. A. Abramson, I. R. Taylor*, The reduction of some adsorbed oxidation-reduction indicators. Jl. of Physical Chem., **40**, 519 (1936). — 3) *D. Ackermann, K. Poller, W. Linneweh*, Trimethylaminoxyd als biol. H-Acceptor. Berichte deutsch. chem. Ges. **59**, 2750 (1926). — 4) *G. A. Adams*, The ultraviolet spectrum of hemoglobin and its derivatives. Biochem. Jl., **30**, 2016 (1936), **32**, 646 (1938). — 5) *C. R. Addinall*, (The story of vitamin $B_1$. Merck & Co., Rahway, N. J., (1937)). — 5a) *E. Adler*, Newer studies on the enzymic oxidation-reduction systems. Skand. Arch. Phys., **77**, 1 (1937). — 6) *E. Adler, H. v. Euler*, Einfluss von Flavinen auf die Atmung von Milchsäure-bacterien. Z. phys. Chem., **225**, 41 (1934). — 6a) *E. Adler, H. v. Euler*, Dehydrierung von Alcohol und Robison Ester. Zs. phys. Chem., **226**, 195 (1934). — 7) *E. Adler, H. v. Euler*, Cozymase und Dihydro-Cozymase in lebenden Zellen. Svensk kem. Tidskr., **48**, 221 (1936). — 8) *E. Adler, H. v. Euler*, c.s. Phosphorylierung und Oxydo-reduktion. Naturwissenschaften, **25**, 282 (1937). — 9) *E. Adler, H. v. Euler, H. Hellström*, Cozymase as the specif. Coenzyme of lactic dehydrogenase from heart muscle. Nature, **138**, 968 (1936). — 10) *E. Adler, H. v. Euler, H. Hellström*, Zur Kenntnis der enzymatischen Wasserstoffüberträger im Muskel. Ark. f. Kemi, Mineral. och Geologi, **12B**, No. 38 (1937). — 11) *E. Adler, H. Hellström, H. v. Euler*, Reduktion der Cozymase. Zs. phys. Chem., **242**, 225 (1936). — 12) *E. Adler, W. L. Hughes*, Enzymatischer Mechanismus der Oxydoreduktion der Triosephosphorsäure. Zt. phys. Chem., **253**, 71, 143 (1938). — 13) *E. Adler, M. Michaelis*, Milchsäuredehydrase aus Herzmuskel. Zs. phys. Chem., **238**, 161 (1936). — 14) *K. Agner*, Reversible Spaltung der Leberkatalase. Zs. phys. Chem., **235**, II (1935). — 14a) *K. Agner*, The preparation and properties of a highly active catalase from horse liver. Biochem. Jl., **32**, 1702 (1938). — 15) *L. Ahlstroem, H. v. Euler*, Sauerstoffaufnahme der Spaltprodukte der Hexosen. Zs. phys. Chem., **200**, 233 (1931). — 16) *G. A. Aikins, A. C. Fay*, Effect of light on the reduction of Methylene-Blue in milk. Jl. of Agr. res., **44**, 85 (1932). — 17) *W. Albach*, Einfl. der Vitalfaerbung auf die Atmung. Protopl., **7**, 395 (1929). — 18) *H. Albers*, Co-ferment-System der Carboxylase. Naturwissenschaften, **24**, 794 (1936). — 19) *H. Albers, F. Schlenk, H. v. Euler*, Nicotinsäureamid aus Cozymase. Ark. f. Kemi, **12B**, Nr. 21 (1936). — 20) *H. A. Alt*, Atmungshemmung durch HCN. Bioch. Zs., **221**, 498 (1930). — 20a) *A. M. Altschul, T. R. Hogness*, Spectroscopic evidence for the existence of carboxycytochrome c. Jl. of Biol. Chem., **124**, 25 (1938). — 21) *N. Alwall*, Wirkung der Dinitrophenole auf die tier. Oxyduktions-Prozesse. Skand. Arch. Phys. **72**, Suppl. (1935). — 22) *P. Ambrus, I. Banga, A. Szent-Györgyi*, Messung des Sauerstoffverbrauchs, des Respirationsquotienten und d. Methylenblaureduktion der Gewebe und der Hefe. Bioch. Zs. **240**, 473 (1931). — 23) *H. Andersag, K. Westphal*, Synthese des antineuritischen Vitamins. Berichte dtsch. chem. Ges. **70**, 2035 (1937). — 24) *R. J. Anderson, M. S. Newman*, The chemistry of the Lipids of Tubercle Bacilli. XXXIV. Isolation of a pigment and of anisic acid. Jl. of Biol. Chem., **101**, 773 (1933). — 25) *R. J. Anderson, M. S. Newman*, The chemistry of the Lipids of Tubercle Bacilli. XXXV. The constitution of iohthiocol. Jl. of Biol. Chem., **103**, 197 (1933). — 26) *R. J. Anderson, M. S. Newman*, The synthesis of Phthiocol. Jl. of Biol. Chem., **103**, 405 (1933). — 27) *K. Ando*, Significance of iron in biol. oxidation I, II. Jl. of Biochem. **9**, 188, 201 (1928). — 28) *E. André, K. Hou*, Oxydase des lipides. Compt. Rend., **194**, 645, 195, 172 (1932). — 29) *E. Annau*, c.s. Acetonkörperbildung aus Brenztraubensäure. Zs. phys. Chem., **224**, 141 (1934). — 30) *E. Annau, I. Banga, B. Goeszy, St. Huszàk, K. Laki, B. Straub, A. Szent-Györgyi*, Über die Bedeutung der Fumarsaure für die tierische Gewebsatmung. 2. Mitt. Zs. phys. Chem., **236**, 1 (1935). — 31) *E. Annau, I. Banga, A. Blazsò, V. Bruckner, K. Laki, F. B. Straub, A. Szent-Györgyi*, Über die Bedeutung der Fumarsäure fuer die tierische Gewebsatmung. 3. Mitt. Zs. phys. Chem. **244**, 105 (1936). — *M. L. Anson*, The estimation of cathepsin with hemoglobin and the partial purification of cathepsin. Jl. of gen. Phys., **20**, 565 (1937). — 32) *M. Anson, A. Mirsky*, Helicorubin. Jl. of Phys., **60**, 221 (1925). — 33) *Ch. A. Ashford*, Glycolytic Mechanism of Brain. Biochem. Jl., **27**, 903 (1933). — 34) *E. Aubel*, Die Oxydations-Reduktions-Potentiale

elbender Zellen und ihre Bedeutung. Zs. angew. Chem., **43**, 939 (1930). — 35) *E. Aubel*, Désamination de l'alanine dans le foie. Ann. de Physiol., **9**, 929 (1933). — 36) *Aubel, E.* Sur la désamination de l'alanine. Soc. Biol. **113**, 37 (1933). — 37) *E. Aubel*, c.s. Potentiel apparent des solutions de sucres reductants. C. R., **184**, 407 (1927). — 38) *E. Aubel, E. Aubertin*, Pot. d'oxido-réduction des milieus ou croissent les anaérobies. Soc. Biol., **97**, 1729 (1927); Ann. de Physiol. et de physicochem. biol. **5**, 1 (1929). — 39) *E. Aubel, L. Genevois*, Potentiel d'oxydo-réduction de la levure. C. R. **184**, 1676 (1927). — 40) *E. Aubel, R. Lévy*, Pot. d'oxydo-réduction dans les chenilles de Galleria mellonella. Soc. Biol., **101**, 756 (1929) **104**, 862, **105**, 358 (1930). — 41) *E. Aubel, P. Mauriac, E. Aubertin*, Sur le potentiels d'oxide-réduction et sur les vitesses des proc. d'oxido-réduction des cellules de mammifères. Ann. de Physiol. et physicochim. biol., **5**, 310 (1929). — 42) *E. Aubel, E. Simon*, Über den Mechanismus der Milchsäurebildung im Muskel. Bioch. Zs., **276**, 309 (1935). — 43) *E. Aubel, R. Wurmser*, Pot. d'oxido-réduction des cellules de mammifères. Soc. Biol., **101**, 880 (1929), — *E. Auhagen*, Weiteres Co-Enzym der alkoholischen Gärung. Naturwissenschaften, **19**, 916 (1931). — 45) *E. Auhagen*, Co-Carboxylase, ein neues Co-Enzym der alkoholischen Gärung .Zs. phys. Chem., **204**, 149 (1932). — 46) *E. Auhagen*, Co-Carboxylase II., Zs. phys. Chem., **209**, 20 (1932). — 47) *E. Auhagen*, Reinigung und Vorkommen von Co-Carboxylase in tierischen Organen. Bioch. Zs., **258**, 330 (1933). — 48) *Fr. Axmacher*, Einwirkung einiger organ. Farbstoffe auf den oxidativen Gasstoffw. überleb. Gewebe. Arch. für exp. Path. **170**, 51 (1933). — 48a) *Axmacher F., H. Bergstermann*, Zur Kenntnis der Carboxylase. Bioch. Zs., **272**, 259 (1934). — 49) *I. K. Baars*, Over sulfat reduction door bacteria. Dissertation, Delft 1930. — 50) *A. Bach*, Mechanismus der Eisenkatalyse bei Autoxydationsprozessen. Berichte deutsch. Chem. Ges. **65**, 1788 (1932). — 51) *A. Bach, A. Kultjugin*, Peroxydasefunktion des Oxyhämoglobins. Bioch. Zs., **167**, 227, 238, 241 (1926). — 52) *A. Bach, D. Michlin*, Enzymatische Umwandlung des Xanthins in Harnsäure. Ber. deutsch. Chem. Ges., **60**, 82 (1927). — 53) *A. Bach, D. Michlin*, Succinodehydrase. Ber. deutsch. Chem. Ges., **60**, 827 (1927). — 54) *A. Bach, K. Nikolajew*, Sind sauerstoffübertragende Enzyme mit wasserstoffübertragenden identisch? Bioch. Zs., **169**, 105 (1926). — 55) *A. Bach, K. Nikolajew*, Chem. Beteiligung des Wassers an der oxydativen Wirk. des Chinons. Ber. deutsch. Chem. Ges., **64**, 2769 (1931). — 56) *H. L. J. Baeckström*, Der Kettenmechanismus bei der Autoxydation von Aldehyden. Zs. physikal. Chem. B **25**, 99 (1934). — 57) *E. G. Ball*, Studies on Oxidation-Reduction. XXI. Phthiocol. the pigment of the hum. tubercle bacillus., Jl. of Biol. Chem., **106**, 515 (1934). — 58) *E. G. Ball*, Studies on Oxidation-Reduction. XXII. Lapachol, Lomatiol, and related compounds. Jl. of Biol. Chem., **114**, 649 (1936). — 58a) *E. G. Ball*, Studies on Oxidation-Reduction. XXIII. Ascorbic Acid. Jl. of Biol. Chem., **118**, 219 (1937). — 58b) *E. G. Ball*, Über die Oxydation und Reduktion der drei Cytochrom-Komponenten. Bioch. Zs., **295**, 262 (1938). — 59) *E. G. Ball*, Xanthine Oxidase: An Alloxazine Proteid. Science, **88**, 131 (1938). — 60) *E. G. Ball, T. T. Chen, (W. M. Clark)*, Studies on oxidation-reduction. XX. Epinephrine and related compounds. Jl. of Biol. Chem., **102**, 691 (1933). — 61) *E. G. Ball, W. M. Clark*, A potentiometric study of epinephrine. Proc. Acad. Nat. Sci., Washington, **17**, 347 (1931). — 62) *W. D. Bancroft, N. F. Murphy*, Oxidation and reduction with $H_2O_2$. Jl. of Physical. Chem., **39**, 377 (1935). — 63) *I. Banga*, Über den Aktivator und Donator der Hauptatmung der Taubenbrustmuskels. Zs. phys. Chem., **249**, 183 (1937). — 64) *I. Banga*, Versuche über die Dehydrase-kuppelung mit Fermentpräparaten. Zs. Phys. Chem., **249**, 200 (1937). — 65) *I. Banga, M. Gerend's, K. Laki, G. Papp, E. Porges, F. B. Straub, A. Szent-Györgyi*, Über die dehydrierende Autoxydation und die biologischen Oxydationen. Zs. phys. Chem., **254**, 147 (1938). — 66) *I. Banga, L. Schneider, A. Szent-Györgyi*, Über den Einfluss der Blausäure auf die Gewebsatmung. Bioch. Zs., **240**, 454 (1931). — 67) *I. Banga, L. Schneider, A. Szent-Györgyi*, Über den Einfluss der arsenigen Säure auf die Gewebsatmung. Bioch. Zs., **240**, 462 (1931). — 68) *I. Banga, A. Szent-Györgyi*, Über die Bedeutung der Fumarsäure für die tierische Gewebsatmung. 4. Mitt. Zs. phys. Chem., **245**, 113 (1937). — 69) *G. Barger, F. Bergel, A. R. Todd*, A crystalline fluoreszent dehydrogenation product from vitamin B₁. Nature, **136**, 259 (1935). — 70) *G. Barkan*, Das Kohlenoxydhämoglobin. Handbuch d. normalen u. pathol. Physiol. VI/I, p. 114, Berlin, 1928. — 70a) *G. Barkan, O. Schales*, Die grünen Derivate des Haemoglobins und die Pseudohaemoglobine. Naturwissenschaften, **25**, 667 (1937). — 71) *H. A Barker*, Biochemistry of the methane fermentation. Arch. für Mikrobiologie, **7**, 403, 420 (1936). — 72) *H. K. Barrenscheen, W. Danzer*, Desaminirung des Glykokolls. Zs. phys. Chem. **220**, 57 (1933). — 73) *E. S. G. Barron*, Eff. of Methylenblue on the O₂-consumption. Jl. of Biol. Chem., **81**, 445; **84**, 83 (1929). — 74) *E. S. G. Barron*, The catalytic effect of methylene blue on the oxygen consumption of tumors and normal tissues. Jl. of Exp. Med., **52**, 447 (1930). — 75) *E. S. G. Barron*, The rate of autoxidation of oxidation-reduction systems and its relation to their free energy. Jl. of Biol. Chem., **97**, 287 (1932). — 76) *E. S. G. Barron*, The oxidation of pyruvic acid by gonococci. Jl. of Biol. Chem., **113**, 695 (1936). — 77) *E. S. G. Barron*, Oxidation-reduction potentials of hemin and some hemochromogens. Science, **85**, 58 (1937). — 78) *E. S. G. Barron*, c.s. Oxidation of ascorbic acid in biol. fluids. Jl. of. Biol. Chem. **116**, 563 (1936). — 79) *E. S. G. Barron, R. H. Demeio, F. Klemperer*, Studies on biological oxidations. 5. Copper and hemochromogens as catalysts for the oxidation of ascorbic acid. Jl. of Biol. Chem., **112**, 625 (1936). — 80) *E. S. G. Barron, L. B. Flexner, L. Michaelis*, Mechanism of cysteine potentials at the Hg-electrode. Jl. of Biol. Chem., **91**, 743 (1929). — 81) *E. S. G. Barron, M. Hamburger*, Effect of cyanide upon the catalytic activity of dyes. Jl. of Biol. Chem., **96**, 299 (1932). — 82) *E. S. G. Barron, G. A. Harrop*, Effect of Methylene-blue upon the $O_2$-cons. of erythrocytes. Jl. of exp. Med., **48**, 207 (1928); Jl. of. Biol. Chem. **79**, 65 (1928). 83) *E. S. G. Barron, A. B. Hastings*, Studies on biological oxidations. II. The oxidation of lactic acid by α-hydroxyoxidase, and its mechanism. Jl. of Biol. Chem., **100**, 155 (1933). — 84) *E. S. G. Barron, A. B. Hastings*, Studies on biological oxidations 3. The oxidation-reduction potential of the system lactate-enzyme-pyruvate. Jl. of Biol. Chem., **107**, 567 (1934). — 85) *E. S. G. Barron, A. B. Hastings*, Oxidation-reduction potentials of cyanide-hemochromogen. Jl. of Biol. Chem., **109**, IV. (1935). — 86) *E. S. G. Barron, L. A. Hoffmann*, Catalytic effect of dyes on the $O_2$-consumption. Jl. of gen. Phys., **13**, 483 (1930). — 87) *E. S. G. Barron, C. P. Miller*, Studies on biological Oxidations. Oxidations produced by gonococci. Jl. of Biol. Chem., **97**, 691 (1932), **113**, 695 (1936). — 88) *I. R. Bates, D. J. Salley*, Production of $H_2O_2$ in Hydrogenation-Oxygen-Reactions. Jl. Amer. Chem. So., **55**, 110 (1933). — 89) *M. A. Battie, J. Smedley-Maclean*, Catalytic action of cupric salts. Biochem. Jl., **23**, 593 (1929). 90) *O. Baudisch*, Der Einfluss von Eisenoxyden und Eisenoxydhydraten auf das Wachstum von Bacterien. Bioch. Zs., **245**, 265 (1932); Svensk. kem. Tidskr., **47**, 115 (1935). — 91) *O. Baudisch*, Aktives Eisen. Hb. der Biochem. Erg. Werke., **1**, Jena 1933. — 92) *K. Baudisch, R. Dubos*, Die Katalasewirkung von Eisenverbindungen in Kulturmedien. Bioch. Zs., **245**, 278 (1932). Earlier papers: *O. Baudisch*, c.s. Jl. of Exp. Med., **42**, 473 (1925); Jl. of Biol. Chem., **71**, 501 (1927). — 93) *J. P. Baumberger, J. J. Jürgensen, K. Bardwell*, The coupled redox potential of the lactate-enzyme-pyruvate system. Jl. of gen. Phys., **16**, 961 (1933). — 94) *F. Beattie*, c.s. Nucleotide composition mammalian cardiac muscle. Biochem. Jl., **28**, 84 (1934). — 95) *F. Bergel, K. Bolz*, Über die Autoxydation von Aminosäurederivaten an Tierkohle und Haemin. Zs. phys. Chem. **215**, 25, **220**, 20, **223**, 66 (1933/34). — 96) *F. Bergel, A. R. Todd*, Structure of aneurin and thiochrome. Nature, **138**, 76 (1936). — 97) *E. Bergmann*, Die chemische Erforschung der Naturfarbstoffe, I. and II. Ergebn. d. Physiologie. **35**, 158 (1933); **36**, 347 (1934). — 98) *F. Bernheim*, The specificity of the dehydrases. The separation of the citric acid dehydrase from liver and of the lactic acid dehydrase from yeast. Biochem. Jl. **22**, 1178 (1928). — 99) *F. Bernheim*, c. s. Oxidation of certain amino acids by resting. B. proteus. Jl. of Biol. Chem. **110**, 165 (1935). — 100) *F. Bernheim, M. L. C. Bernheim*, Pyrrole as a catalyst for certain biological oxidations. Jl. Biol. Chem., **92**, 461 (1931). — 101) *Th. Bersin*, Die Beschleunigung der Dehydrierung von Mercaptoverbindungen durch Metalle. Bioch. Zs., **245**, 466 (1932). — 101a) *Th. Bersin*, Thiolverbindungen und Enzyme. Erg. Enzymforschg., **4**, 68 (1935). — 102) *Th. Bersin*, c.s. Biochemische Beziehungen zwischen Ascorbinsäure und Glutathion. Zs. Phys. Chem., **235**, 12 (1935). — 103) *A. Bertho*, Zur Kenntnis des Essigfermentes. Liebigs Ann., **474**, 1 (1929). — 104) *A. Bertho*, Die Essiggärung. Ergebnisse der Enzymforschung, **1**, 231 (1932). — 105) *A. Bertho*, Mechanismus der Wasserstoffaktivierung. Handbuch der Biochemie, Ergänzungswerk-Jena, **1**, 723 (1933). — 106) *A. Bertho*, Mechanismus der Dehydrierung, Ergebnisse der Enzymforschung, Leipzig, **2**, 204 (1933). — 107) *A. Bertho, H. Glück*, Bildung von $H_2O_2$ durch Milchsäurebacterien. Naturwiss., **19**, 88, 20, 484 (1931/2). — 108) *Bertho, A, H. Glück*, Atmungs-Prozess der Milchsäurebacterien. Liebigs Ann., **494**, 159 (1932). — 109) *A. Bertho, B. v. Zychlinski*, Betrag zur Wirkungsweise der dehydrierenden Enzyme von Milchsäurebakterien. Liebig, Ann. **512**, 81 (1934). — 110) *N. Bezssonoff*, Potentiel de l'acide ascorbique. Jl. de Chim. Physique, **32**, 210 (1935). — 111) *R. Bierich, A. Rosenbohm*, Cytochrom. Zs. phys. Chem., **184**, 246 (1929), **196**, 87 (1931). — 112) *R. Bierich, A. Rosenbohm*, Über reduzierende Substanzen der lebenden Gewebe. Zs. phys. Chem., **223**, 136 (1934). — 113) *R. Bierich, A. Rosenbohm*, Gehalt raschwachsender Gewebe an Ascorbinsäure. Zs. phys. Chem., **231**, 47 (1935). — 114) *E. J. Bigwood* c. s., Forme oxydée du cytochrome de la levure. Ann. de Phys., **9**, 837 (1933). — 115) *E. J. Bigwood c. s.*, Oxydabilité du cytochrome c. Soc. Biol., **117**, 222 (1934). — 116) *E. J. Bigwood c. s.*, Reversibilité de l'action inhibitrice de HCN sur l'oxydase du lait. Ann. de Phys., **12**, 690 (1936); Socl Biol., **123**, 87 (1937). — 117) *E.-J. Bigwood, J. Thomas c. s.*, Action de l'aldéhydéhydrase du lait. Soc. Biol., **118**, 1488, **119**, 337, (1935). — 118) *E. J. Bigwood, J. Thomas*, Action du glutathion sur le cytochrome. Soc. Biol., **120**, 69 (1935). — 119) *E. J. Bigwood, J. Thomas*, Action inhibitrice de HCN et CO sur les oxidations biologiques. Bull. Acad. Méd. Belg., **15**, 439 (1935). — 120) *K. Bingold*, Oxydationssteigernde Wirkung von Blutfarbstoffen. Bioch. Zs., **227**, 457 (1930). — 121) *Th. W. Birch, L. J. Harris*, Titration curve and dissociation constants of vitamin C. Biochem. Jl., **27**, 595 (1933). — 122) *J. Bjerrum, L. Michaelis*, Oxidative faculty of NO. Jl. Amer. Chem. Soc., **57**, 1378 (1935). — 123) *H. Blaschko*, The mechanism of catalase inhibitions. Biochem. Jl., **29**, 2303 (1935). — 124) *H. Blaschko*, Cell respiration and catalase activity. Jl. of Phys., **84**, (1935). — 125) *B. Bleyer, O. Kallmann*, Beiträge zur Kenntnis einiger Inhaltsstoffe der Milch. II. Bioch. Zs., **155**, 54 (1925). — 126) *G. Blix*, Reduktion von Methylenblau in Hexose-Phosphat. Skand. Arch. Phys., **50**, 8 (1927). — 127) *G. Blix*, Adrenalin als Oxydations-Katalysator. Skand. Arch. Phys., **56**, 131 (1929). — 128) *G. Blix*, Oxydo-reduktions-system Homogentisinsäure ⇄ Benzochinon-essigsäure. Zs. phys. Chem. **210**, 87 (1932). — 129) *W. Bockenmüller, Th. Götz*, Autoxydation von Natriumhypophosphit. Liebig Ann. 508, 263 (1934). — 130) *K. Bodendorf*, Zur Kenntnis der Inhibitor-Wirkung. Bre. deutsch. Chem. Ges., **66**, 1608 (1933). — 131) *J. H. Bodine c. s.*, Effect of cyanide on embryonic cells. Jl. of Cell. Phys., **4**, 397, 475, **5**, 97 (1934). **8**, 213 (1936); Proc. Nat. Acad. Nat. Sci., **20**, 640 (1934); Proc. Soc. Exp. Biol., **36**, 21 (1937). — 132) *J. H. Bodine, E. J. Boell*, Effect of MB on respiration of blocked and developing embryonic cells. Proc. Soc. Exp. Biol., **34**, 629 (1936), **36**, 21 (1937). — 133) *J. H. Bodine, E. J. Boell*, Effect of dinitrophenol on öxygen consumption. Proc. Soc. Exp. Biol., **35**, 504 (1936). — 134) *J. H. Bodine, E. J. Boell*, Indophenol oxidase. Jl. of Cell. Phys. **8**, 213 (1936). — 135) *H. Borsook*, Reversible and reversed enzyme reactions. Erg. Enzymforschung, 4, 1 (1935). — 136) *H. Borsook c.s.*, Oxidation of ascorb acid and its reduction in vitro and in vivo. Jl. of Biol. Chem., **117**, 237 (1937). — 137) *H. Borsook, C. E. P. Jeffreys*, Glutathione and ascorbic acid. Science, **83**, 397 (1936). — 138) *H. Borsook, G. Keighley*, The energy of urea synthesis. II The effect of varying hydrogen ion concentration with different metabolites. Proc. of the Nat. Acad. of Sciences, **19**, 720 (1933). — 139) *H. Borsook, G. Keighley*, Oxidation-reduction potentials of ascorbic acid. Proc. Nat. Acad. Sci. Washington, **19**, 875 (1933). — 140) *F. P. Bowden, C. P. Snow*, Photochemistry of vitamins. Nature, **129**, 720 (1932). — 140a) *E. Boyland, M. E. Boyland*, Effect of fumarate on tumores respiration. Biochem. Jl., **30**, 224 (1936). — 141) *P. Boysen-Jensen*, Einwirkung von Monojodessigsäure auf Atmung und Gärung. Bioch. Zs., **236**, 211 (1931). — 142) *L. Brann*, Über katalatische und peroxydatische Wirkungen des Blutfarbstoffs und seiner Derivate. Inaug. Diss. München 1927. — 143) *I. v. Braun, W. Keller*, Autoxydation von Aldehyden in Gegenwart von $MnO_2$. Ber. dentsch. Chem. Ges., **66**, 215 (1933). — 144) *A. E. Braunstein, M. G. Kritzmann*, Ab- und Aufbau von Aminosäuren durch Umaminierung. Enzymologia 2, Heft 3 (1937). — 145) *R. Brdička*, Wielands labiler Wasserstoff bei der Schwermetall katalysierten. Oxydation von SH-verbindungen. Bioch. Zs., **272**, 104 (1934). — 146) *R. Brdička, C. Tropp*, Aktivierung von $H_2O_2$ durch Hemoglobin und Haematin. Bioch. Zs., **289**, 301 (1937). — 147) *F. L. Breusch*, Citronensäure im Gewebsstoffwechsel. Zs. phys. Chem., **250**, 262 (1937). — 148) *F. L. Breusch*, Der Wasserstofftransport im Carcinomgewebe. Bioch. Zs., **295**, 125 (1938). — 149) *F. L. Breusch*, Der Eigenstoffwechsel des Lungengewebes. Bich. Zs., **297**, 24 (1938). — 150) *M. M. Brooks*, Penetration into Valonia of oxidation-reduction indicators. Proc. Soc. Exp. Biol., **23**, 265 (1926); Protopl. 10, 505 (1930). — 151) *R. W. Brown, O. L. Osburn, C. H. Werkman*, Dissimilation of pyruvic acid by Clostridium. Proc. Soc. Exp. Biol., **36**, 203 (1937). — 152) *I. W. Buchanan*, Some limitations of Warburgs theory of the rôle of iron in respiration. Science, **66**, 238 (1927). — 153) *E. R. Buchmann*, Sulfite cleavage of Vitamin $B_1$. Jl. Amer. Chem. Soc., **58**, 1803 (1936). — 154) *E. Bumm*, Glutathion. Handbuch d. Biochem. Erg. Werk. 1, 893 (1933) Jena. — 155) *E. Bumm, H. Appel*, Über die Wirkung von Glutathion auf die Pasteur'sche Reaktion. Zs. phys. Chem., **210**, 79 (1932). — 156) *D. Burk*, Free energy of glycogen-lactic acid breakdown etc. Proc. Roy, Soc., **104**, 153 (1929). — 157) *D. Burk*, Azotase. Ergebnisse der Enzymforschung, **3**, 23 (1934). — 157a) *D. Burk, H. Lineveaver, C. K. Horner*, The physiological nature of humic acid stimulation of azotobacter growth. Soil Science, **33**, 455 (1932). — 157b) *D. Burk*, On the biochemical significance of the Pasteur reaction and Meyerhof Cycle in intermediate carbohydrate metabolism. Occasional Publications of the Amer. Assoc. for the Advancement of Science, Supplement to Science, **85**, 121 (1937). — 158) *Wl. S. Butkewitsch, M. W. Fedoroff*, Umwandlung der Essigsäure durch Mucor. Bioch. Zs., **207**, 302 (1929). — 159) *L. Califano*, Die Verbindung Katalase-CO und ihre Spaltung durch monochromatisches Licht. Naturwissenschaften, **22**, 249 (1934). — 160) *R. K. Cannan*, Electrode potentials of hermidin, the chromogen of mercurialis perennis. Biochem. Jl.

**20**, 927 (1926). — 161) *R. K. Cannan*, Echinochrome. Biochem. Jl., **21**, 184 (1927). — 162) *R. K. Cannan, G. M. Richardson*, The thiol-disulfide system. I. Complexes of thiol-acids with iron. Biochem. Jl. **23**, 1242 (1929). — 163) *L. De Caro*, Antioxidative Wirk. des Thyroxins. Zs. phys. Chem., **219**, 257 (1933). — 164) *L. De Caro, M. Giani*, Oxydations Schutz der Ascorbinsäure durch tierische Gewebe. Zs. phys. Chem., **228**, 13 (1934). — 165) *B. Cavanach*, Enzymatic catalysis of the ionisation of hydrogen. Nature **133**, 797 (1934). — 166) *Fr. Cedrangolo*, Ossidazione dell' acido piruvico. Boll. Soc. Ital. Biol., **9**, 669 (1934); Enzymologia, **1**, 359 (1936). — 167) *R. K. Chakraborty, B. C. Guha*, Ascorbic acid-Oxydase. Ind. Jl. Med. Res., **24**, 839 (1937). Chem. Zbl., 1937, I, 3498. — 168) *B. F. Chow, S. E. Kamerling*, Catalytic effect of ferricyanide in the oxygen absorption of oleic acid. Jl. of Biol. Chem., **104**, 69 (1934). — 169) *E. Ciaranfi*, Abbau der Buttersäure durch Leberschnitte, Bioch. Zs., **285**, 228 (1936). — 169a) *W. M. Clark et al*, Studies on Oxidation-Reduction, I—X. U.S. Public Health Service, Hygienic Lab. Bull. No. 151 (1928). — 170) *W. M. Clark*, The potential energies of oxidation-reduction systems and their biochemical significance. Medicine, **13**, 207 (1934). — 171) *W. M. Clark*, Potential energies of biologically important oxidation-reduction systems. Jl. Applied Physics, **9**, 97 (1938). — 172) *H. T. Clarke, S. Gurin*, Studies of crystalline vitamin B$_1$ : 12. The sulfurcontaining moiety. Jl. Amer. Chem. Soc., **57**, 1876 (1935). — 173) *J. K. Cline, R. R. Williams, J. Finkelstein*, Studies of crystalline vitamin B$_1$ : 17. Synthesis of Vitamin B$_1$. Jl. Amer. Chem. Soc., **59**, 1052 (1937). — 174) *G. H. A. Clowes, M. E. Krahl*, Studies on cell metabolism and cell division. I. On the relation between molecular structures, chemical properties, and biological activities of the nitrophenols. Jl. of gen. Phys., **20**, 145 (1936). — 175) *B. Cohen c. s.*, The reduction-potential of Amoeba dubia by micro-injections of indicators. Jl. of gen. Phys., **11**, 585 (1928). — 176) *B. Cohen c.s.*, II. Reduction potentials of marine ova. Brit. Jl. Exp. Biol., **6**, 229 (1929). — 176a) *E. Cohen, C. A. Elvehjem*, The relation of iron and copper to the cytochrome and oxidase content of animal tissues. Jl. of Biol. Chem., **107**, 97 (1934). — 176b) *J. B. Conant*, The electrochemical formulation of the irreversible reduction and oxidation of organic compounds. Chem. Reviews, **3**, 1 (1927). — 177) *J. B. Conant, B. F. Chow, E. B. Schönbach*, The oxidation of hemocyanin. Jl. of Biol. Chem. **101**, 463 (1933). — 178) *J. B. Conant, F. Dersch, W. E. Mydans*, The prosthetic group of Limulus Hemocyanin. Jl. of Biol. Chem., **107**, 755 (1934). — 179) *J. B. Conant, A. M. Pappenheimer*, A redetermination of the oxidation potential of the hemoglobin-methemoglobin system. Jl. of Biol. Chem., **98**, 57 (1932). — 180) *J. B. Conant, C. O. Tongberg*, The oxidation-reduction potentials of hemin and related substances. I. The potentials of various hemins and hematins in the absence and presence of pyridine. Jl. of Biol. Chem., **86**, 733 (1930). — 181) *R. P. Cook, J. B. S. Haldane, L. W. Mapson*, The relationship between the respiratory catalysts of B. Coli. Biochem. Jl., **25**, 534, 880 (1931). — 182) *R. P. Cook, K. Harrison*, Isolation of acetic acid from mammalian liver. Biochem. Jl., **30**, 1640 (1936). — 183) *S. F. Cook*, Metallic ions as oxidation catalysts. Jl. of gen. Phys., **10**, 289 (1926). — 184) *Th. B. Coolidge*, Cytochrome and yeast iron. Jl. of Biol. Chem., **98**, 755 (1932). — 185) *C. F. Cori, G. T. Cori*, Formation of hexosemonophosphate. Proc. Soc. Exp. Biol., **34**, 702 (1936). — 186) *E. G. Cox*, Crystal structure of hexuronic acid. Nature, **130**, 205 (1932). — 187) *M. Cozic*, Etude biochimique de B. xylinum. Rev. génér. bot., **46**, 1, 48, 209 (1934/36). — 188) *M. Cozic*, Pot. d'oxydo-réduction provoqués par le métabolisme des bact. acétiques. Rev. gén. Bot., **48**, 141 (1936). — 189) *H. P. Créac'h*, Respiration de la cellule de levure. Dissert. Bordeaux 1937. — 190) *W. Cremer*, Reduktion von Haemin durch Cystein. Bioch. Zs., **192**, 426 (1928). — 191) *W. Cremer*, CO-Verbindung des Ferro-Cysteins. Bioch. Zs., **194**, 231, **201**, 490, **206**, 228 (1928/9). — 192) *N. B. Das*, Nachweis der Adenosin-triphosphorsäure in Unterhefe. Svensk kem. Tidskr., **48**, 22 (1936). — 183) *N. B. Das*, The Metabolism of Amino Acids. Dissertation, Szeged (1937). — 194) *N. B. Das*, Studies on the inhibition of lactic and malic dehydrogenases. Biochem. Jl., **31**, 1124 (1937). — 195) *J. G. Davis*, Atmung und Gärung von Milchsäurebacterien. Bioch. Zs., **265**, 90 (1933). — 196) *J. G. Davis*, Atmung und Gärung von Milchsäurebacterien. II. Bioch. Zs., **267**, 357 (1933). — 197) *D. R. Davies, J. H. Quastel*, Dehydrogenations by brain tissue. Biochem. Jl., **26**, 1672 (1932). — 198) *E. F. Degering, F. W. Upson*, Catalytic oxidation of d-glucose. Jl. of Biol. Chem., **94**, 423 (1931), **95**, 409 (1932). — 199) *J. G. Dewan, D. E. Green*, Coenzyme factor. Biochem. Jl., **32**, 626 (1938). — 200) *N. R. Dhar*, Iron and cerium compounds as inductors. Jl. of Physical Chem., **35**, 2043 (1931). — 201) *Ch. Dhéré*, L'absorption des rayons UV par les acides nucléiques. Soc. Biol., **101**, 1129 (1929). — 202) *F. Dickens*, Phenosafranine as an anticatalyst of the Pasteur-Effect. Nature, **135**, 762 (1935). — 203) *F. Dickens*, The metabolism of normal and tumor tissue. Action of some oxidation-reduction systems. Biochem. Jl., **30**, 1064, 1233 (1936). — 204) *Fr. Dickens, G. D. Geville*, Metabolism of normal and tumour tissue. IX. Ammonia and urea formation. Biochem. Jl., 27, 1123 (1933). — 205) *H. Dieterle, E. Kruta*, Inhaltsstoff von Drosera. Arch. der Pharmac., 274, 457 (1936). — 206) *Z. Dische*, Koppelung zwischen Resynthese der Adenosintri-phosphorsäure und Oxydoreduktion zwischen Brenztraubensäure und dem Diaceton-Phosphorsäureester. Naturwissenschaften, 1934, S. 855. — 207) *K. Dixon, E. Holmes*, Mechanism of the Pasteur-effect. Nature, 135, 995, 137, 742, 138, 462 (1935/6). — 208) *M. Dixon*, Effect of CN' on the Schardinger Enzyme. Biochem. Jl., 21, 480 (1927). — 209) *M. Dixon*, The action of carbon monoxide on certain oxidising enzymes. Biochem. Jl., 21, 1211 (1927). — 210) *M. Dixon*, The action of carbon monoxide on the autoxidation of sulphydryl compounds. Biochem. Jl., 22, 902 (1928). — 211) *M. Dixon*, Crystalline tripeptide from liver cells. Nature, 124, 512 (1929). — 212) *M. Dixon*, Oxidation mechanisms in animal tissues. Biol. Reviews, 4, 352 (1929). — 213) *M. Dixon, K. A. C. Elliott*, The effect of cyanide on the respiration of animal tissues. Biochem. Jl., 23, 812 (1929). — 214) *M. Dixon, R. Hill, D. Keilin*, Absorption spectrum of the component c of cytochrome. Proc. Roy. Soc. B. 109, 29 (1931) — 215) *M. Dixon, D. Keilin*, The action of cyanide and other respiratory inhibitors on xanthine oxidase. Proc. Roy. Soc., London, B. 119, 159 (1936). — 216) *M. Dixon, C. Lutwak-Mann*, Aldehyde mutase. Nature, 139, 548 (1937). — 217) *M. Dixon, H. E. Tunnicliffe*, On the reducing power of glutathione and cysteine Biochem. Jl., 21, 844 (1927). — 218) *E. C. Dodds, G. D. Greville*, Acceleration of tissue respiration by a nitrophenol. Nature, 132, 966 (1933). — 219) *E. C. Dodds, G. D. Geville*, Effect of a dinitrophenol on tumour metabolism. The Lancet, 1, 398 (1934). — 220) *H. L. Dube, N. R. Dhar*, Induced oxidation of glucose in presence of insulin. Jl. of Physical Chem., 36, 444 (1932). — 221) *R. Dubos*, Rôle of carbohydrate in biological oxidation. Jl. of Exp. Med., 50, 143 (1929). — 222) *R. Dubrisay, A. Saint-Maxen*, Autoxidation de l'hydroquinone. C. R., 189, 694, 191, 212 (1929/30). — 223) *Ch. Dufraisse c. s.*, Sur la catalyse d'autoxydation. C. R., 191, 1126 (1930), 194, 880 (1932). — 224) *G. Dupont c. s.*, Mécanisme de l'action des catalyseurs dans l'autoxydation. C. R., 190, 1302 (1930). — 225) *G. Dupont, J. Allard*, Mécanisme de l'action antioxygène. C. R., 190, 1419 (1930). — 226) *S. Edlbacher, J. Kraus*, Chemie der Adrenalinwirkung. Zs. phys. Chem., 178, 239 (1928). — 227) *S. Edlbacher, M. Neber*, Intermediärer Stoffwechsel des Histidins. Zs. phys. Chem., 224, 261 (1934). — 228) *S. Edlbacher, A. v. Segesser*, Grünes Derivat des Hämoglobins. Naturwissenschaften, 25, 461 (1937). — Bioch. Zs., 290, 370 (1937). — 229) *N. L. Edson, L. F. Leloir*, Metabolism of ketone bodies. Biochem. Jl., 30, 2319 (1936). — 230) *E. Ehrenfest, E. Ronzoni*, Effect of dinitrophenol on oxidations of tissues. Proc. Soc. Exper. Biol., 31, 318 (1933). — 231) *O. Ehrismann*, Pyocyanin und Bakterien-Atmung. Zs. Hyg., 116, 209 (1934). — 232) *F. Eichholtz*, System biologischer Schwermetallreagentien. Arch. für exp. Path., 148, 369 (1929). — 233) *B. Elema*, Studies on the oxidation-reduction of pyocyanine Part II. Redox potentials of pyocyanine. Rec. Trav. Chim., 50, 807 (1931). — 234) *B. Elema*, Oxidation-reduction potentials of chlororaphine. Réc. Trav. chim. 52, 569 (1933). — 235) *B. Elema*, Theory of the reversible two-step oxidation. Jl. of Biol. Chem., 100, 149 (1933). — 236) *B. Elema*, Some remarks concerning the theory of semiquinone formation. Trav. Chim. Pays-Bas, 54, 76 (1935). — 237) *B. Elema, A. J. Kluyver, J. W. van Dalfsen*, Über die Beziehungen zwischen den Stoffwechselvorgängen der Mikroorganismen und dem Oxydoreduktionspotential im Medium. I. Bioch. Zs., 270, 317 (1934). — 238) *B. Elema, A. C. Sanders*, Studies on the oxidation-reduction of pyocyanine. Part 1. The biochemical preparation of pyocyanine. Trav. chem. des pays-bas, 50, 796 (1931). — 239) *Ph. Ellinger, W. Koschara*, Über eine neue Gruppe tierischer Farbstoffe (Lyochrome). I. Vorl. Mitt. Ber. deutsch. Chem. Ges., 66, 315 (1933). — 240) *K. A. C. Elliott*, Reduction of the S.-S-group by enzyme systems. Biochem. Jl., 22, 1410 (1928). — 241) *K. A. C. Elliott*, Catalysis of oxidation of cysteine and thioglycollic acid by Fe and Cu. Biochem. Jl., 24, 310 (1930). — 242) *K. A. C. Elliott*, Behaviour of glutathoine in yeast. Biochem. Jl., 24, 1421 (1930). — 243) *K. A. C. Elliott*, Oxidations catalysed by horseradish and milk-peroxidases. Biochem. Jl., 26, 1281 (1932). — 244) *K. A. C. Elliott c. s.*, Metabolism of lactic acid. Biochem. Jl., 28, 1920 (1934). — 245) *K. A. C. Elliott, Z. Baker*, The effects of oxidation-reduction indicator dyes on the metabolism of tumour and normal tissues. Biochem. Jl., 29, 2396 (1935). — 246) *K. A. C. Elliott, Z. Baker*, The respiratory quotients of normal and tumour tissue. Biochem. Jl., 29, 2433 (1935). — 247) *K. A. C. Elliott, D. Keilin*, The haematin content of horseradish peroxidase Proc. Roy. Soc. B 114, 210 (1934). — 248) *K. C. Elliott, H. Sutter*, Über die Einwirkung von Kohlenoxyd auf Peroxydase. Zs. phys. Chem., 205, 47 (1932). — 249) *G. W. Ellis*, Contribution to the chemistry of drying oils. Jl. of Soc. Chem. Ind., 45, 193 T (1926). — 250) *G. W. Ellis*, Autoxidation of the fatty acids. Biochem. Jl., 26, 791 (1932), 30, 753 (1936). — 251) *C. A. Elvehjem*, Catalytic action of Cu. Biochem. Jl., 24, 415 (1930). — 252) *C. A. Elvehjem*, The so-called autoxidation of cysteine. Science II, 568 (1931). — 253) **R. Emerson**, Effect of certain respiratory inhibitors on the respiration of Chlorella. Jl. gen. Phys., **10**, 469 (1927). — 254) **W. A. Engelhardt**, Atmung von Vogel-Erythrocyten. Bioch. Zs., **251**, 343 (1932). — 255) **H. Erlenmeyer, Epprecht, v. Meyenburg**, Potential der thiazol-5-carbonsäure. Helv. Chim. Acta, **20**, 310, 514 (1937). — 256) **A. Eucken, K. Fajans**, Empfehlungen bestimmter Formelzeichen seitens der Deutschen Bunsen-Gesellschaft. Zs. Elektrochem., **38**, 681 (1932). — 257) **H. v. Euler**, Biol. aktivierende und hemmende Stoffe. Ark. f. Kemi, **A 11**, Nr. 12 (1933). — 258) **H. v. Euler**, Vitamine A und C. Ark. f. Kemi, **11B**, Nr. 18 (1933). — 259) **H. v. Euler**, Chemische Untersuchungen an hochgereinigter Cozymase. Zs. phys. Chem., **240**, 113 (1936). — 260) **H. v. Euler**, Die Cozymase. Ergebnisse Phys., **38**, 1 (1936). — 261) **H. v. Euler c. s.**, Biolog. Abbau- und Veratmungs-Vorgänge an verschiedenen Stoffgruppen. Zs. phys. Chem. **169**, 123 (151) (1927); Arch. internat. Pharm. **43**, 67 (1932). — 262) **H. v. Euler, c. s.**, Enzymatische Inaktivierung der Cozymase. Zs. phys. Chem., **183**, 60 (1929). — 263) **H. v. Euler c. s.**, Katalatische Wirkung einiger eisenhaltigen Verbindungen. Svensk. kem. Tidskr., **41**, 85 (1929). — 264) **H. v. Euler, c.s.** Zur Kenntnis des Vitamins C und verwandter Stoffe. Akr. f. Kemi, **11B**, Nr. 9, Nr. 13 (1933). — 265) **H. v. Euler c. s.**, Sauerstoffaufnahme durch Gluco-reducton. Zs. phys. Chem., **217**, 1 (1933). — 266) **H. v. Euler, c.s.**, Carotinoidgehalt eines Evertebraten. Zs. phys. Chem. **228**, 77 (1934). — 267) **H. v. Euler c. s.**, Fluorescenz-mikroskopische Studien über das Flavin in Augen. Zs. vergl. Phys., **21**, 739 (1935). — 268) **H. v. Euler c. s.**, Cozymase als Wasserstoffüberträger. Svensk kem. Tidskr., **47**, 290 (1935). — 269) **H. v. Euler c. s.**, Z. K. hochger. Cozymase-präp. Ark. f. Kemi, **12B**, Nr. 4 (1935). — 270) **H. v. Euler c. s.**, Versuche zur Abgrenzung von Codehydrase und Cophosphorylase. Ark. f. Kemi, **12B**, Nr. 25 (1936). — 271) **H. v. Euler c. s.**, Beobachtungen über Enzymsyst. und Vitamingehalt bei Ratten. Ark. f. Kemi, **12B**, 30 (1936). — 272) **H. v. Euler c. s.**, Cozymase, das wasserstoffübertragende Coenzym bei der Muskel-Glykolyse. Zs. phys. Chem., **245**, 217 (1937). — 273) **H. v. Euler c. s.**, Enzymatische Inaktivierung der Codehydrase II. Zs. physiol. Chem., **247**, IV (1937). — 274) **H. v. Euler, E. Adler**, Hemmungs-Versuche an Dehydrasen. Zs. phys. Chem., **232**, 10 (1935). — 275) **H. v. Euler, E. Adler**, Dehydrierung von Hexosen unter Mitwirkung von Adenylpyrophosphorsäure. Zs. phys. Chem., **235**, 122 (1935). — 276) **H. v. Euler, E. Adler**, Über die Komponenten der Dehydrasesysteme: 9. Die Co-Zymase und C-Dehydrase II. Co-Zymase als Wasserstoffüberträger. Zs. phys. Chem., **238**, 233 (1936). — 277) **H. v. Euler, E. Adler, G. Günther, H. Heiwinkel, R. Vestin**, Cozymase und Cophosphorylase. I. Zur Frage des Co-Enzyms der Umphosphorylierung. Arkiv. f. Kemi, **12B**, Nr. 24 (1936). — 278) **H. v. Euler, E. Adler, G. Günther, H. Hellström**, Cozymase, das wasserstoffübertragende Co-Enzym bei der Muskelglykolyse. Zs. phys. Chem., **245**, 217 (1937). — 279) **H. v. Euler, E. Adler, H. Hellström**, Mechanismus der Dehydrierung von Alkohol und Triosephosphaten und der Oxydereduktion. Zs. phys. Chem., **241**, 239 (1936). — 280) **H. v. Euler, E. Adler, S. Kyrning**, Über die Komponenten der Dehydrasesysteme 13. Verschiedenheit von Alkohol- und Triosephosphorsäure-Apodehydrase. Zs. phys. Chem. **242**, 215 (1936). — 281) **H. v. Euler, L. Ahlström**, Oxydations-Grösse Vitamin-A-haltiger Stoffe. Zs. phys. Chem. **204**, 168 (1931). — 282) **H. v. Euler, H. Albers, F. Schlenk**, Über die Co-Zymase. (Vorl. Mitt.). Zs. phys. Chem., **237**, I (1935). — 283) **H. v. Euler, H. Albers, F. Schlenk**, Co-Zymase. Bioch. Zs., **286**, 140 (1936). — 284) **H. v. Euler, E. Bauer**, Umwandlung von Codehydrase I in Codehydrase II. Ber. deutsch. Chem. Ges., **71**, 411 (1938). — 285) **H. v. Euler, D. Burström**, Ascorbinsäure-Bestimmung im Harn durch Titration Bioch. Zs., **283**, 153 (1935). — 286) **H. v. Euler, H. Fink**, Cytochrom in Hefezellen. Zs. phys. Chem., **164**, 69 (1927). — 287) **H. v. Euler, H. Fink, H. Hellström**, Cytochrom in Hefezellen. Zs. phys. Chem., **169**, 10 (1927). — 288) **H. v. Euler, W. Franke, R. Nilsson, K. Zeile**, Chemie der Enzyme II, 3: Katalasen un die Enzyme der Oxydation und Reduktion. München 1934. — 289) **H. v. Euler, G. Günther**, Zur Kenntnis der Temperaturstabilität und Bildung der Cozymase. Ark. f. Kemi, **11B**, Nr. 50 (1935). — 290) **H. v. Euler, G. Günther**, Stabilität des durch Erhitzen der Cozymase entstehenden Aktivators der Glykolyse und des Warburg'schen Coferments. Svensk kem. Tidskr., **47**, 285 (1935); Zs. phys. Chem., **237**, 221 (1935). — 291) **H. v. Euler, G. Günther**, Aktivatoren der Glykolyse III. Zs. phys. Chem., **239**, 83 (1936). — 292) **H. v. Euler, H. Heiwinkel**, Enzymatische Inaktivierung der Cozymase. Naturwissenschaften, **25**, 269 (1937). — 293) **H. v. Euler, H. Hellström**, Cytochrom und die katalatische Wirkung der Hefe. Zs. phys. Chem., **190**, 189 (1930). — 294) **H. v. Euler, H. Hellström**, Zur Kenntnis der enzymatischen Wasserstoffüberträger im Muskel. II. Zs. phys. Chem., **252**, 31 (1938). — 295) **H. v. Euler, B. Jansson**, Katal. $H_2O_2$ Spaltung durch Metallverbindungen. Monats-Hefte der Chemie **53/54**, 1014 (1929). — 296) **H. v. Euler, K. Josephson**, Katalase I, II. Liebigs Ann., **452**, 158, **455**, 1 (1927). — 297) **H. v.**

*Euler, K. Josephson*, Katalytische Spaltung des $H_2O_2$ durch Haemin. Liebigs Ann. **456**, 111 (1927). — 298) *H. v. Euler, P. Karrer c. s.*, Synthese des Lactoflavins (Vitamin $B_2$) und anderer Flavine. Helv. Chim. Acta, **18**, 522 (1935). — 299) *H. v. Euler, P. Karrer, B. Becker*, Konstitution der Pentosephosphorsäure aus Cozymase. Helv. Chim. Acta, **19**, 1060 (1936). — 300) *H. v. Euler, E. Klussmann*, Biochem. Vers. an C-Vitamin und Zuckerderiv. Ark. f. Kemi, **11B**, Nr. 7 (1933). — 301) *H. v. Euler, E. Klussmann*, Reduktions versuche an Zuckerderivaten. Ark. f. Kemi, **11B**, Nr. 8 (1933). — 302) *H. v. Euler, E. Klussmann*, Hochreduzierende Zwischenprodukte (Reduktione) bei der alkalischen Umwandlung einfacher Zuckerarten. Ark. f. Kemi **11B**, Nr. 12 (1933). — 303) *H. v. Euler, E. Klussmann*, Physiologische Versuche über Vitamin C (Ascorbinsäure) und Reduktion (Enol-Tartronaldehyd). Zs. phys. Chem., **217**, 167 (1933). — 304) *H. v. Euler, M. Malmberg*, Vitamin C. Ark. f. Kemi, **B11**, Nr. 9 (1933). — 305) *H. v. Euler, C. Martius*, Ein hochred. Zuckerderiv. (Redukton). Svensk kem. Tidskr., **45**, 73 (1933); Ark. f. Kemi, **11B**. — 306) *H. v. Euler, C. Martius*, Reduktion (Enol-tartronaldehyd) und Ascorbinsäure. Liebigs Ann., **505**, 73 (1933). — 307) *H. v. Euler, K. Myrbäck*, Enzymatischer Abbau und Aufbau der Kohlenhydrate. Svensk. Kem. Tidskr., **37**, 173 (1925). — 308) *H. v. Euler, K. Myrbäck*, Cozymase XIV. Zs. phys. Chem. **169**, 102 (1927). — 309) *H. v. Euler, K. Myrbäck*, Cozymase XV. Zs. phys. Chem., **177**, 237 (1928); Naturwissenschaft, **17**, 291 (1929). — 310) *H. v. Euler, K. Myrbäck*, Cozymase XVII. Zs. phys. Chem., **190**, 93 (1930). — 311) *H. v. Euler, K. Myrbäck*, Cozymase XVIII. Zs. phys. Chem., **198**, 219 (1931). — 312) *H. v. Euler, K. Myrbäck, R. Nilsson*, Molekulargewicht der Cozymase. Zs. phys. Chem., **168**, 177 (1927). — 313) *H. v. Euler, R. Nilsson*, Zur Kenntnis der Reduktase der Hefe. Zs. phys. Chem., **155**, 31 (1926). — 314) *H. v. Euler, R. Nilsson*, Zur Kenntnis der Reduktase der Hefe. Zs. phys. Chem., **162**, 72 (1926). — 315) *H. v. Euler, F. Schlenk*, Einw. vom UV-Licht auf Cozymase. Arkiv. f. Kemi, **12B**, Nr. 19 (1936). — 316) *H. v. Euler, F. Schlenk*, Zusammensetzung des Cozymase-Moleküls. Svensk kem. Tidskr., **48**, 135 (1936). — 317) *H. v. Euler, F. Schlenk*, Cozymase. Zs. phys. Chem., **246**, 64 (1936). — 318) *H. v. Euler, F. Schlenk, R. Vestin*, Adenosin-diphosphorsäure aus Cozymase. Naturwissenschaften, **25**, 318 (1937). — 319) *H. v. Euler, R. Vestin*, Zur Kenntnis der Wirkung der Co-Zymase. Zs. phys. Chem., **237**, 1 (1935). — 320) *H. v. Euler, R. Vestin*, Enzymatische Synthese von Cocarboxylase aus Vitamin $B_1$ und Phosphat. Naturwissenschaften, **25**, 416 (1937). — 321) *H. v. Euler, R. Vestin*, Versuche über enzymatische Umwandlung der Codehydrase I in Codehydrase II. Ark. Kem. Miner. Geol., **12B**, No. 44 (1938). — 322) *H. v. Euler, R. Vestin, H. Heiwinkel*, Umphosphorylierung in Gegenwart von Cozymase. Svensk kem. Tidskr., **48**, 176 (1936). — 323) *H. v. Euler, K. Zeile, H. Hellström*, Zur Kenntnis der aktiven Gruppe der Katalase. Sved. Kem. Tidskr., **42**, 74 (1930). — 324) *H. v. Euler, B. Zondeck*, Über die Stabilität des Prolans, Hinweis auf seine enzymatische Natur. Skand. Arch. Phys., **68**, 232 (1934); Ark. f. Kemi, Å**11**, Nr. 12 (1933). — 325) *U. S. v. Euler*, Beeinflussung der Gewebs-Atmung durch Insulin und Adrenalin. Skand. Arch. Phys., **59**, 123 (1930). — 326) *U. S. v. Euler*, Influence du dinitro-α-napthol sur la consommation d'oxygène. Arch. internat. Pharm., **44**, 464 (1933). — 327) *U. S. v. Euler, H. v. Euler, H. Hellström*, A-Vit.-Wirk. der Lipochrome. Bioch. Zs., **203**, 370 (1928). — 328) *W. L. Evans c. s.*, Mechanism. of carbohydrate oxidation. Jl. Amer. Chem. Soc., **50**, 2267 (1928). — 329) *J. G. Eymers, K. L. van Schouwenburg*, Quantitative data regarding spectra connected with bioluminiscence. Neuberg Festschrift, Enzymologia, **3**, 235 (1937). — 330) *R. Fabre, H. Simonnet*, Contribution a l'étude du pouvoir oxydo-réductif des tissues. Bull. Soc. Chim. Biol., **12**, 777, 800 (1930); C. R., **190**, 1233 (1930). — 331) *W. O. Fenn c. s.*, Stimulating effect of CO on mucle metabolism. Amer. Jl. Phys., **101**, 34, 102, 393 (1932). — 332) *J. Field c. s.*, Effect of 2,4-Dinitrophenol on the $O_2$-consumption of yeast. Jl. of cell. Phys., **4**, 405 (1934); Jl. of Pharm., **53**, 314; Proc. Soc. Exp. Biol., **32**, 1043, 1285, 1342 (1935). — 33) *J. Field, S. M. Field*, Cyanide-stable respiration. Proc. Soc. Exp. Biol., **28**, 995 (1931). — 334) *J. Field, E. G. Tainter*, 2,4-Dinitrophenol and 4,6-Dinitro-o-cresol have a common site of action on the yeast cell. Arch. internat. Pharm., **54**, 184 (1936). — 335) *S. Filitti*, Sur l'équilibre d'oxydorédution des oxypurines. Jl. de Chimie Physique, **32**, 1 (1935); Nature 1935 I, 35; älterer Wert: C. R., **198**, 930 (1934). — 336) *H. Fink*, Klassifizierung von Kulturhefen mit Hilfe des Cytochromspektrums. Zs. phys. Chem., **210**, 197 (1932). — 337) *F. G. Fischer*, Succinodehydrase. Ber. deutsch. Chem. Ges., **60**, 2257 (1927). — 338) *F. G. Fischer, A. Marschall*, Beschleunigung von Aldolkondensationen durch Aminosäuren. Ber. deutsch. Chem. Ges., **64**, 2825 (1931). — 339) *F. G. Fischer et al.*, Biochemische Hydrierung. Liebigs Ann., **529**, 84, 87, **530**, 99 (1937). — 340) *F. G. Fischer, O. Wiedemann*, Hydrierung ungesättigter α-Ketosäuren etc. Liebigs Ann., **513**, 260, **520**, 52, **522**, 1 (1934/36). — 341) *H. Fischer*, Chlorophyll. Chem. Rev., **20**, 41 (1937). — 342) *H. Fischer*,

*R. Bäumler*, Überführung von Chlorophyll derivaten in Phyllocrythrin. Sitzungsber. d. Bayr. Akad. d. Wissenschaft., (Math.-Naturwiss. Abt.) 1929, 77. — 343) *H. Fischer, C. v. Seemann*, Die Konstitution des Spirographishämins. Zs. phys. Chem., 242, 133 (1936). — Zs. angew. Chem., 49, 461 (1936). — 344) *H. Fischer, A. Treibs, K. Zeile*, Farbstoffe mit Pyrrolkernen. Hb. der Biochem., Erg.Band, Jena 1930, S. 72 (114). Erg. Werk Bd. I, 247, Jena 1933. — 345) *H. Fischer, K. Zeile*, Synthese von Hämatoporphyrin, Protoporphyrin und Hämin. Liebigs Ann., 468, 98 (1929). — 346) *E. H. Fishberg, B. T. Dolin*, Homogentisic acid: A physiol, ox. red. system. Jl. of Biol. Chem. 97, LXXXVIII (1932), 101, 159 (1933). — 346a) *K. C. Fisher, L. Irving*, The effect of CO upon the enbryonic fish heart. Proc. Amer. Phys. Soc. 46th Ann. Meet. Amer. Jl. Phys., 109, 36 (1934). — 347) *B. Flaschenträger c. s.*, Neuartiger Abbau der aliphatischen Kette. Zs. phys. Chem., 225, 157, (1934); Helv. Chim. Acta, 18, 962 (1935). — Zs. phys. Chem., 238, 221 (1936). — 348) *W. Fleischmann*, Hemmung biologischer Oxydationen durch Gifte. Bull. Biol. Méd. U.S.S.R., 1, 332 (1936). — 348a) *W. Fleischmann et al.*, Über Kohlenoxydvergiftung von Insektenlarven. Bioch. Zs., 294, 281 (1937). — 349) *W. Fleischmann, S. Kann*, Wirkung des Methylenblau auf die Zellatmung. Bioch. Zs., 257, 293 (1933). — 350) *F. W. Fox, L. F. Levy*, Reversible oxidation of ascorbic acid. Biochem. Jl., 30, 208 (1936). — 351) *H. M. Fox*, Chlorocruorin. Proc. Roy. Soc., B99, 199 (1926). — 351a) *J. Franck, F. Haber*, Autoxydation der Alkalisulfite. Naturwiss., 19, 450 (1931); Sitz.-Ber. Preuss. Akad. d. Wiss., 1931, 250. — 352) *W. Franke*, Gleichgewicht im System Fe₃-Hydrochinon. Liebig Ann. 480, 1, 486, 242 (1930/1). — 353) *W. Franke*, Einfache und komplexe Ferrosalze von Carbonsäuren. Liebig Ann., 491, 30 (1931). — 354) *W. Franke*, Autoxidation der ungesättigten Fettsäuren. I, II. Liebigs Ann., 498, 129 (1932); Zs. phys. Chem., 212, 234 (1932). — 355) *W. Franke*, Zur Energetik von Dehydrierungs-Reaktionen biologischen Interesses. Bioch. Zs., 258, 280 (1933). — 356) *W. Franke*, Wärmtönungen und maximale Nutzarbeiten biochemisch wichtiger Reaktionen. Tab. biol. period., 5, 120 (1935). — 357) *W. Frankenburger*, Fermentreaktionen unter dem Gesichtspunkt der heterogenen Katalyse. Ergebnisse der Enzymforschung, 3, 1 (1934). — 358) *W. Frei*, Atmungsfarbstoffe bei pflanzlichen Mikroorganismen. Festschrift Zangger, 1, 805 (1934). — 359) *L. Fresenius c. s.*, Katalytische Eigenschaften der Mineralwässer. Zs. Anorg. Chem., 160, 273 (1927). — 359a) *K. Freudenberg, Th. Wegmann*, Der Schwefel des Insulins. Zs. phys. Chem., 233, 159 (1935). — 360) *E. A. H. Friedheim*, Potentiel d'oxydo-reduction d'extraits embryoniques. Soc. Biol. 101, 1039 (1929). — 361) *E. A. H. Friedheim*, Pot. d'oxydo-réduction de tissues des mammifères. C. R., 189, 266 (1929). — 362) *E. Friedheim*, La pyocyanine et les oxydations biologiques. C. R. Soc. phys. Genève, 48, Nr. 3 (1931). — 363) *E. A. H. Friedheim*, Pyocyanine, an accessory respiratory enzyme. Jl. exp. Med., 54, 207 (1931). — 364) *E. A. H. Friedheim*, Sur la fonction respiratoire du pigment de Bacillus violaceus. C. R. Soc. Phys. Genève, 49, 125 (1932); Soc. Biol., 110, 353 (1932). — 365) *E. H. A. Friedheim*, Sur deux ferments respiratoires accessoires d'origine animale. C. R. Soc. Phys. Genève, 49, 179 (1932). — 366) *E. A. H. Friedheim*, Significance biol. de la melanogenèse C. R. Soc. Phys. Genève, 50, 20 (1933). — 367) *E. A. H. Friedheim*, Système d'oxydo-réd. réversible biologique. C. R. Soc. Phys. Genève, 50, 162 (1933). — 368) *E. A. H. Friedheim*, Deux systèmes d'oxydoréduction reversibles. C. R. Soc. Phys. Genève, 50, 231 (1933). — 369) *E. A. H. Friedheim*, Fonction respiratoire du pigment rouge de Penic. phoeniceum. C. R. Soc. Biol., 112, 1030 (1933). — 370) *E. A. H. Friedheim*, Ein natürliches, die Zellatmung katalysierendes Redox-System. Naturwissenschaften, 21, 177 (1933). — 371) *E. A. H. Friedheim*, Das Pigment von Halla parthenopea, ein akzessorischer Atmungs-Katalysator. Bioch. Zs., 259, 257 (1933). — 372) *E. A. H. Friedheim*, The effect of pyocyanine on the respiration of some normal tissues and tumours. Biochem. Jl., 28, 173 (1934). — 373) *E. A. H. Friedheim*, Natural reversible oxidation-reduction systems as accessory catalysts in respiration: Juglon and Lawson. Biochem. Jl., 28, 180 (1934). — 374) *E. A. H. Friedheim*, Atmungskatalyse durch ein natürliches Redox-System. Zwischenprodukte der Melaninbildung. Schweiz. med. Wochenschrift, 65, 256 (1935). — 375) *E. A. H. Friedheim, J. G. Bär*, Untersuchungen über die Atmung von Diphyllobothrium latum (L.) Bioch. Zs., 265, 329 (1933). — 376) *E. A. H. Friedheim, L. Michaelis*, Potentiometric study of Pyocyanine. Jl. of Biol. Chem., 91, 355 (1931). — 377) *Cl. Fromageot*, Etude physicochimique du metabolism. Jl. de Chim. Physique, 24, 513, 623 (1927). — 378) *Cl. Fromageot, J. Roux*, Bildung von H₂O₂ durch bacterium bulgaricum. Bioch. Zs., 267, 202 (1934). — 379 *J. S. Fruton*, Oxidation-reduction potentials of ascorbic acid. Jl. of Biol. Chem., 105, 79 (1934). — 380) *O. Fürth, H. Kaunitz*, Oxydation einer physiologischen Substanz durch Tierkohle. Sitzungsberichte der Akademie Wien (IIb), 138, 127 (1929); see also Amer. Jl. Physol., 90, 353 (1928). — 381) *A. Fujita*, Wirkung des CO auf den Stoffwechsel der weissen Blutzellen. Bioch. Zs., 197, 189 (1932). — 382) *A. Fujita, T. Kodama*,

Untersuchungen über Atmung und Gärung pathogener Bakterien. III. Über ytochrom und das sauerstoffübertragende Ferment. Bioch. Zs., **273**, 186 (1934). — 383) *A. Fujita, T. Kodama,* Sauerstoffatmung von Pneumococcus. Bioch. Zs., **277**, 17 (1935). — 384) *R. Gaddie, C. P. Stewart,* Rôle of glutathione in muscle glycolysis. Biochem. Jl., **29**, 2101 (1935). — 385) *H. Gaffron,* Sauerstoff-Übertragung durch Chlorophyll und das photochemische Äquivalent-Gesetz. Ber. deutsch. Chem. Ges., **60**, 755 (1927). — 386) *H. Gaffron,* Über den Mechanismus der Sauerstoffaktivierung durch belichtete Farbstoffe. Bioch. Zs., **264**, 251 (1933). — 386a) *H. Gaffron,* Über die Unabhängigkeit der Kohlensäureassimilation von der Anwesenheit kleiner Sauerstoffmengen und über eine reversible Hemmung der Assimilation durch Kohlenoxyd. Bioch. Zs., **280**, 337 (1935). — 387) *P. H. Gallagher,* Substances, which are capable of behaviour as peroxydases. Biochem. Jl., **18**, 29 (1927). — 388) *D. Ganassini,* Funzione dello zinco in biologia. Arch. ist. biochem. Ital. **3**, 131 (1931). — 389) *A. Geiger,* Rôle of glutathione in anaerobic tissue glycolysis. Biochem. Jl., **29**, 811 (1935). — 390) *M. Geiger-Huber,* Beeinflussung der Hefeatmung durch Neutralrot. Akademie Wet. Amsterdam Proc., **33**, 1059 (1930). — 391) *L. Genevois,* Coloration vitale et respiration. Protopl., **4**, 67 (1928). — 392) *L. Genevois, R. Saric,* Action des dinitrophénols sur la levure. Soc. Biol. 111, 181, 117, 368 (1932/4). — 393) *I. D. Georgescu,* Sur le potentiel d'oxydoréduction des acides hexuroniques. Jl. de Chim. Physique, 29, 217 (1932). — 394) *R. W. Gerard,* Metabolism of Sarcina lutea. Biol. Bull., **60**, 227 (1931). — 395) *R. W. Gerard,* Influence of Metyleneblue on glycolysis. Amer. Jl. Phys., **97**, 523 (1931). — 396) *R. W. Gerard,* Cyanide insensivity of Paramecium. Amer. Jl. Phys., **97**, 524 (1931). — 397) *M. Gerendas,* Autoxydation des Sulfits. Zs. phys. Chem., **254**, 184 (1938). — 398) *H. Gershinowitz,* Free energy and the rate of chemical reactions. Jl. of Chem. Physics, 4, 363 (1936). — 399) *E. G. Gerwe,* Spontaneous oxidation of cysteine. Jl. of Biol. Chem. 92, 399, 525 (1931). — 440) *R. W. Getchell, I. H. Walton,* Activity of peroxydase. Jl. of Biol. Chem., **91**, 419 (1931). — 401) *J. Ch. Ghosh, P. L. Rama Char,* Das Oxydations-Reduktions-Potential der Ascorbinsäure. (Vitamin C). Zs. phys. Chem., **246**, 115 (1937). — 402) *J. Ch. Ghosh, P. C. Rakshit,* Autoxydation von Ascorbinsäure. Bioch. Zs., **289**, 15 (1936). — 403) *L. J. Gillespie, Tsun Hsien Lin*: The reputed dehydrogenation of hydroquinone by Palladium black. Jl. Amer. Chem. Soc., **53**, 3969 (1931). — 404) *A. Görner,* Oxidation of glucose by air in the presence of iron pyrophosphate. Jl. of Biol. Chem., **105**, 705 (1934). — 404a) *B. Gözsy, S. Szent-Györgyi,* Über den Mechanismus der Hauptatmung des Taubenbrustmuskels. Zs. phys. Chem., **224**, 1 (1934). — 405) *H. Goldhammer,* Oxydation von Phenol mit $H_2O_2$ und Eisensalzen. Bioch. Zs., **189**, 81 (1927). — 406) *St. Goldschmidt c. s.,* Mechanismus der Oxydations-reaktionen des $H_2O_2$. Ber. deutsch. Chem. Ges., **61**, 223 (1928); Liebigs Ann. **502**, 1 (1933). — 407) *St. Goldschmidt, K. Freudenberg,* Autoxydation von Linolensäure. Ber. deutsch. chem. Ges., **67**, 1589 1934). — 408) *St. Goldschmidt, St. Pauncz,* Peroxydatische und katalatische Wirkungen von Ferrosalzen. Liebigs Ann. 502, 1 (1933). — 409) *M. Gompel,* Action des anesthés. sur l'oxydation de l'acide oxalique en présence de charbon. Ann. de Phys., **5**, 761 (1929). — 410) *G. Gorr,* Acetessigsäurebildung durch Leberbrei. Bioch. Zs., **254**, 9 1931). — 411) *M. Gradwohl,* Einfluss der Reaktion auf die Oxydation von Aminosäuren an Tierkohle. Bioch. Zs., **219**, 136 (1930). — 412) *F. Grande,* Vorkommen von Palmitico-Dehydrase in Pflanzensamen. Skand. Arch. Phys., **69**, 189 (1934). — 413) *D. E. Green,* The reduction potentials of cysteine, glutathione and glycylcysteine. Biochem. Jl., **27**, 678 (1933). — 414) *D. E. Green,* The potentials of ascorbic acid Biochem. Jl., 27, 1044 (1933). — 415) *D. E. Green,* The oxidation-reduction potentials of Cytochrome c. Proc. Roy. Soc., B. **114**, 423 (1934). — 416) *D. E. Green,* The malic dehydrogenase of animal tissues. Biochem. Jl., **30**, 2095 (1936). — 417) *D. E. Green, J. G. Dewan,* The reversible oxidation and reduction of coenzyme I. Biochem. Jl., **31**, 1069 (1937). — 418) *D. E. Green, J. G. Dewan,* Coenzyme factor of yeast. Biochem. Jl., 32, 1200 (1938). — 418a) *D. E. Green, M. Dixon,* Studies on Xanthine oxidase. XI. Xanthine oxidase and lactoflavine. Biochem. Jl., **28**, 237 (1934). — 418b *D. E. Green, D. Richter,* Adrenochrome. Biochem. Jl. **31**, 596 (1937). — 419) *G. D. Greville,* Fumarate and tissue respiration. I. Effect of dicarboxylic acid on the oxygen consumption. Biochem. Jl., **30**, 877 (1936). — 420) *G. D. Greville,* Fumarate and tissue respiration. II. The respiration of pigeon breast muscle dispersions. Biochem. Jl., **31**, 2274 (1937). — 421) *G. D. Greville, K. G. Stern,* The reduction of dinitrophenols by redox indicators and enzymes. Biochem. Jl., **29**, 487 (1935). — 422) *R. Grewe,* Über das antineuritische Vitamin V. Zs. phys. Chem., **242**, 89 (1936). — 423) *R. Grewe,* Konstitution des Aneurins (Vitamin $B_1$). Naturwissenschaften, **24**, 657 (1936). — 424) *R. Grewe,* Das Aneurin (Vitamin $B_1$). Ergebnisse Phys., **39**, 252 (1937). — 425) *M. Gutstein,* Wasserlösliches Phosphatid und Oxydasereaktionen. Bioch. Zs., **207**, 177 (1929). — 426) *P. György,* Wachstumswirkung synthetischer Flavinpräparate. Zs. f.

Vitaminforschung 4, 223 (1935). — 427) *P. György, R. Kuhn, Th. Wagner-Jauregg*, Vitamin B$_2$. Naturwissenschaften, 21, 560 (1933). — 428) *P. György, R. Kuhn, Th. Wagner-Jauregg*, Über das Vitamin B$_2$. Klin. Ws., 12, 1241 (1933). — 429) *P. György, R. Kuhn, Th. Wagner-Jauregg*, Verbreitung des Vitamins B$_2$ im Tierkörper. Zs. phys. Chem., 223, 21 (1934). — 430) *P. György, R. Kuhn, Th. Wagner-Jauregg*, Darstellung von Vitamin B$_2$ Konzentraten. Zs. phys. Chem., 223, 27 (1934). — 430a) *E. Haas*, Cytochrom. Naturwiss., 22, 207 (1934). — 430b) *E. Haas*, Wirkungsweise des Proteins des gelben Ferments. (Kurze Mitteilung). Bioch. Zs., 290, 291 (1937). — 431) *E. Haas*, Absorptions-spektra der Dihydropyridinverbindungen. Bioch. Zt., 288, 123 (1936). — 432) *E. Haas*, Isolierung eines neuen gelben Ferments. Bioch. Zs., 298, 378 (1938). — 433) *P. Haas, T. G. Hill*, Oxidative and reductive properties of hermidin. Ann. of Bot. 40, 709 (1926). — 434) *F. Haber*, Radikalketten in Lösung. Naturwiss., 20, 468 (1932). — 435) *F. Haber, H. Sachsse*, Reaktion dampfförmigen Na mit O$_2$. Zs. physik. Chem., Bodenstein-Festband, 1931, p. 831. — 436) *F. Haber, J. Weiss*, Katalyse des Hydroperoxyds. Naturwiss., 20, 948 (1932). — 437) *F. Haber, R. Willstätter*, Unpaarigkeit und Radikalketten im Reaktionsmechanismus organischer und enzymatischer Vorgänge. Ber. deutsch. Chem. Ges., 64, 2844 (1931). — 438) *A. Hahn c.s.*, Einfluss des gelben Atmungs-Ferments auf die Dehydrierung. Zs. Biol., 96, 453 (1935). — 439) *A. Hahn, W. Haarmann*, Abbau von Fructosediphosphat im Muskel. Zs. Biol., 90, 231, 92, 364 (1931/32). — 440) *A. Hahn, W. Haarmann c. s.*, Dehydrierung der Bernsteinsäure (Äpelsäure). Zs. Biol., 87, 107, 465, 88, 91, 587, 89, 159, 90, 231, 92, 355 (1927/31). 441) *J. B. S. Haldane*, CO as a tissue poison. Biochem. Jl., 21, 1068 (1927). — 442) *J. B. S., Haldane*, Chain reactions in enzymatic catalysis. Nature, 130, 61 (1932). — 443) *J. B. S. Haldane, K. G. Stern*, Allgemeine Chemie der Enzyme, Dresden, 1932. — 443a) *L. P. Hammett*, Some relations between reaction rates and equilibrium constants. Chem. Rev., 17, 125 (1935). — 444) *D. B. Hand*, Peroxidase. A comparison with other iron-porphyrin catalysts in cells. Ergebnisse der Enzymforschung, 2, 272 (1933). — 445) *H. Handovsky*, Oxidation of phenols by tissues. Biochem. Jl., 20, 1114 (1926). — 446) *H. Handovsky*, Die Oxydations-katalytische Wirkung des Eisens. Zs. phys. Chem., 176, 79 (1928). — 447) *H. Handovsky c. s.*, Action stimulante des nitrodérivatives. Soc. Biol., 115, 1388, 118, 369; Arch. internat. Pharm., 51, 397 (1935). — 448) *M. E. Hanke, J. A. Tuta*, Studies on the oxidation-reduction-potential of blood. Jl. of Biol. Chem., 78, XXXVI (1928). — 449) *A. Hantzsch*, Über meri-chinoide Salze. Ber. deutsch. Chem. Ges., 49, 511 (1916). — 450) *A. Harden*, Alcoholic fermentation, 4. Auflage, London, 1932. — 451) *M. L. C. Hare*, Tyramine oxidase. Biochem. Jl., 22, 968 (1928). — 542) *B. K. Harned*, Oxidations induced by sugars. Jl. of Biol. Chem., 74, XLVII (1927). — 453) *D. C. Harrison*, Catal. act. of traces of iron and copper on the anaerobic oxidation of SH-compounds. Biochem. Jl., 21, 335 (1927). — 454) *D. C. Harrison*, Oxidations by H$_2$O$_2$ in presence of SH-compounds. Biochem. Jl., 21, 507 (1927). — 455) *D. C. Harrison*, Autocatalytic oxidation of SH-compounds. Biochem. Jl., 21, 1404 (1927). — 456) *D. C. Harrison*, Indophenol oxidase. Biochem. Jl., 23, 982 (1929). — 457) *D. C. Harrison*, The oxidation of hexosediphosphoric acid by an enzyme from animal tissues. Biochem. Jl., 25, 1011 (1931). — 458) *D. C. Harrison*, Glucose dehydrogenase: a new oxidising enzyme from animal tissues. Biochem. Jl., 25, 1016 (1931). — 459) *D. C. Harrison*, The product of the oxydation of glucose by glucose dehydrogenase. Biochem. Jl., 26, 1295 (1932). — 460) *D. C. Harrison*, The action of vitamin C on the oxidation of tissues in vitro. Biochem. Jl., 27, 1501 (1933). — 461) *D. C. Harrison*, Dehydrogenases of animal tissues. Ergebnisse der Enzymforschung, 4, 297 (1935). — 462) *D. C. Harrison, S. Thurlow*, Secundary oxidations of some substances of physiological interest. Biochem. Jl., 20, 217 (1926). — 463) *H. Hartmann*, Über das Verhalten von Kohlenoxyd zu Metallverbindungen des Glutathions. Bioch. Zs., 223, 489 (1930). — 464) *F. Haurowitz*, Chemie des Blutfarbstoffes. 7. Zs. phys. Chem., 169, 235 (1927). — 465) *F. Haurowitz*, Chromoproteide, Hb, der Biochem. Erg. Werk. 1, 364 (1930). — 466) *F. Haurowitz*, Über die katalatische Wirkung des Blutfarbstoffs. Zs. phys. Chem., 198, 9 (1931). — 467) *F. Haurowitz*, Über Methämoglobin und seine Verbindungen mit Wasserstoffperoxyd, mit Cyaniden, Fluoriden und Sulfiden. Zs. phys. Chem., 232, 159 (1935). — 468) *F. Haurowitz*, Die Katalatische und peroxydatische Wirkung der Hämine. Enzymologia, 2, 9 (1937). — 469) *F. Haurowitz*, Die Reaktion zwischen Hämin und Wasserstoffperoxyd. Enzymologia, 4, 139 (1937). — 470) *W. N. Haworth, E. L. Hirst c. s.*, Constitution of ascorbic acid. Jl. of Chem. Soc. London, 1933, 1270, 1419, 1934, 62, 1556. — 471) *I. Hayashi*, Einfluss der Farbstoffe auf die Gewebsatmung. Trans. Soc. Path. Jap. 24, 312, 25, 276, 26, 701 (1934/6). — 472) *G. Hecht, F. Eichholtz*, Pharmakol. Analyse des Carcinom Stoffwechsels. Bioch. Zs., 206, 282 (1929). — 473) *H. Hellström*, Die Auslöschung der Methylenblaufluoreszens durch Eisen. Naturwiss., 24, 76 (1936). — 474) *H. Hellström*, Zum

Mechanismus der Hydrosulfitreduktion der Cozymase. Zs. phys. Chem., **246**, 155 (1937). — 475) **St. B. Hendricks**, Refractive Indices of l-ascorbic acid. Nature, **133**, 178 (1934). — 475a) **S. Hennichs**, Aktivität und Eisen-Gehalt hochaktiver Katalase Präparate. Ber. deutsch. Chem. Ges., **59**, 218 (1926). — 476) **M. Henze c. s.**, Umwandlung der Acetessigsäure durch Methylglyoxal. Zs. phys. Chem., **189**, 121, **193**, 88, **195**, 248, **200**, 101, **206**, 1, **212**, 111 (1930/32). — 477) **A. W. H. van Herk**, Beeinflussung der Atmung durch Farbstoffe. Arch. néerl. Phys., **18**, 578 (1933). — 478) **W. Heubner**, Baudischs Befunde und ihre Bedeutung. Med. Klinik, **23**, 1806 (1927). — 479) **G. Hevesy, T. Baranowski, J. Guthke, P. Ostern, J. K. Parnas**, Acta Biol. Exper., In Press, quoted by P. Ostern, T. Baranowski and J. Terszakowec, Über die Phosphorylierung des Adenosins durch Hefe, II., Zs. phys. Chem., **251**, 258 (1938). — 480) **L. F. Hewitt**, Oxido-reduction potential of hemolytix streptococci (pneumococci), Biochem. Jl., **24**, 512, 1551 (1930), — 481) **W. E. van Heyningen**, Inhibition of respiration by cyanide. Biochem. Jl., **29**, 2036 (1935). — 482) **F. F. Heyroth, J. R. Loofbourow**, Chem. constitution of vitamin B₁ as deduced from UV- absorption spectra. Nature, **134**, 461 (1934). — 483) **F. F. Heyroth, J. R. Loofbourow**, Chemical nature of vitamin B₁ from UV absorption spectra. Biochem. Jl., **30**, 651 (1936). — 484) **E. S. Hill**, Spontaneous oxidation of dialuric acid. Jl. of Biol. Chem., **85**, 713, **92**, 471, **95**, 197 (1930/32). — 484a) **E. S. Hill, L. Michaelis**, The effect of iron on the establishment of the oxidation-reduction potential of alloxantin. Science, **78**, 485 (1933). — 484b) **R. Hill**, Chemical nature of hemochromogen and its carbon monoxide compound. Proc. Roy. Soc. London, B, **100**, 419 (1926). — 485) **R. Hill**, Reduced hematin and hemochromogen. Proc. Roy. Soc., B, **105**, 112 (1929). — 486) **R. Hill, D. Keilin**, Porphyrin of component c of cystochrome and its relation to other porphyrins. Proc. Roy. Soc. **B 107**, 286 (1930). — 487) **R. Hill, D. Keilin**, Estimation of haematin iron and the ox--red. equivalent of cytochrome c. Proc. Roy. Soc. B **114**, 104 (1933). — 488) **H. Hillemann**, Beiträge zur Kenntnis des Phenazins, III. Mitt.: Über die Stellung der Methylgruppe im Pyocyanin. Ber. deutsch. Chem. Ges., **71**, 46 (1938). — 489) **K. Hinsberg, G. Holland**, Autoxydation der ungesättigten Fettsäuren. Zs. exp. Med., **94**, 471 (1934). — 489a) **T. R. Hogness, F. P. Zscheile, A. E. Sidwell, E. S. G. Barron**, Cyanide hemochromogen. The ferriheme hydroxide-cyanide reaction: its mechanism and equilibrium as determined by the spectrophotoelectric method. Jl. of Biol. Chem., **118**, 1 (1937). — 490) **E. R. Holiday, K. G. Stern**, Über das spektrale Verhalten des Photo-flavins, des Alloxazins und verwandter Verbindungen: Einfluss der Wasserstoff-Ionen-Konzentration und der zweistufigen Reduktion. Ber. deutsch. Chem. Ges., **67**, 1352 (1934). — 491) **P. Holtz**, Induktion der Ascorbinsäure auf die Oxydation von Zucker. Arch. für exp. Path. **182**, 82, 109 (1936). — 492) **P. Holtz**, Ascorbinsäure als Oxydations-katalysator ungesättigte Fettsäuren. Arch. für exp. Path., **182**, 98 (1936). — 493) **P. Holtz**, Reduktions- und Oxydations-Wirkung bestrahlter Zucker. Arch. für exp. Path., **182**, 141, 160 (1936). — 494) **P. Holtz**, Über den Mechanismus des Histidinabbaus durch Ascorbinsäure und Thioglykolsäure. Zs. phys. Chem., **250**, 87 (1937). — 495) **P. Holtz (G. Triem)**, Histaminbildung durch Ascorbinsäure. Naturwiss., **25**, 14, 251 (1937). — 496) **J. C. Hoogerheide**, La réaction de Pasteur. Ann. de Ferm., **1**, 385 (1935). — 497) **F. G. Hopkins**, On current views concerning the mechanism of biological oxidation. Skand. Arch. Phys., **49**, 33 (1926). — 498) **F. G. Hopkins, K. A. C. Elliott**, Relation of glutathione to cell respiration. Proc. Roy. Soc. B **109**, 58 (1931). — 499) **F. G. Hopkins, E. J. Morgan**, Some relations between ascorbic acid and glutathione. Biochem. Jl., **30**, 1446 (1936). — 500) **T. Imai**, Konstitution des Oryzanins (Vitamin B₁). Zs. phys. Chem., **243**, II (1936). — 500a) **I. M. Innes**, Rôle of the 4 carbon dicarboxylic acids, in muscle respiration. Biochem. Jl., **30**, 2040 (1936). — 501) **R. Itoh**, Xanthinoxidase and catalase, Jl. of Biochem., **22**, 139 (1935). — 502) **B. C. P. Jansen, W. F. Donath**, Isolation of the anti-beri-beri Vitamin. Proc. Akad. Wet. Amsterdam, **29**, 1390 (1926). — 502a) **Jeu, Kia-Khwe, H. N. Alyea**, A comparison of organic inhibitors in chain reactions. Jl. Amer. Chem. Soc., **55**, 575 (1933). — 503) **Br. Jirgensons**, Allgemeine Principien und Wesen chemischer Bindung. Zs. Elektrochem., **35**, 352, 473, 477 (1929). — 504) **Cl. A. Johnson**, Peroxidase activity of hematine. Arch. of Path., **16**, 667 (1923). — 505) **S. W. Johnson, S. S. Zilva**, Oxidation of l-ascorbic acid. Biochem. Jl., **31**, 438 (1937). — 506) **T. B. Johnson, A. Litzinger**, Research on Pyrimidines. Science, **1936**, II, 25; Jl. Amer. Chem. Soc., **58**, 1936 (1936). — 507) **W. P. Jorissen, P. A. A. van der Beek**, Oxydation de la benzaldehyde. Réc. Trav. Chim., **45**, 245, **46**, 42, 47, 286 (1926/8). — 508) **W. R. Jorissen, A. H. Belinfante**, Inducierte Oxydation der Milchsäure. Cjem. Weeklbad, **30**, 618 (1933). — 509) **E. Jorpes, H. v. Euler, R. Nilsson** Cozymase VIII. Zs. phys. Chem., **155**, 137 (1926). — 510) **H. Jost**, Milchsäurebildung in der atmenden Zelle. D. Phys. Ges., **1936**. — 511) **M. Jowett, J. H. Quastel**, Studies in fat meta bolism: I. Biochem. Jl., **29**, 2143 (1935). — 512) **M. Jowett, J. H. Quastel**, Studies in fat Metabolism: II. The oxidation of normal saturated fatty acids in the presence of liver slices. Biochem. Jl., **29**, 2159 (1935). — 513) *M. Jowett, J. H. Quastel,* Studies in fat metabolism: III. The formation and breakdown of acetoacetic acid in animal tissues. Biochem. Jl., **29**, 2181 (1935). — 514) *Ph. Joyet-Lavergne,* Catalyse des Oxydo-réductions dans la celle vivante. Protopl., **23**, 50 (1935). — 515) *Kalmus, H.,* Atmung von Paramäcium. Zt. vergl. Phys., **7**, 314 (1928). — 516) *P. Karrer c. s.,* Vitamin C. Helv. Chim. Acta, **12**, 302 (1933), **17**, 58 (1934). — 517) *P. Karrer c. s.,* Zur Kenntnis des Vitamins C. Helv. Chim. Acta, **16**, 1161 (1933). — 518) *P. Karrer c. s.,* Synthesen von Flavinen. Helv. Chim. Acta, **18**, 69, 72, 426 (1935); Ber. deutsch. Chem. Ges., **68**, 216 (1935). — 519) *P. Karrer, H. Bendas,* Über das Verhalten der Ascorbinsäure gegen Nitrate und Nitrite. Helv. Chim. Acta, **17**, 743 (1934). — 520) *P. Karrer, F. Benz,* Über Reduktionsprodukte des Nicotinsäure-amid-jodmethylats. Helv. Chim. Acta, **19**, 1028 (1936). — 521) *P. Karrer, P. Frei, H. Meerwein,* Zur Konstitution der Lacto-Flavin-phosphorsäure aus Leber. Helv. Chim. Acta, **20**, 79 (1937). — 522) *P. Karrer, P. Frei, B. H. Ringier, H. Bendas,* Lactoflavin-phosphorsäure-adenin-nucleotid aus Leber und das Coferment der d-Alanindehydrase. Helv. Chim. Acta, **21**, 826 (1938). — 523) *P. Karrer, T. Köbner, H. Salomon, F. Zehender,* Über den Lichtabbau der Flavine. Helv. Chim. Acta, **18**, 266, 480 (1935), **19**, 26 (1936). — 523a) *P. Karrer, L. Löwe, H. Hübner,* Konstitution des Astacins. Helv. Chim. Acta, **18**, 96 (1934). — 524) *P. Karrer, H. F. Meerwein,* Synthese des Lactoflavins und 6, 7-Dimethyl-9-(l'-arabityl)-iso-alloxazins. Helv. Chim. Acta, **19**, 264 (1936). — 525) *P. Karerr, B. H. Ringier, J. Büchi, H. Fritzsche, U. Solmssen,* Modellversuche betreffend die wasserstoffübertragenden Wirkungsgruppen der Cofermente. Helv. Chim. Acta, **20**, 55 (1937). — 525a) *P. Karrer, H. Salomon, K. Schöpp,* Isolierung des Hepaflavins. Helv. Chim. Acta, **17**, 419 (1934). — 526) *P. Karrer, H. Salomon, K. Schöpp, E. Schlittler, H. Fritzsche,* Ein neues Bestrahlungsprodukt des Lactoflavins: Lumichrom. Helv. Chim. Acta, **17**, 1010 (1934). — 527) *P. Karrer, H. Salomon, K. Schöpp, E. Schlittler,* Synthese Lactoflavin-ähnlicher Verbindungen. Helv. Chim. Acta, **17**, 1165 (1934). — 528) *P. Karrer, F. Schlenk, H. v. Euler,* Über die Einwirkung von Hypojodit auf einige Pyridiniumbasen. Ark. f. Kemi, **12B**, Nr. 26 (1936). — 529) *P. Karrer, E. Schlittler, K. Pfähler, F. Benz,* Weitere Synthesen Lactoflavin-ähnlicher Verbindungen. II. Helv. Chim. Acta, **17**, 1516 (1934). — 530) *P. Karrer, G. Schwarzenbach, F. Benz, U. Solmssen,* Über Reduktionsprodukte des Nicotinsäure-amid-jodmethylats. Helv. Chim. Acta, **19**, 811 (1936). — 530a) *P. Karrer, G. Schwarzenbach, K. Schöpp,* Redoxpotentials of Ascorbic Acid. Helv. Chim. Act. **16**, 302 (1933). — 531) *P. Karrer, G. Schwarzenbach, G. E. Utzinger,* Dihydro-pyridinverbindungen: 4. N-Phenyl-o-dihydro-pyridin und N-p- Methoxyphenyl-o-dihydro-pyridin. Helv. Chim. Acta, **20**, 72 (1937). — 532) *P. Karrer, O. Warburg,* Jodmethylat des Nicotinsäureamids. Bioch. Zs., **285**, 297 (1936). — 533) *H. Katz,* Magnet. Unterss. an organ. Radikalen. Zs. Physik, **87**, 238 (1933/34). — 534) *W. Keil,* β-Oxydation der δ-Amino-valeriansäure. Zs. phys. Chem., **172**, 310 (1927). — 535) *D. Keilin,* On cytochrome, a respiratory pigment. Proc. Roy, Soc., B, **98**, 312 (1925). — 536) *D. Keilin,* Comparative study of haematin and its bearing on Cytochrome. Proc. Roy. Soc. B, **100**, 129 (1926). — 537) *D. Keilin,* Influence of Co and light on indophenol oxidase of yeast cells. Nature, **119**, 670 (1927). — 538) *D. Keilin,* Cytochrome and respiratory enzymes. Proc. Roy. Soc., **B 104**, 206 (1929). — 539) *D. Keilin,* Cytochrome and intracellular oxidase. Proc. Roy. Soc. B. **106**, 418 (1930). — 540) *D. Keilin,* Cytochrome and intracellular respiratory enzymes. Ergebn. der Enzymforschung, **2**, 239 (1933). — 541) *D. Keilin,* Supposed direct spectroscopic observation of the ,,oxygentransporting ferment''. Nature, **132**, 783 (1933). — 542) *D. Keilin,* Cytochrome and the supposed direct spectroscopic observation of oxidase. Nature, **133**, 290 (1934). — 543) *D. Keilin,* Action of $N_3Na$ on catalytic oxidation reactions. Proc. Roy. Soc. B **121**, 165 (1936). — 544) *D. Keilin,* Mécanism de la respiration intracellulaire. Bull. Soc. Chim. Biol. **18**, 96 (1936). — 545) *D. Keilin, E. F. Hartree,* The combination between methaemoglobin and peroxides: Hydrogen peroxide and ethyl hydroperoxide. Proc. Roy. Soc. London **B 117**, 1 (1935). — 546) *D. Keilin, E. F. Hartree,* Uricase, amino acid-oxidase, and xanthine oxidase. Proc. Roy. Soc. B, **119**, 114 (1936). — 547) *D. Keilin, E. F. Hartree,* Coupled oxidation of alcohol. Proc. Roy. Soc. B **119**, 141 (1936). — 548) *D. Keilin, E. F. Hartree,* On some properties of catalase haematin. Proc. Roy. Soc. London, **B 121**, 173 (1936). — 548a) *D. Keilin, E. F. Hartree,* Preparation of pure cytochrome c from heart muscle and some of its properties. Proc. Roy Soc. London, **B 122**, 298 (1937). — 549) *D. Keilin, E. F. Hartree,* On the mechanism of the decomposition of hydrogen peroxide by catalase. Proc. Roy Soc. Lond. **B 124**, 397 (1938). — 550) *D. Keilin, E. F. Hartree,* Cytochrome a and cytochrome oxidase. Nature, **141**, 870 (1938). — 550a) *D. Keilin, E. F. Hartree,* Cytochrome oxidase. Proc. Roy. Soc. London, **B 125**, 171 (1938). — 551) *D. Keilin, T. Mann,* On the haematin compound of peroxidase. Proc. Roy. Soc. London, **B 122,** 119 (1937). — 552) *D. Keilin,* *T. Mann,* Polyphenol oxidase. Proc. Roy. Soc. Lond. **B 125,** 187 (1938). — 553) *R. A. Kekwick,* *K. O. Pedersen,* Physico-chemical characteristics of the yellow enzyme. Biochem. Jl., 30, 2201 (1936). — 554) *A. E. Kellie, S. S. Zilva,* The catalytic oxidation of ascorbic acid. Biochem. Jl., 29, 1028 (1935). — 555) *W. Kempner,* Wirkung von Blausaure und Kohlenoxyd auf die Buttersäure-gärung. Bioch. Zs., 257, 41 (1933). — 556) *W. Kempner,* Chemical nature of the oxygen-transferring ferment of respiration in plants. Plant Phys., 11, 605 (1936). — 556a) *W. Kempner,* Effect of oxygen tension on cellular metabolism. Jl. Cell. and Comp. Phys., 10, 339 (1937). — 557) *W. Kempner, F. Kubowitz,* Wirkung des Lichtes auf die Kohlenoxydhemmung der Buttersäure-gärung. Bioch. Zs., 265, 245 (1933); 274, 285 (1934). — 558) *E. C. Kendall, J. E. Holst,* Oxidation of cobaltous cysteine. Jl. of Biol. Chem., 91, 435 (1931). — 559) *E. C. Kendall, F. F. Nord,* Reversible oxidation-reduction systems of cysteine-cystine and reduced and oxidized glutathione. Jl. of Biol. Chem., 69, 295 (1926). — 560) *E. C. Kendall, E. I. Witzemann,* Potentiometric study of epinephrine. Jl. of Biol. Chem., 74, XLIX (1927). — 561) *J. Kenner,* Zur Theorie der Oxydationsprozesse. Ber. deutsch. Chem. Ges., 65, 705 (1932). — 561a) *J. Kenner,* Correlation of the yellow oxidation ferment with Warburg's co-ferment. Nature, 139, 25 (1937). — 562) *M. S. Kharasch c.s.,* Metal catalysis in biological oxidation. Jl. of Biol. Chem., 113, 537 (1936). — 563) *W. Kiessling, O. Meyerhof,* Über eine Dinucleotidphosphorsäure der Hefe. Naturwiss., 26, 13 (1938). — 563a) *H. W. Kinnersley, J. O'Brien, R. A. Peters,* Improved yields of vitamin $B_1$. Biochem. Jl., 29, 716 (1935). — 564) *H. W. Kinnersley, J. R. O'Brien, R. A. Peters,* The properties of blue fluorescent substances formed by oxidation of vitamin $B_1$ (Quinochromes), Biochem. Jl., 29, 2369 (1935). — 565) *H. W. Kinnersley, R. A. Peters,* Vitamin $B_1$ and Co-carboxylase. Chem. and Ind., 56, 447 (1937). — 566) *Br. Kisch,* Omegakatalyse der Dioxyphenylalanin-oxydation. Bioch. Zs., 220, 92 (1930). — 567) *Br. Kisch,* Beeinflussung der Gewebs-Atmung durch Amino-säuren. Bioch. Zs., 238, 351, 242, 26, 436, 244, 451, 459, 247, 365 (1932). — 568) *Br. Kisch,* Oxydative Desaminierung der Aminosäuren durch Methylglyoxal. Bioch. Zs., 257, 334 (1933). — 569) *Br. Kisch,* Cyanempfindlichkeit der Atmung verschiedener Gewebe. Bioch. Zs., 263, 75 (1933). — 570) *Br. Kisch,* Steigerung der Gewebs-Atmung durch kleine Cyanmengen. Bioch. Zs., 263, 187 (1933). — 571) *Br. Kisch,* Nicht enzymatische Zwischenkatalysatoren. Hb. der Biochem. Erg. Werk, (Jena), 1, 563 (1933). — 572) *Br. Kisch,* Beinflussung des Gewebs-Stoffwechsels durch optisch-aktive Aminosäuren. Bioch. Zs., 280, 41 (1935). — 573) *Br. Kisch,* Aminodehydrasen im Tierkörper. Enzymologia, 1, 97 (1936). — 574) *Br. Kisch c. s.,* Autokatalyse der Adrenalinoxi-dation. Bioch. Zs., 220, 84, 92, 370 (1930). — 575) *Br. Kisch et al.,* O-Chinone als Ferment-modelle. Bioch. Zs., 242, 1, 21, 244, 440, 247, 371, 249, 63, 250, 135, 252, 380, 254, 148, 257, 89, 334, 259, 455, 263, 98, 268, 158, 271, 424 (1931/4). — 576) *H. Kleinfeller,* Lyochrome. Chem. Ztg., 59, 445 (1935). — 577) *A. J Kluyver,* Chemical activity of micro-organisms. London 1931. — 578) *A. J. Kluyver,* Nieuwe Onderz.... Chemisme der Ademhaling. Chem. Weekblad., 31, H. 18 (1934). — 579) *A. J. Kluyver, J. C. Hoogerheide,* Influence of monojodoacetic acid on yeast. Proc. Akad. Wet. Amsterdam, 36, 596 (1933). — 580) *A. J. Kluyver, J. C. Hoogerheide,* Beziehungen zwischen den Stoffwechselvorgängen von Hefen und Milchsäurebakterien und dem Redoxpotential im Medium. Enzymologia, 1, 1 (1936). — Akad. Wet. Amsterdam Proc. 39, 3 (1936). — 581) *F. Knoop,* Oxydationen im Tierkörper. Stuttgart 1931. — 582) *F. Knoop,* Angeblicher Nachweis einer α-Oxydation von Fettsäuren. Zs. phys. Chem., 209, 277 (1932). — 583) *F. Knoop, C. Martius,* Bildung von Zitronensäure. Zs. phys. Chem., 242, I (1936). — 584) *R. Kobert,* Beiträge zur Kenntnis der Methämoglobine. Arch. Ges. Physiol., 82, 603 (1900). — 585) *G. Kögel,* Reduktions-Beschleunigung des Methylenblaus am Licht durch Eiweiss. Strahlentherap. 42, 384 (1931). — 586) *F. Kögl,* Pilz- und Bakterienfarbstoffe. Handbuch d. Pflanzenanalyse, Wien, 3, 1410 (1932). — 586a) *F. Kögl, J. J. Postowsky,* Über das grüne Stoffwechselprodukt des Bacillus Chlororaphis. Liebigs Ann., 480, 280 (1930). — 587) *I. M. Korr,* Oxidation-reduction potentials in heterogeneous systems. Jl. Cell. and Comp. Physiol., 11, 233 (1938). — 588) *W. Koschara,* Über die Einwirkung von Licht auf Lyochrome. Zs. phys. Chem., 229, 103 (1934). — 589) *E. Krah,* Über die Schwermetallnatur der Zellfermente. Bioch. Zs., 219, 432 (1930). — 590) *M. E. Krahl, G. H. A. Clowes,* Dinitrophenol stimulation of respiration. Proc. Soc. Exp. Biol. 32, 226 (1934). — 591) *M. E. Krahl, G. H. A. Clowes,* Act. of dinitrocresol on yeast. Jl. Amer. Chem. Soc., 57, 1144 (1935). — 592) *M. E Krahl, G. H. A. Clowes,* Studies on cell metabolism and cell division. II. Stimulation of cellular oxidation and reversible inhibition of cell division by dihalo and trihalophenols. Jl. gen. Phys., 20, 173 (1936). — 593) *H. A. Krebs,* Über die Rolle der Schwermetalle bei der Autoxydation von Zuckerlösungen. Bioch. Zs., 180, 377 (1927). — 594)

**H. A. Krebs,** Über die Wirkung von Kohlenoxyd und Licht auf Häminkatalysen. Bioch. Zs., **193,** 347 (1928). — 595) **H. A. Krebs,** Über die Wirkung des Kohlenoxyds auf Häminkatalysen nach M. Dixon. Bioch. Zs., **201,** 489 (1928). — 596) **H. A. Krebs,** Über die Wirkung von Kohlenoxyd und Blausäure auf Hämatinkatalysen. Bioch. Zs., **204,** 322 (1929). — 597) **H. A. Krebs,** Wirk. der Schwermetalle auf die Autoxydation der Alkalisulfide und des $H_2S$. Bioch. Zs., **204,** 343 (1929). — 598) **H. A. Krebs,** Über Hemmung einer Hämatinkatalyse durch Schwefelwasserstoff. Bioch. Zs., **209,** 32 (1929). — 598a) **H. A. Krebs,** Versuche über die proteolytische Wirkung des Papains. Bioch. Zs., **220,** 289 (1930), **238,** 174 (1931). — 599) **H. A. Krebs,** Über die Wirkung der Monojodessigsäure auf die Zellatmung. Bioch. Zs., **234,** 278 (1931). — 600) **H. A. Krebs,** Untersuchungen über den Stoffwechsel der Aminosäuren im Tierkörper. Zs. phys. Chem., **217, 191** (1933). — 601) **H. A. Krebs,** Weitere Untersuchungen über den Abbau der Aminosäuren im Tierkörper. Zs. phys. Chem., **218,** 157 (1933). — 602) **H. A. Krebs,** Grösse der Atmung. Hb. der Biochem. Erg. Werk, **1,** 863, (Jena 1933). — 603) **H. A. Krebs,** Metabolism of amino acids. Biochem. Jl., **29,** 1620, 2077 (1935). — 604) **H. A. Krebs,** Metabolism of amino acids and related substances. Ann. Rev. of Biochemistry, **5,** 247 (1936). — 605) **H. A. Krebs,** Intermediary metabolism of carbohydrates. Nature, **138,** 288 (1936). — 606) **H. A. Krebs,** Dismutation of pyruvic acid in Gonococcus. Biochem. Jl., **31,** 661 (1937). — 607) **H. A. Krebs,** The intermediate metabolism of carbohydrates. Lancet, **233,** 736 (1937). — 608) **H. A. Krebs, c. s.,** Stoffwechsel der Aminosäuren. Klin. Ws., **1932,** 1744. — 609) **H. A. Krebs, W. A. Johnson,** Metabol. of ketonic acids. Biochem. Jl., **31,** 645 (1937). — 610) **H. A. Krebs, W. A. Johnson,** Aceto-pyruvic acid as an intermediary metabolite. Biochem. Jl., **31,** 772 (1937). — 611) **H. A. Krebs, W. A. Johnson,** The role of citric acid in intermediate metabolism in animal tissues. Enzymologia, **4,** 148 (1937). — 612) **F. Kubowitz,** Die Hemmung der Buttersäuregärung durch Kohlenoxyd. Bioch. Zs., **274,** 285 (1934). — 612a) **F. Kubowitz,** Kohlenoxyd-ferroglutathion. Bioch. Zs., **282,** 277 (1935). — 613) **F. Kubowitz,** Über die chemische Zusammensetzung der Kartoffeloxydase. Bioch. Zs., **292,** 221 (1937). — 614) **F. Kubowitz,** Resynthese der Phenoloxydase aus Protein und Kupfer. Bioch. Zs., **296,** 443 (1938). — 615) **F. Kubowitz,** Spaltung und Resynthese der Polyphenoloxydase und des Hämocyanins. Bioch. Zs., **299,** 32 (1938). — 616) **F. Kubowitz, E. Haas,** Ausbau der photochemischen Methoden zur Untersuchung des sauerstoffübertragenden Ferments. (Anwendung auf Essigbakterien und Hefezellen). Bioch. Zs., **255,** 247 (1932). — 617) **A. Th. Küchlin,** Neue Theorie der Zelloxydation. Chem. Weekblad, **28,** 374 (1931). — 618) **A. Th. Küchlin,** Die Fenton'sche Reaktion. Rec. Trav. Chim. **51,** 887 (1932) Bioch. Zs., **261,** 411 (1933). — 619) **A. Th. Küchlin, J. Böseken,** Oxidation of some carbohydrates by $H_2O_2$. Réc. Trav. Chim. **47,** 1011 (1928). — 620) **Th. A. Küchlin, J. Boeseken,** Die Bedeutung der $Fe_2$ und $Fe_3$-Komplexe von Kohlehydraten und Polyalkoholen für den Mechanismus der Fentonschen Reaktion. Proc. Roy. Acad. Amsterdam, **32,** 1218 (1929). — 621) **J. Kühnau,** Über den Abbau der $\beta$-Oxybuttersäure durch Fermente der Leber. Bioch. Zs., **200,** 29 (1928). — 622) **J. Kühnau,** Über den Mechanismus der Verknüpfung von Fett- und Kohlenhydratabbau in der Leber. (Ein Beitrag zur Kenntnis biologischer Redoxpotentiale. Bioch. Zs., **243,** 14 (1931). — 623) **J. Kühnau,** Bildung von Dicarbonsäure aus Acetessigsäure. Intern. Physiol. Kongr. Roma 1932. — 624) **J. Kühnau,** Fette im Stoffwechsel. Hb. der Biochem. Erg. Werk, **3,** 660, Jena 1935. — 625) **F. M. Kün,** Zuckeroxydation mit Luft-$O_2$ und $H_2O_2$. Bioch. Zs., **215,** 12 (1929). — 626) **R. Kuhn,** Über die Wirksamkeit und Specifität von Eisenkatalysatoren. Zs. Angew. Chem., **45,** 353 (1932). — 627) **R. Kuhn,** Natural colouring matters related to vitamins. Jl. of Soc. Chem. Ind., **52,** 981 (1933). — 628) **R. Kuhn,** Flavine. IX. Congr. internac. de Quim. Madrid 1934. — 629) **R. Kuhn,** Synthetic compounds with vitamin $B_2$ action. Nature, **135,** 185 (1935). — 630) **R. Kuhn,** Les flavines. Bull. Soc. Chim. Biol., **17,** 905 (1935). — 631) **R. Kuhn,** Lactoflavin (Vitamin $B_2$). Zs. Angew. Chem., **49,** 6 (1936). — 632) **R. Kuhn,** Reduktions-Stufen des Lactoflavins. Zs. Angew. Chem., **50,** 221 (1937). — 633) **R. Kuhn,** „Hilfs-stoffe" der organischen Synthese. Chem. Ztg., **61,** 17 (1937). — 634) **R. Kuhn,** Wirkstoffe in der belebten Natur. Naturwiss., **25,** 225 (1937). — 635) **R. Kuhn, F. Bär,** Zum photochemischen Verhalten des Lacto-flavins; Modell-Versuche in der Chinoxalin-Reihe. Ber. deutsch. Chem. Ges., **67,** 898 (1934). — 636) **R. Kuhn, P. Boulanger,** Beziehungen zwischen Reduktions-Oxydations-Potential und chemischer Konstitution der Flavine. Ber. Deutsch. Chem. Ges., **69,** 1557 (1936). — 637) **R. Kuhn, P. Boulanger,** Über die Giftigkeit der Flavine. Zs. phys. Chem., **241,** 233 (1936). — 638) **R. Kuhn, L. Brann,** Abhängigkeit der katalatischen und peroxydatischen Wirksamkeit des Fe von seiner Bindung. Ber. Deutsch. Chem. Ges., **59,** 2370 (1926). — 639) **R. Kuhn, L. Brann,** Über die katalatische Wirksamkeit verschiedener Blutfarbstoffderivate. Zs. phys. Chem., **168,** 27 (1927). — 640) **R. Kuhn, P. Desnuelle,** Über die Aminosäuren des gelben Ferments. Ber. Deutsch. Chem. Ges., **70,** 1907

(1937). — 641) *R. Kuhn, P. Desnuelle*, Zur Bestimmung von Sulfhydrylgruppen in Proteinen., Zs. phys. Chem., **251**, 14 (1938). — 642) *R. Kuhn, P. Desnuelle*, Über die Isolierung von Arginin Histidin, Lysin, Glutaminsäure und Asparaginsäure aus gelbem Ferment. Zs. phys. Chem., **251**, 19 (1938). — 643) *R. Kuhn, P. Desnuelle*, Über die Bindung von Silberionen durch das gelbe Ferment und dessen Eiweisskomponente. Zs. phys. Chem., **251**, 23 (1938). — 644) *R. Kuhn, W. Franke*, Über das Redox-Potential des Porphyrexids und des Porphyrindins. Ber. Deutsch. Chem. Ges., **68**, 1528 (1935). — 645) *R. Kuhn, P. György, Th. Wagner-Jauregg*, Über eine neue Klasse von Naturfarbstoffen. (Vol. Mitteil.) Ber. Deutsch. Chem. Ges., **66**, 317 (1933). Mitteil. der Kaiser Wilhelm-Gesellschaft, **2**, Nr. 1/4 (1933). — 646) *R. Kuhn, P. György, Th. Wagner-Jauregg*, Über Ovoflavin, den Farbstoff des Eiklars. Ber. Deutsch. Chem. Ges., **66**, 576 (1933). — 647) *R. Kuhn, P. György, Th. Wagner-Jauregg*, Über Lactoflavin, den Farbstoff der Molke. Ber. Deutsch. Chem. Ges., **66**, 1034 (1933), **67**, 1770 (1934). — 648) *R. Kuhn, D. B. Hand, M. Florkin*, Die Fermente der Zellatmung. Naturwiss., **19**, 771 (1931). — 649) *R. Kuhn, D. B. Hand, M. Florkin*, Über die Natur der Peroxydase. Zs. phys. Chem., **201**, 255 (1931). — 650) *R. Kuhn, F. Köhler, L. Köhler*, Methyloxydationen im Tierkörper. Zs. phys. Chem., **242**, 171 (1936). — 651) *R. Kuhn, J. C. Lyman*, Über das Redox-Potential des Murexids. Ber. Deutsch. Chem. Ges., **69**, 1547 (1936). — 652) *R. Kuhn, K. Meyer*, Autoxydation des Benzaldehyds. Naturwiss., **16**, 1028 (1928). — 653) *R. Kuhn, K. Meyer*, Über katalytische Oxydationen mit Hämin. Zs. phys Chem., **185**, 193 (1929). — 653a) *R. Kuhn, G. Moruzzi*, Über die Dissoziationskonstanten der Flavine; ph-Abhängigkeit der Fluoreszens. Ber. Deutsch. Chem. Ges., **67**, 888 (1934). — 654) *R. Kuhn, G. Moruzzi*, Über das Reduktions-Oxydations-Potential des Lacto-flavins und seiner Derivate. Ber. Deutsch. Chem. Ges., **67**, 1220 (1934). — 655) *R. Kuhn, K. Reinemund*, Über die Synthese des 6.7.9.-Trimethyl-flavins (Lumilactoflavins), Ber. Deutsch. Chem. Ges., **67**, 1932 (1934). — 656) *R. Kuhn, K. Reinemund, F. Weygand*, Synthese des Lumi-lactoflavins. Ber. Deutsch. Chem. Ges., **67**, 1460 (1934). — 657) *R. Kuhn, K. Reinemund, H. Kaltschmitt, R. Ströbele, H. Trischmann*, Synthetisches 6.7.-Dimethyl-9-d-ribo-flavin. Naturwiss., **23**, 260 (1935). — 658) *R. Kuhn, K. Reinemund, F. Weygand, R. Ströbele*, Über die Synthese des Lactoflavins. Ber. Deutsch. Chem. Ges., **68**, 1765 (1935). — 659) *R. Kuhn, H. Rudy*, Über den alkali-labilen Ring des Lacto-flavins. Ber. Deutsch. Chem. Ges., **67**, 892 (1934). — 660) *R. Kuhn, H. Rudy*, Über den alkali-labilen Ring des Lacto-flavins; Monomethyl- und Dimethylverbindungen. Ber. Deutsch. Chem. Ges., **67**, 1125 (1934). — 661) *R. Kuhn, H. Rudy*, Über die Konstitution des Lumilactoflavins (Vorl. Mitt.). Ber. Deutsch. Chem. Ges., **67**, 1298 (1934). — 662) *R. Kuhn, H. Rudy*, Synthetische Vitamin-$B_2$-Phosphorsäure. Ber. Deutsch. Chem. Ges., **68**, 383 (1935). — 663) *R. Kuhn, H. Rudy*, Katalytische Wirkung der Lactoflavin-5'-phosphorsäure; Synthese des gelben Fermentes. Ber. Deutsch. Chem. Ges., **69**, 1974 (1936). — 664) *R. Kuhn, H. Rudy*, Lactoflavin als Co-Ferment; Wirkstoff und Träger. Ber. Deutsch. Chem. Ges., **69**, 2557 (1936). — 665) *R. Kuhn, H. Rudy*, Wachstums-Wirkung von Flavinphosphorsäure. Zs. phys. Chem., **239**, 47 (1936). — 666) *R. Kuhn, H. Rudy, Th. Wagner-Jauregg*, Über Lactoflavin. (Vitamin $B_2$). Ber. Deutsch. Chem. Ges., **66**, 1950 (1933). — 667) *R. Kuhn, H. Rudy, F. Weygand*, Über die zucker-ähnliche Seitenkette des Lactoflavins. Ber. Deutsch. Chem. Ges., **68**, 625 (1935). — 668) *R. Kuhn, H. Rudy, F. Weygand*, Synth. der Lactoflavin-5'-phosphorsäure. Ber. Deutsch. Chem. Ges., **69**, 1543, 1974 (1936). — 669) *R. Kuhn, H. Rudy, F. Weygand*, Über die Bildung eines künstlichen Fermentes aus 6,7-Dimethyl-9-l-Araboflavin-5'-Phosphorsäure. Ber. Deutsch. Chem. Ges., **69**, 2034 (1936). — 670) *R. Kuhn, K. Schön*, Pyocyaninium-perchlorat. Ber. Deutsch. Chem. Ges., **68**, 1537 (1935). 671) *R. Kuhn, R. Ströbele*, Über Vero-, Chloro- und Rhodo-flavine. Ber. Deutsch. Chem. Ges., **70**, 753 (1937). — 672) *R. Kuhn, E. Valkó* quoted in *R. Kuhn + K. Schön*, Pyocyaninium perchlorat, Ber. Deutsch. chem. Ges., **68**, 1538 (1935). — 673) *R. Kuhn, H. Vetter*, Thiochrom. Ber. Deutsch. Chem. Ges., **68**, 2375 (1935). — 674) *R. Kuhn, Th. Wagner-Jauregg*, Über die aus Eiklar und Milch isolierten Flavine. Ber. Deutsch. Chem. Ges., **66**, 1577 (1933). — 675) *R. Kuhn, Th. Wagner-Jauregg*, Über das Reduktions-Oxydations-Verhalten und eine Farbreaktion des Lacto-flavins (Vitamin $B_2$). Ber. Deutsch. Chem. Ges., **67**, 361 (1934). — 676) *R. Kuhn, Th. Wagner-Jauregg, H. Kaltschmitt*, Über die Verbreitung der Flavine im Pflanzenreich. Ber. Deutsch. Chem. Ges., **67**, 1452 (1934). — 677) *R. Kuhn, Th. Wagner-Jauregg, F. W. van Klaveren, H. Vetter*, Über einen gelben, schwefelhaltigen Farbstoff aus Hefe. Zs. phys. Chem., **234**, 196 (1935). — 678) *R. Kuhn, A. Wassermann*, Die Abhängigkeit der katalatischen und oxydativen Wirkung des Eisens von seinem Adsorptions-Zustand. Ber. Deutsch. Chem. Ges., **61**, 1550 (1928). — 679) *R. Kuhn, A. Wassermann*, Komplexbildung und Katalyse. Liebigs Ann. **503**, 203 (1933). — 680) *R. Kuhn, A. Wassermann*, Einfluss von Graphit auf die Hydroperoxydzersetzung durch

Eisen. Liebigs Ann. **503**, 232 (1933). — 681) *R. Kuhn, F. Weygand*, Synthese des 9-Methyl-isoalloxazins. Ber. Deutsch. Chem. Ges., **67**, 1409 (1934). — 682) *R. Kuhn, F. Weygand*, Bedingungen und Geltungsbereich der Flavin-Synthese. Ber. Deutsch. Chem. Ges., **67**, 1459 (1934). — 683) *R. Kuhn, F. Weygand*, Synthetische Verbindungen der Lacto-flavin-Gruppe. (Vorl. Mitteil.). Ber. Deutsch. Chem. Ges., **67**, 1939 (1934). — 684) *R. Kuhn, F. Weygand*, Synthese des 6'7-Dimethyl-9-n-amyl-flavins. Ber. Deutsch. chem. Ges., **67**, 1941, (1934). — 685) *R. Kuhn, F. Weygand*, Synthetisches Vitamin B$_2$. Ber. Deutsch. Chem. Ges., **67**, 2084 (1934). — 686) *R. Kuhn, F. Weygand*, Verbesserung der Flavin-Synthese; Borsäure-Verfahren. Ber. Deutsch. Chem. Ges., **68**, 1282, 166 (1935). — 687) *R. Kuhn, F. Weygand*, O-Nitrophenylhydroxylamin. Ber. Deutsch. Chem. Ges., **69**, 1969 (1936). — 688) *R. Labes, H. Freisburger*, Alloxan als Oxydations-Mittel fuer Thiolgruppen Arch. f. exper. Path., **156**, 226 (1930). — 689) *K. Laki*, Das Ox-Redoxpotential der Ascorbinsäure. Zs. phys. Chem., **217**, 54 (1933). — 690) *K. Laki*, Oxydation und Reduktion der C 4 Dicarbonsäuren. Zs. physiol. Chem., **236**, 31 (1935). — 691) *K. Laki*, Die Oxydation des reduzierten gelben Fermentes. Zs. phys. Chem., **249**, 61 (1937). — 692) *K. Laki*, Über das Redoxpotential des Systems: Oxalessigsäure-l-Apfelsäure. Zs. phys. Chem., **249**, 63 (1937). — 693) *K. Laki*, Malic-Ddehydrogenase. Biochem. Jl., **31**, 1113 (1937). — 694) *K. Laki*, Über die Cytochrome des Taubenbrustmuskels. Zs. phys. Chem., **254**, 27 (1938). — 695) *K. Lang*, Rhodanbildung im Tierkörper. Bioch. Zs., **259**, 243, 263, 261 (1933). — 696) *W. Langenbeck*, Organische Katalysatoren. Ähnlichkeiten in der Wirkung von Fermenten und von definierten organischen Stoffen. Habilitationsschrift, Münster, 1928. — 697) *W. Langenbeck*, Fermentmodelle. Ergebnisse der Enzymforschung, **2**, 314 (1933). — 698) *W. Langenbeck*, Synthèse des diastases arteficielles. Bull. Soc. Chim. Biol., **17**, 627 (1935). — 699) *W. Langenbeck*, Über die Bedeutung der synthetischen organischen Katalysatoren für die Theorie der Enzymwirkung. Chem. Ztg., **60**, 953 (1936). — 700) *W. Langenbeck c. s.*, Isatin als Katalysator der Dehydrierung von Aminosäuren. Ber. Deutsch. Chem. Ges., **60**, 930, **61**, 942 (1927/8), **70**, 672 (1937). — 701) *W. Langenbeck, O. Goedde*, Über organische Katalysatoren, 15. Mitteilg. Künstliche Carboxylasen Ber. Deutsch. Chem. Ges., **70**, 669 (1937). — 702) *W. Langenbeck, R. Hutschenreuter, W. Rottig*, Über die katalatische Wirkung von Imidazolhäminen. Ber. Deutsch. Chem. Ges., **65**, 1750 (1932). — 703) *W. Langenbeck, R. Jüttemann, O. Schäfer, H. Wrede*, Über Carboxylase. I. Zs. phys. Chem., **221**, 1 (1933). — 704) *W. Langenbeck, H. Wrede, W. Schlockermann*, Über Carboxylase. II. Zs. phys. Chem., **227**, 263 (1934). — 705) *H. Laser*, Weitere Untersuchungen über Stoffwechsel und Anaerobiose von Gewebekulturen. Bioch. Zs., **268**, 451 (1934). — 705a) *H. Laser*, Tissue metabolism under the influence of low oxygen tension. Biochem. Jl., **31**, 1671 (1937). — 705b) *H. Laser*, Metabolism of Retina. Nature **136**, 184 (1935). — 706) *A. Lebedew*, Trennung der Oxydoreduktase vom Zymasekomplex. I. Zs. phys. Chem., **156**, 153 (1926). — 707) *E. Lederer, R. Glaser*, Sur l'échinochrome et le spinochrome. Compt. rend. des séances de l'Acad. des Sciences, **207**, 454 (1938). — 708) *J. Lehmann*, Zur Kenntnis biologischer Oxydations-Reduktions-Potentiale. Skand. Arch. Phys., **59**, 173 (1930). — 709) *J. Lehmann*, Zur Kenntnis biologischer Oxydations-Reduktions-Potentiale. Bioch. Zs., **274**, 321 (1934). — 710) *J. Lehmann, E. Martenson*, Über den Sauerstoffverbrauch der vitalen Bernsteinsäureoxydation in Abhängigkeit von Sukzinodehydrogenase, Fumarase, Cytochromoxidase und Katalase. Skand. Arch. Phys., **75**, 61 (1936). — 711) *L. F. Leloir, M. Dixon*, Act. of cyanide and pyrophosphate on dehydrogen. Enzymologia, **2**, H. 2 (1937). — 711a) *R. Lemberg*, Transformation of Haemins into Bile Pigments. Biochem. Jl., **29**, 1322 (1935). — 711b) *R. Lemberg, B. Cortis-Jones, M. Norrie*, Coupled oxidation of ascorbic acid and haemochromogens. Biochem. Jl., **32**, 149 (1938). — 711c) *R. Lemberg, R. A. Wyndham*, Some observations on the occurrence of bile pigment haemochromogens in nature and on their formation from haematin and haemoglobin. Jl. and Proc. Roy Soc. of New South Wales, **70**, 343 (1937). — 712) *A. Lennerstrand, J. Runnstroem*, Oxydation, Phosphorylierung u. Gärung. Bioch. Zs., **283**, 12, 287, 172, 289, 104 (1935/6); Naturwiss., **25**, 347 (1937). — 713) *G. N. Lewis, M. Randall*, Thermodynamics and the free energy of chemical substances. New York 1923. — 714) *Fr. Lieben, V. Getreuer*, System Aminokörper-Aldehyd-H$_2$-Acceptor. Bioch. Zs., **252**, 420 (1932) und Intern. Physiol. Kongr. Rom 1932, S. 155. — 715) *F. Lieben, V. Getreuer*, Verhalten des Systems Aminosäure-Aldehyd. Bioch. Zs., **269**, 69 (1934). — 716) *Fr. Lieben, E. Molnar*, Ferment-freie Hydrierung von Methylenblau. Bioch. Zs., **232**, 209 (1931). — 717) *F. Lipmann*, Über die oxydative Hemmbarkeit der Glykolyse und den Mechanismus der Pasteurschen Reaktion. Bioch. Zs., **265**, 133 (1933). — 718) *F. Lipmann*, Über die Hemmung der Mazerationssaftgärung durch Sauerstoff in Gegenwart positiver Oxydoreduktionssysteme. Bioch. Zs., **268**, 205 (1934). — 719) *F. Lipmann*, Versuche zur potentiometrischen Erfassung der Oxydo-Reduktions-Vorgänge in gärendem

Hefeextrakt. Bioch. Zs., **274**, 329 (1934). — 720) *F. Lipmann*, Fermentation of phosphogluconic acid. Nature, **138**, 588 (1936). — 721) *F. Lipmann*, Hydrogenation of Vitamin $B_1$. Nature, **138**, 1097 (1936). — 722) *F. Lipmann*, Pyruvic acid dehydrogenation, Vitamin $B_1$ and Cocarboxylase. Nature, **140**, 25 (1937). — 723) *F. Lipmann*, A coloured intermediate on reduction of vitamin $B_1$. Nature, **140**, 849 (1937). — 724) *F. Lipmann*, Photochemische Reduction von $Fe_3$ und Methylenblau durch Brenztraubensäure. Skand. Arch. Phys., **76**, 186, 193 (1937). — 725) *F. Lipmann*, Über den Umsatz der Brenztraubensäure und den Mechanismus der Vitamin $B_1$ Wirkung. Skand. Arch. Phys., **76**, 255 (1937). — 726) *F. Lipmann*, Dehydrierung der Brenztraubensäure. Enzymologia, **4**, 65 (1937). — 726a) *F. Lipmann, G. Perlmann*, Hydrogenation of vitamin $B_1$. Jl. Am. Chem. Soc., **60**, 2574 (1938). — 727) *R. O. Löbel*, Atmung und Glykolyse tierischer Gewebe. Bioch. Zs., **161**, 219 (1925). — 728) *E. Löffler, R. Rigler*, Wachstumshemmungen durch HCN. Bioch. Zs., **173**, 449 (1926). — 729) *K. Lohmann*, Darstellung der Adenyl-pyro-phosphorsäure aus Muskulatur. Bioch. Zs., **233**, 460 (1931). — 730) *K. Lohmann*, Chemische Natur des Cofermentes der Milchsäurebildung. Bioch. Zs., **237**, 445 (1931). — 731) *K. Lohmann*, Enzymatische Umwandlung von synthetischem Methylglyoxal in Milchsäure. Bioch. Zs., **254**, 332 (1932). — 732) *K. Lohmann*, Abbau der Zucker. Handbuch der Biochem. Erg. Werk, **1**, 926 (1933). — 733) *K. Lohmann*, Aufspaltung der Adenyl-pyro-phosphorsäure im Krebsmuskel. Bioch. Zs., **282**, 109 (1935). — 734) *K. Lohmann*, Konstitution der Adenyl-pyrophosphorsäure und Adenosin-diphosphorsäure. Bioch. Zs., **282**, 120 (1935). — 735) *K. Lohmann*, Co-Carboxylase. Zs. Angew. Chem., (Vortragsreferat) **50**, 221 (1937). — 736) *K. Lohmann, Ph. Schuster*, Über das Vorkommen der Adenin-Nucleotide in den Geweben. Bioch. Zs., **282**, 104 (1935). — 737) *K. Lohmann, Ph. Schuster*, Co-Carboxylase. Bioch. Zs., **294**, 188 (1937). — 738) *K. Lohmann, Ph. Schuster*, Über die Co-Carboxylase. Naturwiss., **25**, 26 (1937). — 739) *J. W. H. Lugg c. s.*, Colouring matters of Drosera Whittackeri. Jl. of Chem. Soc. (1936) 1457ss. — 740) *E. Lundsgaard*, Einwirkung der Monojodessigsäure auf den Spalt- und Oxydations-Stoffwechsel. Bioch. Zs., **220**, 8 (1930), **250**, 61 (1932). — 741) *C. Lutwak-Mann*, Decomposition of adenine compounds by bacteria. Biochem. Jl., **30**, 1410 (1936). — 741a) *C. M. Lyman, E. S. G. Barron*, Studies on biological oxidations. 8. The oxidation of glutathione with copper and hemochromogens as catalysts. Jl. of Biol. Chem., **121**, 275 (1937). — 742) *Th. F. Macrae*, Autoxydation von Leucomethylenblau. Ber. Deutsch. Chem. Ges., **64**, 133 (1931). — 743) *Th. F. Macrae*, Formation of $H_2O_2$ in catalytic dehydrogenations. Biochem. Jl., **27**, 1248 (1933). — 744) *J. Madinaveitia*, 2-Methyl-1,4-naphtho quinone. An. Soc. espan. fisic. y quim., **31**, 750 (1933). — 745) *J. Magat*, Action ox.-réd. de la lécithine colloidale. Soc. Biol., **116**, 1367 (1934). — 746) *H. Magne c. s.*, Intoxication par le dinitrophénol. Ann. de Phys., **8**, 1 (1933). — 747) *K. Makino*, Konstitution der Adenosin-tri-phosphorsäure. Bioch. Zs., **278**, 161 (1935). — 748) *K. Makino, T. Imai*, Chemie des antineuritischen Vitamins. Zs. phys. Chem., **239**, I (1936). — 749) *A. M. Malkov, N. Zwetkowa*, Die Rolle der Phosphate im Oxydations-Prozess. Bioch. Zs., **246**, 191 (1932). — 750) *W. Manchot, G. Lehmann*, Einwirkung von $H_2O_2$ auf Ferrosalz. Liebigs Ann. **460**, 179 (1928). — 751) *W. Manchot, W. Pflaum*, Mechanismus der Oxydations-Vorgänge. Zs. Anorg. Chem., **211**, 1 (1933). — 752) *W. Manchot, H. Schmid*, Mechanismus der Oxydations-Vorgänge. Ber. Deutsch. Chem. Ges., **65**, 98 (1932). — 753) *P. J. G. Mann*, Kinetics of peroxydase action. Biochem. Jl., **25**, 918 (1931). — 754) *Ph. I. Mann*, Reduction of glutathione by a liver system. Biochem. Jl., **26**, 785 (1932). — 755) *C. Martius*, Abbau der Citronensäure. Zs. phys. Chem., **247**, 104 (1937). — 756) *C. Martius, F. Knoop*, Phys. Abbau der Citronensäure. Zs. phys. Chem., **246**, I (1937). — 757) *H. L. Mason*, Spontaneous cleavage of glutathione. Jl. of Biol. Chem., **90**, 25 (1931). — 758) *H. A. Mattill*, Antioxidants and the autoxidation of fats. Jl. of Biol. Chem., **90**, 141 (1931). — 759) *H. A. Mattill, H. I. Mattill*, Further studies of anti-oxydants. Amer. Jl. Phys., **90**, 447 (1929). — 760) *C. A. Mawson*, The influence of animal tissues on the oxidation of ascorbic acid. Biochem. Jl., **29**, 569 (1935). — 761) *A. Mayer, L. Plantefol*, Pouvoir hydrogénant des tissues des végétaux et de leur constituents solubles. Ann. de phys. et physicochim. biol., **4**, 297 (1928). — 762) *A. Mayer, R. Wurmser*, L'oxydation des corps organiques à la temperature ordinaive. Ann. Phys. Physicochim. Biol. **2**, 329 (1926). — 763) *N. Mayer*, Oxidation-réduction potentiel de l'acide réductinique. C. r., **204**, 115 (1937); Jl. de Chim. Physique, **34**, 109 (1937). — 764) *N. Mayer-Reich*, Action de l'oxygène sur les glucides. C. R., **196**, 1337 (1933). — 765) *F. P. Mazza*, Deidrog. di acidi fenil-alifat. Arch. di Sci. Biol., **21**, 320 (1935). — 766) *F. P. Mazza*, Oss. biol. d. acidi bibasici alif., Arch. di. Sci. Biol., **22**, 1 (1936). — 767) *F. P. Mazza c. s.*, S. deidrogen. d. acidi grassi. Atti Accad. Linc. (6), **17**, 476, **18**, 461, **20**, 113 (1933/34). — 768) *F. P. Mazza, O. Carera*, Deidrogenazione della glicerina con palladio. Boll. Soc. Ital. Biol. sper. **9**, 177 (1934). — 769) *F. P. Mazza, A. Cimmino*, Ossid. dell' ac. piruv . . .

d. B. coli. Arch. di Sci. Biol., **20**, 486 (1934). — 770) *F. P. Mazza, G. Stolfi*, Pigm. di Halla par-thenopaea Costa. Boll. Soc. Ital. Biol., **5**, 74 (1930). — 771) *F. P. Mazza, G. Stolfi*, Contrib. allo stud. della psicol. degli esercizi fisizi. Arch. di Sci. Biol., **16**, 1 (1931). — 772) *G. Medes*, Is cystine sulfoxide an intermediate in the oxidat. metabolism of cystine? Jl. of Biol. Chem., **109**, LXIV (1935). — 773) *R. Meier, K. Ballowitz*, Die Bedeutung gekoppelter Reaktionen niederer alipha-tischer C-Verbindungen für Kohlenhydrat und Fettabbau. Zs. phys. Chem., **230**, 122 (1932). — 774) *R. H. De Meio, E. S. G. Barron*, Effect of 1-2-4 Dinitrophenol cn cellular respiration. Proc. Soc. Exp. Biol. and Med., **32**, 36 (1934). — 775) *R. H. de Meio, M. Kissin, E. S. G. Barron*, Catalytic effect of reversible dyes on cellular respiration. Jl. of Biol. Chem., **107**, 579 (1934). — 776) *E. Meirowsky*, Pigmentbildung durch oxydiertes Adrenalin. Arch. für Derm., **163**, 135 (1931). — 777) *N. U. Meldrum*, Reduction of glutathion in mammalian erythrocytes. Biochem. Jl., **26**, 817 (1932). — 778) *N. U. Meldrum, M. Dixon*, Properties of pure glutathione. Biochem. Jl., **24**, 472 (1930). — 779) *N. U. Meldrum, H. L. A. Tarr*, The reduction of glutathione by the Warburg-Christian system. Biochem. Jl., **29**, 108 (1935). — 779a) *V. La Mer, J. W. Temple*, The autoxi-dation of hydroquinone. Proc. Nat. Acad. Sci., **15**, 191 (1929). — 730) *A. De Merrit Welch*, Epinephrine oxidation and stabilisation. Amer. Jl. Phys. **108**, 360 (1934). — 781) *W. K. Mertens, A. G. van Veen*, Die Bongkrek-vergiftung in Banjumas. Med. Dienst Volksgezondheid in Ned.-Indie, **4**, 209 (1933). — 782) *K. Meyer*, Oxidation of benzaldehyde. Jl. of Biol. Chem., **103**, 25 (1933). — 783) *K. Meyer*, Oxidation of pyruvic acid. Jl. of Biol. Chem., **103**, 39 (1933). — 784) *K. Meyer*, Oxidation of ergosterol. Jl. of Biol. Chem., **103**, 607 (1933). — 784a) *O. Meyerhof*, Untersuchungen zur Atmung getöteter Zellen. I. Die Wirkung des Methylenblaus auf die Atmung lebender und getöteter Staphylokokken nebst Bemerkungen über den Einfluss des Milieus, der Blausäure und Narkotika. Pflüg. Arch., **169**, 87 (1917). — 785) *O. Meyerhof*, Intermediärvor-gänge bei der biologischen Kohlenhydratspaltung. Ergebnisse der Enzymforschung, **4**, 208 (1935). — 786) *O. Meyerhof*, Neuere Versuche über zellfreie alkoholische Gärung. Naturwiss., **24**, 689 (1936). — 787) *O. Meyerhof*, Über die Synthese der Kreatinphosphorsäure im Muskel und die „Reaktonsform" des Zuckers. Naturwiss., **25**, 443 (1937). — 788) *O. Meyerhof*, Über die Inter-mediärvorgänge der enzymatischen Kohlehydratspaltung. Ergebnisse Physiologie, **39**, 10 (1937). — 789) *O. Meyerhof, E. Boyland*, Über den Atmungs-vorgang jodessigsäure- vergifteter Muskeln. Bioch. Zs., **237**, 406 (1931). — 790) *O. Meyerhof, W. Kiessling*, Über Cozymasepyrophosphat. Naturwiss., **24**, 557 (1936). — 791) *O. Meyerhof, W. Kiessling, W. Schulz*, Über die Reaktions-gleichung der alkoholischen Gärung. Bioch. Zs., **292**, 25 (1937). — 792) *O. Meyerhof, K. Loh-mann*, Enzymatische Milchsäurebildung im Muskelextrakt. Bioch. Zs., **185**, 113 (1927). — 793) *O. Meyerhof, K. Lohmann*, Über freiwillige enzymatische Spaltungen mit negativer Wärme-tönung. Naturwiss., **22**, 452 (1934). — 794) *O. Meyerhof, K. Lohmann, Ph. Schuster*, Adol-kondensation von Dioxyaceton-phosphorsäure. Bioch. Zs., **286**, 301, 319 (1936). — 795) *O. Meyer-hof, P. Ohlmeyer*, Über die Rolle der Co-Zymase bei der Milchsäurebildung im Muskelextract. Bioch. Zs., **290**, 334 (1937). — 796) *O. Meyerhof, P. Ohlmeyer, W. Gentner, H. Maier-Leibnitz*, Studium der Zwischenreaktionen der Glykolyse mit Hilfe von radioaktivem Phosphor. Bioch. Zs., **298**, 396 (1938). — 797) *O. Meyerhof, P. Ohlmeyer, W. Möhle*, Die Cozymase als Ampholyt. Naturwiss., **25**, 172 (1937). — 798) *O. Meyerhof, P. Ohlmeyer, W. Möhle*, Über die Koppelung zwischen Oxydoreduktion und Phosphatveresterung bei der anäroben Kohlenhy-dratspaltung. Bioch. Zs., **297**, 90, 113 (1938). — 799) *O. Meyerhof, W. Schulz*, Reduktion von NO durch das Oxydationsferment. Bioch. Zs., **275**, 147 (1934). — 800) *L. Michaelis*, Effect of iron on cysteine oxidation. Jl. of Biol. Chem., **84**, 777 (1929). — 801) *L. Michaelis*, The formation of semiquinones as intermediary reaction products from pyocyanine and some other dyestuffs. Jl. of Biol. Chem., **92**, 211 (1931). — 802) *L. Michaelis*, Potentiometric study of Wurster's red and blue. Jl. Amer. Chem. Soc., **53**, 2953 (1931). — 803) *L. Michaelis*, Chemische physikochemische und biologische Eigenschaften des Farbstoffes des Bac. pyocyaneus. Verh. 14. internat. Kongr. Phys. 181 (1932). — 804) *L. Michaelis*, Oxydations-Reduktionspotentiale. 2. Auflage, Berlin, 1933. — 805) *L. Michaelis*, Semiquinones, the intermediate steps of reversible organic oxidation-reduction. Chem. Reviews, **16**, 243 (1935). — 806) *L. Michaelis*, Potentiometric study of beta-naphthoqui-none sulfonate. A futher contribution to the semiquinone problem. Jl. Amer. Chem. Soc., **58**, 873 (1936). — 807) *L. Michaelis*, Oxidation and catalysis of organic compounds. Collect. Net (Marine biol. Lab. Woods Hole), **12**, 101 (1937). — 808) *L. Michaelis*, Review of the semiquinone problem. 71. Meet. of the electrochem. Soc. Jl. Amer. Electrochem. Soc., **71**, 185 (1937). — 809) *L. Michaelis et al.*, Complexes of cysteine. Jl. of Biol. Chem. **83**, 191, **84**, 777 (1929). — 810) *L. Michaelis, E. S. G. Barron*, Reducing effects of cysteine induced by free metals. Jl. of. Biol. Chem. **81**, 29 (1929). —

811) **L. Michaelis, G. F. Böker, R. K. Reber,** Paramagnetism of semiquinone of phenantrene-quinone-3-sulfonate. Jl. Am. Chem. Soc., 60, 202 (1938). — 812) **L. Michaelis, E. S. Fetcher Jr.,** The equilibrium of the semiquinone of phenanthrene-3-sulfonate with its dimeric compound. Jl. Am. Chem. Soc., 59, 2460 (1937). — 813) **L. Michaelis, L. B. Flexner,** Reduction potential of cysteine. Jl. of. Biol. Chem. 89, 689 (1928). — 814) **L. Michaelis, E. F. H. Friedheim,** Potentiometric studies on complex iron systems. Jl. of Biol. Chem., 91, 343 (1931). — 815) **L. Michaelis, E. S. Hill,** Potentiometric studies on semiquinones. Jl. Amer. Chem. Soc., 55, 1481 (1933). — 816) **L. Michaelis, E. S. Hill, M. P. Schubert,** Die reversible zweistufige Reduktion von Pyocyanin und α-Oxyphenazin. Bioch. Zs., 255, 66 (1932). — 817) **L. Michaelis, H. Pechstein,** Untersuchungen über die Katalase der Leber. Bioch. Zs., 53, 320 (1913). — 818) **L. Michaelis, R. K. Reber, J. A. Kuck,** Supplement to a recent paper on the paramagnetism of semiquinones. Jl. Amer. Chem. Soc., 60, 214 (1938). — 819) **L. Michaelis, K. Salomon,** Methämoglobin-Erzeugung und Atmungssteigerung durch organische Farbstoffe. Bioch. Zs., 234, 107 (1931). — 820) **L. Michaelis, M. P. Schubert,** Cobalt complexes of cysteine. Jl. of Biol. Chem., 83, 367, 84, 777 (1929). — 821) **L. Michaelis, M. P. Schubert,** Some problems in two-step oxidation treated for the case of phenanthrenequinone-sulfonate. Jl. of. Biol. Chem., 119, 133 (1937). — 822) **L. Michaelis, M. P. Schubert, C. V. Smythe,** Potentiometric study of the flavins. Jl. of Biol. Chem., 116, 587 (1936). — 823) **L. Michaelis, M. P. Schubert, C. V. Smythe,** The semiquinone of the flavine dyes, including vitamin B$_2$. Science, 84, 138 (1936). — 824) **L. Michaelis, G. Schwarzenbach,** The intermediate forms of oxidation-reduction of the flavins. Jl. of Biol. Chem., 123, 527 (1938). — 825) **L. Michaelis, C. V. Smythe,** Correlation between rate of oxidation and potential in iron system. Jl. of. Biol. Chem., 94, 329 (1931/2). — 826) **L. Michaelis, C. V. Smythe,** Influence of certain dyestuffs on fermentation and respiration of yeast extract. Jl. of. Biol. Chem., 113, 717 (1936). — 827) **L. Michaelis, C. V. Smythe,** The pentacyano-aquo-complexes of iron. Compt. rend. Lab. Carlsberg (Sørensen Volume), 22, 347 (1937). — 827a) **L. Michaelis, K. G. Stern,** Einfluss von Schwermetallen und Metallkomplexen auf proteolytische Vorgänge. Bioch. Zs., 240, 192 (1931). — 828) **F. Micheel,** Vitamin C. Zt. Angew. Chem., 47, 550 (1934). — 829) **F. Micheel, F. Jung,** Oxytetronsäure, einfachster Stoff vom Typ der Ascorbinsäure. Ber. Deutsch. Chem. Ges., 66, 1291 (1933). — 830) **F. Micheel, K. Kraft,** Die Konstitution des Vitamins C. VIII. Mitt. Zs. phys. Chem., 215, 215 (1933); Nature, 131, 274 (1933). — 831) **F. Michaelis, R. Mittag,** Zur Kenntnis des Vitamin C. Zs. phys. Chem., 247, 34 (1937). — 832) **D. Michlin, B. Severin,** Pflanzliche Aldehydrase und Mutase. Bioch. Zs., 237, 339 (1931). — 833) **N. A. Milas,** Studies in autoxidation reactions. Jl. Amer. Chem. Soc., 53, 221 (1931), 56, 486 (1934); Jl. of Physical Chem., 33, 1204 (1929), 38, 411 (1934); Chem. Rev., 10, 295 (1932). — 833a) **A. E. Mirsky, M. L. Anson,** Sulfhydril and disulfide groups of proteins, Jl. Gen. Phys., 19, 427 (1936). — 833b) **A. E. Mirsky, M. L. Anson,** On pyridine hemochromogen. Jl. Gen. Phys., 12, 581 (1929). — 833c) **M. L. Anson, A. E. Mirsky,** The reactions of cyanide with globin hemochromogen., Jl. Gen. Phys., 14, 43 (1930). — 834) **B. R. Monaghan, F. O. Schmitt,** Effect of carotene on the oxidation of linoleic acid. Jl. of Biol. Chem., 96, 387 (1932). — 835) **J. Mosters,** Blutesterasesteigerung nach peroral. Ascorbinsäure-gaben. Klin. Ws., 15, 1557 (1936). — 836) **Ch. Moureu, Ch. Dufraisse c. s.,** Autoxydation et action antioxygène. Rec. Trav. Chim., 43, 645 (1924); C. R. 180, 993 (1925), 186, 1673, 187, 157 (1928). — 837) **Ch. Moureu, Ch. Dufraisse c. s.,** Autoxydation et action antioxygène. C. R., 182, 949, 183, 823 (1926). — 838) **Ch. Moureu, Ch. Dufraisse c. s.,** Considérations sur l'autoxydation. II Cons. de Chim. Bruxelles, 24, 4, 1925, Paris 1926. — 839) **Ch. Moureu, Ch. Dufraisse c. s.,** Catalysis and auto-oxidation. Chem. Rev., 3, 113 (1926). — 840) **Ch. Moureu, Ch. Dufraisse c. s.,** Autoxidation et action antioxygene. Compt. Rend., 183, 408 (1926); Jl. of. Soc. Chem. Ind., 47, 819 (1928); Rec. Trav. Chim., 48, 826 (1929). — 841) **D. Moyle-Needham,** Succinic acid in muscle. Biochem. Jl., 21, 739, 24, 208 (1927/1930). — 842) **D. Mueller,** Alkoholdehydrase aus Hefe losgelöst. Bioch. Zs., 262, 239 (1933). — 843) **E. Müller,** Electrolytische Oxydation der Ameisensäure. Zs. Elektrochem., 33, 561, 34, 170 (1927/8). — 844) **E. Muntwyler,** Effect of dinitrophenol on oxygen uptake. Proc. Soc. Exp. Biol., 32, 1060 (1935). — 845) **E. Muntwyler, D. Binns,** Effect of cyanide on the oxygen uptake. Amer. Jl. Phys., 108, 80 (1934). — 846) **K. Myrbäck,** Cozymase und ihre Bestimmung. Zs. phys. Chem., 177, 158 (1928). — 847) **K. Myrbaeck,** Cozymase und Phosphatase. Zs. phys. Chem., 217, 249 (1933). — 848) **K. Myrbäck,** Versuche über die Cozymase der Hefe. Zs. phys. Chem., 219, 173 (1933). — 849) **K. Myrbäck,** Cozymase. Ergebnisse der Enzymforschung, 2, 139 (1933). — 850) **K. Myrbäck,** Zur Kenntnis der Cozymase I. Zs. phys. Chem., 225, 125 (1934). — 851) **K. Myrbäck,** Titrations-kurve der Cozymase. Zs. phys. Chem., 225, 199 (1934). — 852) **K. Myrbäck,** Zur Kenntnis der Cozymase V. Zs., phys.

Chem., **233**, 95 (1935). — 853) *K. Myrbäck*, Cozymase aus tierischen Organen. Zs. phys. Chem., **233**, 154 (1935). — 854) *K. Myrbäck*, Oxydation und Reduktion der Cozymase. Zs. phys. Chem., **234**, 259 (1935). — 855) *K. Myrbäck*, Natur der reducierenden Gruppe der Cozymase. Zs. phys. Chem., **241**, 223 (1936). — 856) *K. Myrbäck*, Co-Zymase. Tab. Biol., **14**, 110 (1937). — 857) *K. Myrbäck, H. v. Euler*, Über Eigenschaften hochgereinigter CO-Zymasepräparate. Zs. phys. Chem., **198**, 236 (1931). — 858) *K. Myrbäck, H. v. Euler*, Chemische Natur der Cozymase Zs. phys. Chem., **203**, 143 (1931). — 859) *K. Mybäck, H. v. Hellström*, Hefen-Cozymase. Zs. phys. Chem., **212**, 7 (1932). — 860) *K. Myrbäck, H. v. Euler, H. Hellström*, Weitere Untersuchungen über die Hefen-Cozymase. Zs. phys. Chem., **214**, 184 (1933). — 861) *K. Myrbäck, E. Jorpes*, Freie Diffusion von Nucleinsäure als Mittel zur Bestimmung ihrer Mol. Grösse. Zs. phys. Chem., **237**, 159 (1935). — 862) *K. Myrbäck, H. Larsson*, Isolierung der Cozymase. Zs. phys. Chem., **225**, 131 (1934). — 863) *K. Myrbäck, R. Nilsson*, Cozymase XI. Zs. phys. Chem., **165**, 140 (1927). — 864) *K. Myrbaeck, B. Oertenblad*, Zur Kenntnis der Cozymase IV. Zs. phys. Chem., **233**, 87 (1935). — 865) *K. Myrbaeck, B. Oertenblad*, Darstellung und Eigenschaften eines hochgereinigten Cozymase-Präparates. Zs. phys. Chem., **233**, 148 (1935). — 866) *K. Myrbäck, B. Oertenblad*, Über die Einwirkung der Adenyl-säuredes-aminase auf Cozymase. Zs. phys. Chem., **234**, 254 (1935). — 867) *K. Myrbäck, B. Oertenblad*, Phosphatase und Cozymase. Zs. phys. Chem., **241**, 148 (1936). — 868) *W. M. McCord*, Effect of 2,4-Dinitrophenol on the oxygen uptake. Amer. Jl. Phys., **109**, 232 (1934). — 869) *W. D. McFarlane*, Reduction of iron by ascorb. acid. Biochem. Jl., **30**, 1472 (1936). — 869a) *Needham*, Dorothy Moylen, see Moylen-Needham. — 870) *J. Needham c.s.*, Mechanism of carbohydrate breakdown. Nature, **138**, 462 (1936), **139**, 368 (1937). — 871) *J. Needham, H. Lehmann*, Glucolysis without phosphorylation in the chick embryo. Nature, **139**, 368 (1937). — 872) *J. Needham, D. M. Needham*, H-concentration and the oxidation-reduction-potentials of the cell-interior. Proc. Roy Soc. London, **B 98**, 259 (1925). — 873) *J. Needham, D. M. Needham*, The oxidation-reduction-potential of protoplasma: A review. Protoplasma, **1**, 255 (1926). — 874) *J. Needham, D. M. Needham*, Further micro-injection studies on the oxidation-reduction potential of the cell-interior. Proc. Roy. Soc. London, **B 99**, 383 (1926). — 875) *E. Negelein*, Über die Wirkung des Schwefel-wasserstoffs auf chemische Vorgänge in Zellen. Bioch. Zs., **165**, 203 (1925). — 876) *E. Negelein*, Verbrennung von Kohlenoxyd zu Kohlensäure durch grüne und mischfarbene Hämine. Bioch. Zs., **243**, 386 (1931). — 877) *E. Negelein*, Kryptohämin. Bioch. Zs., **248**, 243 (1932). — 878) *E. Negelein*, Über Kryptohämin. Bioch. Zs., **250**, 577 (1932). — 879) *E. Negelein*, Extraktion eines von Bluthämin verschiedenen Hämins aus Herzmuskel. Bioch. Zs., **266**, 412 (1933). — 879a) *E. Negelein*, Methode zur Gewinnung des A-Proteins der Gärungsfermente. Bioch. Zs., **287**, 329 (1936). — 880) *E. Negelein, W. Gerischer*, Nachweis des sauerstoffübertragenden Ferments in Azotobakter. Naturwiss., **21**, 884 (1933). Bioch. Zs., **268**, 1 (1934). — 881) *E. Negelein, W. Gerischer*, Verbesserte Methode zur Gewinnung des Zwischenferments aus Hefe. Bioch. Zs., **284**, 289 (1936). — 882) *E. Negelein, E. Haas*, Über die Wirkungsweise des Zwischenferments. Bioch. Zs., **282**, 206 (1935). — 883) *E. Negelein, H. J. Wulff*, Diphosphopyridine protein. Alkohol, Acetaldehyd. Bioch. Zs., **293**, 351 (1937). — 884) *C. Neuberg, A. Hildesheimer*, Über zuckerfreie Hefegärungen. I. Bioch. Zs., **31**, 170 (1910/11). — 885) *C. Neuberg, M. Kobel*, Abbau von Aminosäuren durch Methylglyoxal. Bioch. Zs., **185**, 477, **188**, 197 (1927). — 886) *M. S. Newman, J. A. Crowder, R. J. Anderson*, The chemistry of the Lipids of Tubercle Bacilli. XXXVIII. A new synthesis of phthiocol. Jl. of Biol. Chem., **105**, 279 (1934). — 887) *M. Nicloux*, Oxydation du glucose. C. R., **186**, 1218 (1928); Bull. Soc. Chim. Biol., **10**, 1135 (1928). — 888) *M. Nicloux, H. Nebenzahl*, Oxydation des sucres par $O_2$ gazeuse. Soc. Biol., **101**, 189 (1929). — 889) *R. Nilsson*, Reinigungsversuche an Cozymase. Ark. f. Kemi, **9**, Nr. 31 (1926). — 890) *R. Nilsson*, Glykolytischer Kohlenhydratabbau. Bioch. Zs., **258**, 198 (1933). — 891) *F. F. Nord*, Influence of heat and hydrogen ion concentration on biological transportation systems containing sulfur. Jl. of Physical. Chem., **31**, 867 (1927). — 891a) *S. Ochoa*, Darstellung reiner Cozymase aus Warmblütermuskulatur. Bioch. Zs., **292**, 68 (1937). — 892) *S. Ochoa*, Vitamin B$_1$ and Cocarboxylase. Nature, **141**, 831 (1938). — 893) *A. Oerstroem*, Analyse der Atmungssteigerung bei der Befruchtung des Seeigel-Eis. Protopl., **15**, 566 (1932). — 894) *A. Oerstroem*, Einfluss des Kohlenoxyds auf die Atmung der Hefezelle in verschiedenen Substraten. Protoplasma, **24**, 177 (1935). — 895) *F. J. Ogston, D. E. Green*, The mechanism of the reaction of substrates with molecular oxygen. I. Biochem. Jl., **29**, 1983 (1935); II. Biochem. Jl., **29**, 2005 (1935). — 896) *H. Ohle*, Vitamin C; l-Ascorbinsäure. Hb. der Biochem. Erg. Werk., **3**, 800 (1935) (Jena). — 897) *H. Ohle c. s.*, D-Gluco-saccharosonsäure, ein Isomeres der Ascorbinsäure. Ber. Deutsch. Chem. Ges., **67**, 324 (1934). — 898) *P. Ohlmeyer*, Über die Beteiligung des Adenylsäuresytems und der Cozymase an der alkoholischen Gärung. Bioch. Zs., **287**, 212 (1936). — 899) *D. Okuyama*, Redox-potential of the living tissues. Jl. of Biochem., **14**, 69 (1931/2). (Japan). — 900) *H. S. Olcott*, Antioxidants and the autoxidation of fat. Jl. Amer. Chem. Soc., **56**, 2492 (1934). — 901) *A. Oparin*, Oxydations-Vorgänge in der lebenden Zelle. Bioch. Zs., **182**, 155 (1927). — 902) *C. Oppenheimer*, Die Fermente, Leipzig, 1925—1926. — 903) *C. Oppenheimer*, Desmolasen, Allgemeiner Teil. Handbuch der Biochemie, Ergänzungswerk, Jena, **1**, 490 (1933). — 904) *C. Oppenheimer*, Desmolyse, Allgemeiner Teil. Handbuch der Biochemie, Ergänzungswerk, Jena, **1**, 823 (1933). — 905) *C. Oppenheimer*, Chemische Grundlagen der Lebensvorgänge, Leipzig 1933 (In abridged from: Einführung in die Allgemeine Biochemie, Leiden 1936). — 906) *C. Oppenheimer*, Die Fermente und ihre Wirkungen, Supplement, Den Haag, 1936—1938. — 907) *P. Ostern*, Di-Adenosin-pentaphosphorsäure (Herznucleotid). Bioch. Zs., **270**, 1 (1934). — 908) *P. Ostern*, Enzymatische phosphorylierung von Stärke. Enzymologia, **3**, 5 (1937). — 909) *P. Ostern, T. Baranowski*, Verkettung der chemischen Vorgänge im Muskel X. Bioch. Zs., **281**, 157 (1935). — 910) *Z. Otani, K. Ichihara*, Desaminierung von Alanin. Fol. Jap. Pharm., **1**, 397 (1925). — 911) *E. Ott c. s.*, Eine neben Ascorbinsäure vorkommende Verbindung. Zs. phys. Chem., **243**, 199 (1936). — 912) *I. H. Page, M. Bülow*, Sauerstoffaufnahme von Phosphatiden etc. Zs. phys. Chem., **231**, 10 (1934). — 913) *C. C. Palit*, Induced oxidation in biological phenomena. Jl. of Physical Chem., **36**, 2504 (1932). — 914) *C. C. Palit, N. R. Dhar*, Catalytic and induced oxidation of some carbohydrates. Jl. of Physical Chem., **30**, 939 (1926). — 915) *C. C. Palit, N. R. Dhar*, Slow and induced oxidation of glycogen. Jl. of Physical Chem., **34**, 711 (1930). — 916) *W. v. Pantschenko-Jurewicz, H. Kraut*, Über den Zusammenhang zwischen Ascorbinsäure u. Leberesterase. Bioch. Zs., **285**, 407 (1936). — 917) *J. K. Parnas*, Über den Mechanismus der Glykogenolyse im Muskel. Ergebnisse der Enzymforschung, **6**, 57 (1937). — 918) *L. M. B. Patterson*, Effect of cyanide on fermentation by yeast preparations. Biochem. Jl., **25**, 1593 (1931). — 918a) *L. Pauling, Ch. D. Coryell*, The magnetic properties and structure of hemoglobin, oxyhemoglobin and Carbonmonoxyhemoglobin. Proc. Nat. Acad. of Sciences, **22**, 210 (1936). — 919) *V. A. Pawlow, M. M. Issakowa-Keo*, Studien über Reduktions-Potentiale in biologischen Systemen. I.: Reduktions Potentiale im Hühnerei vor und während der Entwicklung. Bioch. Zs., **216**, 19 (1929). — 920) *Fr. Perrin*, Desactivation induite des molecules et la théorie des antioxygènes. C. R., **184**, 1121 (1927). — 921) *R. A. Peters*, Oxygen consumption of Colpidium. Jl. of Phys., **68**, II (1929). — 922) *R. A. Peters*, The vitamin B Complex. British Med. Jl., **1936**, II, 903. — 923) *R. A. Peters*, Biochemical lesion in vitamin $B_1$ deficiency. Lancet, **1936**, I, 1161, 1165. — 924) *R. A. Peters*, Pyruvic acid. oxidation in brain. Biochem. Jl., **30**, 2206 (1936). — 925) *R. A. Peters, I. St. Philpot*, UV absorption of crystalline preparations of vitamin $B_1$. Proc. Roy. Soc., B **113**, 48 (1933). — 925a) *L. B. Pett*, Change in the flavin content of yeast. Ark. f. Kemi, **11 B**, No. 53 (1935). — 926) *E. Pfankuch*, Enzymatic Reduction von Dehydro-ascorbinsäure. Naturwiss., **22**, 821 (1934). — 927) *E. Pietsch, E. Josephy, W. Roman*, Topochemie der Korrosion und Passivität. Zs. physikal Chem. A **157**, 363 (1931); Korrosion Metallschutz, **8**, 57 (1932). — 928) *N. W. Pirie*, Cuprous derivatives of some SH-compounds. Biochem. Jl., **25**, 614 (1931). — 929) *N. W. Pirie*, Oxidation of SH-compounds by $H_2O_2$. Biochem. Jl., **25**, 1565 (1931). — 930) *N. W. Pirie*, Formation of sulphate from cystine and methionine by tissues in vitro. Biochem. Jl., **28**, 305 (1934). — 931) *R. F. Pitts*, Effect of cyanide on Colpidium. Proc. Soc. Exp. Biol., **29**, 542 (1932). — 932) *L. Plantefol*, Action du dinitrophénol sur la respiration. Ann. de Phys., **8**, 124, 11, 32 (1935); Soc. Biol., **113**, 147 (1933). — 933) *B. St. Platt*, Peroxide formation by pneumoccus. Biochem. Jl., **21**, 19 (1927). — 934) *B. St. Platt, A. Wormall*, Nature and reactions of the substance „Tyrin". Biochem. Jl., **21**, 26 (1927). — 935) *G. Possenti*, Ric. dell' ac. piruv. n. retina. Riv. Pat. Sper., **4**, 229 (1935). — 936) *P. W. Preisler*, Ox.-red. potential and the possible respiratory significance of the pigment of nudibranch. Chromodoris zebra. Jl. of gen. Phys., **13**, 349 (1929/30). — 937) *P. W. Preisler, L. H. Hempelmann*, Oxidation-reduction potentials of Beta-Hydroxyphenazine and N-Methyl-Beta-Oxyphenazine. Jl. Amer. Chem. Soc., **59**, 141 (1937). — 938) *C. E. M. Pugh*, Activity of certain oxidase preparations. Biochem. Jl., **23**, 456 (1929). — 939) *C. E. M. Pugh*, On the supposed oxidase activity of cobaltammines, with reference to tyrosinase. Biochem. Jl., **27**, 480 (1933). — 940) *A. Purr*, The influence of vitamin C on intracellular enzyme action. Biochem. Jl., **27**, 1703 (1933). — 941) *G. Quagliariello*, Pres. n. bile di un enz. deidr. l'ac. stear. Atti Accad. Linc. (6), **16**, 387, 552 (1932). — 942) *G. Quagliariello*, Biologische Oxydation der höheren Fettsäuren. Zs. Angew. Chem., **47**, 370 (1934). — 943) *J. H. Quastel*, Dehydrogenations produced by resting bacteria. A theory of the mechanism of oxidations and reductions in vitro. Biochem. Jl., **20**, 166 (1926), **21**, 148 (1927). — 944) *J. H. Quastel*, Effect of CO on biological reduction of nitrate. Nature, **130**, 207

(1932). — 945) *J. H. Quastel c. s.*, Some reactions of resting bacteria. Biochem. Jl., **19**, 304 (1925). 946) *J. H. Quastel, A. H. M. Wheatley*, Biological oxidations in the succinic acid series. Biochem. Jl., **25**, 117 (1931). — 947) *J. H. Quastel, A. H. M. Wheatley*, Oxidations by the brain. Biochem. Jl., **26**, 725 (1932). — 948) *J. H. Quastel, A. H. M. Wheatley*, Hepatic oxidation of fatty acids. Biochem. Jl., **27**, 1753 (1933). — 949) *J. H. Quastel, A. H. M. Wheatley*, Effect of ascorbic acid on fatty acid oxydation Biochem. Jl., **28**, 1014 (1934). — 950) *J. H. Quastel, A. H. M. Wheatley*, Studies in fat metabolis,. 4. Acetoacetic acid breakdown in the kidney. Biochem. Jl., **29**, 2773 (1935). — 951) *J. H. Quastel, M. D. Whetham*, Equilibria existing between succinic and fumaric acid in presence of resting bacteria. Biochem. Jl., **18**, 519 (1924). — 952) *J. H. Quastel, W. R. Wooldridge*, Effect of chemical changes in environment on resting bacteria. Biochem. Jl., **21**, 148 (1927). — 953) *J. H. Quastel, W. R. Wooldridge*, Some properties of the dehydrogenating enzymes of bacteria. Biochem. Jl., **22**, 689 (1928). — 954) *J. H. Quastel, W. R. Wooldridge*, Red. potential energy exchange and cell growth. Biochem. Jl., **23**, 115 (1929). — 955) *N. Raikow*, Eine neue Theorie über den Mechanismus der Oxydation. Zs. anorg. Chem., **168**, 297 (1928), **189**, 36 (1930). — 956) *S. Ranganathan, G. Sankaran*, Migration of ascorbic acid (vitamin C) in an electric field. Indian Jl. of med. Res., **24**, 213 (1936). — 957) *H. S. Raper*, The aerobic oxidases. Phys. Reviews, **8**, 245 (1928). — 958) *H. S. Raper, C. E. M. Pugh*, Action of tyrosinase on phenols. Biochem. Jl., **21**, 1370 (1927). — 959) *L. Rapkine*, Potentiel de réduction et les oxydations. Soc. Biol., **96**, 1280 (1927). — 960) *L. Rapkine*, Energy du développement. Ann. de Phys., **7**, 242, 382 (1931). — 961) *L. Rapkine. R. Wurmser*, Potentiel de réduction des cellules vertes. Soc. Biol. **94**, 1347 (1926). — 962) *L. Rapkine, R. Wurmser*, Potentiel de réduction des cellules. Soc. Biol. **95**, 604 (1926). Vortr. 12. Internat. Phys.-Kongr., Stockholm 3.—6. VIII. 1926, S. 139; BPh. **38**, 363. — 963) *L. Rapkine, R. Wurmser*, On the intracellular oxidation-reduction potential. Proc. Roy. Soc., B **102**, 128 (1927). — 964) *E. Raymond*, Oxydation de l'aldehyde benzoique. C. R., **191**, 616 (1930). — 965) *E. Redslob, P. Reiss*, Potentiel d'ox.-réd. du corps vitré. Soc. Biol., **102**, 1060 (1929). — 966) *L. Reichel*, Schardingersches Ferment der Milch. Naturwiss., **23**, 260 (1935). — 967) *L. Reichel*, Anthocyane als biol. Wasserstoff-acceptoren. Naturwiss., **25**, 318 (1937). — 968) *L. Reichel, H. Köhle*, Aldehydrase der Leber. Zs. phys. Chem., **236**, 145, 158 (1935). — 969) *T. Reichstein et al.*, Synthese der d- und l-Ascorbinsäure (Vit. C). Helv. Chim. Acta, **16**, 1019 (1933)., **17**, 311 (1934). — 970) *T. Reichstein, R. Oppenauer*, Reduktinsäure, ein stark reduzierendes Abbauprodukt aus Kohlehydrat. Helv. Chim. Acta, **16**, 988 (1933). — 971) *A. Reid*, Oxydation scheinbar autoxydabler Leucobasen durch molekularen Sauerstoff. Ber. Deutsch. Chem. Ges., **63**, 1920 (1930). — 972) *A. Reid*, Oxydation des Leuco-Methylenblaus. Bioch. Zs., **228**, 487 (1930). — 973) *A. Reid*, Manometrische Messung der sauerstofflosen Atmung. Bioch. Zs., **242**, 159 (1931). — 974) *A. Reid*, Das sauerstoffübertragende Ferment der Atmung. Ergebnisse der Enzymforschung, **1**, 325 (1932). — 975) *A. Reid*, Fermenthämine. Zs. angew. Chem., **47**, 515 (1934). — 976) *L. Reiner, C. V. Smythe, J. T. Pedlow*, On the glucose metabolism of trypanosomes (trypanosoma equiperdum and trypanosoma lewisi). Jl. of Biol. Chem., **113**, 75 (1936). — 977) *I. Remesow c. s.*, Katalytische Eigenschaften von Cholesterin. Bioch. Zs., **246**, 431, **266**, 330, **269**, 63 (1932/4). — 978) *M. S. Resnitschenko*, Mehrphasenwirkung des KCN auf die lebende Zelle. Bioch. Zs., **191**, 345 (1927). — 979) *Fr. Reuter, H. Willstädt, K. L. Zirm*, Zur Kenntnis der peroxydatischen Wirkung III. Bioch. Zs., **261**, 353 (1933). — 979a) *G. M. Richardson, R. K. Cannan*, The dialuric acid-alloxan equilibrium. Biochem. Jl., **23**, 68 (1929). — 980) *D. Richter*, Die Wirkung des Ferro-Eisens bei induzierten Reaktionen. Ber. Deutsch. Chem. Ges., **64**, 1240 (1931). — 981) *D. Richter*, Chain reactions in enzymatic catalysis. Nature, **130**, 97, **129**, 870 (1932). — 982) *E. K. Rideal, W. M. Wright*, Low temperature oxidation at charcoal surfaces. Jl. of Chem. Soc. **1925**, 1347, **1926**, 1813, 3182, **1927**, 2323. — 983) *A. Rieche*, Oxyalkylhydroperoxyde. Ber. Deutsch. Chem. Ges., **64**, 2328 (1931). — 984) *J. Roche*, Propriétés phys.-chemiques de la chlorocruorine. Soc. Biol., **115**, 776 (1933). — 985) *J. Roche*, Rech. sur les globines II (Helicorubin). Bull. Soc. Chim. Biol., **15**, 110 (121) (1933). — 986) *J. Roche*, Sur l'hémérythrine du Siponcle. Bull. Soc. Chim. Biol., **15**, 1415 (1933). — 987) *J. Roche*, Pigments hematies des actinies. Bull. Soc. Chim. Biol., **18**, 825 (1936). — 988) *J. Roche. M. Th. Bénévent*, Constitution du cytochrome c. Bull. Soc. Chim. Biol., **17**, 1473 (1935). — 989) *J. Roche, M. Th. Bénévent*, Les hématines des tissues. Bull. Soc. Chim. Biol., **18**, 1650 (1936). — 990) *J. Roche, M. Th. Bénévent*, Les hématines des cytochromes a. C. R., **203**, 128 (1936). — 991) *J. Roche, M. Th. Bénévent*, Hématines et cytochrome c. Soc. Biol., **123**, 18 (1937). — 992) *J. Roche, M. Th. Bénévent.*, Hématines cellulaires et cytochromes. Soc. Biol., **123**, 20 (1937). — 993) *J. Roche, (H. M. Fox)*, Cryst. Chlorocruorin. Proc. Roy. Soc., B **114**, 161 (1933). — 994) *J. Roche, J.*

*Morena*, Hélicorubine. Compt. Rend. Soc. Biol., **123**, 1215, 1218 (1936). — 994a)*J. H. Roe, G. L. Barnum*, The anti-scorbutic potency of reversibly oxidized ascorbic acid and the observation of an enzyme in blood which reduces the reversibly oxidized vitamin. Jl. of Nutrition, **11**, 359 (1936). — 995) *M. H. Roepke, J. M. Ort*, Formation of the action red. of serveral sugars. Jl., of physical Chem., **35**, 3596 (1931). — 996) *J. Roest*, Oxydations-Katalysen durch Adrenalin. Bioch. Zs., **176**, 17 (1926). — 997) *W. Roman*, Bedeutung der Phasengrenzflächen-Reaktion für den Organismus. Nederl. Tijdschr. Geneesk., **78**, 1467 (1934). — 998) *W. Roman*, Deskriptive und physikalische Chemie der Proteine. Tab. Biol., **14**, 274 (330) (1937). — 999) *P. Rona c. s.*, Die Oxydations-Katalysatoren der Insekten. Bioch. Zs., **223**, 205 (1930). — 1000) *P. Rona c. s.*, Autoxydation von Doppelbindungen. Bioch. Zs., **250**, 149 (1932). — 1001)*A. Roncato*, Stud. polarograf. appl. alla biochimica. Arch. di Sci. Biol., **20**, 146 (1934). — 1002) *E. Ronzoni, E. Ehrenfest*, Effect of dinitrophenol on the metabolism of frog muscle. Jl. Biol. Chem., **115**, 749 (1936). — 1003) *Rosenberg, H.* Ascorbinsäure-Oxydation. Skand. Arch. Phys., **76**, 119 (1937). — Chem. Zentralblatt, **1937**, II, 240. — 1004) *A. Rosenbohm*, Vorkommen und Nachweis eines hämochromogenähnlichen Pigmentes in tierischen Nebennieren. Zs. phys. Chem., **178**, 250 (1928). — 1005) *S. M. Rosenthal, C. Voegtlin*, The action of sulphydryl, iron and cyanide compounds on the oxygen consumption of the living cells. Public health reports, **46**, 521 (1931). — 1006)*S. M. Rosenthal, C. Voegtlin*, The action of heavy metals on cysteine and on sulphydryl groups of proteins. Public health reports, **48**, 347 (1933). — 1007) *E. Ross*, Action of respiratory catalycts on $O_2$ consumption by Nitella. Proc. Soc. Exp. Biol., **32**, 64 (1934). — 1008) *O. T. Rotini, E. Damman, F. F. Nord*, Dehydrierung durch Fusarium. Bioch. Zs., **288**, 414 (1936). — 1009) *H. Rudy*, Enzymatische Spaltung von Lacto-flavin-Phosphorsäure. Zs. phys. Chem., **242**, 198 (1936). — 1010) *H. Rudy*, Absorptionsspektren im Dienste der Vitaminforschung. Naturwiss., **24**, 497 (1936). — 1011) *J. Runnström*, Atmungs-mechanismus und Entwicklungs-erregung bei dem Seeigel-Ei. Protopl., **10**, 106 (1930), **15**, 532 (1932). — 1012)*J. Runnström*, Influence of pyocyanine on sea urchin eggs. Biol. Bull., **68**, 327 (1935). (Woods Hole.) — 1013) *J. Runnström, A. Lennerstrand, H. Borei*, Oxydation und Phosphatbindung im Hämolysat der Pferdeblutkörperchen. Bioch. Zs., **271**, 15 (1934). — 1014) *J. Runnström, L. Michaelis*, Correlation of oxidation and phosphorylation in hemolyzed blood. Jl. gen. Phys., **18**, 717 (1935). — 1015) *A. Schäffner, E. Bauer*, Einfluss von SH-Gruppen auf Phosphatasen verschiedener Herkunft. Zs. phys. Chem., **225**, 245 (1934). — 1016) *F. Schlenk*, Pentosephosphorsäure aus Cozymase. Ark. f. Kemi, **12 B**, Nr. 17 (1936). — 1017) *F. Schlenk*, Triphospho-pyridin-nucleotid (Codehydrase II; Warburg's Coferment). Tab. Biol., **14**, 186 (1937). — 1018)*F. Schlenk*, Co-Carboxylase. Tab. Biol., **14**, 193 (1937). — 1019) *F. Schlenk*, Adenosin-5-phosphorsäuren (Cophosphorylasen). Tab. Biol., **14**, 354 (1937). — 1020) *F. Schlenk*, Die Einwirkung von Phosphoroxychlorid auf Cozymase. Naturwiss., **25**, 668 (1937). — 1021) *F. Schlenk*, Säurehydrolyse der Cozymase. Naturwiss., **25**, 270 (1937). — 1022) *F. Schlenk, H. v. Euler*, Cozymase. Naturwiss., **24**, 794 (1936). — 1023) *F. Schlenk, H. v. Euler*, Über das Verhalten der Cozymase gegen Alkali. Arkiv. f. Kemi, **12 B**, Nr. 20 (1936). — 1024)*F. Schlenk, H. v. Euler c. s.*, Einw. von alkali auf Cozymase. Zs. phys. Chem., **247**, 23 (1937). — 1025) *F. Schlenk, G. Günther, H. v. Euler*, Einwirkung von Phosphatase auf Cozymase. Arkiv f. Kemi, Miner. och Geol., **12 B**, (1938) No. 56. — 1025a)*F. Schlenk, H. Hellström, H. v. Euler*, Desaminocozymase. Ber. Deutsch. Chem. Ges., **71**, 1471 (1938). — 1026)*H. Schmalfuss c. s.*, Über die Entstehung von Pigmenten in Pflanzen. Bioch. Zs., **190**, 424 (1927). — 1027) *F. O. Schmitt, M. G. Scott*, Effect of CO tissue respiration. Amer. Jl. Phys., **107**, 85 (1934). — 1028) *E. Schneider*, Carotinoide der Purpurbakterien. Rev. Fac. Sci. Univ. d'Istanbul, **1**, 74 (1936). — 1029) *A. Schöberl*, Sulfhydrylverbindungen als Antikatalysatoren bei Oxydationen mit molekularem Sauerstoff. II Mitt. Zs. phys. Chem., **201**, 161, 167 (1931). — 1030) *A. Schöberl*, Cystein und Glutathion als Antikatalysator bei Oxydationen mit molekularem Sauerstoff. Ber. Deutsch. Chem. Ges., **64**, 546 (1931). — 1031) *A. Schöberl, H. Eck*, Hydrolytische Aufspaltung der Disulfid-bindung. Liebigs Ann. **522**, 97 (1936). — 1032) *A. Schöberl, M. Wiesner*, Modellversuche zum oxydativen Abbau biologisch wichtiger organischer Schwefelverbindungen. Liebigs Ann., **507**, 111 (1933). — 1032a) *O. v. Schönebeck, [C. J. Neuberg]*, Versuche zur Reinigung von Carboxylase. Bioch. Zs., **272**, 42 (1934), **275**, 330 (1934/35), **277**, 451 (1935). — 1033)*S. A. Schou*, Absorption des rayons ultraviolets par les aldéhydes. C. R., **182**, 965 (1926). — 1034) *H. Schröder*, Hemmung der Dopa-Reaktion durch Vitamin C. Klin. Ws. **13**, I, 553 (1934). — 1035)*M. P. Schubert*, Complex types involved in the catalytic oxidation of thiol acids. Jl. Amer. Chem. Soc., **54**, 4077 (1932). — 1036) *M. P. Schubert*, A new complex of cobalt and cysteine and its behavior with hydrogen peroxide. Jl. Amer. Chem. Soc., **55**, 3336 (1933).

— 1037) *H. Schüler*, Über die Oxydation des Hämoglobineisens durch Ferricyankalium und das Gleichgewicht der Reaktion. Biochem. Zs., **255**, 474 (1932). — 1038) *K. Schuwirth*, Chinone als Fermentmodell. Bioch. Zs., **271**, 427 (1934). — 1039) *G. M. Schwab*, Die Katalyse vom Standpunkt der chemischen Kinetik. Berlin, 1931. — 1040) *G. M. Schwab c. s.*, Kettencharakter der Katalase-Wirkung. Ber. Chem. Ges., **66**, 661 (1933). — 1041) *C. Schwob*, Catalytic properties of charcoal. Jl., Amer. Chem. Soc., **58**, 1115 (1936). — 1042) *K. C. Sen*, Effect of narcotics on some dehydrogenases. Biochem. Jl., **25**, 849 (1931). — 1043) *K. Ch. Sen*, Reduction of glutathione in liver. Ind. Jl., Med. Res., **20**, 1051 (1933). — 1044) *H. K. Sen, S. Bannerjee*, Synthetische Enzyme. Trans. Nat. Inst. Sci, India, **1**, 83 (1935); Chem. Zbl., **1937**, I, 4378. — 1045) *M. G. Sevag, (L. Maiweg)*, Atmungsmechanismus der Pneumokokken. Ann. Chem. Pharm. (liebig), **507**, 92; Bioch. Zs., **267**, 211 (1933); Jl. of exp. Med., **60**, 95 (1934). — 1046) *Ph. A. Shaffer*, Intermediary peroxides in the oxidation of ferrous salts. Jl. of Biol. Chem., **74**, XLVI (1927). — 1047) *P. A. Shaffer, B. K. Harned*, Oxidations induced by sugars. Jl. of Biol. Chem., **93**, 311 (1931). — 1048) *M. Sherif, E. G. Holmes*, Oxygen consumption of nerve in presence of glucose and galactose. Biochem. Jl., **24**, 400 (1930). — 1049) *K. Shibata*, Wirkungs- Mechanismus der oxydoreduzierenden Enzyme. Acta phytochim., **4**, 373 (1929). — 1050) *K. Shibata*, Cytochrom und Zellatmung. Ergebn. der Enzymforschung, **4**, 348 (1935). — 1051) *K. Shibata, Y. Shibata*, Oxidaseartige Wirkung gewisser Metallkomplexe. Acta phytochim., **4**, 363 (1929). — 1052) *K. Shibata, Y. Shibata c. s.*, Die katalytische Wirkung der Metallkomplexverbindungen. Iwata Inst. of Plant. Biochem. Publ. 2, Tokyo 1936. — 1053) *K. Shibata, H. Tamiya*, Bedeut. des Cytochroms. Acta Phytochim., **5**, 23 (1930), **7**, 191 (1933). — 1054) *Ch. S. Shoup, I. P. Boykin*, Insensitivity of Paramecium to HCN. Jl. of gen. Phys., **15**, 107 (1931/32). — 1055) *P. E. Simola*, Co-Zymase und Co-Carboxylasegehalt des Rattenorganismus bei B-Avitaminose. Bioch. Zs., **254**, 229 (1932). — 1056) *A. Simon, (K. Koetschau)*, Aktives Eisen. Zs. Anorg. Chem., **164**, 101, 168, 129, 194, 89 (1927/30). — 1057) *A. Simon, Th. Reetz*, Aktives Eisen. Liebigs Ann., **485**, 73 (1931). — 1058) *A. Smakula*, Optische Untersuchungen des antineuritischen Vitamins. Zs. phys. Chem., **230**, 231 (1935). — 1059) *I. Smedley-Maclean et al.*, Catal. action of cupric salts etc. Biochem. Jl., **23**, 593 (1929), **25**, 1252 (1931), **28**, 486, 892 (1934). — 1060) *I. H. C. Smith, H. A. Spöhr*, Kinetics of the oxidation with Na-Ferro-pyrophosphate. Jl. Amer. Chem. Soc., **48**, 107 (1926). — 1061) *C. V. Smythe*, The mechanism of iron catalysis in certain oxidation. Jl. of Biol. Chem., **90**, 251 (1931). — 1062) *C. V. Smythe*, Die Wirkung der Blausäure auf Methylglyoxal. Bioch. Zs., **257**, 371 (1933). — 1063) *B. Sokoloff*, Influence de la quinone sur l'oxydation. Soc. Biol., **107**, 115 (1931). — 1064) *H. A. Sphör, I. H. C. Smith*, Studies on atmospheric oxidation. II. Jl. Amer. Chem. Soc., **48**, 236 (1926). — 1064a) *F. J. Stare*, A potentiometric study of hepatoflavin. Jl. of. Biol. Chem., **112**, 223 (1935). — 1064b) *F. J. Stare, C. A. Baumann*, Effect of fumarate on respiration. Proc. Roy. Soc. B **121**, 338 (1936); Biochem. Jl., **30**, 2257 (1936). — 1065) *F. J. Stare, C. A. Elvehjem*, Respiration of animal tissues. Amer. Jl. Phys., **105**, 655 (1933). — 1066) *C. C. Steele*, Recent progress. in determining the structure of Chlorophyll. Chem. Rev., **20**, 1 (1937). — 1067) *H. N. Stephens*, Studies in autoxidations. Jl. of Physical Chem., **37**, 209, **38**, 419 (1933/4). — 1068) *M. Stephenson*, Lactic dehydrogenase. Biochem. Jl., **22**, 605 (1928). — 1069) *M. Stephenson, L. H. Stickland*, Hydrogenase. Biochem. Jl., **25**, 205, 215 (1931). — 1070) *M. Stephenson, L. H. Stickland*, Biochem. Jl., **26**, 712, **27**, 1517, 1528 (1932/3). — 1071) *K. G. Stern*, Über die Katalase farbloser Blutzellen. Zs. phys. Chem., **204**, 259 (1932). — 1072) *K. G. Stern*, Der isoelektrische Punkt der Katalase. 2. Mitteilung über Katalase. Zs. phys. Chem., **208**, 86 (1932). — 1073) *K. G. Stern*, Über die Hemmungstypen und den Mechanismus der katalatischen Reaktion. III. Mitt. über Katalase. Zs. phys. Chem., **209**, 176 (1932). — 1074) *K. G. Stern*, Über das optische Verhalten der Katalase. 4. Mitteilung über Katalase. Zs. phys. Chem., **212**, 207 (1932). — 1075) *K. G. Stern*, Über die katalatische Aktivität von Häminkomplexen. 5. Mitt. über Katalase. Zs. phys. Chem., **215**, 35 (1933). — 1076) *K. G. Stern*, Über die Teilchengrösse und das Molekulargewicht der Katalase. 6. Mitteilung über Katalase. Zs. phys. Chem., **217**, 237 (1933). — 1076a) *K. G. Stern*, Synthetische Haeminkatalasen. Zs. phys. Chem., **219**, 105 (1933). — 1077) *K. G. Stern*, Über den Zusammenhang zwischen Leber-Lyochrom und Katalase. Ber. Deutsch. Chem. Ges., **66**, 555 (1933). — 1077a) *K. G. Stern*, Mechanismus des Pyocyanineffektes auf die Atmung., Naturwiss., **21**, 350 (1933). — 1077b) *K. G. Stern*, Isolation of Hepatoflavin Nature, **132**, 784 (1933). — 1078) *K. G. Stern*, On the question of the flavin potentials. Ber. Deutsch. Chem. Ges., **67**, 654 (1934). — 1079) *K. G. Stern*, Uroflavin, Maltoflavin, and redox potentials of lyochromes. Nature, **133**, 178 (1934). — 1080) *K. G. Stern*, Potentiometric study of photoflavin. Biochem. Jl., **28**, 949 (1934). — 1081) *K. G. Stern*, Redox-Systeme. Tab. biol. period., **4**, 1 (1935). — 1081a) *K. G. Stern*, Constitution of the

Prosthetic Group of Catalase. Nature, **136**, 302 (1935). — 1082) *K. G. Stern*, Spectroscopy of an enzyme reaction. Nature, **136**, 335 (1935). — 1083) *K. G. Stern*, Oxidation-reduction potentials of toxoflavin. Biochem. Jl., **29**, 500 (1935). — 1084) *K. G. Stern*, The constitution of the prosthetic group of catalase. Jl. of Biol. Chem., **112**, 661 (1936). — 1085) *K. G. Stern*, On the mechanism of enzyme action. Jl. of Biol. Chem., **114**, 473 (1936). — 1086) *K. G. Stern*, On the constitution and the mode of action of catalase. Biochimia (USSR), **2**, 198 (1937). — 1087) *K. G. Stern*, Spectrography of the reaction of catalase with ethyl hydrogen peroxide. Enzymologia, **4**, 145 (1937). — 1088) *K. G. Stern*, On the absorption spectrum of catalase. Jl. of Biol. Chem., **121**, 561 (1937). — 1089) *K. G. Stern*, Spectroscopy of catalase. Jl. of gen. Phys., **20**, 631 (1937). — 1090) *K. G. Stern, D. Dubois*, A photoelectric method for recording fast chemical reactions. Jl. of Biol. Chem., **116**, 575 (1936). — 1091) *K. G. Stern, D. Dubois*, A spectroscopic method for the kinetic study of rapid chemical reactions. Jl. of Biol. Chem., **121**, 573 (1937). — 1092) *K. G. Stern, J. W. Hofer*, Synthesis of Co-Carboxylase from Vitamin $B_1$. Enzymologia (Neuberg-Festschrift), **3**, 82 (1937). — 1093) *K. G. Stern, J. W. Hofer*, Synthesis of Co-Carboxylase from Vitamin $B_1$. Science, **85**, 483 (1937). — 1094) *K. G. Stern, E. R. Holiday*, Zur Konstitution des Photo-flavins; Versuche in der Alloxazin-Reihe (Vorläuf. Mitteil.) Ber. Deutsch. Chem. Ges., **67**, 1104 (1934). — 1095) *K. G. Stern, E. R. Holiday*, Die Photo-flavine, eine Gruppe von Alloxazin-Derivaten. Ber. Deutsch. Chem. Ges., **67**, 1442 (1934). — 1096) *K. G. Stern, G. I. Lavin*, Ultra-Violet absorption spectrum of catalase. Science, **88**, 263 (1938). — 1096a) *K. G. Stern, K. Salomon*, Ovoverdin. Jl. of Biol. Chem., **122**, 461 (1938). — 1097) *K. G. Stern, A. White*, Studies on the constitution of insulin. I. Properties of reduced insulin preparations. Jl. of Biol. Chem., **117**, 95 (1937). — 1098) *K. G. Stern, R. W. G. Wyckoff*, An ultracentrifugal study of catalase. Jl. of Biol. Chem., **124**, 573 (1938). — 1099) *K. G. Stern, R. W. G. Wyckoff*, An ultracentrifugal study of catalase. Science, **87**, 18 (1938). — 1100) *L. Stern*, Specificité des accepteurs d'hydrogène. Bull. Biol. Méd. U.S.S.R., **1**, 330 (1936). — 1101) *L. H. Stickland*, Bacterial decomposition of formic acid. Biochem. Jl., **23**, 1187 (1929). — 1102) *L. H. Stickland, D. E. Green*, Negative oxidation-reduction system of B. Coli. Nature, **133**, 573 (1934). — 1103) *R. D. Stiehler, H. M. Huffmann*, Heat capacity, entropy and free energy of adenine. Jl. Amer. Chem. Soc., **57**, 1741 (1935). — 1104) *R. Stöhr*, Schicksal der Acetessigsäure im tierischen Organismus. Zs. phys. Chem., **237**, 165 (1935). — 1105) *F. M. Stone, C. B. Coulter*, Porphyrin compounds derived form Bacteria. Jl. of gen. Phys., **15**, 629 (1932). — 1105a) *E. Stotz, C. J. Harrer, C. G. King*, „Ascorbic Acid Oxidase" in Relation to Copper. Jl. of Biol. Chem., **119**, 511 (1937). — 1105b) *E. Stotz, C. J. Harrer, M. O. Schultze, C. G. King*, Tissue respiration studies on normal and scorbutic guinea pig liver and kidney. Jl. of Biol. Chem., **120**, 129 (1937). — 1106) *F. B. Straub*, Dehydrasekoppelungen in der $C_4$-Dicarbonsäurekatalyse. Zs. phys. Chem., **249**, 189 (1937). — 1107) *F. B. Straub*, Coenzyme of the d-Amino-acid Oxidase. Nature, **141**, 603 (1938). — 1107a) *F. B. Straub*, Beitrag zur Existenz der Ascorbinsäureoxydase. Zs. phys. Chem., **254**, 205 (1938). — 1107b) *J. B. Sumner, A. L. Dounce*, Crystalline catalase. Jl. of Biol. Chem., **121**, 417 (1937). — 1108) *J. B. Sumner, N. Gralen*, The molecular weight of crystalline catalase. Jl. of Biol. Chem., **125**, 33 (1938). — 1109) *J. B. Sumner, N. Gralen*, The molecular weight of crystalline catalase. Science, **87**, 284 (1938). — 1110) *J. B. Sumner, S. F. Howell*, Hematin and the peroxidase of fig. sap. Enzymologia, **1**, 133 (1936). — 1111) *The Svedberg, A. Hedenius*, Molecular weights of the blood pigments. Nature, **131**, 325 (1933). — 1112) *J. L. Svirbely, A. Szent-Györgyi*, Hexuronic acid as the antiscorbutic factor. Nature, **129**, 690 (1932). — 1113) *A. Szent-Györgyi*, Zellatmung II, III. Bioch. Zs., **157**, 50, 67 (1925). — 1114) *A. v. Szent-Györgyi*, Zellatmung IV. Bioch. Zs., **250**, 399 (407) (1925). — 1115) *A. v. Szent-Györgyi*, Die Rolle des $H_2O_2$ bei der biologischen Oxydation. Bioch. Zs., **178**, 75 (1926). — 1116) *A. v. Szent-Györgyi*, Zellatmung. Bioch. Zs., **162**, 399 (1925), **181**, 425 (1927). — 1117) *A. Szent-Györgyi*, Description of a new carbohydrate derivative. Biochem. Jl., **22**, 1387 (1928). — 1118) *Szent-Györgyi*, Function of peroxidase systems. Biochem. Jl., **22**, 1387 (1928). — 1119) *A. Szent-Györgyi*, Function of hexuronic acid in the respiration. Jl. of Biol. Chem., **90**, 385 (1931). — 1120) *A. Szent-Györgyi*, Die freie Energie der Milchsäureoxydation. Zs. phys. Chem., **217**, 51 (1933). — 1121) *A. Szent-Györgyi*, Identification of vitamin C. Nature, **131**, 225 (1933). — 1122) *A. Szent-Györgyi*, Non-enzymic catalysts of cellular oxidation. Arch. exp. Zellforschung, **15**, 29 (1934). — 1123) *A. v. Szent-Györgyi*, Mechanism of respiration. Nature, **135**, 305 (1935). — 1124) *A. v. Szent-Györgyi*, Oxydations, Fermente. Schweizer Med. Ws., 1936, II, 885. — 1125) *A. v. Szent-Györgyi*, Studies on Biological Oxidation Acta Med. Szeged, **9**. 1 (1937). — (Also in bookform, Budapest-Leipzig, 1937). — 1126) *A. v. Szent-Györgyi c. s.*, Mechanismus der Hauptatmung d. Taubenbrustmuskels. Zs. phys. Chem., **224**, 1 (1934), **236**, 1, **244**, 105 (1936), **245**,

113 (1937). — 1126a) *A. v. Szent-György et al.*, Über die dehydrierende Autoxydation und die biologischen Oxydationen. Zs. phys. Chem., 254, 147 (1938). — 1127) *A. v. Szent-Györgyi, I. Banga*, Über Co-Fermente, Wasserstoffdonatoren und Arsenvergiftung der Zellatmung. Bioch. Zs., 246, 203 (1932). — 1128) *H. Tait, E. J. King*, Oxidation of lecithin in presence of glutathione. Biochem. Jl., 30, 285 (1936). — 1129) *T. Takahashi, T. Asai*, Acids formed by Rhizopus. Proc. Imp. Acad. Tokyo, 3, 86 (1927). — 1130) *B. Tamamushi, c. s.*, Oxydation der Bernsteinsäure an Kohle. Zs. Elektrochem., 41, 761 (1935). — 1131) *B. Tamamushi, H. Umezawa*, Dehydrierung der Bernsteinsäure an Kohle. Acta phytochimica, 8, 221 (1934/5). — 1132) *H. Tamiya*, Cytochrom in Schimmelpilz-zellen. Acta Phytochim, 4, 215 (1928). — 1133) *H. Tamiya c. s.*, Einfluss des CO auf die MB-Reduktion. Acta Phytochim, 5, 119 (1930). — 1134) *H. Tamiya, Y. Ogura*, Wirkungs-mechanismus der einzelnen Cytochromkomponenten in der Zellatmung. Acta phytochim., 9, 123 (1937). — 1135) *H. Tamiya, S. Yamagutchi*, Cytochromspectrum von verschiedenen Mikroorganismen. Acta Phytochim., 7, 233 (1933). — 1136) *K. Tanaka*, Physiologie der Essiggärung. Acta phytochim., 5, 239 (1930). — 1137) *K. Tanaka*, Gluconsäuregärung der Essigbacterien. Acta Phytochim., 7, 265 (1933). — 1138) *H. Tangl, N. Berend*, Fettresorption durch die Desaturierung der Fettsäuren. Bioch. Zs., 220, 234, 232, 181, 260, 490 (1930/33). — 1139) *H. L. A. Tarr*, Enzymatic formation of $H_2S$ by bacteria. Biochem. Jl., 27, 1869, 28, 192 (1933/4). — 1140) *H. Tauber*, Interaction of peroxidase and ascorbic, acid (vitamin C) in biological oxidations and reductions. Enzymologia, 1, 209 (1936). — 1141) *H. Tauber*, Enzymic synthesis of cocarboxylase form vitamin $B_1$. Enzymologia, II, 171 (1937). — 1142) *H. Tauber*, Enzymic synthesis of co-carboxylase. Science, 86, 180 (1937). — 1143) *H. Tauber*, Synthesis of Co-Carboxylase from Vitamin $B_1$. Jl. Amer. Chem. Soc., 60, 730 (1938). — 1144) *H. Tauber*, Preparation of Cocarboxylase. Proc. Soc. Exp. Biol. and Med., 38, 888 (1938). — 1144a) *H. Tauber*, The interaction of ascorbic acid (vitamin C) with Enzymes. Ergebnisse der Enzymforschung, 7, 301 (1938). — 1145) *H. Tauber, J. S. Kleiner*, Chemical nature of catalase. Proc. Soc. Exp. Biol. and Med., 33, 391 (1935). — 1146) *K. Terai*, Einfluss des Redoxsystems auf die Adrenalinwirkung. Fol. pharm. Jap., 19, 79, 248 (1934). — 1147) *H. Theorell*, Kristallinisches Myoglobin. I, V. Bioch. Zs., 252, 1, 268, 46, 55, 64, 73 (1931/34). — 1148) *H. Theorell*, Reindarstellung (Kristallisation) des gelben Atmungsfermentes und die reversible Spaltung desselben. Bioch. Zs., 272, 155 (1934). — 1149) *H. Theorell*, Kataphoretische Studien über das Atmungs-Co-Ferment der roten Blutzellen. Bioch. Zs., 275, 11 (1934); Kongr. f. Phys. Helsinki 1934; Skand. Arch. Phys., 71, (1934). — 1150) *H. Theorell*, Bestimmung der Anzahl von sauren Gruppen des Atmungs-Co-Ferments durch Diffusionsmessungen. Bioch. Zs., 275, 19 (1934). — 1151) *H. Theorell*, Über die Wirkungsgruppe des gelben Ferments. Bioch. Zs., 275, 37 (1934). — 1152) *H. Theorell*, Elektrophoretische Studien über das Atmungskoferment aus roten Blutzellen. Naturwiss., 22, 290 (1934). — 1153) *H. Theorell*, Reindarstellung der Wirkungsgruppe des gelben Ferments. Bioch. Zs., 275, 344 (1935). — 1154) *H. Theorell*, Das gelbe Oxydationsferment. Bioch. Z., 278, 263 (1935). — 1155) *H. Theorell*, Quantitative Bestrahlungsversuche am gelben Ferment, Flavinphosphorsäure und Lactoflavin. Bioch. Zs., 279, 186 (1935). — 1156) *H. Theorell*, Reines Cytochrom C. (Vorl. Mitt.) Bioch. Zs., 279, 463 (1935). — 1157) *H. Theorell*, Reines Cytochrom c. II. Bioch. Zs., 285, 207 (1936). — 1158) *H. Theorell*, Die physiologische Reoxydation des reduzierten gelben Ferments. Bioch. Zs., 288, 317 (1936). — 1159) *H. Theorell*, Keilin's cytochrome c and the respiratory mechanism of Warburg and Christian. Nature 138, 687 (1936). — 1160) *H. Theorell*, Präparative Versuche über das „unmodifizierte Porphyrin c". Enzymologia, 4, 192 (1937). — 1161) *H. Theorell*, Das gelbe Ferment: seine Chemie und Wirkungen. Ergebn. der Enzymforschung, 6, 111 (1937). — 1162) *H. Theorell*, Die freie Eiweisskomponente des gelben Ferments und ihre Kupplung mit Lactoflavinphosphorsäure. Bioch. Zs., 290, 293 (1937). — 1163) *H. Theorell*, Über die chemische Konstitution des Cytochroms C. Bioch. Zs., 298, 242 (1938). — 1164) *H. Theorell, P. Karrer, K. Schöpp, P. Frei*, Flavinphosphorsäure aus Leber. Helv. Chim. Acta, 18, 1022 (1935). — 1165) *E. J. Theriault, C. T. Butterfield, P. D. McNamee*, Catalysis of air oxidation by iron salts. Jl. Amer. Chem. Soc., 55, 2012 (1933). — 1166) *T. Thunberg*, Intermediärer Stoffwechsel. Skand. Arch. Phys., 40, 1 (1920). — 1167) *T. Thunberg*, Acceptormethoden, in Oppenheimer-Pincussen, Fermentmethoden p. 1118 (1134) (1929). — 1168) *T. Thunberg*, Die biologischen Reduktions-Oxydations-Potentiale. Handbuch der Biochemie, Ergänz.-Band Jena 1930, p. 213. — 1169) *T. Thunberg*, Der jetzige Stand der Lehre vom biologischen Oxydations-Mechanismus. Handbuch der Biochemie, Ergänzungswerk, Bd. I. p. 248, Jena 1933. — 1170) *T. Thunberg*, Über das Schicksal des Sauerstoffs in den biologischen Oxydations-Prozessen. Arch. internat. Pharm., 38, 89 (1930). — 1171) *T. Thunberg*, Die Dehydrasen. Handbuch der Biochemie, Ergänzungswerk, Jena, 1, 518 (1933). — 1172) *T.*

*Thunberg*, Zn und Cd als stimulierende Mittel für die Oxydations Prozesse. Skand. Arch. Phys., **69**, 247, **72**, 283 (1934/35). — 1173) *T. Thunberg*, Zur Kenntnis der wasserstoffaktivierenden Fähigkeit der Hefe. Skand. Arch. Phys., **75**, 248 (1937). — 1174) *T. Thunberg*, Die Dehydrogenasenforschung der letzten Jahre. Ergeb. Enzymforschung, **7**, 163 (1938). — 1175) *S. Thurlow*, Studies in xanthine oxidase IV. Biochem. Jl., **19**, 175 (1925). — 1176) *W. C. Tobie*, Pigment of Bac. violaceum. Jl. of Bact., **29**, 223 (1935); Proc. Soc. Exp. Biol., **34**, 620 (1936). — 1177) *S. Toda*, Die Wirkung von Blausäureäthylester auf Schwermetallkatalysen. Bioch. Zs., **172**, 17 (1926). — 1178) *Sh. Toda*, Wasserstoffaktivierung durch Eisen. Bioch. Zs., **172**, 34 (1926). — 1179) *A. R. Todd, F. Bergel*, A synthesis of aneurin. Jl. Chem. Soc. Lond., **1937**, 364. — 1180) *A. R. Todd, F. Bergel, Fränkel-Conrat, Jacob*, Aneurin. VI. Synthesis of thiochrome. Jl. Chem. Soc. Lond., **1936**, 1601. — 1181) *A. R. Todd, F. Bergel, A. Jacob*, Aneurin. Ber. Deutsch. Chem. Ges., **69**, 217 (1936); Jl. of Chem. Soc., **1936**, 1555, 1557, 1559, **1937**, 364. — 1182) *E. Töniessen, E. Brinkmann*, Oxydativer. Abbau der Kohlehydrate. Zs. phys. Chem., **187**, 137 (1931). — 1183) *G. de Toeuf*, Potentiel d'oxydo-reduction de cytochrome c. Journ. Chim. Phys., **34**, 740 (1937). — 1184) *I. Torres*, Restatmung in Blausäure. Bioch. Zs., **280**, 114 (1935). — 1185) *W. Traube, W. Lange*, Oxydations-, Reduktions Vorgänge. Ber. Deutsch. Chem. Ges., **58**, 2773, **59**, 2860 (1925/26). — 1186) *K. Trautwein c. s.*, ph-Empfindlichkeit der atmenden und gärenden Bierhefe. Bioch. Zs., **236**, 35, **240**, 423 (1931). — 1187) *M. Tsuchihashi*, Zur Kenntnis der Blutkatalase. Bioch. Zs., **140**, 63 (1923). — 1188) *M. Tsukano*, Redox potentials of biozucker plus its dehydrogenase. Jl. of Biochem., **15**, 491 (1932). — 1189) *H. Ucko*, Peroxydasenatur des aktiven Eisens. Klin. Ws., **7**, 1515 (1928). — 1190) *F. Urban*, Spectrographic study of the cytochromes. Jl. of Biol. Chem., **109**, XCIII (1935). — 1191) *H. J. Ustvedt*, Die Rolle des Eisens bei der Glykolyse. Bioch. Zs., **265**, 154 (1933). — 1192) *A. G. van Veen, J. K. Baars*, Constitution of Toxoflavin. Proc. Acad. Sci. Amsterdam, **40**, 498 (1937). — 1193) *A. G. van Veen, W. K. Mertens*, On the isolation of a toxic bacterial pigment. Kon. Akad. Wetenschap. Amsterdam., **36**, 666 (1933). — 1194) *A. G. van Veen, W. K. Mertens*, Die Giftstoffe der sogenannten Bongkrek-Vergiftung auf Java. Rec. Trav. Chim. Pays-Bas, **53**, 257 (1934). — 1195) *A. G. van Veen, W. K. Mertens*, Das Toxoflavin, der gelbe Giftstoff der Bongkrek. Rec. Trav. Chim. Pays-Bas, **53**, 398 (1934). — 1196) *E. Vellinger*, Recherche potentiométrique sur le ph intérieur et sur le potentiel d'ox-réd. de l'oeuf d'oursin. Arch. de Physique Biol. **6**, 141 (1928) BPh. **45**, 321. — 1197) *P. E. Verkade et al.*, Fettstoffwechsel. Akad. Wet. Amsterdam, **35**, 251, **36**, 314, 876 (1932/33); Zs. phys. Chem., **215**, 225, **225**, 230, **230**, 207, **234**, 21 (1934/35); Biochem. Jl., **28**, 31 (1934). — 1198) *R. Vesterberg, W. Braun*, Hydrierung und Dehydrierung als gekoppelte Reaktion. Ber. Deutsch. Chem. Ges., **65**, 1473, (1932). — 1199) *R. Vestin*, Cozymase als Phosphatüberträger. Zs. phys. Chem., **240**, 99 (1936). — 1200) *R. Vestin*, Enzymatische Umwandlung von Codehydrase I in Codehydrase II. Naturwiss., **25**, 667 (1937). — 1201) *R. Vestin, H. v. Euler*, Leicht hydrolysierbares Phosphat aus Cozymase. Zs., phys. Chem., **245**, I (1936). — 1202) *R. Vestin, F. Schlenk, H. v. Euler*, Ein Beitrag zur Konstitutionsermittelung der Cozymase; Isolierung des Spaltstückes Adenosin-di-phosphorsäure. Ber. Deutsch. Chem. Ges., **70**, 1369 (1937). — 1203) *K. Vietorisz*, Einfluss von As und HCN auf die Atmung maliguer Geschwülste. Bioch. Zs., **240**, 488 (1931). — 1204) *A. I. Virtanen, L. Pulkki*, Formation of citric acid in nature. Ann. Acad. Sci. Fenn. A **33**, Nr. 3 (1930). — 1204a) *C. Vögtlin et al.*, The relation of protein metabolism to malignant growth. Occasional publications of the Amer. Ass. f. the Advancement of Science. No. 4. June, 1937. Supplement to Science, Vol. 85. — 1204b) *C. Vögtlin, J. M. Johnson, H. A. Dyer*, Quantitative estimation of the reducing power of normal and cancer tissue. Jl. of Pharm., **24**, 305 (1924). — 1205) *C. Vögtlin, J. M. Johnson, H. A. Dyer*, Biological significance of cystine an glutathone. I. On the mechanism of the cyanide action. Jl. of Pharm. and exp. Therap. **27**, 467 (1926). — 1206) *C. Vögtlin, J. M. Johnson, S. M. Rosenthal*, The oxidation catalysis of crystalline glutathione with particular reference to copper. Jl. of Biol. Chem. **93**, 435 (1931); Jl. of Pharm., **42**, 256 (1931). — 1207) *Th. Wagner-Jauregg*, Flavine. Zs. Angew. Chem., **47**, 318 (1934). — 1208) *Th. Wagner-Jauregg*, B-Vitamine. Zs. Angew. Chem., **47**, 547 (1934). — 1209) *Th. Wagner-Jauregg*, Lactoflavin (Vitamin B$_2$) und Fermentvorgänge. Ergebnisse der Enzymforschung, **4**, 333 (1935). — 1210) *Th. Wagner-Jauregg, E. F. Möller*, Über die Milchsäureredehydrase. Zs. phys. Chem., **236**, 216 (1935). — 1211) *Th. Wagner-Jauregg, E. F. Moeller*, Aktivierung der enzymatischen Dehydrierung des Alkohols durch Glutathion. Zs. phys. Chem., **236**, 232 (1935). — 1212) *Th. Wagner-Jauregg, H. Rauen*, Über die enzymatische Dehydrierung der Citronensäure. Zs phys. Chem., **233**, 215 (1935). — 1213) *Th. Wagner-Jauregg, H. Rauen, F. F. Möller*, Über die Wirkungsweise des Vitamins B$_2$ und die Beteiligung von Flavoproteinen an enzymatischen Dehydrierungen. (Vorl. Mitt.). Zs. phys. Chem., **228**, 273

(1934). — 1214) *Th. Wagner-Jauregg, H. Ruska*, Flavine als biologische Wasserstoff-Acceptoren. Ber. Deutsch. Chem. Ges. 66, 1298 (1933). — 1215) *Th. Wagner-Jauregg, H. Wollschitt*, Bemerkungen zu einer Mitteilung ,,Über Urochrom und die Teilnahme von Lyochromen an der Zellatmung. Naturwiss., 22, 107 (1934). — 1216) *O. Warburg*, Über die Geschwindigkeit der photochemischen Kohlensäurezersetzung in lebenden Zellen. Bioch. Zs., 100, 230 (1919). — 1216a) *O. Warburg*, Über Eisen, den sauerstoffübertragenden Bestandteil des Atmungsferments. Bioch. Zs., 152, 479 (1924). — 1217) *O. Warburg*, Eisen, der sauerstoffübertragende Bestandteil des Atmungsferments. Ber. Deutsch. Chem. Ges., 58, 1001 (1925). — 1218) *O. Warburg*, Über die Wirkung der Blausaure auf die alkoholische Gärung. Bioch. Zs., 165, 196, 172, 432, 189, 354 (1925/7). — 1219) *O. Warburg*, Über die Assimilation der Kohlensäure. Bioch. Zs., 166, 386 (1925). — 1220) *O. Warburg*, Wirkung von Blausäureäthylester (Aethylcarbylamin) auf die Pasteur'sche Reaktion. Bioch. Zs., 172, 432 (1926). — 1221) *O. Warburg*, Wirkung des CO auf den Stoffwechsel der Hefe. Bioch. Zs., 177, 471 (1926). — 1222) *O. Warburg*, Methode zur Bestimmung von Kupfer und Eisen und über den Kupfergehalt des Blutserums. Bioch. Zs., 187, 255 (1927). — 1223) *O. Warburg*, Über die Wirkung von CO und Stickoxyd auf die Atmung und Gärung. Bioch. Zs., 189, 354 (1927). — 1224) *O. Warburg*, Über Kohlenoxydwirkung ohne Hämoglobin und einige Eigenschaften des Atmungsferments. Naturwiss., 15, 546 (1927). — 1225) *O. Warburg*, Über die chemische Konstitution des Atmungsferments. Naturwiss., 16, 345 (1928). — 1225a) *O. Warburg*, Katalytische Wirkungen der lebendigen Substanz. Berlin, 1928. — 1226) *O. Warburg*, Über die oxydationskatalytische Wirkung des Eisens nach Handovsky. Bioch. Zs., 198, 241 (1928). — 1227) *O. Warburg*, Wie viele Atmungsfermente gibt es? Bioch. Zs., 201, 486 (1928). — 1228) *O. Warburg*, Bemerkung zu einer Arbeit von D. Keilin. Bioch. Zs., 207, 494 (1929). — 1229) *O. Warburg*, Über die chemische Konstitution des Atmungsferments. Zs. Elektrochemie, 35, 549 (1929). — 1230) *O. Warburg*, Über Nicht-Hemmung der Zellatmung durch Blausäure. Bioch. Zs., 231, 493 (1931). — 1231) *O. Warburg*, Wirkung der Blausäure auf die katalytische Wirkung des Mangans. Bioch. Zs., 233, 245 (1931). — 1232) *O. Warburg*, Phaeophorbid-b-Eisen. Ber. deutsch. Chem. Ges., 64, 682 (1931). — 1233) *O. Warburg*, Das sauerstoffübertragende Ferment der Atmung. Zs. Angew. Chem., 45, 1 (1932) (Nobelvortrag). — 1234) *O. Warburg*, Sauerstoffübertragende Fermente. Naturwiss., 22, 441 (1934). — 1235) *O. Warburg*, Chemische Konstitution von Fermenten. Ergebn. Enzymforschg., 7, 210 (1938). — 1236) *O. Warburg, W. Christian*, Phaeohämin b. Bioch. Zs., 235, 240 (1931). — 1237) *O. Warburg, W. Christian*, Über die Aktivierung von Kohlenhydrat in roten Blutzellen. Bioch. Zs., 238, 131 (1931). — 1238) *O. Warburg, W. Christian*, Über die Aktivierung der Robisonschen Hexose- Mono-Phosphorsaure in roten Blutzellen. Bioch. Zs., 242, 206 (1931). — 1239) *O. Warburg, W. Christian*, Über ein neues Oxydationsferment und sein Absorptionsspektrum. Bioch. Zs., 254, 438 (1932). — 1240) *O. Warburg, W. Christian*, Ein zweites sauerstoffübertragendes Ferment und sein Absorptionsspektrum. Naturwiss., 20, 688 (1932). — 1241) *O. Warburg, W. Christian*, Über das neue Oxydationsferment. Naturwiss., 20, 980 (1932). — 1242) *O. Warburg, W. Christian*, Über das gelbe Oxydationsferment. Bioch. Zs., 257, 492 (1933). — 1243) *O. Warburg, W. Christian*, Über das gelbe Oxydationsferment. Bioch. Zs., 258, 496 (1933). — 1244) *O. Warburg, W. Christian*, Sauerstoffübertragendes Ferment in Milchsäurebazillen. Bioch. Zs., 260, 499 (1933). — 1245) *O. Warburg, W. Christian*, Über das gelbe Oxydationsferment. Bioch. Zs., 263, 228 (1933). — 1246) *O. Warburg, W. Christian*, Über das gelbe Ferment und seine Wirkungen. Bioch. Zs., 266, 377 (1933). — 1247) *O. Warburg, W. Christian*, Coferment-Problem. Bioch. Zs., 274, 112 (1934). — 1248) *O. Warburg, W. Christian*, Coferment-Problem. Bioch. Zs., 275, 464 (1934— 1935). — 1249) *O. Warburg, W. Christian*, Zerstörung des wasserstoffübertragenden Co-Ferments durch ultraviolettes Licht. Bioch. Zs., 282, 221 (1936). — 1250) *O. Warbrug, W. Christian*, Gärungs-Coferment. Bioch. Zs., 285, 156 (1936). — 1251) *O. Warburg, W. Christian*, Optischer Nachweis der Hydrierung und Dehydrierung des Pyridins im Gärungs-Coferment. Bioch. Zs., 286, 81 (1936). — 1252) *O. Warburg, W. Christian*, Pyridin als Wirkungs-Gruppe dehydrierender Fermente. Bioch. Zs., 286, 142 (1936). — 1253) *O. Warburg, W. Christian*, Pyridin, der wasserstoffübertragende Bestandteil von Gärungs-fermenten. (Pyridin-Nucleotide). Bioch. Zs., 287, 291 (1936). — 1254) *O. Warburg, W. Christian*, Verbrennung von Robison-Ester durch Triphospho-Pyridin-Nucleotid. Bioch. Zs., 287, 440 (1936). — 1255) *O. Warburg, W. Christian*, Pyridin, der wasserstoffübertragende Bestandteil von Gärungsfermenten. Helv. Chim. Acta, 19, E 79 (1936). — 1256) *O. Warburg, W. Christian*, Co-ferment der d-alanin-oxydase. Bioch. Zs., 296, 294, 298, 150 (1938). — 1257) *O. Warburg, W. Christian*, Bemerkung über gelbe Fermente. Bioch. Zs., 298, 368 (1938). — 1258) *O. Warburg, W. Christian, A. Griese*, Wirkungs-Gruppe des Coferments aus roten Blutzellen. Bioch. Zs., **279**, 143 (1935). — 1259) *O. Warburg, W. Christian, A. Griese,* Wasserstoffübertragendes Co-Ferment, seine Zusammensetzung und Wirkungsweise. Bioch. Zs., **282**, 157 (1935). — 1260) *O. Warburg, E. Haas,* Über eine Absorptionsbande im Gelb in Bäckerhefe. Naturwiss., **22**, 207 (1934). — 1261) *O. Warburg, F. Kubowitz,* Atmung bei sehr kleinen Sauerstoffdrucken. Bioch. Zs., **214**, 5 (1929). — 1262) *O. Warburg, F. Kubowitz,* Über katalatische Wirkung von Bluthäminen und von Chlorophyll-Häminen. Bioch. Zs., **227**, 184 (1930). — 1263) *O. Warburg, F. Kubowitz, W. Christian,* Über die katalytische Wirkung von Methylenblau in lebenden Zellen. Bioch. Zs., **227**, 245 (1930). — 1264) *O. Warburg, F. Kubowitz, W. Christian,* Über die Wirkung von Phenylhydrazin und von Phenylhydroxylamin auf die Atmung roter Blutzellen. Bioch. Zs., **233**, 240 (1931), **242**, 170 (1931). — 1265) *O. Warburg, E. Negelein,* Verteilung des Atmungsferments zwischen CO und $O_2$. Bioch. Zs., **193**, 334 (1928). — 1266) *O. Warburg, E. Negelein,* Einfluss der Wellenlänge auf die Verteilung des Atmungsfermentes. Bioch. Zs., **193**, 339 (1928). — 1267) *O. Warburg, E. Negelein,* Über die photochemische Dossoziation von Eisencarbonylverbindungen (Kohlenoxyd-Hämochromogen, Kohlenoxyd-Ferrocystein) und das photochemische Aequivalentgesetz. Bioch. Zs., **200**, 414 (1928). — 1268) *O. Warburg, E. Negelein,* Über die photochemische Dissoziation bei intermittierender Belichtung und das absolute Absorptionsspektrum des Atmungsferments. Bioch. Zs., **202**, 202 (1928). — 1269) *O. Warburg, E. Negelein,* Über die photochemische Spaltung einer Eisencarbonylverbindung und das photochemische Aequivalentgesetz. Naturwiss., **16**, 387 (1928). — 1270) *O. Warburg, E. Negelein,* Absolutes Absorptionsspektrum des Atmungsferments. Bioch. Zs., **204**, 495 (1929). — 1271) *O. Warburg, E. Negelein,* Absolutes Absorptionsspektrum des Atmungsferments. Bioch. Z., **214**, 64 (1929). — 1272) *O. Warburg, E. Negelein,* Über das Absorptionsspektrum des Atmungsferments der Netzhaut. Bioch. Zs., **214**, 101 (1929). — 1272a) *O. Warburg, E. Negelein,* Grünes Hämin aus Blut-Hämin. Ber. Deutsch. Chem. Ges., **63**, 1816 (1930). — 1273) *O. Warburg, E. Negelein,* Über die Hauptabsorptionsbanden der Mac Munnschen Histohämatine. Bioch. Zs., **233**, 486 (1931). — 1274) *O. Warburg, E. Negelein,* Photographische Abbildung der Hauptabsorptionsbanden der Mac Munnschen Histohämatine. Bioch. Zs., **238**, 135 (1931). — 1275) *O. Warburg, E. Negelein,* Über das Hämin des sauerstoffübertragenden Ferments der Atmung, über einige künstliche Hämoglobine und über Spirographis-Porphyrin. Bioch. Zs., **244**, 9 (1932). — 1276) *O. Warburg, E. Negelein,* Notiz über Spirographishämin. Bioch. Zs., **244**, 239 (1932). — 1277) *O. Warburg, E. Negelein,* Direkter spektroskopischer Nachweis des sauerstoffübertragenden Ferments in Essigbakterien. Bioch. Zs., **262**, 237 (1933). — 1278) *O. Warburg, E. Negelein, W. Christian,* Carbylamin-Hämoglobin und die photochemische Dissoziation seiner Kohlenoxydverbindung. Bioch. Zs., **214**, 26 (1929). — 1279) *O. Warburg, E. Negelein, E. Haas,* Spirographishämin. Bioch. Zs., **227**, 171 (1930). — 1280) *O. Warburg, E. Negelein, E. Haas,* Spektroskopischer Nachweis des sauerstoffübertragenden Ferments neben Cytochrom. Bioch. Zs., **266**, 1 (1933). — 1281) *O. Warburg, A. Reid,* Oxydation von Hämoglobin zu Methämoglobin. Bioch. Zs., **242**, 149 (1931). — 1282) *A. Wassermann,* Übertragung von peroxydisch gebundenen Sauerstoff auf $H_2S$. Ber. Deutsch. Chem. Ges., **65**, 704 (1932). — 1283) *A. Wassermann,* Kinetik und Hemmbarkeit der eisenkatalysierten $H_2O_2$-$H_2S$-Reaktion. Liebigs Ann., **503**, 249 (1933). — 1284) *A. Wassermann,* Benztraubensäure-$H_2O_2$-Reaktion. Bioch. Zs., **263**, 1 (1933). — 1285) *A. Wassermann,* Specificity of iron as a catalyst. Jl. of Chem. Soc., **1935**, 826. — 1286) *A. Watanabe,* Beeinflussung der Atmung von einigen grünen Algen durch KCN und MB. Acta Phytschin., **6**, 315 (1932). — 1286a) *J. Weijlard, H. Tauber,* Synthesis, Isolation and identification of cocarboxylase. Jl. Amer. Chem. Soc., **60**, 2263 (1938). — 1287) *H. Weil-Malherbe,* Brain metabolism I. II. Biochem. Jl., **30**, 665, **31**, 299 (1936/37). — 1288) *H. Weil-Malherbe c. s.,* Carbohydrate metabolism. Nature, **138**, 551 (1936); Biochem. Jl., **31**, 299 (1937). — 1289) *J. Weiss,* Reaction mechanism of oxidation-reduction processes. Nature, **133**, 648 (1934). — 1289a) *J. Weiss,* Reaction mechanism of the enzymes catalase and peroxidase in the light of the theory of chain reactions. Jl. of Physical Chem., **41**, 1107 (1937). — 1290) *E. Weitz et al.,* Über freie Ammonium-Radikale. Ber. Deutsch. Chem. Ges., **55**, 395, **57**, 153, **59**, 423, 2307, **64**, 2909 (1922/1931); Zs. f. Elektrochem., **34**, 538 (1928). — 1291) *W. B. Wendel,* Oxidation of lactate by methemoglobin in erythrocytes with regeneration of hemoglobin. Proc. Soc. Exp. Biol., **28**, 401 (1931). — 1292) *W. B. Wendel,* Oxidations by erythrocytes and the catalytic influence of methylene blue. Jl. of Biol. Chem., **102**, 373 (1933). — 1293) *W. B. Wendel,* Oxidations by erythrocytes and the catalytic influence of methylene blue. II. Methemoglobin and the effect of cyanide. Jl. of Biol. Chem., **102**, 385 (1933). — 1294) *C. H. Werkman et al.,* Dissimilation of phosphate esters. Enzymologia, **4**, 24 (1937). — 1295) *E. Wertheimer,* Autoxydations System als Modell einer Schwermetallkatalyse. Fermentforschung,

8, 497 (1926). — 1296) *F. Weygand, H. Stocker*, Reindarstellung von gelbem Ferment aus Hefe. Zs. phys. Chem., 247, 167 (1937). — 1297) *M. Wheldale-Onslow, M. E. Robinson*, Mechanism of plant oxidases. Biochem. Jl., 20, 1138 (1926). — 1298) *A. White, K. G. Stern*, Studies on the constitution of insulin. II. Further experiments on reduced insulin preparations. Jl. of Biol. Chem., 119, 215 (1937). — 1299) *H. Wieland*, Über den Mechanismus der Oxydationsvorgänge. 9. Liebigs Ann., 445, 181 (1925). — 1300) *H. Wieland*, Neuere Untersuchungen über die biologische Oxydation. Zs. Angew. Chem., 44, 579 (1931). — 1301) *H. Wieland*, Dehydrierende Enzyme, Helv. Chim. Acta, 15, 521 (1932). — 1302) *H. Wieland*, Verlauf der Oxydationsvorgänge. Stuttgart 1933. — 1302a) *H. Wieland*, L'oxygène, ses relations chimiques et biologiques. Paris, Gauthier-Villars, 1935 (V. Conseil de Chimie Solvay, Bruxelles, 1934). — 1303) *H. Wieland c. s.*, Hydrolytische Spaltung von Aminosäuren durch Tierkohle. Liebigs Ann. 513, 203 (1934). — 1304) *H. Wieland c. s.*, Anaerobe Vergärung der Fumarsäure. Liebigs Ann. 525, 119 (1936). — 1305) *H. Wieland, A. Bertho*, Über den Mechanismus der Oxydationsvorgänge. 15. Das Wesen der Essigsäuregärung. Liebigs Ann., 467, 95 (1928). — 1306) *H. Wieland, K. Bossert*, Eisenkatalyse von Diaethylperoxyd. Liebigs Ann. 509, 1 (1934). — 1307) *H. Wieland, F. Chrometzka*, Die Katalytische Zersetzung von Diäthylperoxyd durch Eisen. Ber. Deutsch. Chem. Ges., 63, 1028 (1930). — 1308) *H. Wieland, B. Claren*, Dehydrierung durch Hefe. Liebigs Ann. 509, 182 (1933). — 1309) *H. Wieland, O. B. Claren, B. N. Pramanik*, Enz. Dehydrierung von Milchsäure etc. durch Hefe. Liebigs Ann., 507, 203 (1933). — 1310) *H. Wieland, F. G. Fischer*, Über den Mechanismus der Oxydationsvorgänge. 11. Katalytische Dehydrierung. Ber. Deutsch. Chem. Ges., 59, 1180 (1926). — 1311) *H. Wieland, K. Frage*, Über den Mechanismus der Oxydationsvorgänge. 20. Beiträge zur Kenntnis der Bernsteinsäure-Dehydrase. Liebigs Ann. 477, 1 (1929). — 1312) *H. Wieland, W. Franke*, Über den Mechanismus der Oxydationsvorgänge. 12. Die Aktivierung des Hydroperoxyds durch Eisen. Liebigs Ann., 457, 1 (1927). — 1313) *H. Wieland, W. Franke*, Über den Mechanismus der Oxydationsvorgänge. 14. Die Aktivierung des Sauerstoffs durch Eisen. Liebigs Ann., 464, 101 (1928). — 1314) *H. Wieland, W. Franke*, Über den Mechanismus der Oxydationsvorgänge. 16. Über das Rosten des Eisens. Liebigs Ann. 469, 257 (1929). — 1315) *H. Wieland, W. Franke*, Verhältnis der Oxydationsgeschwindigkeit von molekularen $O_2$ und $H_2O_2$. Liebig Ann., 473, 289 (1929). — 1316) *H. Wieland, W. Franke*, Über den Mechanismus der Oxydationsvorgänge. 18. Weitere Versuche über die Aktivierung des Hydroperoxyds durch Eisen. Liebigs Ann., 475, 1 (1929). — 1317) *H. Wieland, W. Franke*, Über den Mechanismus der Oxydationsvorgänge. 19. Über kombinierte Autoxydationssysteme. Liebigs Ann. 475, 19 (1929). — 1318) *H. Wieland, A. Lawson*, Über den Mechanismus der Oxydationsvorgänge 27. Weitere Beiträge zur Kenntnis der Dehydrasen des Muskelgewebes. Liebigs Ann. 485, 193 (1931). — 1319) *H. Wieland, Th. F. Macrae*, Über den Mechanismus der Oxydationsvorgänge. 26. Weitere Studien über die dehydrierenden Enzyme der Milch. Liebigs Ann., 483, 217 (1930). — 1320) *H. Wieland, W. Mitchell*, Über den Mechanismus der Oxydationsvorgänge. 29. Dehydrogenating enzymes of milk. Liebigs Ann., 492, 156 (1932). — 1321) *H. Wieland, H. J. Pistor*, Dehydrierendes Enzym-System von Acetobacter peroxydans. Liebigs Ann., 522, 116 (1936). — 1322) *H. Wieland, K. Rauch, A. F. Thompson*, Die Hemmbarkeit von Atmung und Gärung. Liebigs Ann., 521, 214 (1936). — 1323) *H. Wieland, D. Richter*, Autoxydation der Aldehyde. Liebigs Ann., 486, 226 (1931). — 1324) *H. Wieland, D. Richter*, Autoxydation der Aldehyde. Liebigs Ann., 495, 284 (1932). — 1325) *H. Wieland, B. Rosenfeld*, Über den Mechanismus der Oxydationsvorgänge. 21. Über die dehydrierenden Enzyme der Milch., Liebigs Ann., 477, 32 (1929). — 1326) *H. Wieland, M. G. Sevag*, Über den Mechanismus der Oxydationsvorgänge. 33. Dehydrierungsreaktionen mit Buttersäurebakterien. Liebigs Ann., 501, 151 (1933). — 1327) *H. Wieland, R. Sonderhoff*, Enzym. Oxyd. von Essigsäure durch Hefe. Liebigs Ann., 499, 213, 503, 61 (1932/33). — 1328) *H. Wieland, H. Sutter*, Beiträge zur Wirkungsweise von Oxydasen und Peroxydasen. Ber. Deutsch. Chem. Ges., 61, 1060, 63, 66 (1928/30). — 1329) *H. Wieland, F. Wille*, Ärobe Dehydrierung von Alkohol durch Hefe. Liebigs Ann., 503, 70 (1933). — 1330) *H. Wieland, A. Wingler*, Katalytische Spaltung von Ketosäuren. Liebigs Ann., 436, 229 (1924). — 1331) *R. R. Williams*, Structure of vitamin $B_1$. Jl. Amer. Chem. Soc., 58, 1063 (1936). — 1332) *R. R. Williams, T. D. Spies*, Vitamin $B_1$ and its use in medicine. New York, 1938. — 1332a) *R. R. Williams*, Chemistry of vitamin $B_1$ (Thiamin). Erg. Vitamin- und Hormonforschg. 1, 213 (1938). — 1333) *R. R. Williams c. s.*, Structure of vitamin $B_1$. Jl. Amer. Chem. Soc., 57, 229, 536, 1093, 1751, 1849, 1856 (1935). — 1334) *R. R. Williams, J. K. Cline*, Synthesis of vitamin $B_1$. Jl. Amer. Chem. Soc., 58, 1504 (1936). — 1334a) *R. R. Williams, A. E. Rühle*, Studies of crystalline vitamin $B_1$: 11. Presence of quaternary nitrogen. Jl. Amer. Chem. Soc., 57, 1856 (1935). — 1335) *H. Will-*

*städt,* Die Vitamine. Klin. Ws., **14**, 1665, 1705 (1935). — 1336) *R. Willstätter,* Sauerstoff-Übertragung in der lebenden Zelle. Ber. Deutsch. chem. Ges., **59**, 1871 (1926). — 1336a) *R. Willstätter,* Advances in the isolation of enzymes. Ber. Deutsch. Chem. Ges., **59**, 1 (1926). — 1336b) *R. Willstätter, A. Pollinger,* Über Peroxydase. Liebigs Ann., **430**, 290 (1923). — 1337) *R. Willstätter, M. Rohdewald,* Aufbau des Glykogens durch Leukocyten. Zs. phys. Chem., **247**, 115, 269 (1937). — 1338) *R. Willstätter, H. Weber,* Quantitative Bestimmung der Peroxydase. Liebigs Ann., **449**, 156, 175 (1926). — 1339) *M. O. P. Wiltshire,* Influence of tissues and amino acids on adrenaline, Jl. of Phys., **72**, 88 (1931). — 1340) *H. Winberg, K. M. Brandt,* Zwischenstufen bei der Vergärung der Maltose. Svensk. kem. Tidskr., **48**, 213 (1936). — 1341) *A. Windaus, R. Tschesche, R. Crewe,* Über antineuritisches Vitamin aus Hefe. Zs. phys. Chem., **204**, 123 (1932); **228**, 27 (1934), **237**, 98 (1935). — 1342) *O. Wintersteiner, R. R. Williams, Rühle,* Elementary composition and UV-Absorption of Vitamin B$_1$. Jl. Amer. Chem. Soc., **57**, 517 (1935). — 1343) *H. G. Wood, Ch. H. Werkman,* Pyruvic acid breakdown by propionic acid bacteria. Biochem. Jl., **28**, 745 (1934), **30**, 618 (1936), **31**, 349 (1937); Iowa state Coll. Jl. Sci., **11**, 287, (1937). — 1344) *D. D. Woods,* Synthesis of formic acid by bacteria (Hydrogenlyases). Biochem. Jl., **30**, 515 (1936). — 1345) *B. Woolf,* Enzymes in B. coli.-Addition compound theory of enzyme action. Biochem. Jl., **23**, 472, 25, 342 (1929/31). — 1346) *F. Wrede, A. Rothaas,* Über das Violacein, den violetten Farbstoff des Bazillus violaceus. I. Zs. phys. Chem., **223**, 113 (1934). — 1347) *F. Wrede, E. Strack,* Pyocyanin, der blaue Farbstoff des Bac. Pyocyaneus I—IV. Zs. phys. Chem., **140**, 1, **142**, 103, **177**, 177, **181**, 58 (1924/29). — 1348) *F. Wrede, E. Strack,* Zur Synthese des Pyocyanins und einiger Homologen. Ber. Deutsch. Chem. Ges., **62**, 2051 (1929). — 1349) *G. P. Wright, M. v. Alstyne,* Oxidation of linseed oil by hematin. Jl. of Biol. Chem., **93**, 71 (1931). — 1350) *G. P. Wright, I. B. Conant, S. E. Kamerling,* Catalytic. effect of ferricyanide. Jl. of Biol. Chem., **94**, 411 (1931). — 1351) *K. Wunderly,* Aminolyse am Kohlekontakt. Helv. Chim. Acta, **16**, 515 (1933). — 1352) *K. Wunderly,* Das Verhalten einiger Harnstoffderivate. (Hexosen) an Tierkohle. Helv. Chim. Acta, **16**, 1009, 1018 (1933). — 1353) *R. Wurmser,* Sur les syntheses effectués dans les cellules et le potentiel de reduction 12. Internat. Phys. Kongr., Stockholm 3.—6. VIII. 1926, S. 175; BPh., **38**, 363. — 1354) *R. Wurmser,* Oxydations et réductions. Paris 1930. — 1355) *R. Wurmser,* Etude de poudres d'organes sur glutathion. Bull. Soc. Chim. Biol., **15**, 801 (1933). — 1356) *R. Wurmser et al.,* Potentiel apparant des solutions de glucose. C. R., **185**, 1038 (1927), **192**, 680, **194**, 888 (1931/2); Jl. de Chim. Physique, **25**, 641, **26**, 424, 565 (1928/9); Soc. Biol., **104**, 135 (1930). — 1357) *R. Wurmser, S. Filitti,* Réversibilité des oxydations en présence de l'alcool-déhydrase. Soc. Biol., **118**, 1027 (1935); Jl. de Chim. Physique, **33**, 577 (1936). — 1358) *R. Wurmser, S. Filitti-Wurmser,* Le potentiel d'oxydoréduction du cytochrome. C. rend. Soc. Biol., **127**, 471 (1938). — 1359) *R. Wurmser, S. Filitti-Wurmser,* Le potentiel d'oxydoréduction du système cytochrome c $\rightleftarrows$ oxycytochrome c. Jl. de Chim. Physique, **35**, 81 (1938). — 1360) *R. Wurmser, J. A. de Loureiro,* La réversibilité de l'oxydation de certains dérivés des glucides, en particulier de l'acide ascorbique. Jl. de Chim. Physique, **31**, 419 (1934). — 1361) *R. Wurmser, J. A. de Loureiro,* Réversibility des systèmes oxydored. dérivés des glucides. C. R., **198**, 738 (1934); Soc. Biol., **113**, 543, **116**, 101 (1933—1934). — 1362) *R. Wurmser, N. Mayer,* Réversibilité du système oxyd.-réd. des glucides. C.R., **194**, 888 (1932), **196**, 1337 (1933) Lit. — 1363) *R. Wurmser, N. Mayer,* L'équilibre entre les acides lactique et pyruvique. C. R., **195**, 81 (1932). — 1364) *R. Wurmser, N. Mayer,* Sur la réversibilité de l'oxydation de l'acide ascorbique. C. rend. seances Soc. de Biol., **121**, 3 (1936). — 1365) *R. Wurmser, N. Mayer, O. Crépy,* Potentiel d'oxydore-duction de la reductone. Jl. de Chim. Physique, **33**, 101 (1936). — 1366) *S. Yamagutchi,* Über die Beeinflussung der Sauerstoffatmung von verschiedenen Bakterien durch Blausäure und Kohlenoxyd. Acta Phytochimica, **8**, 157 (1934). — *S. Yamagutchi,* Untersuchungen über die intrazelluläre Indophenolreaction bei Bakterien. Acta Phytochimica, **8**, 263 (1935). — 1368) *A. Yamamoto,* Einfluss einiger Gifte auf die Atmung, Energie etc. der Schimmelpilze. Acta phytochim., **7**, 65 (1933). — 1369) *M. Yamamoto,* Stabilisierung des Vit. C durch Adrenalin. Zs. phys. Chem., **243**, 266 (1936). — 1370) *K. Zeile,* Zur Kinetik der Hydroperoxydspaltung durch Porphyrin-Metallkomplexsalze. Zs. phys. Chem., **189**, 127 (1930). — 1371) *K. Zeile,* Über die aktive Gruppe der Katalase. II. Zs., phys. Chem., **195**, 39 (1931). — 1372) *K. Zeile,* Synthetische Beiträge zur Konstitution des Cytochroms. I. Zs. phys. Chem., **207**, 35 (1932). — 1372a) *K. Zeile,* Hämin-haltige Fermente, Erg. Phys., **35**, 498 (1933). — 1373) *K. Zeile,* Schwermetallhaltige tierische Oxydasen. Handbuch der Biochemie, Ergänzungswerk, Jena, **1**, 544 (1933). — 1374) *K. Zeile,* Mechanismus der Sauerstoffaktivierung, Schwermetallkatalyse. Handbuch der Biochemie, Ergänzungswerk, Jena, **1**, 708 (1933). — 1375) *K. Zeile,* Über die Bestimmung von Diffusionskoeffizienten hochmolekularer Körper und über einige Beobachtungen bei der Diffusion von Katalase., Bioch. Z., **258**, 347 (1933). Notiz zur obenstehenden Mitteilung Bioch. Z. **261**, 156 (1933). — 1376) *K. Zeile*, Über Cytochrom c. (IV Mitt.). Zs. phys. Chem. **236**, 212 (1935). — 1377) *K. Zeile, H. Hellström*, Über die aktive Gruppe der Leberkatalase. Zs. phys. Chem., **192**, 171 (1930). — 1378) *K. Zeile, P. Piutti*, Synthetische Beiträge zur Konstitution des Cytochroms II. Zs. phys. Chem., **218**, 52 (1933). — 1379) *K. Zeile, F. Reuter*, Über Cytochrom C. III. Zs. phys. Chem., **221**, 101 (1933). — 1380) *N. D. Zelinsky*, Autoxydation von Cyclohexan durch Sauerstoff. Ber. Deutsch. Chem. Ges., **63**, 2362 (1930). — 1381) *R. v. Zeynek*, Über krystallisiertes Cyanhämoglobin. Zs. phys. Chem., **33**, 426 (1901). — 1382) *S. S. Zilva et al.*, Reversible enzymatic oxidation of Vitamin C. Biochem. Jl., **28**, 663, **30**, 1215, **31**, 438 (1934/7). — 1383) *F. Zuckerkandl*, (*Klebermass*), Eisen bei der alkoholischen Gärung. Bioch. Zs., **261**, 55, **271**, 435 (1933).

# INDEX